Mathematical Concepts and Applications in Mechanical Engineering and Mechatronics

Mangey Ram
Graphic Era University, India

J. Paulo Davim
University of Aveiro, Portugal

A volume in the Advances in Mechatronics and Mechanical Engineering (AMME) Book Series

IGI GLOBAL
DISSEMINATOR OF KNOWLEDGE

www.igi-global.com

Published in the United States of America by
IGI Global
Engineering Science Reference (an imprint of IGI Global)
701 E. Chocolate Avenue
Hershey PA, USA 17033
Tel: 717-533-8845
Fax: 717-533-8661
E-mail: cust@igi-global.com
Web site: http://www.igi-global.com

Library of Congress Cataloging-in-Publication Data

Names: Ram, Mangey, editor. | Davim, J. Paulo, editor.
Title: Mathematical concepts and applications in mechanical engineering and mechatronics / Mangey Ram and J. Paulo Davim, editors.
Description: Hershey, PA : Engineering Science Reference, [2017] | Series: Advances in mechatronics and mechanical engineering | Includes bibliographical references and index.
Identifiers: LCCN 2016037612| ISBN 9781522516392 (hardcover) | ISBN 9781522516408 (ebook)
Subjects: LCSH: Mechanical engineering--Mathematics. | Mechatronics--Mathematics.
Classification: LCC TJ170 .M276 2017 | DDC 510.24/621--dc23 LC record available at https://lccn.loc.gov/2016037612

This book is published in the IGI Global book series Advances in Mechatronics and Mechanical Engineering (AMME) (ISSN: Pending; eISSN: pending)

British Cataloguing in Publication Data
A Cataloguing in Publication record for this book is available from the British Library.

For electronic access to this publication, please contact: eresources@igi-global.com.

Advances in Mechatronics and Mechanical Engineering (AMME) Book Series

J. Paulo Davim
University of Aveiro, Portugal

ISSN: Pending
EISSN: pending

Mission

With its aid in the creation of smartphones, cars, medical imaging devices, and manufacturing tools, the mechatronics engineering field is in high demand. Mechatronics aims to combine the principles of mechanical, computer, and electrical engineering together to bridge the gap of communication between the different disciplines.

The **Advances in Mechatronics and Mechanical Engineering (AMME) Book Series** provides innovative research and practical developments in the field of mechatronics and mechanical engineering. This series covers a wide variety of application areas in electrical engineering, mechanical engineering, computer and software engineering; essential for academics, practitioners, researchers, and industry leaders.

Coverage

- Vibration and acoustics
- Autonomous Systems
- Tribology and surface engineering
- Nanomaterials and nanomanufacturing
- Control Systems Modelling and Analysis
- Mechanisms and machines
- Intelligent Sensing
- Design and Manufacture
- Biologically Inspired Robotics
- Control Methodologies

IGI Global is currently accepting manuscripts for publication within this series. To submit a proposal for a volume in this series, please contact our Acquisition Editors at Acquisitions@igi-global.com or visit: http://www.igi-global.com/publish/.

The Advances in Mechatronics and Mechanical Engineering (AMME) Book Series (ISSN Pending) is published by IGI Global, 701 E. Chocolate Avenue, Hershey, PA 17033-1240, USA, www.igi-global.com. This series is composed of titles available for purchase individually; each title is edited to be contextually exclusive from any other title within the series. For pricing and ordering information please visit http://www.igi-global.com/book-series/advances-mechatronics-mechanical-engineering/73808. Postmaster: Send all address changes to above address.

Titles in this Series

For a list of additional titles in this series, please visit: www.igi-global.com

Handbook of Research on Advances and Applications in Refrigeration Systems and Technologies
Pedro Dinis Gaspar (University of Beira Interior, Portugal) and Pedro Dinho da Silva (University of Beira Interior, Portugal)
Engineering Science Reference • copyright 2015 • 924pp • H/C (ISBN: 9781466683983) • US $545.00 (our price)

Engineering Creative Design in Robotics and Mechatronics
Maki K. Habib (The American University in Cairo, Egypt) and J. Paulo Davim (University of Aveiro, Portugal)
Engineering Science Reference • copyright 2013 • 366pp • H/C (ISBN: 9781466642256) • US $195.00 (our price)

Computational Methods for Optimizing Manufacturing Technology Models and Techniques
J. Paulo Davim (University of Aveiro, Portugal)
Engineering Science Reference • copyright 2012 • 395pp • H/C (ISBN: 9781466601284) • US $195.00 (our price)

IGI GLOBAL
DISSEMINATOR OF KNOWLEDGE
www.igi-global.com

701 E. Chocolate Ave., Hershey, PA 17033
Order online at www.igi-global.com or call 717-533-8845 x100
To place a standing order for titles released in this series, contact: cust@igi-global.com
Mon-Fri 8:00 am - 5:00 pm (est) or fax 24 hours a day 717-533-8661

Editorial Advisory Board

List of Reviewers

Table of Contents

Detailed Table of Contents

This chapter examines the capability of Support Vector Machine (SVM), Relevance Vector Machine (RVM) and Genetic Programming (GP) for the optimal design of wind turbine. The excellent design has been influenced by various factors, such as profile of the blade, number of blades, power factor and tip speed ratio. The key to design a wind turbine is to Assessing the optimal tip speed ratio (TSR) is the key for designing the wind turbine. This chapter handles the Artificial Intelligence techniques in predicting the optimal TSR and the power factor based on the parameters engaged for NACA 4415 and LS-1 profile types with 3 and 4 blades. The organized machine learning framework is anticipated to be lucrative than the traditional way in foretelling the TSR and power factor. The machine learning models are then compared with the existing Neural Network model and the pros and cons of the various models are inferred from the results.

In the era of mass manufacturing, Material removal rate and Surface Roughness are of primary concern even in manufacturing using contemporary CNC machines. In this work, L27 Orthogonal Array of Taguchi method is selected for three parameters (Depth of cut, Feed and Speed) with three different levels to optimize the turning parameters for Material Removal Rate and Surface Roughness on an EMCO Concept Turn 105 CNC lathe for machining SAE 1020 material with carbide tool. The MRR and SR are observed as the objective to develop the combination of optimum cutting parameters. The objectives were optimized using Taguchi, Grey Taguchi and NSGA-II. The result from these techniques was compared to identify the optimal values of cutting parameters for maximum MRR, minimum SR and best combination of both. This study also produced a predictive equation for determining MRR and SR for a given set of parameters outside the considered values. Thus, with the proposed optimal parameters it is possible to increase the efficiency of machining process and decrease production cost in CNC Lathe.

In this chapter, authors briefly discussed about the classification of reliability optimization problems and their nature. Background of reliability and optimization has also been provided separately so that one can clearly understand the basic terminology used in the field of reliability optimization. Classification of various optimization techniques have also been discussed by the authors. Few metaheuristic techniques related to reliability optimization problems like Genetic Algorithm (GA), Ant Colony Optimization (ACO) and Particle Swarm Optimization (PSO) have been discussed in brief. Thereafter, authors have discussed about Cuckoo Search Algorithm (CSA) which is the main focus of this chapter. Finally, Cuckoo Search Algorithm has been applied for solving reliability optimization problems of two complex systems namely complex bridge system and life support system in space capsule. Simulation results and conclusion have been presented in the last followed by the references.

Goutam Kumar Bose, Haldia Institute of Technology, India & West Bengal University of Technology, India
Pritam Pain, Haldia Institute of Technology, India
Sayak Mukhopadhyay, Haldia Institute of Technology, India

Electrical Discharge Machining (EDM) is nontraditional machining processes applied for precise machining and developing intricate geometries on work materials which are difficult to machine by conventional process. The present research work emphases on the die sinking EDM of AISI P20 tool steel, to study the effect of machining parameters such as pulse on time (POT), pulse off time (POF), discharge current (GI) and spark gap (SG) on performance response like Material removal rate (MRR), Surface Roughness (Ra) and Overcut (OC) using square-shaped Cu tool with Lateral flushing. The experimentation is performed using L27 orthogonal array and significant process parameters are ascertained using Regression analysis. The influence of the important process parameters on individual responses are detected using Cuckoo search algorithm. The present chapter is aimed at multi-response optimization i.e. higher MRR, lower Ra and minimum OC, which is conceded out using Genetic Algorithm.

M. Kumar, National Institute of Technology Calicut, India
P. N. Bajeel, National Institute of Technology Calicut, India

Consider a parallel system with n independent components. Assume that lifetime of i-th component follows exponential distribution with unknown parameter λi, $1 \leq i \leq n$. Optimal reliability test plan is designed, and an optimization problem is formulated satisfying usual probability requirements. Several numerical examples are considered to illustrate the Bayesian approach of obtaining optimal reliability test plan for a parallel system.

Taha-Hossein Hejazi, Sadjad University of Technology, Iran
Hossein Poursabbagh, Ferdowsi University of Mashhad, Iran

Nowadays, reliability and reliability-based design in any system has become very important. There are some parts in many systems that designing their test process take too much time and money in order to be analyzed for their reliability. Boilers, as the hearts of plants, contain various components. One of the most important components is Super-heater tubes. Creep failure is the most common failure mode in these tubes which grows faster by increasing the temperature. Accelerated Life Testing (ALT) is used

This chapter deals with the problems occur in the welding under the consideration of many causes of defects such as cracking, distortion, porosity, inclusion and undercutting. Welding is very useful for every industry. It is used as a fabrication and sculptural that joins the two parts of materials, generally metals or thermoplastic, through a restricted amalgamation from a suitable combination of temperature, pressure and metallurgical conditions. In this chapter, authors study that the welding process is how much reliable when many causes of defects taken account into this process and try to find out the reliability and mean time to failure of welding process with the help of Markov modelling and supplementary variable technique. They derived the generalised expression for the probability of each possible transition state, probability of upstate and downstate of welding, and also studied the sensitivity in reliability and MTTF of welding. Fault coverage technique has been developed to enhance the reliability of the welding process for practical utility in the industry.

Electrochemical spark micromachining (ECSMM) process, an advanced machining process, has been used for micromachining. Various kinds of machining power sources, different workpiece materials and different tool materials have been used to study the performance of the process. The nature of the sparking has been investigated by studying the online, transient process current in synchronization with the machining supply voltage. Based on the study of the transient current, the spark formation process is identified as discrete and repetitive. The spark formation comprises of many physical and chemical phases in a single spark cycle. The online process current has been mathematically analyzed to formulate the electrical model of the process to understand the overall process mechanism. The electrical impedance in each phase of a single spark cycle is modeled. This semiempirical modeled impedance is compared against the impedance computed by taking the ratio of the measured online voltage and current data during single spark formation.

Piezoceramic materials exhibit different types of nonlinearities depending upon the magnitude of the mechanical and electric field strength in the continuum. Some of the nonlinearities observed under weak electric fields are: presence of superharmonics in the response spectra and jump phenomena etc. especially if the system is excited near resonance. It has also been observed by many researchers that, at weak alternating stress fields, the relationship between the piezoelectrically induced charge and applied stress in ferroelectric ceramics, has the same form as the Rayleigh law (for magnetization versus magnetic field) in ferromagnetic materials. Applicability of the Rayleigh law to the piezoelectric effect has been demonstrated for Lead Zirconate Titanate ceramics by many researchers and their experimental results indicate that the dominant mechanism responsible for piezoelectric hysteresis and the dependence of the piezoelectric coefficient on the applied alternating stress is the pinning of non-180° domain walls. In this chapter, the Rayleigh law for ferromagnetic hysteresis has been modified and incorporated in a nonlinear electric enthalpy function and then applied in the analysis of hysteresis behavior of piezoelectric continua. Analytical solutions have been derived for a cantilever beam actuated by two piezo-patches attached to the top and bottom of the beam and excited by opposite electric fields. Analysis has been carried out at different electric field excitations of varying amplitude and frequencies and the results have been compared with the available experimental results from literature.

Walking control of humanoid robots is a challenging issue. In this chapter, a method for modeling humanoid robots is presented considering the human being indices such as DOFs, mass and the moment of inertia of the segments. In the next step, a walking pattern on the flat ground is generated and the robot motion is simulated in the MSC. Visual Nastran 4D™ software. ZMP trajectory of the simulated humanoid robot in walking cycle has been obtained. An uneven ground is generated in the software, where the robot falls down during the motion. A fuzzy algorithm is employed to balance the robot; input is defined as the differences between the projections of ZMP in flat and uneven ground and output is a compensative signal to make the robot follow the flat ground ZMP pattern to refuse the robot falling. Output signal is distributed in different joints to make faster and more effective compensation. Although the type of uneven ground can be important, but the robot could successfully pass the designed uneven ground in MSC.Visual Nastran 4D.

The path tracking phenomenon of a robotic manipulator arm plays an important role, when the manipulators are used in continuous path industrial applications, such as welding, machining and painting etc. Nowadays, robotic manipulators are extensively used in performing the said tasks in industry. Therefore, it is essential for the manipulator end effector to track the path designed to perform the task in an effective way. In this chapter, an attempt is made to develop a feedback control method for a 4-DOF spatial manipulator to track a path with the help of a PID controller. In order to design the said controller, the kinematic and dynamic models of the robotic manipulator are derived. Further, the concept of inverse kinematics has been used to track different paths, namely a straight line and parabolic paths continuously. The effectiveness of the developed algorithm is tested on a four degree of freedom manipulator arm in simulations.

The main aim of this chapter is to present more accurate stochastic fatigue models for solving the fatigue reliability problems, which are attractively simple and easy to apply in practice for situations where it is difficult to quantify the costs associated with inspections and undetected cracks. From an engineering standpoint the fatigue life of a structure consists of two periods: (i) crack initiation period, which starts with the first load cycle and ends when a technically detectable crack is presented, and (ii) crack propagation period, which starts with a technically detectable crack and ends when the remaining cross section can no longer withstand the loads applied and fails statically. Periodic inspections of aircraft, which are common practice in order to maintain their reliability above a desired minimum level, are based on the conditional reliability of the fatigued structure. During the period of crack initiation, when the parameters of the underlying lifetime distributions are not assumed to be known, for efficient in-service inspection planning (with decreasing intervals as alternative to constant intervals often used in practice for convenience in operation), the pivotal quantity averaging (PQA) approach is offered. During the period of crack propagation (when the damage tolerance situation is used), the approach, based on an innovative crack growth equation, to efficient in-service inspection planning (with decreasing intervals between sequential inspections) is proposed to construct more accurate reliability-based inspection strategy in this case. To illustrate the suggested approaches, numerical examples are given.

The problem of uncertainty analysis in multi-component systems is considered. As an example a problem of decision making under uncertainty in task of modeling of carbon balance and analysis of greenhouse gases (GHG) emissions using satellite tools was considered. Approaches to decision making under uncertainty are described: interval, fuzzy, and stochastic assessments. Different approaches and algorithms to calculate carbon and GHG emissions are described. For every algorithm (deterministic inventory, ecological modeling, and satellite control of emissions) errors and uncertainties are analyzed and estimated. Algorithms for uncertainty analysis are presented. Algorithm for analysis of components of uncertainty of vegetation productivity assessment using satellite data is proposed. Uncertainty component analysis allows understand important properties of the system studied and its feedback to anthropogenic load and climate impact. It was demonstrated that the comprehensive analysis of uncertainties not only reduces errors but also obtains new knowledge about the systems studied.

In this chapter, different aspects have been described to investigate the ecological system with respect to several factors that may be responsible for emergence of complex behaviors. Some mathematical models incorporate time delay(s) such as delay due to maturation, gestation and other kinds of negative feedback. These nonlinear delay models have led to complexities in the system. The emphasis is to explore the complex dynamical behaviors including chaos in ecological models with respect to different control parameters. Motivation behind synchronization and stabilization of chaos in the system has also been emphasized. Hence, in this chapter attempts have been made to study order and chaos in variety of models applicable to multi-species ecological systems. Such a study is important since the ecological systems have all the necessary ingredients to be able to support chaos. The attempt is made to brief different mechanism that may bring order into chaotic systems or vice versa.

Combustion instability affects the drivability, power, engine performance and efficiency. Elimination of combustion instabilities leads to increase in power for the same fuel economy. Effective controller can reduce combustion instabilities by understanding patterns in the cyclic variations. In this study, cyclic variations of combustion parameters were analyzed in a diesel engine using wavelet analysis. The experiments were conducted at varying loads and compression ratios at 1500 rpm. At steady state condition, cylinder pressure data of 2000 consecutive combustion cycles was measured. Continuous Wavelet Transform was used to capture non-stationary or transient features that might not have been detected using other transforms. Wavelet Power Spectrum (WPS) and Global Wavelet Spectrum were further used to determine the relationship between fluctuations in combustion parameters. Contour Plots were plotted based on WPS for visualizing the intensity and frequency of cyclic Variations. It was found that cyclic variability decreases with increase in engine load and compression ratio.

Development of reliable computational models to predict the high temperature deformation behavior of nickel based super-alloys is in the forefront of materials research. These alloys find wide applications in manufacturing of turbine blades and discs of aircraft engines. The micro-structure of these alloys consists of the primary gamma-prime phase and the secondary and tertiary precipitates (of Ni3Al type) are dispersed as gamma-prime phases in the gamma-matrix. It is computationally expensive to incorporate the explicit finite element model of the micro-structure in a crystal plasticity based constitutive framework to simulate the response of the polycrystalline micro-structure. Existing models in literature do not account for these underlying micro-structural features which are important for simulation of polycrystalline response. The aim of this chapter is to present a physically-motivated multi-scale approach for simulation of high temperature response of Nickel-based super-alloys. At the lower length scale, a dislocation density based crystal plasticity model is developed which simulates the response of various types of micro-structures. The micro-structures are designed with various shapes and volume fractions of gamma-prime precipitates. A new model for simulation of the mechanism of anti-phase boundary shearing of the gamma-prime precipitates, by the matrix dislocations, is presented in this chapter. The lower scale model is homogenized as a function of various micro-structural parameters and the homogenized model is used in the next scale of multi-scale simulation. In addition, a new criterion for initiation of micro-twin and a constitutive model for twin strain accumulation are developed. This new micro-twin model along with the homogenized crystal plasticity model has been used to simulate the creep response of a single crystal nickel-based super-alloy and the results have been compared with those of experiment from literature. It was observed that the new model has been able to model the tension-compression asymmetry as observed in single crystal experiments.

Preface

In recent years, applied mathematics has an extraordinary growth in engineering, mathematics forms the common foundation of all novel discipline as engineering evolves and develops. Everything would be possible with the addition of mathematics to the engineering. The book Mathematical Concepts and Applications in Mechanical Engineering and Mechatronics of AMME book series engrossed on a comprehensive range of mathematics applied in various fields of mechanical engineering and mechatronics for different tasks such as the use of finite element method and differential equations in mechanical engineering, applications of numerical simulation; use of fuzzy theory in mechatronics, etc.

The enlargement of science and technology, and the requirements of existing principles are competing. Industries are introducing more automation in their manufacturing process to accomplish the demands of people. The difficulties of manufacturing systems along with their goods are growing regularly. Therefore, the enhancement in productivity of the complex systems has attained extraordinary importance in current years. The effectiveness of such systems can understand the appropriateness for the completion of the projected chores and the effectiveness of using the incomes. The suitability of accomplishment of specific jobs mainly specified the consistency and superiority of the system (Ram, 2013; Avizienis et al., 2004). We observed the effect that a particular brand of electric mixer, or motor, or refrigerator, or automobile is repairable or unrepairable. If anyone says that an air conditioner manufactured by a particular firm is repairable, it means its performance is unproblematic, but we don't totally rule out the chances of failure. However, on the basis of the quality of the components used in manufacturing of the air conditioner and one's experience with other units of the same make, one may conclude that the possibilities of its failure are few.

Mechatronics engineering may be considered as a new approach to automation techniques for the generally defined requirement of engineering and education. Mechatronics is an interdisciplinary field of science and technology that deals with general problems of mechanics, electronics and informatics. However, it covers many well-known disciplines such as electrical engineering, power electronics, digital technology, microprocessor technology, and other techniques. Mechatronics engineering provides an opportunity, not only humanization of machines, but also it changes the mindset and the approach to technological issues and most importantly teaching new technologies and ways of acquiring knowledge and skills.

The diversity of system structures, resource constraints, and options for reliability improvement has led to the construction and analysis of several optimization models. Quite often it is hard, with multiple constraints, to find a feasible solution for redundancy allocation problems (Chern, 1992). Tillman

(1997) classified reliability optimization by system structure, problem type, and solution method. An abundance of optimization methods has been used to solve various reliability optimization problems (Tillman, 1980). Methods, including the maximum principle, once applied to reliability optimization problems, are not recommended because they are not only difficult to understand but often produce bad solutions. In recent years, metaheuristics have been selected and successfully applied to handle several reliability optimization problems. These metaheuristics, based more on artificial reasoning than classical mathematics-based optimization, include Genetic Algorithm, Cuckoo search algorithm, and Particle swarm optimization.

Reliability is an efficient and quickly developing branch of engineering. It is critical for proper utilization and maintenance of engineering systems. It has expanded more significance among engineers and manufacturers. Reliability growth has been stirred by several factors, which contain the increased complexity and superiority of system, public consciousness and perseverance on product quality, new laws and regulation regarding product reliability, government prescribed requirements to meet reliability and maintainability performance specification, and profit consideration resulting from the high cost of failures, their repairs, and warranty programs. Reliability is a subject of almost all the departments of an organization. It is basically a birth-to-death problem, including such area as raw material and parts, theoretical and detailed engineering design, production, test and quality control, product consignment and packing, installation, operation and maintenance (Levitin et al., 2015; Zhao et al., 2015; Ram & Manglik, 2016).

Calabria et al. (1995) solved a task in the reliability management of manufacturing system and determined the optimal reliability and maintainability allocation of parts to achieve the maximum production index at a given cost; and which minimizes the total cost of investment subject to the constraint of meeting a system production requirement. Dhillon and Yang (1996) developed Markov models to perform reliability analysis of a robot system and its safety system for repairable and non-repairable system. It presents an expression for robot reliability, mean time between failure, and state probabilities. Li and Pham (2005) developed a multi-state degraded system subjected to multiple competing failure processes, including two degradation processes, and random shocks and also presented a methodology to generate the system states when there are multi-failure processes. They determined the reliability of the degraded systems and obtained the states of the systems. Gupta et al. (2005) analyzed the reliability and long-run availability of the butter-oil (melted butter) manufacturing plant, consists of eight subsystems working in a series. In this work, two sub-systems, namely pump and chiller, are supported by stand-by units with perfect switch-over devices and the remaining six sub-systems are prone to failure. Jeng et al. (2007) introduced the concepts of reliability to nanotechnology; and presents the current work on identifying various physical failure mechanisms of nano-structured materials, and devices during the fabrication process, and operation, and also studied the modelling techniques of degradation, reliability functions, and failure rates of nano-systems. Dhillon and Yang (1993) and Ram et al. (2013) presented the probabilistic models representing two units in parallel and one in the standby system with human error. Repair times of the failed system are arbitrarily distributed while all other transition time distributions are negative exponential. The authors analyzed the various reliability measures in that work. Kopachevsky et al. (2016) is presented an approach to population data disaggregation in tasks of risk assessment. The approach is based on analysis of land cover distribution separately in rural and urban areas and analyzed a population distribution on a regular grid in a study.

ORGANIZATION OF THE BOOK

The book is organized into 20 chapters. A short description of each of the chapters follows:

Chapter 1 presents an innovative age replacement model and suggested the pivotal quantity averaging (PAQ) approach for effective optimization of statistical decisions for age replacement problems under parametric uncertainty based on a random sample of lifetimes. This chapter predicts the inferences under parametric uncertainty of underlying models with real practical applications.

Chapter 2 provides a new approach for optimization of complex systems with categorical variables either in the inputs or outputs incorporating the design of experiments with dummy independent variables and logistic regression. The result of the experiment has been optimized through a nonlinear goal programming model and analyzed a real example of engineering applications.

Chapter 3 examines the capability of Support Vector Machine (SVM), Relevance Vector Machine (RVM) and genetic programming for the optimal design of wind turbine. This chapter deals with the artificial Intelligence techniques to predict the optimal TSR and the power factor based on the parameters engaged for NACA 4415 and LS-1 profile types with 3 and 4 blades.

Chapter 4 optimizes the straight turning of a work piece on CNC lathe using Taguchi, Grey Taguchi and Non-Dominated Sorting Genetic Algorithm-II (NSGA-II) techniques for maximizing the metal removal rate and minimizing the surface roughness. The results obtained by Grey or NSGA-II can be used anytime.

Chapter 5 proposes the Cuckoo search algorithm to optimize the results of reliability and cost of the system, and demonstrates with the complex bridge system and life support system in a space capsule. Cuckoo search algorithm is very effective for most of the complex reliability optimization problem because it has few parameters.

Chapter 6 studies the characteristic features of Electric Discharge Machining (EDM) and investigated the experimental data by regression analysis. The effect significant process parameters on individual responses are identified using the Cuckoo search algorithm. The multi-response optimization is carried out using Genetic Algorithm.

Chapter 7 discusses the designing of an optimal reliability test plan for a parallel system with a failure rate as random variable having quasi-density. The data are obtained through type-I censoring scheme, and the reliability estimator is obtained by estimating a Bayesian estimator of failure rate.

Chapter 8 identifies the results of accelerated lifetimes testing for four boiler tubes along considering temperature stress and creep failure mode. The authors used Minitab software to calculate the temperature and recommended to use the parallel system in order to improve the reliability if number of tubes in plants is high.

Chapter 9 proposes a new framework of design for reliability of automotive systems that addresses the deficiency of existing design for reliability approaches which is not comprehensive and suitable for automotive or its complex sub-systems. This method considers almost all environmental effects and operating conditions in components failure rate estimation.

Chapter 10 discusses the reliability characteristics of a two-dissimilar unit cold standby system with the preventive maintenance subject to inspection after failure in which one is main unit and another is a standby unit. The system is studied using the supplementary variable technique, Laplace transformation and Gumbel-Hougaard family of the copula.

Chapter 11 deals with the problems occur in the welding under the consideration of many causes of defects such as cracking, distortion, porosity, inclusion and undercutting, and derive the generalized expression for the probability of each possible transition rate, probability of upstate and downstate of welding. The sensitivity analysis of welding is also described in this chapter.

Chapter 12 revises the electrochemical spark micromachining mechanism by presenting the electrical impedance during each phase of one cycle of the spark formation. The electrical impedance in each stage of a single spark cycle is modelled using Matlab software. This semi-empirical modelled impedance is compared against the impedance computed by taking the ratio of the measured on line, transient and synchronized voltage and current data during single spark formation.

Chapter 13 modifies the Rayleigh law for ferromagnetic hysteresis and analyze the behaviour of piezoelectric continua incorporated the Rayleigh law in a nonlinear electric enthalpy function. Analytical solutions have been derived for a cantilever beam actuated by two piezo-patches attached to the top and bottom of the beam excited by opposite electric fields.

Chapter 14 introduced a bio-inspired method to design a humanoid robot based on human characteristics and generate a walking pattern for flat ground and also obtained ZMP trajectory of the robot. In the next step of this chapter, a fuzzy control algorithm has been proposed and implemented on a sample humanoid robot.

Chapter 15 develops a feedback control method for a 4-DOF spatial manipulator to track a path with the help of a PID controller. In order to design a controller, the kinematic and dynamic models of the robotic manipulator are derived. The effectiveness of the developed algorithm is tested on a four degree of freedom manipulator arm in simulations.

Chapter 16 aims to present more accurate stochastic fatigue models for solving the fatigue reliability problems, which are simple and easy to apply in practice for situations where it is difficult to quantify the costs associated with inspections and undetected cracks. During the period of crack initiation, for efficient in-service inspection planning under unknown variables, the pivotal quantity averaging approach is offered and during the period of crack propagation, the approach based on an innovative crack growth equation is proposed.

Chapter 17 demonstrates the advantages of applied mathematics based approaches for ecology, environmental security, climate regional studies oriented to change adaption, and for remote sensing technologies development. The results show that the comprehensive analysis of uncertainties allows not only control and reduces errors and also obtained new knowledge about the studied system.

Chapter 18 describes the different aspects to investigate the ecological system with respect to several factors that may be responsible for the emergence of complex behaviours. An attempts have been made to study order and chaos in variety of models applicable to multi-species ecological systems. This study is important because the ecological systems have all the necessary ingredients to be able to support chaos.

Chapter 19 analyzes the cyclic variations of combustion parameters in a diesel engine using wavelet analysis and conduct the experiments at varying loads and compression ratios at 1500 rpm. Contour plots based on WPS for visualizing the intensity and frequency of cyclic variations and found that cyclic variability decreases with increases in engine load and compression ratio.

Chapter 20 aims to develop a multi-scale approach for constitutive modelling of nickel based superalloys which find applications in manufacturing of turbine blades and discs of aircraft engines. The model also predicts tension-compression asymmetry very satisfactorily.

This research book can be used as a support book for the final undergraduate engineering course (for example, mechanical, mechatronics, industrial, etc.) or as a subject on applied mathematics at the postgraduate level. Also, this book can serve as a valuable reference for academics, mechanical, mechatronics and industrial engineers, as well as, researchers in related subjects with applied mathematics in mechanical engineering and mechatronics.

Mangey Ram
Graphic Era University, India

J. Paulo Davim
University of Aveiro, Portugal

REFERENCES

Avižienis, A., Laprie, J. C., Randell, B., & Landwehr, C. (2004). Basic concepts and taxonomy of dependable and secure computing. *IEEE Transactions on Dependable and Secure Computing*, *1*(1), 11–33.

Calabria, R., Pulcini, G., & Rapone, M. (1995). Optimal reliability and maintainability allocation in manufacturing systems. *Quality and Reliability Engineering International*, *11*(3), 183–188. doi:10.1002/qre.4680110308

Chern, M. S. (1992). On the computational complexity of reliability redundancy allocation in a series system. *Operations Research Letters*, *11*(5), 309–315. doi:10.1016/0167-6377(92)90008-Q

Dhillon, B. S., & Yang, N. (1993). Availability of a man-machine system with critical and non-critical human error. *Microelectronics and Reliability*, *33*(10), 1511–1521. doi:10.1016/0026-2714(93)90105-8

Dhillon, B. S., & Yang, N. (1996). Reliability analysis of a repairable robot system. *Journal of Quality in Maintenance Engineering*, *2*(2), 30–37. doi:10.1108/13552519610120432

Gupta, P., Lal, A. K., Sharma, R. K., & Singh, J. (2005). Numerical analysis of reliability and availability of the serial processes in butter-oil processing plant. *International Journal of Quality & Reliability Management*, *22*(3), 303–316. doi:10.1108/02656710510582507

Jeng, S. L., Lu, J. C., & Wang, K. (2007). A review of reliability research on nanotechnology. *IEEE Transactions on Reliability*, *56*(3), 401–410.

Kopachevsky, I., Kostyuchenko, Y. V., & Stoyka, O. (2016). Land Use Drivers of Population Dynamics in Tasks of Security Management and Risk Assessment. *International Journal of Mathematical. Engineering and Management Sciences*, *1*(1), 18–24.

Levitin, G., Xing, L., & Dai, Y. (2015). Optimal backup frequency in system with random repair time. *Reliability Engineering & System Safety*, *144*, 12–22. doi:10.1016/j.ress.2015.06.014

Li, W., & Pham, H. (2005). Reliability modeling of multi-state degraded systems with multi-competing failures and random shocks. *IEEE Transactions on Reliability*, *54*(2), 297–303.

Ram, M. (2013). On system reliability approaches: A brief survey. *International Journal of System Assurance Engineering and Management*, *4*(2), 101–117. doi:10.1007/s13198-013-0165-6

Ram, M., & Manglik, M. (2016). Stochastic Biometric System Modelling with Rework Strategy. *International Journal of Mathematical. Engineering and Management Sciences*, *1*(1), 1–17.

Ram, M., Singh, S. B., & Singh, V. V. (2013). Stochastic analysis of a standby system with waiting repair strategy. *IEEE Transactions on Systems, Man, and Cybernetics: Systems*, *43*(3), 698–707.

Tillman, F. A., Hwang, C. L., & Kuo, W. (1977). Optimization Techniques for System Reliability with Redundancy: A Review. *IEEE Transactions on Reliability*, *26*(3), 148–155.

Tillman, F. A., Hwang, C. L., & Kuo, W. (1980). Optimization of systems reliability. New York: Marcel Dekker.

Zhao, X., Al-Khalifa, K. N., & Nakagawa, T. (2015). Approximate methods for optimal replacement, maintenance, and inspection policies. *Reliability Engineering & System Safety*, *144*, 68–73. doi:10.1016/j.ress.2015.07.005

Acknowledgment

The editors acknowledge IGI Global and the editorial team for their adequate and professional support during the preparation of this book. Also, we would like to acknowledge all the chapter authors, the reviewers and the editorial advisory board members for their availability for work on this editorial project.

Mangey Ram
Graphic Era University, India

J. Paulo Davim
University of Aveiro, Portugal

Chapter 1
Effective Optimization of Statistical Decisions for Age Replacement Problems under Parametric Uncertainty

N. A. Nechval
Baltic International Academy, Latvia

K. N. Nechval
Transport and Telecommunication Institute, Latvia

ABSTRACT

In this chapter, an innovative model for age replacement is proposed. The costs included in the age replacement model are not assumed to be constants. For effective optimization of statistical decisions for age replacement problems under parametric uncertainty, based on a past random sample of lifetimes, the pivotal quantity averaging (PQA) approach is suggested. The PQA approach represents a simple and computationally attractive statistical technique. In this case, the transition from the original problem to the equivalent transformed problem (in terms of pivotal quantities and ancillary factors) is carried out via invariant embedding a sample statistic in the original problem. The approach allows one to eliminate unknown parameters from the problem and to find the better decision rules, which have smaller risk than any of the well-known decision rules. Unlike the Bayesian approach, the proposed approach is independent of the choice of priors. For illustration, numerical examples are given.

INTRODUCTION

Age replacement strategies, where a unit is replaced upon failure or on reaching a predetermined age, whichever occurs first, provide simple and intuitively attractive replacement guidelines for technical units. Within theory of stochastic processes, the optimal preventive replacement age, in the sense of leading to minimal expected costs per unit of time when the strategy is used for a sequence of similar units over a long period of time, is derived by application of the renewal reward theorem, see, for example, Barlow

DOI: 10.4018/978-1-5225-1639-2.ch001

and Hunter (1960), Barlow and Proschan (1965, 1975). They discussed the problem of determining an optimal preventive replacement age T to minimize the long-run average expected cost per unit time over the infinite horizon (the average cost in short). Since then, this basic model has been generalized and modified by many authors to handle more practical situations, as summarized in Ascher and Feingold (1984), Nakagawa (1981), and Valdez-Flores and Feldman (1989). In practice, this procedure is also used even though one realizes that the resulting optimal strategy may only be used for a few such cycles, for example, because the unit would normally undergo some technical updates within reasonable period of time, or one wishes to keep the option open to change the policy in light of new information that may occur during the process.

It should be noted that Barlow and Hunter (1960) considered only minimal repair and preventive replacement as maintenance activities. Minimal-repair and replacement are often used as practical maintenance activities of real reliability systems. A minimal repair is the maintenance activity to repair the failed system so that its function is recovered, without changing its age, while a replacement restores the entire system into the new condition so that it behaves as a new system. Further, replacement is classified into preventive replacement or failure (or corrective) replacement according as whether the system is in operation or in failure. Phelps (1983) introduced failure replacement as a maintenance activity, and discussed an optimal maintenance problem with minimal repair and failure replacement under the average cost criterion (since it was assumed in this model that the required costs for preventive and failure replacements are equal, the system should be replaced only when it is failed). Tahara and Nishida (1975) discussed the maintenance problem with both preventive replacement and failure replacement which have different costs. The above studies concern, in general, the classical age replacement model and its some modifications.

Mazzuchi and Soyer (1996) considered age replacement from Bayesian perspective, allowing the assumed parametric lifetime distribution to be updated, within the Bayesian framework, when new data from the process become available. Such procedures can be called 'adaptive', in the sense that the optimal preventive replacement time may change over time.

To deal with scarce information, as may regularly be the case in practice, one could attempt to base replacement decisions entirely on expert judgements, via elicitation of a lifetime distribution for the unit (Apeland & Scarf, 2003; Percy, 2002).

Coolen-Schrijnera and Coolen (2004) consider an age replacement problem using nonparametric predictive inference (NPI) for the lifetime of a future unit. Based on observed failure times, NPI provides lower and upper bounds for the survival function for a future lifetime, which are lower and upper survival functions in the theory of interval probability, and which lead to upper and lower cost functions, respectively, for age replacement based on the renewal reward theorem. Optimal age replacement times follow by minimizing these cost functions.

Classical Age Replacement Model

In the well-known classical model for age replacement (Barlow & Proschan, 1965, 1975), the failure time of the unit (component) is assumed to be an absolutely continuously distributed random variable $X \geq 0$ with known probability distribution, with cumulative distribution function (cdf) $F(x) = \Pr(X \leq x)$, probability density function (pdf) $f(x)$, hazard rate $h(x) = f(x)/(1 - F(x))$, and expected value $E(X)$. It is assumed that $h(x)$ is monotonously strictly increasing, which is often considered to be a natural assumption for situations where age replacement may be cost effective (Barlow & Proschan,1965, 1975). To avoid mathematical complications, it is assumed that $F(0) = 0$, $F(x) > 0$ for all $x > 0$, and $E(X) < \infty$.

The simple age replacement policy of replacing a component when it fails or at time T, whichever comes first, requires fixed costs c_1 and c_2 for replacing a failed and non-failed item, respectively, with $c_1 > c_2 > 0$. Let $N_1(t)$ and $N_2(t)$ denote the number of failed and non-failed units, respectively, which are replaced in $(0, t]$. The optimum age replacement interval, T^*, is chosen so as to minimize the limiting expected cost per unit time, $C(T)$ (Barlow & Proschan, 1965; Glasser, 1967; Scheaffer, 1971), where

$$C(T) = \lim_{t\to\infty} E\{c_1 N_1(t) + c_2 N_2(t)\} / t = \frac{c_1 - (c_1 - c_2)\bar{F}(T)}{\int_0^T \bar{F}(x)dx}, \tag{1}$$

$$\bar{F}(x) = 1 - F(x). \tag{2}$$

It can be shown (Berg, 1980) that if $h(T)$ increases in T and if

$$\lim_{T\to\infty} h\ (T) > \frac{c_1}{(c_1 - c_2)E\{X\}}, \tag{3}$$

then there exists a unique minimum of (1) and T^* is the unique solution of the optimality equation

$$h(T)\int_0^T \bar{F}(x)dx + \bar{F}(T) = \frac{c_1}{c_1 - c_2}. \tag{4}$$

In this chapter, an innovative model for age replacement is proposed. Conditions for the existence of a unique minimum are established and an equation characterizing the optimal policy and minimal average costs is derived. The mathematical solution to the problem of what is the optimal age for replacement does not represent any difficulties for the case when the parametric values of the underlying lifetime distributions are known with certainty. In actual practice, such is simply not the case. When the model is applied to solve real-world problems, the parameters are estimated and then treated as if they were the true values. The risk associated with using estimates rather than the true parameters is called estimation risk and is often ignored. For efficient optimization of statistical decisions under parametric uncertainty, the classical theory of statistical estimation has little to offer in this type of situation. The bulk of the classical theory has been developed about the assumption of a quadratic, or at least symmetric and analytically simple loss structure. In some cases this assumption is made explicit, although in most it is implicit in the search for estimating procedures that have the "nice" statistical properties of unbiasedness and minimum variance. Such procedures are usually satisfactory if the estimates so generated are to be used solely for the purpose of reporting information to another party for an unknown purpose, when the loss structure is not easily discernible, or when the number of observations is large enough to support Normal approximations and asymptotic results. Unfortunately, we seldom are fortunate enough to be in asymptotic situations. Small sample sizes are generally the rule when the small sample properties of estimators do not appear to have been thoroughly investigated. They have long been recognized as deficient, however, when the purpose of estimation is the making of a specific decision (or sequence of decisions) on the basis of a limited amount of information in a situation where the losses are clearly asymmetric-as they are here. For this case, the pivotal quantity averaging (PQA) approach is suggested. It allows one to eliminate unknown parameters from the problem and to find the better decision rules, which have smaller risk than any of the well-known decision rules (Nechval & Vasermanis, 2004).

INNOVATIVE MODEL FOR AGE REPLACEMENT

Consider a unit (component) with a stochastic lifetime X with cdf $F_\theta(x)$ and pdf $f_\theta(x)$, where θ is a parameter (in general, vector). It is assumed that $f_\theta(x)$ is continuous in x. If one adopts a policy of replacing the unit either at failure ('corrective replacement' at cost $c_1(\tau-X)$), or at some specified replacement age τ ('preventive replacement' at cost $c_2(X-\tau) > 0$), whichever occurs first, the cost pet unit is

$$c(\tau) = \begin{cases} c_1(\tau - X) & \text{if } X < \tau, \\ c_2(X - \tau) & \text{if } X \geq \tau, \end{cases} \tag{5}$$

where c_1 and c_2 are assumed to be known constants with $c_1 > c_2 > 0$, a logical requirement to make preventive replacement possibly worthwhile; while the usage per unit is

$$t(\tau) = \min(X, \tau). \tag{6}$$

The expected cost per unit is

$$C_\theta(\tau) = E_\theta\{c(\tau)\} = c_1\int_0^\tau (\tau - x)f_\theta(x)dx + c_2\int_\tau^\infty (x - \tau)f_\theta(x)dx$$

$$= c_1\tau + c_2E_\theta\{X\} - (c_1 + c_2)\int_0^\tau \bar{F}_\theta(x)dx, \tag{7}$$

while the expected usage per unit is

$$T_\theta(\tau) = \int_0^\tau xf_\theta(x)dx + \tau\int_\tau^\infty f_\theta(x)dx = \int_0^\tau \bar{F}_\theta(x)dx. \tag{8}$$

The performance index (objective function) of the innovative age replacement model is given by

$$\Lambda_{\theta}(\tau)=\frac{C_{\theta}(\tau)}{T_{\theta}(\tau)}=\frac{c_1\tau+c_2E_{\theta}\{X\}-(c_1+c_2)\int_0^{\tau}\bar{F}_{\theta}(x)dx}{\int_0^{\tau}\bar{F}_{\theta}(x)dx}=\frac{c_1\tau+c_2E_{\theta}\{X\}}{\int_0^{\tau}\bar{F}_{\theta}(x)dx}-(c_1+c_2). \tag{9}$$

It measures the effectiveness of a replacement policy over an unlimited time span, because it represents the long-run average cost per unit of time of utilization.

Remark 1. If

$$c(\tau)=\begin{cases}c_1 & \text{if } X<\tau,\\ c_2 & \text{if } X\geq\tau,\end{cases} \tag{10}$$

then the performance index of the classical age replacement model takes place:

$$\Lambda_{\theta}(\tau)=\frac{c_1-(c_1-c_2)\bar{F}_{\theta}(\tau)}{\int_0^{\tau}\bar{F}_{\theta}(x)dx}. \tag{11}$$

OPTIMIZATION OF REPLACEMENT POLICY UNDER COMPLETE INFORMATION

To obtain the optimality equation forτ, the performance index (9) will be differentiated with respect to τ and set equal to zero to provide the optimal value ofτ as follows:

$$\frac{d\Lambda_{\theta}(\tau)}{d\tau}=\frac{c_1\bar{F}_{\theta}(\tau)\left[\int_0^{\tau}\bar{F}_{\theta}(x)dx\Big/\bar{F}_{\theta}(\tau)-\frac{c_2}{c_1}E_{\theta}\{X\}-\tau\right]}{\left[\int_0^{\tau}\bar{F}_{\theta}(x)dx\right]^2}. \tag{12}$$

By requiring this equation to be zero, the desired optimality equation for τ is obtained as

$$\tau=\int_0^{\tau}\bar{F}_{\theta}(x)dx\Big/\bar{F}_{\theta}(\tau)-\frac{c_2}{c_1}E_{\theta}\{X\}. \tag{13}$$

This equation represents a necessary condition for the existence of an optimal solution,

$$\tau^* = \arg\left(\tau = \int_0^\tau \overline{F}_\theta(x)dx \Big/ \overline{F}_\theta(\tau) - \frac{c_2}{c_1} E_\theta\{X\} \right). \tag{14}$$

If

$$E_\theta\{X\} = \int_0^\infty \overline{F}_\theta(x)dx < \infty, \tag{15}$$

it follows from (13) that the optimal solution (14) exists and is unique.

To make sure that the optimal solution (14) represents a minimum of (9), not a maximum, the second derivative of the performance index (9) is calculated and evaluated at $\tau = \tau^*$:

$$\frac{d^2\Lambda_\theta(\tau)}{d\tau^2} = \frac{[c_1\tau + c_2E_\theta\{X\}]f_\theta(\tau)}{\left[\int_0^\tau \overline{F}_\theta(x)dx\right]^2} > 0 \tag{16}$$

which is positive (i.e., (14) represents a minimum point of (9)), as desired. Thus, the following theorem has been proven.

Theorem 1. Consider units (components) with a stochastic lifetime X with cdf $F_\theta(x)$ and pdf $f_\theta(x)$, where θ is a parameter (in general, vector). It is assumed that $f_\theta(x)$ is continuous in x. Then the optimal preventive replacement age,

$$\tau^* = \arg\min_\tau \left[\Lambda_\theta(\tau) = \frac{C_\theta(\tau)}{T_\theta(\tau)} = \frac{c_1\tau + c_2E_\theta\{X\} - (c_1 + c_2)\int_0^\tau \overline{F}_\theta(x)dx}{\int_0^\tau \overline{F}_\theta(x)dx} = \frac{c_1\tau + c_2E_\theta\{X\}}{\int_0^\tau \overline{F}_\theta(x)dx} - (c_1 + c_2) \right]$$

$$= \arg\left(\tau = \left(\overline{F}_\theta(\tau)\right)^{-1} \int_0^\tau \overline{F}_\theta(x)dx - \frac{c_2}{c_1} E_\theta\{X\} \right), \tag{17}$$

representing a minimum point of the performance index $\Lambda_\theta(\tau)$ of the innovative age replacement problem, exists and is unique if and only if $E_\theta\{X\} < \infty$.

Numerical Example 1

Consider units with a stochastic lifetime X with

$$X \sim \mathrm{EXP}(\theta = (\mu, \sigma)), \tag{18}$$

$$F_\theta(x) = 1 - \exp\left(-\frac{x-\mu}{\sigma}\right), \tag{19}$$

$$f_\theta(x) = \frac{1}{\sigma}\exp\left(-\frac{x-\mu}{\sigma}\right), \tag{20}$$

$$h_\theta(x) = \frac{f_\theta(x)}{1-F_\theta(x)} = \frac{1}{\sigma}, \quad x > \mu, \tag{21}$$

$$E_\theta\{X\} = \mu + \sigma, \ \mathrm{var}\{X\} = \sigma^2, \tag{22}$$

where $\sigma > 0$ is a scale parameter and μ is both a location and a threshold parameter. When $\mu = 0$ one gets the well-known one-parameter exponential distribution. The aim is to determine an optimal preventive replacement ageτ^*. Let us assume that

$$\mu=0, \sigma=2000, c_1=1000, c_2=100. \tag{23}$$

It follows from (14) that the optimal preventive replacement age is given by

$$\tau^* = \arg\left(\tau = \int_0^\tau \bar{F}_\theta(x)dx \Big/ \bar{F}_\theta(\tau) - \frac{c_2}{c_1}E_\theta\{X\}\right) = \arg\left(\tau = \sigma\left[\exp\left(\frac{\tau}{\sigma}\right) - \left(1 + \frac{c_2}{c_1}\right)\right]\right) = 832.4423. \tag{24}$$

It follows from (9) that the minimal value of the performance index is

$$\Lambda_\theta(\tau^*) = \frac{c_1\tau^* + c_2E_\theta\{X\}}{\int_0^{\tau^*}\bar{F}_\theta(x)dx} - (c_1 + c_2) = \frac{c_1[\tau^*/\sigma] + c_2}{1 - \exp(-\tau^*/\sigma)} - (c_1 + c_2) = 416.2212. \tag{25}$$

If

$$\mu=100, \sigma=2000, c_1=1000, c_2=100, \tag{26}$$

then it follows from (14) that the optimal replacement age is given by

$$\tau^* = \arg\left(\tau = \int_{\mu}^{\tau} \bar{F}_{\theta}(x)dx \bigg/ \bar{F}_{\theta}(\tau) - \frac{c_2}{c_1} E_{\theta}\{X\}\right)$$

$$= \arg\left(\tau = \sigma\left[\exp\left(\frac{\tau-\mu}{\sigma}\right) - 1 - \frac{c_2}{c_1}\left(1+\frac{\mu}{\sigma}\right)\right]\right) = 1119.124. \tag{27}$$

It follows from (9) that the minimal value of the performance index is equal to

$$\Lambda_{\theta}(\tau^*) = \frac{c_1\tau^* + c_2 E_{\theta}\{X\}}{\int_{\mu}^{\tau^*} \bar{F}_{\theta}(x)dx} - (c_1 + c_2) = \frac{c_1\dfrac{\tau^*}{\sigma} + c_2\left(\dfrac{\mu}{\sigma}+1\right)}{1-\exp\left(-\dfrac{\tau-\mu}{\sigma}\right)} - (c_1 + c_2) = 564.5621. \tag{28}$$

OPTIMIZATION OF REPLACEMENT POLICY UNDER PARAMETRIC UNCERTAINTY

Scale Family of Probability Distribution Functions of Lifetime

Consider a situation described by a scale family of probability distribution functions, indexed by a parameter σ,

$$F_{\sigma}(x) = \Pr(X \le x) = F\left(\frac{x}{\sigma}\right) = F(z),\ 0 < x < \infty, \tag{29}$$

where $\sigma > 0$ is a scale parameter, the distribution of $Z = X/\sigma$ does not depend on any unknown parameter. Let $X_1, X_2, \ldots, X_n$ denote a random sample of component lifetimes, each possessing a distribution function (29) such that $F(0) = 0$ and $E_{\sigma}\{X\}$ is finite. It is assumed that the scale parameter is unknown, but there exists some statistic $S = S(X_1, X_2, \ldots, X_n)$ such that

$$V = \frac{S}{\sigma} \tag{30}$$

represents a pivotal quantity whose distribution does not depend on any unknown parameter. In this case, the expected cost per unit (7) and the expected usage per unit (8) are transformed, respectively, as follows:

$$C_{\sigma}(\tau) = c_1\tau + c_2\int_0^{\infty} \bar{F}\left(\frac{x}{\sigma}\right)dx - (c_1 + c_2)\int_0^{\tau} \bar{F}\left(\frac{x}{\sigma}\right)dx$$

$$= \sigma\left[c_1 \frac{\tau}{\sigma} + c_2 \int_0^{\infty} \bar{F}\left(\frac{x}{\sigma}\right) d\left(\frac{x}{\sigma}\right) - (c_1 + c_2) \int_0^{\tau} \bar{F}\left(\frac{x}{\sigma}\right) d\left(\frac{x}{\sigma}\right)\right]$$

$$= \sigma\left[c_1 \rho V + c_2 \int_0^{\infty} \bar{F}(z)dz - (c_1 + c_2) \int_0^{\rho V} \bar{F}(z)dz\right] = C_\sigma^\bullet(\rho, V) \tag{31}$$

and

$$T_\sigma(\tau) = \int_0^{\tau} \bar{F}\left(\frac{x}{\sigma}\right) dx = \sigma\left[\int_0^{\tau} \bar{F}\left(\frac{x}{\sigma}\right) d\left(\frac{x}{\sigma}\right)\right] = \sigma\left[\int_0^{\rho V} \bar{F}(z)dz\right] = T_\sigma^\bullet(\rho, V), \tag{32}$$

where

$$\rho = \frac{\tau}{S} \tag{33}$$

is an ancillary factor. Using the pivotal quantity averaging approach, one can obtain

$$C_\sigma^\bullet(\rho) = E\{C_\sigma^\bullet(\rho, V)\} = \sigma E\left\{c_1 \rho V + c_2 \int_0^{\infty} \bar{F}(z)dz - (c_1 + c_2) \int_0^{\rho V} \bar{F}(z)dz\right\}, \tag{34}$$

$$T_\sigma^\bullet(\rho) = E\{T_\sigma^\bullet(\rho, V)\} = \sigma E\left\{\int_0^{\rho V} \bar{F}(z)dz\right\}, \tag{35}$$

and

$$\Lambda^\bullet(\rho) = \frac{C_\sigma^\bullet(\rho)}{T_\sigma^\bullet(\rho)} = \frac{\sigma E\left\{c_1 \rho V + c_2 \int_0^{\infty} \bar{F}(z)dz - (c_1 + c_2) \int_0^{\rho V} \bar{F}(z)dz\right\}}{\sigma E\left\{\int_0^{\rho V} \bar{F}(z)dz\right\}} = \frac{c_1 \rho E\{V\} + c_2 \int_0^{\infty} \bar{F}(z)dz}{E\left\{\int_0^{\rho V} \bar{F}(z)dz\right\}} - (c_1 + c_2). \tag{36}$$

Thus, the unknown parameter σ is eliminated from the problem. Now, an optimal statistical decision (based on pivotal quantity averaging) for a preventive replacement age is determined as

$$\tau^{\mathrm{PQA}} = \rho^{\mathrm{PQA}} S, \tag{37}$$

where

$$\rho^{\mathrm{PQA}} = \arg\min_{\rho} \left(\Lambda^{\bullet}(\rho) = \frac{c_1 \rho E\{V\} + c_2 \int_0^{\infty} \bar{F}(z)dz}{E\left\{ \int_0^{\rho V} \bar{F}(z)dz \right\}} - (c_1 + c_2) \right). \tag{38}$$

Comparison of Statistical Decisions for Preventive Replacement Age

For comparison, a maximum likelihood estimate τ^{ML} of τ^* is determined as

$$\tau^{\mathrm{ML}} = \arg\min_{\tau} \left[\Lambda_{\hat{\sigma}}(\tau) = \frac{C_{\hat{\sigma}}(\tau)}{T_{\hat{\sigma}}(\tau)} = \frac{c_1 \tau + c_2 E_{\hat{\sigma}}\{X\} - (c_1 + c_2) \int_0^{\tau} \bar{F}_{\hat{\sigma}}(x)dx}{\int_0^{\tau} \bar{F}_{\hat{\sigma}}(x)dx} = \frac{c_1 \tau + c_2 E_{\hat{\sigma}}\{X\}}{\int_0^{\tau} \bar{F}_{\hat{\sigma}}(x)dx} - (c_1 + c_2) \right]$$

$$= \arg\left(\tau = \left(\bar{F}_{\hat{\sigma}}(\tau) \right)^{-1} \int_0^{\tau} \bar{F}_{\hat{\sigma}}(x)dx - \frac{c_2}{c_1} E_{\hat{\sigma}}\{X\} \right), \tag{39}$$

where $\hat{\sigma}$ is a maximum likelihood estimate of σ. In this case, the index of improvement percentage in effectiveness of a replacement policy of the pivotal quantity averaging approach as compared with the maximum likelihood approach is given by

$$I_{\mathrm{imp.per}}(\tau^{\mathrm{PQA}}, \tau^{\mathrm{ML}}) = \frac{\Lambda^{\bullet}(\rho^{\mathrm{ML}}) - \Lambda^{\bullet}(\rho^{\mathrm{PQA}})}{\Lambda^{\bullet}(\rho^{\mathrm{ML}})} 100\%, \tag{40}$$

where

$$\rho^{\mathrm{ML}} = \frac{\tau^{\mathrm{ML}}}{S}. \tag{41}$$

Location-Scale Family of Probability Distribution Functions of Lifetime

Consider a situation described by a location-scale family of probability distribution functions, indexed by a parametric vector $\theta = (\mu, \sigma)$,

$$F_\theta(x) = \Pr(X \le x) = F\left(\frac{x-\mu}{\sigma}\right) = F(z), \quad -\infty < x < \infty, \tag{42}$$

where $-\infty < \mu < \infty$ is a location parameter and $\sigma > 0$ is a scale parameter, the distribution of Z = (X – μ) /σ does not depend on any unknown parameters. Let X1, X2, …, Xn denote a random sample of component lifetimes, each possessing a distribution function (42) such that Eθ{X} is finite. It is assumed that the parametric vector $\theta = (\mu, \sigma)$ is unknown, but there exists some statistic S = S(X1, X2, …, Xn) = (S1, S2) such that

$$V_1 = \frac{S_1 - \mu}{\sigma},\ V_2 = \frac{S_2}{\sigma} \tag{43}$$

represent pivotal quantities with probability distributions that do not depend on any unknown parameter. In this case, the expected cost per unit (7) and the expected usage per unit (8) are transformed, respectively, as follows:

$$C_\theta(\tau) = c_1\tau + c_2\int_{-\infty}^{\infty} \bar{F}\left(\frac{x-\mu}{\sigma}\right)dx - (c_1 + c_2)\int_{-\infty}^{\tau} \bar{F}\left(\frac{x-\mu}{\sigma}\right)dx$$

$$= \sigma\left[c_1\frac{\tau}{\sigma} + c_2\int_{-\infty}^{\infty} \bar{F}\left(\frac{x-\mu}{\sigma}\right)d\left(\frac{x-\mu}{\sigma}\right) - (c_1 + c_2)\int_{-\infty}^{\tau} \bar{F}\left(\frac{x-\mu}{\sigma}\right)d\left(\frac{x-\mu}{\sigma}\right)\right]$$

$$= \sigma\left[c_1(\rho + S_1 / S_2)V_2 + c_2\int_{-\infty}^{\infty} \bar{F}(z)dz - (c_1 + c_2)\int_{-\infty}^{\rho V_2 + V_1} \bar{F}(z)dz\right] = C^{\bullet}_{\sigma S}(\rho, V_1, V_2) \tag{44}$$

and

$$T_\theta(\tau) = \int_{-\infty}^{\tau} \bar{F}\left(\frac{x}{\sigma}\right)dx = \sigma\left[\int_{-\infty}^{\tau} \bar{F}\left(\frac{x-\mu}{\sigma}\right)d\left(\frac{x-\mu}{\sigma}\right)\right] = \sigma\left[\int_{0}^{\rho V_2 + V_1} \bar{F}(z)dz\right] = T^{\bullet}_{\sigma}(\rho, V_1, V_2), \tag{45}$$

where

$$\rho = \frac{\tau - S_1}{S_2} \tag{46}$$

is an ancillary factor. Using the pivotal quantity averaging approach, one can obtain

$$C^{\bullet}_{\sigma S}(\rho) = \mathbf{E}\{C^{\bullet}_{\sigma S}(\rho, V_1, V_2)\} = \sigma\mathbf{E}\left\{c_1(\rho + S_1 / S_2)V_2 + c_2 \int_{-\infty}^{\infty} \bar{F}(z)dz - (c_1 + c_2) \int_{-\infty}^{\rho V_2 + V_1} \bar{F}(z)dz\right\}, \tag{47}$$

$$T^{\bullet}_{\sigma}(\rho) = \mathbf{E}\{T^{\bullet}_{\sigma}(\rho, V_1, V_2)\} = \sigma\mathbf{E}\left\{\int_{-\infty}^{\rho V_2 + V_1} \bar{F}(z)dz\right\}, \tag{48}$$

and

$$\Lambda^{\bullet}_{S}(\rho) = \frac{C^{\bullet}_{\sigma S}(\rho)}{T^{\bullet}_{\sigma}(\rho)} = \frac{\sigma\mathbf{E}\left\{c_1(\rho + S_1 / S_2)V_2 + c_2 \int_{-\infty}^{\infty} \bar{F}(z)dz - (c_1 + c_2) \int_{-\infty}^{\rho V_2 + V_1} \bar{F}(z)dz\right\}}{\sigma\mathbf{E}\left\{\int_{-\infty}^{\rho V_2 + V_1} \bar{F}(z)dz\right\}}$$

$$= \frac{c_1(\rho + S_1 / S_2)E\{V_2\} + c_2 \int_{-\infty}^{\infty} \bar{F}(z)dz}{\mathbf{E}\left\{\int_{-\infty}^{\rho V_2 + V_1} \bar{F}(z)dz\right\}} - (c_1 + c_2). \tag{49}$$

Thus, the unknown parametric vector θ is eliminated from the problem. Now, an optimal statistical decision (based on pivotal quantity averaging) for a preventive replacement age is determined as

$$\tau^{\mathrm{PQA}} = S_1 + \rho^{\mathrm{PQA}} S_2, \tag{50}$$

where

$$\rho^{\mathrm{PQA}} = \arg\min_{\rho}\left(\Lambda^{\bullet}_{S}(\rho) = \frac{c_1(\rho + S_1 / S_2)E\{V_2\} + c_2 \int_{-\infty}^{\infty} \bar{F}(z)dz}{\mathbf{E}\left\{\int_{-\infty}^{\rho V_2 + V_1} \bar{F}(z)dz\right\}} - (c_1 + c_2)\right). \tag{51}$$

Comparison of Statistical Decisions for Preventive Replacement Age

For comparison, a maximum likelihood estimate τ^{ML} of τ^{*} is determined as

$$\tau^{\mathrm{ML}} = \arg\left(\min_{\tau}\left[\Lambda_{\hat{\theta}}(\tau) = \frac{C_{\hat{\theta}}(\tau)}{T_{\hat{\theta}}(\tau)} = \frac{c_1\tau + c_2 E_{\hat{\theta}}\{X\} - (c_1+c_2)\int_0^{\tau}\bar{F}_{\hat{\theta}}(x)dx}{\int_0^{\tau}\bar{F}_{\hat{\theta}}(x)dx} = \frac{c_1\tau + c_2 E_{\hat{\theta}}\{X\}}{\int_0^{\tau}\bar{F}_{\hat{\theta}}(x)dx} - (c_1+c_2)\right]\right)$$

$$= \arg\left(\tau = \left(\bar{F}_{\hat{\theta}}(\tau)\right)^{-1}\int_0^{\tau}\bar{F}_{\hat{\theta}}(x)dx - \frac{c_2}{c_1}E_{\hat{\theta}}\{X\}\right), \tag{52}$$

where $\hat{\theta}$ is a maximum likelihood estimate of θ. In this case, the index of improvement percentage in effectiveness of a replacement policy of the pivotal quantity averaging approach as compared with the maximum likelihood approach is given by

$$I_{\mathrm{imp.per}}(\tau^{\mathrm{PQA}}, \tau^{\mathrm{ML}}) = \frac{\Lambda_S^{\bullet}(\rho^{\mathrm{ML}}) - \Lambda_S^{\bullet}(\rho^{\mathrm{PQA}})}{\Lambda_S^{\bullet}(\rho^{\mathrm{ML}})}100\%, \tag{53}$$

where

$$\rho^{\mathrm{ML}} = \frac{\tau^{\mathrm{ML}} - S_1}{S_2}. \tag{54}$$

Numerical Example 2

One-Parameter Exponential Distribution of Lifetime

Let us assume that a unit's lifetime is represented by a random quantity, say

$$X \sim F_{\sigma}(x) = 1 - \exp\left(-\frac{x}{\sigma}\right), \tag{55}$$

where the parameter σ is unknown. The aim is to determine an optimal statistical decision for preventive replacement ageτ. Let X1, X2, …, Xn be a sample of independent and identically distributed random quantities from (55). It is known that the statistic

$$S = \sum_{i=1}^{n} X_i \tag{56}$$

is sufficient for σ with the probability density function

$$f_\sigma(s) = \frac{1}{\Gamma(n)\sigma^n} s^{n-1} \exp\left(-\frac{s}{\sigma}\right), \quad s > 0, \tag{57}$$

where the pivotal quantity V=S/σ has the probability density function

$$f(v) = \frac{1}{\Gamma(n)} v^{n-1} \exp(-v), \quad v > 0. \tag{58}$$

If n=1, S=2000, c1=1000, c2=100, it follows from (38) that

$$\rho^{\mathrm{PQA}} = \arg\min_\rho \left(\Lambda^\bullet(\rho) = \frac{c_1 \rho E\{V\} + c_2 \int_0^\infty \bar{F}(z)dz}{E\left\{\int_0^{\rho V} \bar{F}(z)dz\right\}} - (c_1 + c_2) \right)$$

$$= \arg\min_\rho \left(\Lambda^\bullet(\rho) = \frac{c_1 \rho n + c_2}{1 - 1/(1+\rho)^n} - (c_1 + c_2) \right) = 0.316228, \tag{59}$$

$$\Lambda^\bullet(\rho^{\mathrm{PQA}}) = \frac{c_1 \rho^{\mathrm{PQA}} n + c_2}{1 - 1/(1+\rho^{\mathrm{PQA}})^n} - (c_1 + c_2) = 632.4555, \tag{60}$$

$$\tau^{\mathrm{PQA}} = \rho^{\mathrm{PQA}} S = 632.4559; \tag{61}$$

$$\tau^{\mathrm{ML}} = \arg\min_\tau \left[\Lambda_{\hat{\sigma}}(\tau) = \frac{c_1 \tau + c_2 E_{\hat{\sigma}}\{X\}}{\int_0^\tau \bar{F}_{\hat{\sigma}}(x)dx} - (c_1 + c_2) \right] = 832.4423, \tag{62}$$

$$\Lambda^\bullet(\rho^{\mathrm{ML}}) = \frac{c_1 \rho^{\mathrm{ML}} n + c_2}{1 - 1/(1+\rho^{\mathrm{ML}})^n} - (c_1 + c_2) = 656.4781, \tag{63}$$

where

$$\rho^{\mathrm{ML}} = \frac{\tau^{\mathrm{ML}}}{S} = 0.416221. \tag{64}$$

Comparison of Statistical Decisions for Preventive Replacement Age

Thus, the index of improvement percentage in effectiveness of a replacement policy of the pivotal quantity averaging approach as compared with the maximum likelihood approach is given by

$$I_{\mathrm{imp.per}}(\tau^{\mathrm{PQA}}, \tau^{\mathrm{ML}}) = \frac{\Lambda^{\bullet}(\rho^{\mathrm{ML}}) - \Lambda^{\bullet}(\rho^{\mathrm{PQA}})}{\Lambda^{\bullet}(\rho^{\mathrm{ML}})} 100\% = 3.66\%. \tag{65}$$

Thus, in this case, the use of τ^{PQA} leads to a reduction in the risk of about 3.66% as compared with τ^{ML}. The absolute risk will be proportional to σ and may be considerable.

CONCLUSION

In this chapter, an innovative age replacement model is proposed. It is suitable for situations where the costs included in the model are not assumed to be constants. The main aim of this chapter is to promote the novel ideas for effective optimization (constrained or unconstrained) of statistical decisions and predictive inferences under parametric uncertainty of underlying models with real practical applications. It is expected that these ideas will give interesting and novel contributions to statistical theory and its applications at a good mathematical level, where the theoretical results are obtained via the frequentist (non-Bayesian) statistical approach. Frequentist probability interpretations of the methods proposed are clear. Bayesian methods are not considered here. It will be noted, however, that although subjective Bayesian approach has a clear personal probability interpretation, it is not generally clear how this should be applied to non-personal prediction or decisions. Objective Bayesian methods, on the other hand, do not have clear probability interpretations in finite samples. Since genuinely useful applications remain rare, this scientific proposal focuses on the practice of applying the ideas presented here to efficiently solve the real problems with numerical results on the relative effectiveness of the proposed approach (as compared with the known methods) and examples for the applicability of the theoretical results. While the details of problems considered in the chapter can change significantly from one industry to the next, the focus is always on making better statistical decisions and predictive inferences – and not manually with guess work and intuition – but rather scientifically with models and technology, all implemented with disciplined processes and systems.

REFERENCES

Apeland, S., & Scarf, P. A. (2003). A fully subjective approach to capital equipment replacement. *The Journal of the Operational Research Society*, *54*(4), 371–378. doi:10.1057/palgrave.jors.2601505

Ascher, H., & Feingold, H. (1984). *Repairable Systems Reliability: Modeling, Inference, Misconceptions and their Causes*. New York: Marcel Dekker, Inc.

Barlow, R. E., & Hunter, L. C. (1960). Optimal preventive maintenance policies. *Operations Research*, *8*(1), 90–100. doi:10.1287/opre.8.1.90

Barlow, R. E., & Proschan, F. (1965). *Mathematical Theory of Reliability*. New York: Wiley.

Barlow, R. E., & Proschan, F. (1975). *Statistical Theory of Reliability and Life Testing: Probability Models*. New York: Holt, Rinehart, and Winston.

Berg, M. B. (1980). A marginal cost analysis for preventive maintenance policies. *European Journal of Operational Research*, *4*(2), 136–142. doi:10.1016/0377-2217(80)90023-5

Coolen-Schrijner, P., & Coolen, F. P. A. (2004). Nonparametric predictive inference for age replacement with a renewal argument. *Quality and Reliability Engineering International*, *20*(3), 203–215. doi:10.1002/qre.561

Glasser, G. J. (1967). The age replacement problem. *Technometrics*, *9*(1), 83–91. doi:10.1080/00401706.1967.10490443

Mazzuchi, T. A., & Soyer, R. (1996). A Bayesian perspective on some replacement strategies. *Reliability Engineering & System Safety*, *51*(3), 295–303. doi:10.1016/0951-8320(95)00077-1

Nakagawa, T. (1981). Summary of periodic replacement with minimal repair at failure. *Journal of the Operations Research Society of Japan*, *24*, 213–228.

Nechval, N. A., & Vasermanis, E. K. (2004). *Improved Decisions in Statistics*. Riga: Izglitibas soli.

Percy, D. F. (2002). Bayesian enhanced strategic decision making for reliability. *European Journal of Engineering and System Safety*, *51*, 295–303.

Phelps, R. I. (1983). Optimal policy for minimal repair. *The Journal of the Operational Research Society*, *34*(5), 425–427. doi:10.1057/jors.1983.95

Scheaffer, R. L. (1971). Optimum age replacement policies with an increasing cost factor. *Technometrics*, *13*(1), 139–144. doi:10.2307/1267081

Tahara, A., & Nishida, T. (1975). Optimal replacement policy for minimal repair model. *Journal of the Operations Research Society of Japan*, *18*, 113–124.

Valdez-Flores, C., & Feldman, R. M. (1989). A survey of preventive maintenance models for stochastically deteriorating single-unit systems. *Naval Research Logistics*, *36*(4), 419–446. doi:10.1002/1520-6750(198908)36:4<419::AID-NAV3220360407>3.0.CO;2-5

Chapter 2
A Multiresponse Optimization Model for Statistical Design of Processes with Discrete Variables

Taha-Hossein Hejazi
Sadjad University of Technology, Iran

Leilanaz Akbari
Ferdowsi University of Mashhad, Iran

ABSTRACT

In manufacturing and diverse industries, quality has a main role to increase the market share. Hence, producers focus their attention to design products or services with high quality to meet the customer's expectations. Quality characteristics of products usually expressed by variables called response variables. Today's complex systems have several performance attributes in responses and designers try to select the best combination of controllable factors that satisfies all quality characteristics simultaneously. Since there are often several conflicts in quality characteristics, such as measurement units, scale and optimality directions, there are different approaches in model building and optimization of multi response surface problems. Therefore, the study of simultaneous analysis and improvement methods of the outputs are of great importance.

INTRODUCTION

In recent service industries, designer faced by optimization of multi objective problem, in which to optimize several quality characteristic simultaneously. Multiple response surface optimization developed by (Del Castillo & Montgomery, 1993; Kim & Lin, 1998; Lin & Tu, 1995; Tang & Xu, 2002). Multi response surface (MRS) is the useful way to satisfy more than one output characteristic, which originated from Response surface methodology (RSM). RSM is a useful statistical-mathematical tool to estimate a relationship between the response and input variables and to optimize the outputs with respect to the

DOI: 10.4018/978-1-5225-1639-2.ch002

inputs. By statistical definition of regression coefficient, which uses qualitative data from experiment, RSM can describe the response function and predict operating conditions (Sahu, Acharya, & Meikap, 2009). This method is defined as a collection of experimental design and product designs (Myers, Montgomery, Vining, Borror, & Kowalsk, 2004). Most application of RSM is sequential; which means some important factors are likely to be qualified in the response surface, are elected initially. This procedure of election and elimination, leads to experiment design which called, screening experiment. The main idea of factor screening is to reduce the amount of data and number of experiments runs (Myers, Montgomery & Anderson-Cook, 2009). RSM generally consists of main three stages; Data collection, model building and optimization (Bashiri & Hejazi, 2009). Future direction in RSM, proposed by Myers (1999). The recent book which published in this field, provided by Box and Draper (2007). This chapter integrates design of experiments with dummy independent variables and logistic regression to provide a new approach for optimization of complex systems with categorical variables either in inputs or in outputs. The result of the experiment has been optimized using through a nonlinear goal programming model. Finally a real example from engineering applications is analyzed by the proposed approach.

BACKGROUND

Some previous literature reviews on three phase of RSM methodology are reviewed separately as below.

Step 1: Data Collection

There are several approaches proposed in designing experiments. Response surface designs such as central composite design (CCD) and Box–Behnken can be used to fit the quadratic response surface model, which has linear and curvilinear effects for each of the factors along with all pairwise interaction effects. Box–Behnken design is known as economical design which requires only 3 levels for each factor and consists of a particular subset of the factorial design (khuri & Mukhopadhyay, 2010). Factorial design is applied to evaluate two or more factors simultaneously. This method divided into three parts such as; one-factor-at-a-time, full and fractional design which used to find the linear response (Montgomery, 2005). In order to find optimal design, several criteria proposed for multi response experiments. There are many methods of achieving optimal design such as simultaneous experiment design or sequential design (Ramachandran & Tsokos, 2015). Sequential design proposed by (Sitter & Forbes, 1997; Sitter & Wu, 1999). The most common criterion for multi response design is D-optimality, which expanded for linear multiresponse designs. (Atashgah & Seifi, 2007; Bischoff, 1993; Chang, 1994, 1997; Imhof, 2000; Kovach, 2009). Other criteria presented to find optimal design are; A-optimal, E-optimal (Wong, 1994), block design (Box & Draper, 1972; Das, Mandal & Sinha, 2003), and Bayesian design (Chaloner & Larntz, 1992; Chaloner & Vardinelli, 1995).

Step 2: Statistical Analysis and Model Fitting

Statistical model between the responses and factors determine settings on the design variable for analysis such as optimization, which produce optimal or at least acceptable values of the responses. There are several estimation techniques in RSM. Ordinary least square (OLS) is the main approach where there is no dependency among the variables. Principal component regression is another method for fitting where there is a meaningful correlation either among responses or factors. Seemingly unrelated regression (SUR) is used to find the correlated multiple response surfaces by a single-step calculation (Castillo, 2009; Hejazi, Seyyed-Esfahani & Mahootchi, 2014; Peterson, Miró-Quesada & Del Castillo, 2009; Shah, Montgomery, & Carlyle, 2004). Also principal component analysis (PCA) is the most common method to solve problem with property of correlation among the responses (Augusto & Ferreira, 2010; Mehrjoo & Bashiri, 2013; Riberio, Teófilo, Augusto, & Ferreira, 2010; Salmasnia, Baradaran Kazemzadeh, & Niaki, 2012). In particular, logistic regressions, probit model, Linear Probability Model (LPM) are presented as flexible methods for statistical modeling of categorical dependent variables. OLS consider as an inappropriate method for qualitative and limited dependent variable. Hence logit and probit models extended as common approaches, which applied in qualitative analysis (Feinstein & Thomas, 2002). Method Logit analysis is determined by prediction of probability of the event, which transformed to the logistic regression based on a ratio of chances (Kollár, 2014). Logistic regression is employed almost in situation with discrete value of response variable. In this case, classical regression which used for continuous responses is not applicable. Logit and probit models are antecedent (Spuchľáková & Cúg, 2014). Whereas probit assume normal distribution of independent variable, logistic regression does not consider normal distribution. Logistic regression eliminates disadvantage of discriminant analysis (Kollár & Cisko, 2014). Logistic regression has become the standard model for evaluating the functional relationship between the dependent variable and a set of independent variable, among the many binary and multi-level categorical response models. Three types of logistic regression model applied, according to the type of responses (ordinal, nominal, binary); (Ezgi, Anaguan, & Koksal, 2009; Huang, Biro, & Dorn, 2009; Lawson & Montgomery, 2006; Zhou, Wu, Yang, & Jiao, 2008). Recent logistic regression method explains the transformation of discrete variable into a continuous function (Kotu & Deshpande, 2015).

Step 3: Optimization

Desirability function, Taguchi loss function and classical optimization approaches are formally applied to optimize multi response problems in certain condition. Several studies have presented approaches addressing multiple quality characteristics. Desirability function is one of the common methods which convert multiple response optimization (MRO) problems into a single objective optimization problem, introduced by Harrington (1965) and proved by (Del castillo, 1996; Derringer & suich, 1980; Kim & Lin, 2006; Wu, 2005). Taguchi loss function recognized as a method which reduce the quality variation and bring the mean close to the target value, simultaneously (Tong, Wang, & Chen, 2005). Considering correlation among responses in providing single response for optimization, proposed by (Khuri & Conlon, 1981; Vining, 1998). In order to consider correlation structure of the responses in multiple optimization problem, Weighted Principle Component applied by Taguchi method (Routara, Mohanty, Datta, Bandyopadhyay, & Mahapatra, 2010).Utilizing goal programming method for obtaining non-dominated solutions which ranked by TOPSIS approach, proposed by (Ramezini, Bashiri, & Atkinson, 2011). In study of (Bashiri & Hejazi, 2012; Hejazi, Bashiri, Díaz-García, & Noghondarian, 2012; Hejazi et al., 2014), Mathematical programing constructed for considering correlation among response, based on PCA and goal programming model to handle multi objective problem. General framework for optimization of multi response problem based on goal programming model proposed by (Kazemzadeh, Bashiri, Atkinson, & Noorossana, 2008). In some cases, independent, dependent variables and some parameters have stochastic character. This uncertainty discussed in various stage of multi objective optimization process (Chiao & Hamada, 2001; Hejazi et al., 2012; Kazemzadeh et al., 2008).

The main idea of this chapter is to address the last two phases of RSM where the inputs and outputs are measured by categorical ways. There are some difficulties in optimization of business processes due to the essence of output variables that are binary, nominal, ordinal. Therefore logistic regression presented to model these types of data (Lawsonl et al., 2006). Measuring and assessment of categorical variables lead to inclusion in of one of the predetermined categories. If data analyzed as independent variable, commonly analysis of variance approach or dummy variable utilized (Hosmer & Lemeshow, 2000).

MATERIAL AND METHODS

Phase 1: Design

In the first phase of RSM method, designers conduct experiments as a series of runs, in order to identify cause – and - effect relationships of the process, whenever some input variables change. Experimental design known as substantial tool in both scientific and engineering fields, which implies in designing and formulating a new product, manufacturing process development, reducing total cost and developing time, marketing and service operations. Hence, applying experimental design influences on final products, which manufactured them easier and more reliable.

There are two aspects of experimental design:

- The design of experiments and
- Statistical design of experiments.

In the existence of experimental errors, statistical analysis consider as an efficient method. Using statistical design of experiments for developing products that are robust to environmental factors and other sources of variability is very important in the field of engineering problems. Robust design approach tries to select the levels of controllable factors, to attach a level of confidence to a statement. Whereas, statistical methods, only measure the errors arises from uncontrollable and unavoidable process, and doesn't involve in a process of choosing a specific factor.

In recent years, designed of experiments have been implied in real case study of financial services, telecommunications, health care, e-commerce, legal services, marketing, logistics and transportation (Montgomery, 2012).

Phase 2: Modeling

In some recent complex systems, the relation between dependent and independent variables are unknown, in terms of lack of information related to the engineering, chemical or physical principles. In this situation, experimenters approximate the unknown function with empirical model, which referred to response surface model. Fitting response surface model is based on experimental design, model building techniques and optimization methods, which known as RSM (Myers et al., 2009). RSM began with the work of Box and Wilson (1951), and illustrated as an iterative heuristic technique, for optimizing real systems, namely chemical systems. This methodology, consider as multi stage method; first uses linear polynomial models, mainly first degree and second degree model. This method figures out OLS as the best linear estimator, due to the white noise assumption of responses. To estimate first degree polynomials, RSM applies classic design of resolution ɪɪɪ, whereas CCD applies for second degree polynomials. In order to consider in which direction the factors will be changed, gradient implies to fit first degree polynomial through a full or fractional factorial design. At the end, RSM implies mathematical programming to the second degree polynomial to examine the optimal response (Barton & Meckesheimer, 2006; Kleijnen, 2008). RSM has been utilized as an applicable method in recent case studies (khuri, 2006; Merkuryeva, 2005; Rosen, Harmonosky, & Traband, 2007).

Many distribution functions have been proposed for statistical analysis of dichotomous responses, in the process of RSM method. Logistic regression consider as the most practical way, because of the flexible and easily used function and also meaningful interpretation result (Cox & Snell, 1989). For fitting logistic regression, Maximum likelihood method applies, instead of uses least squares in linear regression. After estimating the coefficients of the variable, the significance of the coefficients obtain similar to the linear regression, also by considering likelihood function (Hosmer et al., 2000). In order to check the signification of adding new variables, log likelihood and likelihood ratio tests are standard features for logistic regression software. Likelihood ratio tests became the most beneficial test among wald test and score test (Jennings, 1986).

Phase 3: Optimization

Goal programming is one of the most practical techniques for multiple objective decision analysis, which affected from the field of mathematical programming and multiple criteria decision making (MCDM). This method characterized due to the types of mathematical programming models such as linear programming, integer programming and nonlinear programming (Hillier & Lieberman, 2001). Goal programming model seeks to consider all of the feasible combination of the decision variables, in order to achieve the goals or targets of each objective function, which assigned by decision makers. Objective function tries to minimize any deviation between the target level and the value that is able to be achieved in a given solution, weighted and prioritized (Jones & Tamiz, 2010). Goal programming propound as useful method where the different objectives of the problem cannot reach their optimal values simultaneously, due to the conflicting nature of the multiple goal. Hence, this method cited as satisfying technique rather than optimizing method.

Goal programming models categorized in terms of the mathematical nature of decision variables. For instance, adding binary or integer variables causes goal programming model as a practical technique, for formulating both logical condition and multiple conflicting goals. Modeling integer goal programming makes some limitation on decision variables and constraints, which increased solution time. More over in some issues, objective function, goals and constraints occur in nonlinear format). If nonlinear goal programming model cannot be solved by linearization techniques, heuristic or meta- heuristic method, such as genetic algorithm utilize as an effective way to solve large- scale problem (Deb, 2001; Leung, 2007; Taguchi, Ida, & Gen, 1997). Chebyshev goal programming implies minimization of the maximum deviation from the set of goals. This method has an important feature, which give the most appropriate solution when decision makers desire to balance between the levels of satisfaction of the targets. The preference of chebyshev goal programming is that there is no need to get preference information such as weighs and priorities about decision maker's targets. The most popular variants of goal programming method is, fuzzy goal programming, due to the fuzzy nature of the data, priority structure and target levels regarded to the decision makers (Jones et al., 2010; Ravindran, 2008). To solve fuzzy goal programming problems, several approaches have been provided, according to the classification of the model (Azaron, Furmans, & Modarres, 2010; Chanas & Kuchta, 2002; Yaghoobi, Jones, & Tamiz, 2008; Yaghoobi & Tamiz, 2007).

PROPOSED APPROACH

In presented approach, optimization of multiple responses is discussed by applying mathematical programming approach in which, either input and output variables are categorical. In this case several steps performed by utilizing RSM methodology. The steps can be explained as follows. First, control and response variables determined by experimental result. In statistical modeling section, input variables modified to dummy variables. Interaction between variables considered by defining new variable. Then for constructing functional relationship between input and response variables, a regression equation is employed. In the last step, optimization of multi response problem proposed to maximize the probability of each category. Figure 1 shows a flowchart representing steps of the methodology.

Mathematical Modeling

In this section, definitions of parameters and variables are provided and then, the structure of model building is discussed.

Parameter Definition

Sets:

$N = \{i = 1, \ldots, n\}$ Number of experimental runs

Figure 1. Flowchart view of proposed method

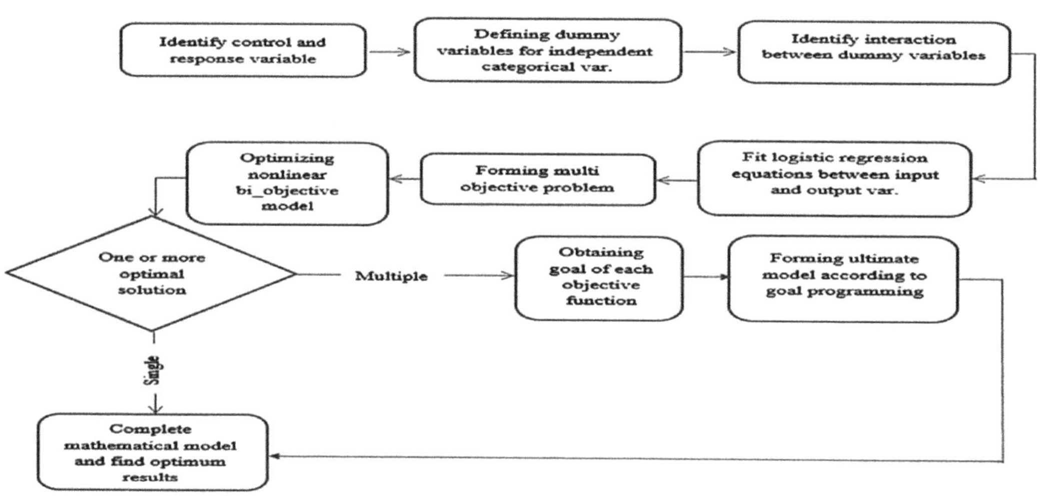

Index:

J The number of categories for response variable

Parameters and Variables:

$y_i = \left(y_{i1}, y_{i2}, \ldots, y_{ij}\right)^{T}$ Vector of response variable

$\beta_j = \left(\beta_{j1}, \beta_{j2}, \ldots, \beta_{jk}\right)$ Coefficients vector of control variables

$D_i = \left(D_{i1}, D_{i2}, \ldots, D_{ik}\right)^{T}$ Vector of independent variable (dummy variables)

Π_j Probability of being in each of the categories

$\Pi_{j_{\max}}$ Function with highest priority

$\eta_{ij} = \theta_j + D_i^{T}\beta_j$ linear combination of multiple explanatory variables

$p\left(y \le j\right) = \pi_1 + \ldots + \pi_j$ Cumulative event probabilities of response variables

$\text{Odds}\left(y \le j\right) = \dfrac{p\left(y \le j\right)}{1 - p\left(y \le j\right)} = \dfrac{\pi_j}{1 - \sum_{j=1}^{j-1} \pi_j}$ The ratio of cumulative event probability for j^{th} category to the summation of category j^{th} to j-1(odds ratio of jth category)

$\ln\left(\dfrac{\pi_j}{1 - \sum_{j=1}^{j-1} \pi_j}\right) = T_m$ Natural logarithm of the last probability odds

T_m The target value of mth objective function

d_m, $d_m^{'}$ Negative and positive deviation from the target

W_m Relative importance of mth response variable

$\widehat{y_m}(D)$ Estimated logistic regression

$F(X)$ Objective function

Equations

In this subsection, authors first consider the analysis of ordinal logistic regression and then discuss binary logistic regression in summary. The probability of being in each of the categories is given by Equation (1).

$$\pi_j = \frac{e^{\eta_{ij}}}{1+e^{\eta_{ij}}} \qquad j=1,..,j\text{-}1 \tag{1}$$

Equation (1) is named as logistic function or logit, but it is nonlinear with respect to parameters of model. The value of cumulative event probabilities of response variables are calculated by Equation (2).

$$p(y \le j) = \pi_1 + \ldots + \pi_j \qquad j=1,..,j\text{-}1 \tag{2}$$

The ratio of cumulative event probability for j^{th} category to the summation of category j^{th} to j-1, known as odds ratio shown in Equation (3).

$$\text{Odds}(y \le j) = \frac{p(y \le j)}{1 - p(y \le j)} = \frac{\pi_j}{1 - \sum_{j=1}^{j-1} \pi_j} = \exp\left(D_i^T \beta_j\right) \qquad j=1,..,j\text{-}1 \tag{3}$$

Defining binary logistic regression is as same as ordinal one, but probability of response variable is calculated as below.

$$p(y_j = 1) = \pi_j$$

$$p(y_j = 0) = 1 - \pi_j$$

Mathematical Programming

In this section, mathematical programming constructed for optimizing multi response problem. In ordinal logistic regression, variables are categorized from lowest to the highest utility. Hence, the maximum probability function must be obtained to construct the final optimization model.

$$\text{Max}\,\Pi_{j_{\max}} \tag{4}$$

Subject to:

$$\pi_{j+1} > \pi_j \;\; \forall j = 1,2,\ldots, j-1 \tag{5}$$

$$\sum_{j=1}^{2} \text{D}_{ij} \leq 1 \quad \text{for all } i \tag{6}$$

$$D_{ij}\text{: Binary} \tag{7}$$

Constraint 5: since the aim of ordinal logistic regression is to maximize the function with highest priority, and variables are categorized from lowest to the highest utility, so this constraint should be established.

Constraint 6: to allocate only one value to independent variable.

Constraint 7: dummy variables consider as binary.

Goal Programming Approach

Goal programming is a multi-objective method developed to minimize both positive and negative deviation of objective values from the corresponding targets, which are determined by Decision Maker (DM). However DM may not be able to determine the exact importance of the goals according to the function. In this study probability of each category is given by DM as previous information. So discussion about calculating the goal of each objective function is presented by Converting the goal to the binary and ordinal logit as shown in below:

$$\text{Odds}\left(y \leq j\right) = \frac{\text{p}\left(y \leq j\right)}{1-\text{p}\left(y \leq j\right)} = \frac{\pi_j}{1-\sum_{j=1}^{j-1}\pi_j} \qquad j=1,.., j\text{-}1 \tag{8}$$

By calculating the natural logarithm of the last probability odds, the goal of mth objective function calculated by Equation (9).

$$\ln\left(\frac{\pi_j}{1-\sum_{j=1}^{j-1}\pi_j}\right) = T_m \qquad j=1,.., j\text{-}1 \tag{9}$$

The general form of goal programming can be developed in terms of defined variables in 2.1.1. In this formulation two types of variables; deviational variables and decision variables are utilized. There are two classes of constraints, system constraints and goal constraints. Goal constraints express both negative and positive deviational variable, which infraction of the goals, will not make the solution infeasible as Equation (11). System or (hard) constraints are restrictions based on decision variables which could be equality or an inequality constraint as Equation (12).

The basic idea of goal programming is to minimize under and over achievements of the objective function; see Equation (10). Equation (11), investigate how well each of the target levels attained by studying the deviational variables. Both negative and positive deviational variables are consider as positive variables as Equation (13).

$$\text{Min } F(x) = \sum_{m=1}^{M} W_m \left(d_m + d_m^{'} \right) \tag{10}$$

Subject to:

$$\widehat{y_m}\left(D\right) + d_m - d_m^{'} = T_m \ \forall m = 1,\ldots, M \tag{11}$$

$$g_i\,(x) \leq 0 \ \forall i = 1,\ldots, k \tag{12}$$

$$d_m, d_m^{'} \geq 0 \tag{13}$$

Real Numerical Example

In this section in order to provide a better interpretation of the proposed method a real problem from engineering applications (friction stir welding process) analyzed. Friction stir welding (FSW) was first invented in 1991 in The Welding Institute (TWI), which produces better weld mechanical properties in some aluminum alloys. In order to use FSW for producing tailor welding blank (TWB) sheets, the factors of this process detected by Box-Behnken design, and applied in an optimized way. It should be noticed that, many researches have been presented for optimizing FSW parameters. RSM method applied to maximize yield strength by optimizing FSW parameters (Ghaffarpour, Aziz, & Hejazi, 2015). The factors and response variables of friction stir welding process have been defined in Table 1. The process of friction stir welding is shown in Figure 2.

Table 1. Variable definition

Factors	Responses	Analyzing Tool for Modeling
Rotational speed (x_1) Linear motion speed (x_2) Shoulder diameter tool (x_3)	Elongation	Ordinal logistic regression
	Tensile strength	Binary logistic regression

Figure 2. Schematic of friction stir welding process

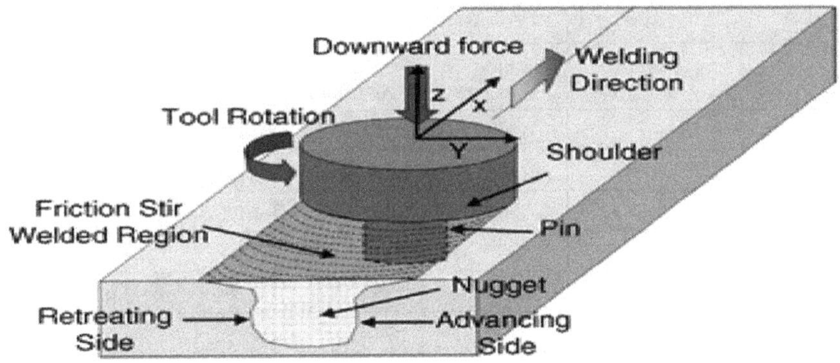

For categorizing independent variable, dummy variable dedicate to them as shown in Table 2.

By discussing the results of Minitab software which used to model binary and ordinal logistic regression, "shoulder diameter tool" is not significant in the both model. So this variable is omitted. Frequency of each response variable after omitting variable X_3, shown in Table 3.

In order to consider the interaction effect between dummy variables in each logit regression, first investigate that if this interaction is significant or not. Results show that this attribute is only significant in ordinal logit regression. Hence, it can be considered by variable D_{1221} defined as below.

$$D_{12}{*}D_{21}=D_{1221}$$

Estimation of ordinal logistic regression and probability of each surface are shown below:

Ordinal Logit

$$\eta_{12} = 3.75076\text{-}1.77667\, D_{11}\text{-}2.7785\, D_{12}\text{-}5.11443\, D_{21} +3.01062\, D_{1221}$$

$$\eta_{11} = \text{-}1.77667\, D_{11}\text{-}2.7785\, D_{12}\text{-}5.11443\, D_{21} +3.01062\, D_{1221}$$

Table 2. Defined dummy variables for independent variables

Value of Independent Variable		Dummy Variable			
		D_{11}	D_{12}	D_{21}	D_{31}
X_1	1 2 3	0 0 1	0 1 0		
X_2	1 2			0 1	
X_3	1 2				0 1

Table 3. Frequency of response surfaces

ID	X_1	X_2	Y_1			Y_2	
			First Category	Second Category	Third Category	First Category	Second Category
1	1	1	0.75	0.125	0.125	1	
2	1	2		0.25	0.75	0.25	0.75
3	2	1		0.875	0.125	0.5	0.5
4	2	2		0.25	0.75		1
5	3	1	0.5	0.25	0.25	0.875	0.125
6	3	2			1	1	

$$p(y \le 2) = \pi_1 + \pi_2 = \frac{e^{3.75076 - 1.77667D_{11} - 2.7785D_{12} - 5.11443D_{21} + 3.01062D_{1221}}}{1 + e^{3.75076 - 1.77667D_{11} - 2.7785D_{12} - 5.11443D_{21} + 3.01062D_{1221}}}$$

$$p(y \le 1) = \pi_1 = \frac{e^{-1.77667D_{11} - 2.7785D_{12} - 5.11443D_{21} + 3.01062D_{1221}}}{1 + e^{-1.77667D_{11} - 2.7785D_{12} - 5.11443D_{21} + 3.01062D_{1221}}} = p_{11}$$

$$p(y \le 2) - p(y \le 1) = p_{12}$$

$$1 - p(y \le 2) = \pi_3 = \frac{1}{1 + e^{3.75076 - 1.77667D_{11} - 2.7785D_{12} - 5.11443D_{21} + 3.01062D_{1221}}} = p_{13}$$

Binary Logit

η_{21} = -2.41424 - 2.7004 D_{11}+2.56225 D_{12} +3.11296 D_{21}

$$p(y_2 = 2) = \frac{e^{-2.41424 - 2.7004D_{11} + 2.56225D_{12} + 3.11296D_{21}}}{1 + e^{-2.41424 - 2.7004D_{11} + 2.56225D_{12} + 3.11296D_{21}}} = P_{22}$$

$$p(y_2 = 1) = \frac{1}{1 + e^{-2.41424 - 2.7004D_{11} + 2.56225D_{12} + 3.11296D_{21}}} = P_{21}$$

Mathematical Programming

$$\text{Max } Z_1 = \frac{1}{1+e^{3.75076-1.77667D_{11}-2.7785D_{12}-5.11443D_{21}+3.01062D_{1221}}}$$

$$\text{Max } Z_2 = \frac{e^{-2.41424\ -\ 2.7004D_{11}+2.56225D_{12}+3.11296D_{21}}}{1+e^{-2.41424\ -\ 2.7004D_{11}+2.56225D_{12}+3.11296D_{21}}}$$

St:

$$\frac{e^{3.75076-1.77667D_{11}-2.7785D_{12}-5.11443D_{21}+3.01062D_{1221}}}{1+e^{3.75076-1.77667D_{11}-2.7785D_{12}-5.11443D_{21}+3.01062D_{1221}}} > 2\frac{e^{-1.77667D_{11}-2.7785D_{12}-5.11443D_{21}+3.01062D_{1221}}}{1+e^{-1.77667D_{11}-2.7785D_{12}-5.11443D_{21}+3.01062D_{1221}}}$$

$D_{11}+D_{12}\leq 1$

D_{ij}: Binary

Optimization of Bi-Objective Model

Optimization of each response surface is solved by nonlinear programming solver built in Lingo 11 optimization software. The results are shown in Table 4.

Goal Programming Model

In this subsection, an agreement answers of two objective functions obtained by goal programming approach. According to the goal ratio of each surface response in Tables 5 and 6, the goal value of each objective functions obtained as shown.

Table 4. Value of objective functions and probability of each category

Objective Functions	Z_1 $D_{11}, D_{21}=1$	Z_2 $D_{12}, D_{21}=1$	P_{11}	P_{12}	P_{13}	P_{21}	P_{22}
Z_1	-3.14.34	1.13155	0.00101	0.04045	0.95852	0.88097	0.11902
Z_2	-2.00168	3.26097	0.00752	0.23635	0.75612	0.03693	0.9603

Table 5. Goal ratio of objective functionY_1

Category	Goal Ratio
1	0.01
2	0.04
3	0.95

Table 6. Goal ratio of objective functionY_2

Category	Goal Ratio
1	0.05
2	0.95

Goal Value of Y_1:

$$\text{Odds ratio} = \frac{0.04+0.01}{1-\left(0.04+0.01\right)} = 0.0526$$

By calculating natural logarithm of odds ratio, the goal value of objective function is -2.9444
Goal Value of Y_2:

$$\text{Odds ratio} = \frac{0.95}{0.05} = 19$$

Finally the goal value of objective function obtained equals to 2.9444.
In this study by considering sensitivity analysis of goal ratio, each time optimal solution is same. So the goal programming of this model is shown in below:

$$\text{MinZ} = d_1 - d_1^{'} - d_2 + d_2^{'}$$

$$\frac{1}{1+e^{3.75076-1.77667D_{11}-2.7785D_{12}-5.11443D_{21}+3.01062D_{1221}}} - d_1 + d_1^{'} = -2.9444$$

$$\frac{e^{-2.41424-2.7004D_{11}+2.56225D_{12}+3.11296D_{21}}}{1+e^{-2.41424-2.7004D_{11}+2.56225D_{12}+3.11296D_{21}}} - d_2 + d_2^{'} = 2.9444$$

$$\frac{e^{3.75076-1.77667D_{11}-2.7785D_{12}-5.11443D_{21}+3.01062D_{1221}}}{1+e^{3.75076-1.77667D_{11}-2.7785D_{12}-5.11443D_{21}+3.01062D_{1221}}} > 2\frac{e^{-1.77667D_{11}-2.7785D_{12}-5.11443D_{21}+3.01062D_{1221}}}{1+e^{-1.77667D_{11}-2.7785D_{12}-5.11443D_{21}+3.01062D_{1221}}}$$

$D_{11}+D_{12}\leq 1$

D_{11}, D_{12}, D_{21}, D_{1221}: Binary

The result of optimal solution of each objective function is shown in below (obtained by Lingo 11 optimization software):

d_1=1. 81285, d_2=0. 31657, D_{11}=0, D_{12}=1, D_{21}=1

So x_1=2, x_2=2 Optimal levels of y_1, y_2 are respectively the third and second ones with 75 and 100 percent of occurrence.

FUTURE RESEARCH DIRECTIONS

The numerical results of friction stir welding process demonstrate the benefits of the proposed approach for modeling and optimization parts of RSM method. The proposed approach is feasible and easy to implement in other manufacturing and service industries. Several aspects of this study can be expanded by considering correlation, dispersion effects between responses by utilizing PCA, OLS, and SUR methods for instance. However, statistical methods make some restrictions which need to comply assumptions such as normality of the data and the same error variance. Hence, utilizing meta-heuristic method such as Artificial Neural Network (ANN) may omit these limitations, and resulting high reliability in recognizing and removing non-random fluctuations in the data, as well as exploring the interactions between variables. In order to optimize response variables, meta-heuristic methods can apply by estimating and improving the regression coefficient. To consider uncertainty in the case of vague nature of systems, in statistical modeling part, fuzzy logistic regression can be applied. In this circumstance, explanatory and response variables are non-precise and measured by linguistic terms.

CONCLUSION

In this chapter, response surface methodology proposed as statistical mathematical tool to find optimal conditions of operation process. The results of experiments gathered as categorical input variables which defined by dummy variables. In model fitting section, Logistic regression is applied to model discrete responses. Also, interaction between dummy variables has been considered by defining a new variable. Afterwards, in order to maximize the probability of each category, mathematical programming proposed based approach nonlinear goal programming method.

Specific strengths of this chapter, consider as proposing mathematical programming for optimization of multi objective problems, which in real world designers seeks to optimize several performance of contradict attributes. Logistic regression and dummy variable provided as a statistical modeling technique.

The weakness of the chapter point as neglecting uncertainty, which can result in deceptive parameter estimates and may cause weak process design. While the dimensional of the problem extended, it is difficult to obtain exact model, due to insufficient knowledge of the systems.

REFERENCES

Atashgah, A. B., & Seifi, A. (2007). Application of semi-definite programming to the design of multi-response experiments. *IIE Transactions*, *39*(7), 63–769. doi:10.1080/07408170701245353

Azaron, A., Furmans, K., & Modarres, M. (2010). Nurse Scheduling by Fuzzy Goal Programming. In D. Jones, M. Tamiz, & J. Ries (Eds.), *New Developments in Multiple Objective and Goal Programming* (pp. 154–163). Heidelberg, Germany: Springer.

Barton, R. R., & Meckesheimer, M. (2006). Metamodel- Based Simulation Optimization. In S.G. Henderson & B.L. Nelson (Eds.), *Handbooks in Operations Research and Management Science* (pp. 535– 574). Amsterdam: Elsevier.

Bashiri, M., & Hejazi, T. (2009). An extension of multi-response optimization in MADM view. *Journal of Applied Sciences*, *9*(9), 1695–1702. doi:10.3923/jas.2009.1695.1702

Bashiri, M., & Hejazi, T. H. (2012). A mathematical model based on principal component analysis for optimization of correlated multiresponse surfaces. *Journal of Quality*, *19*(3), 223–239.

Bischoff, W. (1993). On D-optimal designs for linear models under correlated observations with an application to a linear model with multiple response. *Journal of Statistical Planning and Inference*, *37*(1), 69–80. doi:10.1016/0378-3758(93)90080-P

Box, G. E. P., & Draper, N. R. (2007). Introduction to Response Surface Methodology. In S. Quigley & S. S. Filler (Eds.), *Response Surfaces, Mixtures and Ridge Analyses* (pp. 1–16). Wiley and Sons. doi:10.1002/9780470072769.ch1

Box, G. E. P., & Wilson, K. B. (1951). On the experimental attainment of optimum conditions. *Journal of the Royal Statistical Society. Series B. Methodological*, *13*(1), 1–38.

Box, M. J., & Draper, N. R. (1972). Estimation and design criteria for multiresponse non-linear models with non-homogeneous variance. *Applied Statistics*, *21*(1), 13–24. doi:10.2307/2346599

Castillo, E. (2009). A Bayesian reliability approach to multiple response optimization with seemingly unrelated regression models. *Quality Technology & Quantitative Management*, *6*(4), 353–369. doi:10.1080/16843703.2009.11673204

Chaloner, K., & Larntz, K. (1992). Bayesian design for accelerated life testing. *Journal of Statistical Planning and Inference*, *33*(2), 245–259. doi:10.1016/0378-3758(92)90071-Y

Chaloner, K., & Verdinelli, I. (1995). Bayesian experimental design: A review. *Statistical Science*, *10*(3), 273–304. doi:10.1214/ss/1177009939

Chanas, S., & Kuchta, D. (2002). Fuzzy goal programming- one notion, many meanings. *Control Cybern*, *4*(31), 871–890.

Chang, S. (1997). An algorithm to generate near D-optimal designs for multiple response surface models. *IIE Transactions*, *29*(12), 1073–1081. doi:10.1080/07408179708966426

Chang, S. I. (1994). Some properties of multiresponse D-optimal designs. *Journal of Mathematical Analysis and Applications*, *184*(2), 256–262. doi:10.1006/jmaa.1994.1198

Chiao, C. H., & Hamada, H. (2001). Analyzing Experiments With Correlated Multiple Responses. *Journal of Quality Technology*, *33*(4), 451–465.

Cox, D. R., & Snell, E. J. (1989). *Binary response variables. Analysis of Binary Data* (2nd ed.). London, UK: Chapman & Hall.

Das, K., Mandal, N. K., & Sinha, B. K. (2003). Optimal experimental designs for models with covariates. *Journal of Statistical Planning and Inference*, *115*(1), 273–285. doi:10.1016/S0378-3758(02)00151-9

Del Castillo, E. (1996). Multi response process optimization via constrained confidence regions. *Journal of Quality Technology*, *28*(1), 61–70.

Del Castillo, E., & Montgomery, D. C. (1993). A nonlinear programming solution to the dual response problem. *Journal of Quality Technology*, *25*(3), 199–204.

Derringer, G., & Suich, R. (1980). Simultaneous optimization of several response variables. *Journal of Quality Technology*, *12*(4), 214–219.

Ezgi Aktar, D., Anagun, A. S., & Koksal, G. (2009). Determination of optimal product styles by ordinal logistic regression versus conjoint analysis for kitchen faucets. *International Journal of Industrial Ergonomics*, *39*(5), 866–875. doi:10.1016/j.ergon.2009.06.007

Feinstein, C. H., & Thomas, M. (2002). Logit, probit, and tobit models. In J. Mackenzie (Ed.), *Making History Count: A Primer in Quantitative Methods for Historians* (pp. 384–434). Cambridge, UK: Cambridge University Press. doi:10.1017/CBO9781139164832.014

Ghaffarpour, M., Aziz, A., & Hejazi, T. H. (2015). Optimization of friction stir welding parameters using multiple response surface methodology. *Journal of Materials Design and Applications*, *0*(0), 1–13.

Harrington, E. C. (1965). The desirability function. Industrial Quality Control. *American Journal of Operations Research*, *21*(10), 494–498.

Hejazi, T. H., Bashiri, M., Díaz-García, J. A., & Noghondarian, K. (2012). Optimization of probabilistic multiple response surfaces. *Applied Mathematical Modelling*, *36*(3), 1275–1285. doi:10.1016/j.apm.2011.07.067

Hejazi, T. H., Seyyed-Esfahani, M., & Mahootchi, M. (2014). Optimization of degree of conformance in multiresponse–multistage Systems with a Simulation-based meta heuristic. *Quality and Reliability Engineering International*, *31*(4), 645–658. doi:10.1002/qre.1622

Hillier, F. S., & Lieberman, G. J. (2001). Other Algorithms for Linear Programming. In E.M. Munson. In M. Lorkovic (Ed.), *Introduction to the operational research* (pp. 332–340). New York, NY: Mc Graw-Hill.

Hosmer, D. W., & Lemeshow, S. (2000). *Applied Logistic Regression* (W. A. Shewhart & S. S. Wilks, Eds.). New York, NY: John Wiley & Sons. doi:10.1002/0471722146

Huang, B., Biro, F. M., & Dorn, L. D. (2009). Determination of Relative Timing of Pubertal Maturation through Ordinal Logistic Modeling: Evaluation of Growth and Timing Parameters. *The Journal of Adolescent Health*, *45*(4), 383–388. doi:10.1016/j.jadohealth.2009.02.013 PMID:19766943

Imhof, L. (2000). Optimum designs for a multi response regression model. *Journal of Multivariate Analysis*, *72*(1), 120–131. doi:10.1006/jmva.1999.1841

Jennings, D. E. (1986). Judging inference adequacy in logistic regression. *Journal of the American Statistical Association*, *81*(394), 471–476. doi:10.1080/01621459.1986.10478292

Jones, D., & Tamiz, M. (2010). Practical goal programming. In F. S. Hillier (Ed.), *International Series in Operations Research & Management Science* (pp. 1–170). New York, NY: Springer.

Kazemzadeh, R. B., Bashiri, M., Atkinson, A. C., & Noorossana, R. (2008). A general framework for multiresponse optimization problems based on goal programming. *European Journal of Operational Research*, *189*(2), 421–429. doi:10.1016/j.ejor.2007.05.030

Khuri, A. I. (2006). Response Surface Methods and Their Application in the Treatment of Cancer with Drug Combinations: Some Reflections. In A. I. Khuri (Ed.), *Response Surface Methodology and Related Topics* (pp. 159–173). Singapore: World Scientific Printers. doi:10.1142/5915

Khuri, A. I., & Conlon, M. (1981). Simultaneous optimization of multiple responses represented by polynomial regression functions. *Technometrics*, *23*(4), 363–375. doi:10.1080/00401706.1981.10487681

Khuri, A. I., & Mukhopadhyay, S. (2010). Response Surface Methodology. *Wiley Interdisciplinary Reviews: Computational Statistics*, *2*(2), 128–149. doi:10.1002/wics.73

Kim, K., & Lin, D. (2006). Optimization of multiple responses considering both location and dispersion effects. *European Journal of Operational Research*, *169*(1), 133–145. doi:10.1016/j.ejor.2004.06.020

Kim, K. J., & Lin, D. K. J. (1998). Dual response surface optimization: A fuzzy modeling approach. *Journal of Quality Technology*, *30*(1), 1–10.

Kleijnen, J. P. C. (2008). Simulation optimization. In F. S. Hillier (Ed.), *Design and Analysis of Simulation Experiments* (pp. 101–138). New York, NY: Springer. doi:10.1007/978-0-387-71813-2_4

Kollár, B. (2014). *Credit value at risk and options of its measurement*. Paper presented at International Conference on Economics and Social Science (ICESS), Information Engineering Research Institute, Advances in Education Research, Shenzhen, China.

Kollár, B., & Cisko, Š. (2014). *Credit risk quantification with the use of Credit Risk*. Paper presented at International Conference on Management, Education, Business, and Information Science, Durgapur, India.

Kotu, V., & Deshpande, B. (2015). Regression methods. In S. Elliot (Ed.), *Predictive Analytics and Data Mining* (pp. 165–193). Waltham, MA: Elsevier. doi:10.1016/B978-0-12-801460-8.00005-7

Kovach, J., & Cho, B. R. (2009). A D-optimal design approach to constrained multiresponse robust design with prioritized mean and variance considerations. *Computers & Industrial Engineering*, *57*(1), 237–245. doi:10.1016/j.cie.2008.11.011

Lawsonl, C., & Montgomery, D. C. (2006). logistic regression analysis of customer satisfaction data. *Quality and Reliability Engineering International*, *22*(8), 971–984. doi:10.1002/qre.775

Leung, S. C. H. (2007). A non-linear goal programming model and solution method for the multi-objective trip distribution problem in transportation engineering. *Optimization and Engineering*, *8*(3), 277–298. doi:10.1007/s11081-007-9019-x

Lin, D. K. J., & Tu, W. (1995). Dual response surface optimization. *Journal of Quality Technology*, *27*(1), 34–39.

Mehrjoo, S., & Bashiri, M. (2013). An application of principal component analysis and logistic regression to facilitate production scheduling decision support systems: An automative industry case. *Journal of Industrial Engineering International*, *9*(9), 1–12.

Merkuryeva, G. (2005). Response surface-based simulation meta modeling methods with applications to optimization problems. In A. Dolgui, J. Soldek, & O. Zaikin (Eds.), *Supply chain optimization product / process design, facility location and flow control* (pp. 205–215). Boston, MA: Springer.

Montgomery, D. C. (2005). *Design and Analysis of Experiments*. New York, NY: John Wiley & Sons, Inc.

Montgomery, D. C. (2012). *Design and Analysis of Experiments* (L. Ratts & A. Melhorn, Eds.). John Wiley & Sons, Inc.

Myers, R. H. (1999). Response Surface Methodology: Current Status and Future Directions (with discussion). *Journal of Quality Technology*, *31*(1), 30–44.

Myers, R. H., Montgomery, D. C., & Anderson Cook, C. M. (2009). *Response Surface Methodology: Process and Product Optimization Using Designed Experiments* (3rd ed.). John Wiley and Sons.

Myers, R. H., Montgomery, D. C., Vining, G. G., Borror, C. M., & Kowalski, S. M. (2004). Response surface methodology: A retrospective and literature survey. *Journal of Quality Technology*, *36*(1), 53–77.

Peterson, J. J., Miró-Quesada, G., & Del Castillo, E. (2009). A Bayesian reliability approach to multiple response optimization with seemingly unrelated regression models. *Quality Technology & Quantitative Management*, *6*(4), 353–369. doi:10.1080/16843703.2009.11673204

Ramachandran, K. M., & Tsokos, C. P. (2015). Design of experiment. In J. Cetel & A. Sambamoorthy (Eds.), *Mathematical Statistics with Applications in R* (2nd ed., pp. 459–494). London, UK: Elsevier. doi:10.1016/B978-0-12-417113-8.00009-6

Ramezani, M., Bashiri, M., & Atkinson, A. C. (2011). A goal programming-TOPSIS approach to multiple response optimization using the concepts of non-dominated solutions and prediction intervals. *Expert Systems with Applications*, *38*(8), 9557–9563. doi:10.1016/j.eswa.2011.01.139

Ravindran, A. R. (2008). Multiple criteria decision making. In A. R. Ravindran (Ed.), *operation research methodologies* (pp. 128–135). Washington, DC: CRC press.

Ribeiro, J. S., Teofilo, R. F., Augusto, F., & Ferreira, M. M. C. (2010). Simultaneous optimization of the micro extraction of coffee volatiles using response surface methodology and principal component analysis. *Chemometrics and Intelligent Laboratory Systems*, *102*(1), 45–52. doi:10.1016/j.chemolab.2010.03.005

Rosen, S. C., Harmonosky, C. M., & Traband, M. T. (2007). A simulation optimization method that considers uncertainty and multiple performance measures. *European Journal of Operational Research*, *181*(1), 105–110. doi:10.1016/j.ejor.2006.05.040

Routara, B. C., Mohanty, S. D., Datta, S., Bandyopadhyay, A., & Mahapatra, S. S. (2010). Combined quality loss (CQL) concept in WPCA-based Taguchi philosophy for optimization of multiple surface quality characteristics of UNS C34000 brass in cylindrical grinding. *International Journal of Advanced Manufacturing Technology*, *51*(1-4), 135–143. doi:10.1007/s00170-010-2599-1

Sahu, J., Acharya, J., & Meikap, B. (2009). Response surface modeling and optimization of chromium (VI) removal from aqueous solution using Tamarind wood activated carbon in batch process. *Journal of Hazardous Materials*, *172*(2-3), 818–825. doi:10.1016/j.jhazmat.2009.07.075 PMID:19748729

Salmasnia, A., Baradaran Kazemzadeh, R., & Niaki, S. T. A. (2012). An approach to optimize correlated multiple responses using principal component analysis and desirability function. *International Journal of Advanced Manufacturing Technology*, *62*(5-8), 1–12. doi:10.1007/s00170-011-3824-2

Shah, H. K., Montgomery, D. C., & Carlyle, W. M. (2004). Response Surface Modeling and Optimization in Multiresponse Experiments Using Seemingly Unrelated Regressions. *Quality Engineering*, *16*(3), 387–397. doi:10.1081/QEN-120027941

Sitter, R. R., & Forbes, B. (1997). Optimal two-stage designs for binary response experiments. *Statistica Sinica*, *7*, 941–956.

Sitter, R. R., & Wu, C. F. J. (1999). Two-stage design of quantal response studies. *Biometrics*, *55*(2), 396–402. doi:10.1111/j.0006-341X.1999.00396.x PMID:11318192

Spuchľáková, E., & Cúg, J. (2014). Lost Given Default and the Credit risk. In *Proceedings of ICMEBISInternational Conference on Management, Education, Business, and Information Science*. EDU Gait Press.

Taguchi, T., Ida, K., & Gen, M. (1997). Method for Solving Nonlinear Goal Programming Tasks with Interval Coefficients Using GA. *Computers & Industrial Engineering*, *33*(4), 597–600. doi:10.1016/S0360-8352(97)00201-5

Tang, L. C., & Xu, K. (2002). A unified approach for dual response optimization. *Journal of Quality Technology*, *34*(4), 437–447.

Tong, L. I., Wang, C. H., & Chen, H. C. (2005). Optimization of multiple responses using Principal component analysis and technique for order preference by Similarity to ideal solution. *International Journal of Advanced Manufacturing Technology*, *27*(3), 407–414. doi:10.1007/s00170-004-2157-9

Vining, G. G. (1998). A compromise approach to multiresponse optimization. *Journal of Quality Technology*, *30*(4), 309–313.

Wong, W. K. (1994). Comparing robust properties of A, D, E and G-optimal designs. *Computational Statistics & Data Analysis*, *18*(4), 441–448. doi:10.1016/0167-9473(94)90161-9

Wu, C. F. J. (2005). Optimization of correlated multiple quality characteristics using desirability function. *Quality Engineering*, *17*(1), 119–126. doi:10.1081/QEN-200028725

Yaghoobi, M. A., Jones, D. F., & Tamiz, M. (2008). Weighted additive models for solving fuzzy goal programming problems. *Asia-Pacific Journal of Operational Research*, *25*(5), 715–733. doi:10.1142/S0217595908001973

Yaghoobi, M. A., & Tamiz, M. (2007). A method for solving fuzzy goal programming problems based on MINMAX approach. *European Journal of Operational Research*, *177*(3), 1580–1590. doi:10.1016/j.ejor.2005.10.022

Zhou, F., Wu, D., Yang, X., & Jiao, J. (2008). Ordinal logistic regression for affective product design, In *Proceedings of Conference Industrial Engineering and Engineering Management conference*. Singapore: IEEE. doi:10.1109/IEEM.2008.4738219

KEY TERMS AND DEFINITIONS

Categorical Variable: Categorical variables are also known as discrete or qualitative variables. Categorical variables can be further categorized as nominal, ordinal or dichotomous. These variables defined in both dependent and independent variables. In which, logistic regression and dummy variables distinguished respectively.

Design of Experiments: This branch of applied statistics developed, to determine the factors affecting the process in the case of minimizing the number of tests.

Dummy Variable: Dummy variables are binary numerical variables used in regression analysis which defined to control variables when there are distinguish as categorical variable.

Goal Programming: GP is a multi-objectives approach which devised to minimize any deviations from the target, determined by DM.

Logistic Regression: Logistic regression is a method to model the data where the output is binary, ordinal or nominal.

Mathematical Programming: Mathematical Programming (MP) is the use of mathematical models, particularly optimizing models.

Response Surface Methodology: RSM is statistical-mathematical tool which mostly used in offline quality control associated with the design of experiments (DOE), in order to estimate a relationship between input and output variables.

Chapter 3
Design Optimization of a Wind Turbine Using Artificial Intelligence

Jagan Jayabalan
Galgotias University, India

Dookie Kim
Kunsan National University, South Korea

Dalkilic Yildirim
Erzincan University, Turkey

Pijush Samui
National Institute of Technology Patna, India

ABSTRACT

This chapter examines the capability of Support Vector Machine (SVM), Relevance Vector Machine (RVM) and Genetic Programming (GP) for the optimal design of wind turbine. The excellent design has been influenced by various factors, such as profile of the blade, number of blades, power factor and tip speed ratio. The key to design a wind turbine is to Assessing the optimal tip speed ratio (TSR) is the key for designing the wind turbine. This chapter handles the Artificial Intelligence techniques in predicting the optimal TSR and the power factor based on the parameters engaged for NACA 4415 and LS-1 profile types with 3 and 4 blades. The organized machine learning framework is anticipated to be lucrative than the traditional way in foretelling the TSR and power factor. The machine learning models are then compared with the existing Neural Network model and the pros and cons of the various models are inferred from the results.

INTRODUCTION

The drastic utilization of renewable energy sources due to booming level of pollution in all sources likewise air, water and soil and global iffy. The energy creators have emphasized on wind energy due to its pros likewise size of the machines, easy and manageable ability, endowment by the bureaucracy, tax benefits, etc. The omnipotence availability, spotless and the free of source make the best decision for encouraging wind power. This supremacy has succored the human race for centuries by mobilizing the ships and catalyzing wind turbines, pump water, etc. The popularly priced readiness and plentiful

DOI: 10.4018/978-1-5225-1639-2.ch003

supply of petroleum, the high cost and uncertainty of wind placed it inconvenience to the economy. The recent hike in oil prices made the people to discern that world's oil and gas accumulation would be cast off and that led to the development of other energy sources. In India, the wind power was modernized in early 1990's and indispensably escalated in the last few years. According to the strategy of World Wind Energy Association, 2008, India spotted the position of fifth largest installed wind power capacity in the world. During the year 2009-2010 India competed other dominating countries and accomplished the highest growth rate. India produced 22,644.63 MW through the wind power by the year 2015 which was appreciable.

The range of wind velocity that is utilized by a specific wind turbine for the production of power is called productive wind speed. The power available from wind is proportional to cube of the wind's speed. This depicts that, when the velocity of wind abates, and then there would be a rapid fall in the generation of energy. If we flip that around, the more wind speed may overexert the turbine. Productive wind velocity may vary from 4 m/sec and 35 m/sec. The stipulated least possible speed for most advantageous performance of exorbitant scale wind farms is about 6 m/s. The wind speed makes the propeller like blades around a rotor, which was banded together with the main shaft. It is then whirls a generator to be the source of energy. It was advised to mount the wind turbines on a tower. This is important in order to apprehend the most energy. The wind turbines can do the trick of taking advantages of faster and less wind turbulent at an altitude of 100 feet (30 meters) or more above ground. Wind turbines can be employed for yielding power not only for a building or home but also to a vast dissemination by connecting with the electricity grid.

The outline of the wind turbine blade hinged on different factors, likewise the turbine profile, number of blades, power factor, and tip speed ratio. The optimal Tip Speed Ratio (TSR) is quintessential for devising the wind turbine. This will undeviatingly influence the power generated and, eventually it disturbs the investment made. It was advised to adopt TSR between the range 7 and 8 and during operation it was recommended to 7 for a 3-blade network connected wind turbine. However, the optimal TSR is counted on the profile type adopted and the blade number and could fall out of the boundaries suggested. Tip Speed Ratio can be augured by engaging the classical regression modeling techniques. This chapter employs the recent dominating soft computing techniques like of Support Vector Machine (SVM), Relevance Vector Machine (RVM) and Genetic Programming (GP) for foretelling the Tip Speed Ratio.

The Tip Speed Ratio (TSR) and Power factor (Cp) are the meat and potatoes for the composition of wind turbines. In our literature survey we found certain research in which TSR is calculated using artificial neural network (ANN). ANN, although widely popular has certain drawbacks such as:

1. They are highly specific in nature. One program code is not suitable for multiple purposes.
2. They are termed as "Black Box", i.e. they are unpredictable and hard to train. There exists no knowledge of what is happening inside the code.
3. They are not statistical in nature, thus reducing the overall reliability.

This chapter will be using Support Vector Machine (SVM), Relevance Vector Machine (RVM) and Genetic Programming (GP) which are prevalent machine learning methods. SVM, RVM and GP are found to be more accurate than ANN in real world research. It also compares the results and use optimum TSR and power factor to optimize the design process of a wind turbine.

The objective is to predict the TSR by developing an equation using the database taken from various wind farm locations. The equation will be derived by using soft computing algorithms such as SVM,

RVM and GP. We will also make a comparative study between the above developed models. To prepare a usable code to predict the optimum Cp and TSR thus reducing the human error in the designing phase. Also to find out if the use of machine learning techniques is a superior option to ANN.

BACKGROUND

Soft Computing is used for problems having no known algorithm that can compute an exact solution. It deals with imprecision, uncertainty, approximation to achieve practicability and low solution cost. Example: Neural Networks. Dmitiris (2008) formulated an SVM and RVM algorithm for finding the deposits of micro calcification, which is an early sign of breast cancer and the results were found good compared to other existing techniques. The MATLAB software was used in this study and they concluded that the relevance vector machines for regression are robust technique for function approximation. It was an important research and contribution in the field of medical sciences. Yurdusev (2004) formulated an ANN code for calculating the tip speed ratio for various blade profiles. Data used was an experimental data. It was found that results were more accurate from the ANN approach as compared to that of conventional methods. Manwell (2002) did an excellent study on wind power. His work on calculation of various losses and blade profile are very useful and accurate. The wind engineering mini codes are inundated with the set of computer codes for evaluating wind energy related controversies, specifically in the context of an academic setting.

Numerical weather predictions models were adopted for the most of the general wind prediction systems (Ernst et al., 2007), and velocity of wind is converted to electrical power output. The faultlessness of the forecasting can be elevated significantly when the models are fused to ensembles. Ernst et al. (2009) explained the prominent role of wind and photovoltaic prediction for balancing, and grid operations. Costa et al. (2008) provide a survey of 30 years of short-term prediction differentiating between mathematical, statistical and physical models. Negnevitsky et al. (2009) examined the prediction methods utilized for the power system applications which emphasize electricity load, price foreseeing and wind power prediction. Milligan et al. (2009) explains the property of wind energy. For a single wind turbine, predictions on one or two hours basis can achieve an accuracy level of approximately 5-7% mean absolute error to installed wind capacity, increasing to 20% for day-ahead forecasts.

Theodore and Huseyin Inc. (2006) formulated an SVM algorithm for financial forecasting of top MNCs and the results were found good compared to other existing techniques. They used techniques such as Back propagation and Radial Basis Function. It was mentioned that the SVMs lead to a quadratic programming (QP) problem. The MATLAB software was used in this study and they concluded that the support vector machines for regression are robust technique for function approximation. Samui (2007) determined the settlement of shallow foundations on cohesion-less soil using Support Vector Machine. It must be noted that the existing techniques were unsteady for figuring out the settlement of shallow foundations. Hence the soft computing method SVM was engaged for forecasting the settlement of shallow foundation on cohesion-less soil. It was proved that SVM has its own capability for predicting the settlement. Santos et al. (2015) adopted the SVM technique for determining the fault in the wind turbines. They have compared with the other machine learning techniques which implies that SVM posses less time than the others. Also it has the very low error percentage of 0.02. SVM has homogenized functional form as that of ANN with better generalization achievement and ability to perform accurate predictions (Maria et al., 2014). Based on the wavelet decomposition Maria et al. (2014) has compared with different models wind power. RVM based typecasting approach to take measure the liquefaction capability from the Cone Penetration Test (CPT) test and it was found propitious (Sarat & Samui, 2008). In this study, Gaussian, Polynomial and Spline functions were used as kernel functions. It was also concluded that once developed and trained model perform the simulation in a small factor of the time.

Using the multiclass RVM and SVM, Widodo et al. (2009) examined the defected diagnosis of low speed bearing. Widodo et al., (2009) adopted two different techniques SVM and RVM of multi-class classification methods for fault diagnosis. From the results obtained, both RVM and SVM classification models, it was decided that the RVM produces favorable results for wield in minimal velocity bearing fault diagnosis. RVM was utilized for determining the stability analysis of slope by First Order Moment Method (FOSM) (Samui et al., 2011). RVM hinge on the Bayesian hypothesis and utilize an inductive modeling methodology that allows fusion of prior knowledge in the estimation process. It was concluded that RVM based FOSM may not be perfect replacement but a viable alternative for reliability problems. Utilizing the Empirical Mode Decomposition (EMD) and RVM, Bao et al. (2013) predicted the short-term wind power. RVM successfully predicted the Intrinsic Mode Function (IMF) and the coefficient of correlation and predicted the short-term wind power after combining the IMF and the r. Yan et al. (2013) adopted Optimized RVM (ORVM) for forecasting the wind power interval and its fluctuation range at the favorable level. They described the potential of RVM that outperforms other model in predicting the wind power.

Tipping (2001) paved the way to a general Bayesian structure for procuring sparse solutions to regression and classification assignments utilizing models linear in the parameters. He introduced the RVM method, a model of identical form to the popular SVM. He depicted that, by capitalizing a probabilistic Bayesian learning structure can extract flawless prediction models which routinely make use of minor primitive functions than associated SVM while offering a number of additional advantages. Kiong (2000) determined the size, geometry and topology optimization of structures by adopting and utilizing the Genetic Programming (GP). Adil et al. (2008) adopted GP for predicting the compression strength and the tensile strength of the limestone. The wellness of the GP produced the direct equation for determining the strength in future. Also, the roughness of the surface was predicted and the equations were determined by Brezocnik et al. (2004). Chandwani et al. (2014) in modeling slump of ready mix concrete using genetic algorithms have explained the methods of implementing genetic algorithms and the various error parameters that can be used as the benchmark test.

An encrypting program was bestowed to map between the absolute structures and the genetic programming parse trees. Ochieng (2014) forecasted the value of TSR of a wind turbine and examined its upshot on power coefficient. The power coefficient (Cp) was accomplished at a specific tip speed ratio, which is primitive to device the turbine. Jack (1995) discussed the pros and cons of Artificial Neural Networks versus Regression techniques for predicting the outcomes. It was specifically divulged the disadvantages like Black Box approach, proneness to over fitting, etc.

TECHNIAL SPECIFICATION

This chapter involves in predicting the optimum tip speed ratio and the power factor based on the parameters NACA 4415 and LS-1 profile types for three and four blades. NACA 4415- NACA stands for National Aeronautics Council Advisory (NACA), now NASA, USA. This blade profile was designed by Aeronautical and Astronautical Research Laboratories.

Tip Speed Ratio (λ)

The Tip Speed Ratio or TSR for wind turbines is ratio between the tangential speed of the tip of blade and the actual velocity of the wind. The tip-speed ratio is related to efficiency, with the optimum varying with blade design. When the tip speed ratio was adopted to the minimal limit, then the rotor blade design would be sick enough which in turn produces fruitless power generation. On the other side, when the tip speed ratio is utilized in maximal range the turbine will whirls faster through turbulent air. This leads to resist the optimal extraction of power from the wind and may result in lofty stress and the probability of breakdown of structure. The commencing point of designing the turbine is to determine the relation between the wind speed and the turbine blade tip speed. The most favorable TSR band modifies with the "solidity" of the turbine disc, which defines the number of blades the turbine has.

When contriving at the design of TSR, supreme power can be pulled out from the wind. The power can be reduced eventually when the actual TSR fluctuates up or down in its design point. The single bladed turbine which also has the counter weight on the other hand might be devised for the TSR ranges 10 and 15. TSR goes down when the number of blades are incorporated effectively. For a 3 blade turbine, the TSR value ranges from 4 to 7, this gives the better outcome. Windmill farm usually consists of multi-blade, executes at a TSR lower than one. The 3 blade turbine consists of 2m diameter rotor and a TSR 4.5 with a velocity of 5 m/sec then the optimum tip speed would be approximately 22.5 m/sec. This in turn transfigure to the turbine shaft of 215 rpm. In order to maintain the same TSR, the 20m diameter turbine has the rotational speed of 21.5 rpm. These less velocity do not fit to the most alternators or generators. In the case of the 20 meter turbine, would certainly require stepping up by a gearbox. It must be highlighted that when the velocity of the wind doubles, then the turbine must also doubled in order to allow the extraction of maximum power.

$$\lambda = \frac{Tip\ speed\ of\ blade}{Wind\ speed} \tag{1}$$

Power Coefficient (C_p)

Power coefficient is a quantity that expresses what fraction of the power in the wind is being extracted by the wind turbine. It is generally assumed to be a function of both tip-speed ratio and pitch angle. One would always want the turbine to run at Cp_{max} so as to have max efficiency. But the Betz criterion defines that the maximum power that can be extracted out of a wind limits to 59%. Hence $Cp_{max} = 0.59$. It has been realized that the value of power coefficient much rely upon the tip speed ratio of the turbine. It also converts the length of the rotor blade. The power generated is directly proportional to the power coefficient. When the power coefficient escalates, then the generation of power increases eventually. Another prominent factor directly influence the power production is the angle of inclination (pitch angle) of the turbine blade. The angle of inclination is inversely proportional to the power coefficient. When the pitch angle is equal to 0, then the value of power coefficient is more, and C_p is decreases with increasing pitch angle.

Profile Losses

This can be considered using $\eta = 1 - \frac{\lambda}{\varepsilon}$, where λ is tip speed ratio and ε is the slip and can be expressed as $\varepsilon = \frac{C_L}{C_D}$ where C_L is the coefficient of lift force of chosen profile and C_D is the coefficient of drag force.

End Losses

At the extremity of the blade, air flows from the underneath side of the profile to the greater side. Blending with the airflow moving towards the blade, this airflow spreads by fits and starts. It can be computed by the following equation:

$$\eta_{end} = 1 - \left(\frac{1.84}{z} * \lambda\right) \tag{2}$$

where λ is the number of turbine blades.

Design Approach

There are several factors such as turbine profile, number of blades, power factor and tip speed ratio that influence the design of the wind turbine. The optimal Tip Speed Ratio (TSR) is quintessential for devising the wind turbine. This will undeviatingly influence the power generated and, eventually it disturbs the investment made. It was advised to adopt TSR between the range 7 and 8 and during operation it was recommended to 7 for a 3-blade network connected wind turbine. However, the optimal TSR is counted on the profile type adopted and the blade number and could fall out of the boundaries suggested. This chapter attempts to enrich the comprehensive designing procedures and also the design of the wind turbine, in order to flourish its design and reliability. Using the conventional auto regression model along with the hot off the fire techniques, TSR can be forecasted. This chapter utilizes the algorithm of Support Vector Machine (SVM), Relevance Vector Machine (RVM) and Genetic Programming (GP).

SUPPORT VECTOR MACHINE

Boser *et al.* (1992) is the forerunner, derive Support Vector Machine (SVM) from the impression of statistical learning theory. Vapnik (1995) flourished the Support Vector Machine. More details can be found in literature on SVM (Boser *et al.* 1992; Vapnik 1998). This chapter includes the dominating regression technique SVM by implementing an insensitive loss function ε. Considering a set of training data $\{(x_1, y_1),....,(x_l, y_l)\}$, x R_n, y r . Where x is the input, y is the output, R_n is the N-dimensional vector space and r is the one-dimensional vector space. The SVM has also been used different geotechnical engineering applications (Samui *et al.,* 2008; Das *et al.,* 2011). Neural Network modeling techniques depicts some hardship with generalization, developing models that could over-fit the data. This is the aftereffect of the optimization algorithms utilized for choosing the boundaries and the statistical computations caused to determine the superlative model. The amount and the character of the scrutiny resolve the depiction of any statistical model. In most cases, data is finite and sampled and, typically, sampling is non-uniform and, due to the high-dimensional nature of the problem, the data forms only a sparse distribution in the input space. The maturation of SVM has acquired attractiveness over time due to its encouraging solutions and many captivating attributes. They can be normally be thought as comprising an another possibility training methodology for Multi-Layer Perceptron and Radial Basis Function classifiers, in which the weights of the network are determined by sorting out a Quadratic Programming (QP) problem with linear disparity and equality damper, rather than resolving a non-convex, unrestricted denigration problem, as in typical neural network training methods. The upcoming Figure 1 depicts the process of the SVM.

Details of SVM

Let us consider the typical regression formulation for forecasting erudition (Cherkassky & Mulier, 1998; Hastie, Tibshirani, & Friedman, 2001; Vapnik, 1999). The intention is to assess unknown real-valued function in the relationship

$$y = r\left(x\right) + \delta \tag{1}$$

where x is a multivariate input, y is the output and δ is the noise data which is uniformly distributed zero mean random error. The prediction rely on a definite set of pieces i.e. training data $\left(x_i, y_i\right), \left(i = 1, \ldots, n\right)$. The uniformly distributed training data samples are engendered according to the probability density function.

$$p\left(x, y\right) = p\left(x\right) p\left(y \mid x\right) \tag{2}$$

Figure 1. Flowchart of SVM

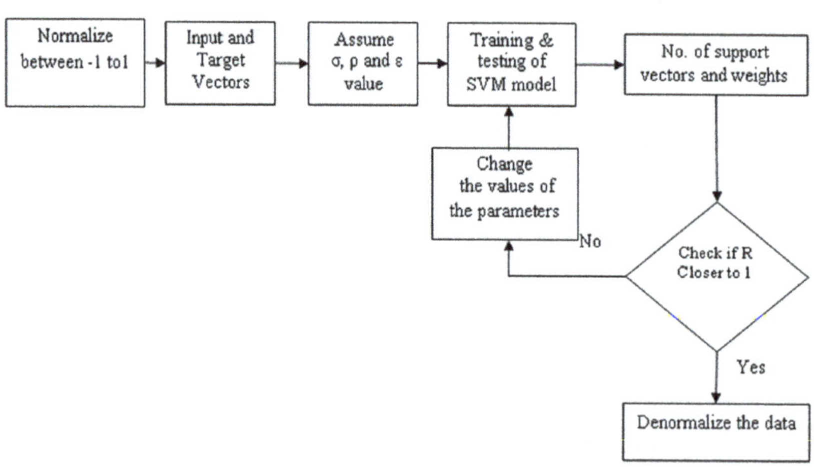

The unknown function in Equation 1 is the mean of the output regression function

$$r(x) = \int yp(y \mid x)dy \tag{3}$$

An estimation procedure identifies the best model $f(x, \varpi_0)$ from the set of approximating functions $f(x, \varpi)$. It is then parameterized by a set of parameters $\varpi \in \Omega$. The eminence of the approximation can be determined by the inconsistency measure $L(y, f(x, \varpi))$. The ultimate aim of learning this is to pick the best model reducing estimation risk.

$$R(\varpi) = \int L(y, f(x, \varpi))p(x, y)dxdy \tag{4}$$

The regression function is Equation 3 is the one minimizing estimation risk (4) with the squared loss function

$$L(y, f(x, \varpi)) = (y - f(x, \varpi))^2 \tag{5}$$

It should be noted that $f(x, \varpi)$, carried by the learning technique may or may not have the regression function (3). Hence we have to determine the regressor function $f(x, \varpi_0)$ that reduces the prediction risk function using only the training data.

$$R(\varpi) = \int (y - f(x, \varpi))^2\, p(x, y)dxdy \tag{6}$$

The risk functional ascertains the accuracy of the unknown target function $r(x)$. SVM maps the input x to the n dimensional space using some non-linear mapping and then the linear mapping will be developed in the feature space (Cherkassky & Mulier, 1998; Smola & Scholkopf, 1998; Vapnik, 1998, 1999).

$$f(x, \varpi) = \sum_{i=1}^{n} \varpi_i g_i(x) + b \tag{7}$$

where $g_i(x)$, i=1,…..,n indicates the non-linear transformations and b is the bias.

The diminution of empirical risk over the training data can be procured. Conventional loss functions utilized for denigration of empirical risk encompass the squared error and absolute value error. Vapnik (1998, 1999) adopts ε-insensitive loss function, a special type of loss function in SVM.

$$L_{\varepsilon}\left(y, f\left(x, \varpi\right)\right) = \begin{cases} 0 & \text{if } \left|y - f\left(x, \varpi\right) \leq \varepsilon\right| \\ \left|y - f\left(x, \varpi\right) \leq \varepsilon\right| & \text{otherwise} \end{cases} \quad (8)$$

The empirical risk is defined in the following Equation 9.

$$R_{emp}\left(\varpi\right) = \frac{1}{n} \sum_{i=1}^{n} L_{\varepsilon}\left(y_i, f\left(x_i, \varpi\right)\right) \quad (9)$$

SVM carry out the linear regression in a large dimensional feature space by utilizing the ε insensitive loss. Also simultaneously it engages to deduce the complexity of the model by minimizing $\left\|\varpi\right\|^2$. This can be illustrated by setting up a positive slack variable ξ_i, ξ_i^* $i = 1, \ldots, n$ in order to determine the divergence training samples away from ε insensitive region. Thus SVM regression is comprised as minimization of the function which is in the following equation.

Minimize: $\frac{1}{2}\left\|\varpi\right\|^2 + C \sum_{i=1}^{n} \left(\xi_i + \xi_i^*\right)$

Subjected to: $\begin{cases} y_i - f\left(x_i, \varpi\right) - b \leq \varepsilon + \xi_i^* \\ f\left(x_i, \varpi\right) + b \leq \varepsilon + \xi_i^* \\ \xi_i, \xi_i^* \geq 0 \end{cases}$ (10)

where C is the regularization parameter which is a non-negative constant. This optimization formulation can be transmuted into the two-fold issues (Vapnik, 1998, 1999).

$$f\left(x\right) = \sum_{i=1}^{n} \left(\alpha_i - \alpha_i^*\right) K\left(x_i, x\right) + b \quad (11)$$

The two-fold variables are subjected to constraint $0 \leq \alpha_i, \alpha_i^* \leq C$ and $K\left(x_i, x\right)$ is the kernel function. The points that come into view with non-zero coefficients are called support vectors. SVM generalization performance depends on the better setting of hyper parameters and the kernel parameters. The problem of determining depends upon the above mentioned hyper and kernel parameters.

The parameter C measures the arbitrate whereas the degrees of deviation are determined by ε. For instance, when C is abdominous i.e. infinity, then the empirical risk alone can be minimized without regards to model complexity. Parameter ε governs the thickness of the ε -insensitive region, utilized to harmonize the training data (Cherkassky & Mulier, 1998; Vapnik, 1998, 1999). The range of ε can influence the amount of support vectors accompanies to organize the regression function. Greater value of ε, upshot in lesser support vectors chosen, and ends in greater 'plane' (fewer complexes) regression assay. Hence, both C and ε -values impinge on the complexity of model, but in an antithetic path.

Radial basis function has been adopted for this study.

$$K\left(x_i, x\right) = \exp\left\{-\frac{\left(x_i - x\right)^T \left(x_i - x\right)}{2\sigma^2}\right\} \tag{12}$$

x_i and x are the training and test patterns, T is the transpose and σ is a the global basis function width.

The following Table 1, composes the dataset involved for the optimal design of wind turbine.

Table 1. Dataset for developing the model

TSR	$C_{pschmitz}$	Z_{end}	Z_{end}	Z_{end}	Z_{end}	$Z_{profile}$	$Z_{profile}$	$Z_{profile}$	C_{popt3}	C_{popt4}	Profile Number
		(Z=1)	(Z=2)	(Z=3)	(Z=4)	(a=1°)	(a=5°)	(a=10°)			
0.5	0.238	-2.68	-0.84	-0.2267	0.08	0.9878	0.9922	0.9908	-0.0534	0.0189	2
1	0.4	-0.84	0.08	0.3867	0.54	0.9756	0.9845	0.9816	0.1523	0.2127	2
1.5	0.475	-0.2267	0.3867	0.5911	0.6933	0.9633	0.9767	0.9724	0.2742	0.3217	2
2	0.515	0.08	0.54	0.6933	0.77	0.9511	0.969	0.9632	0.346	0.3843	2
2.5	0.531	0.264	0.632	0.7547	0.816	0.9389	0.9612	0.954	0.3852	0.4165	2
3	0.537	0.3867	0.6933	0.7956	0.8467	0.9267	0.9535	0.9448	0.4073	0.4335	2
3.5	0.538	0.4743	0.7371	0.8248	0.8686	0.9144	0.9457	0.9355	0.4196	0.4419	2
4	0.541	0.54	0.77	0.8467	0.885	0.9022	0.938	0.9263	0.4296	0.4491	2
4.5	0.544	0.5911	0.7956	0.8637	0.8978	0.89	0.9302	0.9171	0.4371	0.4543	2
5	0.547	0.632	0.816	0.8773	0.908	0.8778	0.9225	0.9079	0.4427	0.4582	2
5.5	0.55	0.6655	0.8327	0.8885	0.9164	0.8655	0.9147	0.8987	0.447	0.461	2
6	0.553	0.6933	0.8467	0.8978	0.9233	0.8533	0.907	0.8895	0.4503	0.4631	2
6.5	0.556	0.7169	0.8585	0.9056	0.9292	0.8411	0.8992	0.8803	0.4528	0.4646	2
7	0.559	0.7371	0.8686	0.9124	0.9343	0.8289	0.8915	0.8711	0.4547	0.4656	2
7.5	0.562	0.7547	0.8773	0.9182	0.9387	0.8166	0.8837	0.8619	0.456	0.4662	2
8	0.565	0.77	0.885	0.9233	0.9425	0.8044	0.876	0.8527	0.457	0.4665	2
8.5	0.568	0.7835	0.8918	0.9278	0.9459	0.7922	0.8682	0.8435	0.4576	0.4665	2
9	0.57	0.7956	0.8978	0.9319	0.9489	0.78	0.8605	0.8343	0.457	0.4654	2
9.5	0.572	0.8063	0.9032	0.9354	0.9516	0.7677	0.8527	0.825	0.4563	0.4641	2
10	0.574	0.816	0.908	0.9387	0.954	0.7555	0.845	0.8158	0.4553	0.4627	2
0.5	0.238	-2.68	-0.84	-0.227	0.08	0.986	0.994	0.996	-0.054	0.019	1
1	0.4	-0.84	0.08	0.387	0.54	0.971	0.989	0.991	0.153	0.214	1
1.5	0.475	-0.227	0.387	0.591	0.693	0.957	0.983	0.987	0.277	0.325	1
2	0.515	0.08	0.54	0.693	0.77	0.943	0.978	0.983	0.351	0.39	1
2.5	0.531	0.264	0.632	0.755	0.816	0.929	0.972	0.978	0.392	0.424	1

continued on following page

Table 1. Continued

TSR	$C_{pschmitz}$	Z_{end} (Z=1)	Z_{end} (Z=2)	Z_{end} (Z=3)	Z_{end} (Z=4)	$Z_{profile}$ (a=1°)	$Z_{profile}$ (a=5°)	$Z_{profile}$ (a=10°)	C_{popt3}	C_{popt4}	Profile Number
3	0.537	0.387	0.693	0.796	0.847	0.914	0.967	0.974	0.416	0.443	1
3.5	0.538	0.474	0.737	0.825	0.869	0.9	0.961	0.97	0.43	0.453	1
4	0.541	0.54	0.77	0.847	0.885	0.886	0.956	0.965	0.442	0.462	1
4.5	0.544	0.591	0.796	0.864	0.898	0.871	0.95	0.961	0.452	0.469	1
5	0.547	0.632	0.816	0.877	0.908	0.857	0.945	0.957	0.459	0.475	1
5.5	0.55	0.665	0.833	0.888	0.916	0.843	0.939	0.952	0.465	0.48	1
6	0.553	0.693	0.847	0.898	0.923	0.829	0.934	0.948	0.471	0.484	1
6.5	0.556	0.717	0.858	0.906	0.929	0.814	0.928	0.944	0.475	0.488	1
7	0.559	0.737	0.869	0.912	0.934	0.8	0.923	0.939	0.479	0.491	1
7.5	0.562	0.755	0.877	0.918	0.939	0.786	0.917	0.935	0.482	0.493	1
8	0.565	0.77	0.885	0.923	0.943	0.771	0.912	0.931	0.486	0.496	1
8.5	0.568	0.784	0.892	0.928	0.946	0.757	0.906	0.926	0.488	0.498	1
9	0.57	0.796	0.898	0.932	0.949	0.743	0.901	0.922	0.49	0.499	1
9.5	0.572	0.806	0.903	0.935	0.952	0.729	0.895	0.918	0.491	0.499	1
10	0.574	0.816	0.908	0.939	0.954	0.714	0.89	0.913	0.492	0.5	1
10.5	0.574	0.825	0.912	0.942	0.956	0.7	0.884	0.909	0.491	0.499	1
11	0.574	0.833	0.916	0.944	0.958	0.686	0.879	0.905	0.49	0.498	1
11.5	0.574	0.84	0.92	0.947	0.96	0.671	0.873	0.9	0.489	0.496	1

In order to develop the SVM model the following steps are:

1. The Schmitz coefficient, end losses, profile losses and a blade number used as an Input Data for the algorithm.
2. The measured Tip speed ratio, power coefficient for three and four blade wind turbine is used as an Output Database.
3. Training dataset consist of 70% of the data which was used for developing the model whereas the remaining 30% of the data are adopted as the Testing dataset which examines the efficiency of the developed model.
4. All the database values are normalized between -1 and 1 for simpler analysis. Range of data varies which can affect the regression analysis thus we normalize the data in one defined range to increase the accuracy. Formula used for normalizing is discussed below:

$$D = \frac{2\left(A - A_{\min}\right)}{\left(A_{\max} - A_{\min}\right)} - 1 \tag{13}$$

where A- Input data

D - Normalised data input

A_{max} - Maximum value in the set A

A_{min} - Minimum value in set A

$$A = \frac{(D+1)*(Amax - Amin) + 2*Amin}{2}$$

5. MATLAB coding is developed using different kernels (for different models) for analyzing the data and to get an equation for tip speed ratio.
6. Correlation coefficient is calculated between the predicted values and the actual values.
7. Output values are denormalized as to compare with the original data. Denormalizing formula is discussed below:

$$A = \frac{(D+1)*(A_{max} - A_{min}) + 2*A_{min}}{2} \quad (14)$$

where A - Renormalized Output data

D - Normalised output

A_{max} - Uppermost value in the set A

A_{min} - Lowermost value in set A

RELEVANCE VECTOR MACHINE

The prominent character of SVM is to decrease the error and concurrently widens the margin between two classes which was implicitly explained by the kernel function. These were endeavored by target functions. Tipping (2001) adopted and explained this eminently emphatic contrivance in order to escape from the over-fitting. Since everything has its own uncertainty, SVM also has its own practical difficulties, despite of SVM's success in various fields.

Tipping (2001) familiarized an innovative technique called Relevance Vector Machine (RVM) in order to escape from the problems faced by SVM. RVM is a distinct case of a sparse kernel model, which stand for a Bayesian treatment of a comprehensive linear model of indistinguishable functional form to support vector machine (SVM). RVM varies with SVM likewise the case of solution, which outcomes the probabilistic interpretation of its output (Wei et al., 2005). RVM avoids the difficulty that establishes the models which have both the composition and the parameterization process altogether to the information index of the available data.

RVM hire a fully probabilistic structured and familiarize a preceding over the model weights governed by asset of hyper parameters, one associated with each weight, whose most probable values are iteratively estimated from the data. RVM has the most favorable feature is that, it uses some kernel functions which are potential for generalization which was similar to SVM. The following Figure 2 reveals the flowchart of RVM.

Details of RVM

RVM is a sparse method for training generalized linear models (Tipping, 2000). Let us consider a set of example of input vectors $\left\{x_i\right\}_{i=1}^{N}$ is given along with a corresponding set of targets $t = \left\{t_i\right\}_{i=1}^{N}$. The RVM constructs a logistic regression model based on a set of sequence features derived from the input patterns, i.e.

$$p\left(C_1/x\right) \approx \sigma\left\{y\left(x;w\right)\right\}, \text{ with } y\left(x;w\right) = \sum_{i=1}^{N} w_i \Phi_i\left(x\right) + w_0 \quad (15)$$

where φ_i is an ith component of the basis vector function

$$\Phi\left(x\right) = \left(\Phi_1\left(x\right), \Phi_2\left(x\right), \ldots, \Phi_N\left(X\right)\right)^T = \left[1, K\left(x_i, x_1\right), K\left(x_i, x_2\right), \ldots, K\left(x_i, x_N\right)\right]^T,$$

Figure 2. Flowchart of RVM

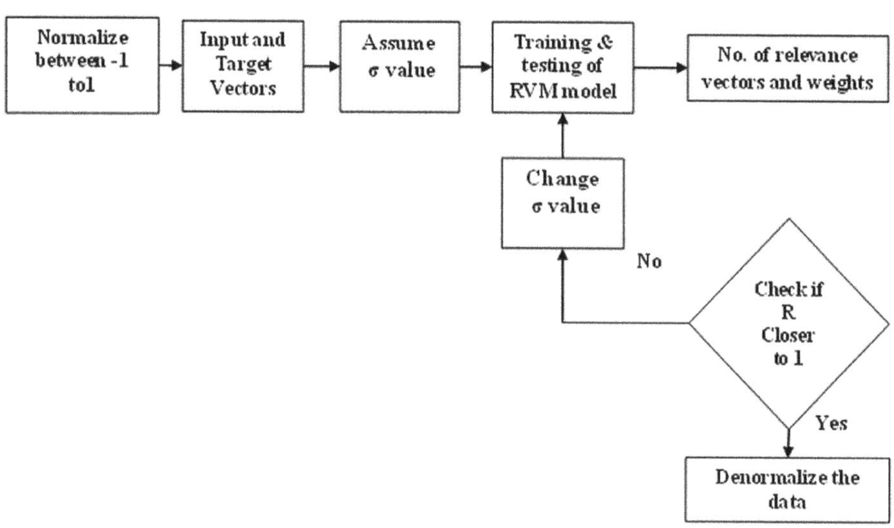

$w = \left(w_0, \ldots, w_N\right)^T$ are a vector of weights, $\tilde{A}\{y\} = \left(1 + \exp\{-y\}\right)^{-1}$ is the logistic sigmoid link function and $K\left(x_i, x_j\right)_{j=1}^{N}$ are kernel terms. Assuming a Bernoulii distribution for $P\left(t/x\right)$, the likelihood is written as (Tipping, 2001):

$$P\left(t/w\right) = \prod_{i=1}^{N} \sigma\left\{y\left(x_i; w\right)\right\}^{t_i} \left[1 - \sigma\left\{y\left(x_i; w\right)\right\}\right]^{1-t_i} \tag{16}$$

We cannot integrate the weights analytically. The RVM adopts the following separable Gaussian prior, with a distinct hyper-parameter, α_i, for each weight,

$$p\left(w/\alpha\right) = \prod_{i=1}^{N} N\left(w_i/0, \alpha_i^{-1}\right) \tag{17}$$

The optimal parameters of the model are then derived by minimizing the penalized negative log-likelihood,

$$\log\left\{P\left(t/w\right)p\left(w/\alpha\right)\right\} = \sum_{i=1}^{N}\left[t_i \log y_i + \left(1 - t_i\right)\log\left(1 - y_i\right)\right] - \frac{1}{2}w^T A w \tag{18}$$

If we differentiate twice equation (15), the expression is given below:

$$\nabla w \nabla w \log p\left(w/t, \alpha\right) = -\left(\Phi^T B \Phi + A\right) \tag{19}$$

where $B = diag\left(\beta_1, \ldots, \beta_N\right)$ is a diagonal matrix with $\beta_n = \sigma\left\{y\left(x_n\right)\right\}\left[1 - \sigma\left\{y\left(x_n\right)\right\}\right]$

The following equation has been used for updating hyper-parameter

$$\alpha_i^{new} = \frac{1 - \alpha_i \Sigma_{ii}}{\mu_i^2} \tag{20}$$

where μ_i is the ith posterior mean weight, $\sum_{ii}$ is the ith diagonal element of the posterior weight covariance. The process is repeated until the ultimate goal is met. The property of this optimization problem is that the value of many w will be zero. The nonzero weights are called relevance vectors.

RVM adopts the same normalization technique, kernel function, training and testing dataset as used in the SVM model. The program of RVM has been done by using MATLAB.

Genetic Programming

Genetic Programming (GP) was proposed by Koza (1992). GP is one of the best parts of the evolutionary algorithm and also the specification of genetic algorithms (GA) (Goldberg, 1989). It helps in producing the clear prediction equations which satisfies the inabilities of some soft computing techniques (Adil et al., 2008). Genetic programming can be used as an invention machine by employing a two-part fitness measure that incorporates both the degree to which an individual in the population satisfies the given technical requirements and the degree to which the individual does not possess the key characteristics of preexisting technology. GP adopts Darwins concept of "survival of the fittest" and also some appropriate combination of computer programs. The computer programs can be conveyed in terms of "parse tree".

Details of GP

This section involves in expressing the details of the Genetic Programming. The primitive step includes the creation of arbitrary population of equations. Later in the next step, the fitness value of each equation was figured out. In this step, 'Parents' were determined. The third step continues with the process of reproduction, mutation and crossover to produce the offspring's. In the next step, the superlative equations are generated.

GP take over the similar normalization technique, kernel function, training and testing dataset which was utilized in the previous SVM and RVM models. MATLAB have been adopted for cultivating the GP model.

RESULTS AND DISCUSSION

The design values have been varied over a specific corresponding range in order to get better Coefficient of Correlation value (R). When the value of R is near to one, then it can be agreed to be a valuable model. The following equation is used to determine the R value.

$$R = \frac{\sum_{i=1}^{n}\left(h_i - \bar{h}_i\right)\left(t_i - \bar{t}_i\right)}{\sum_{i=1}^{n}\left(h_i - \bar{h}_i\right)^2 \sum_{i=1}^{n}\left(t_i - \bar{t}_i\right)^2} \tag{21}$$

Figures 3, 4, and 5 depict the achievement of testing dataset of the SVM model.

The above figures depicted about the performance of SVM.

The following figures show the performance of RVM for predicting C_{p3}, C_{p4} and TSR. The Figures 6, 7, and 8 are depicted below.

Figure 3. Regression plot of actual verses predicted for Cp3 by SVM

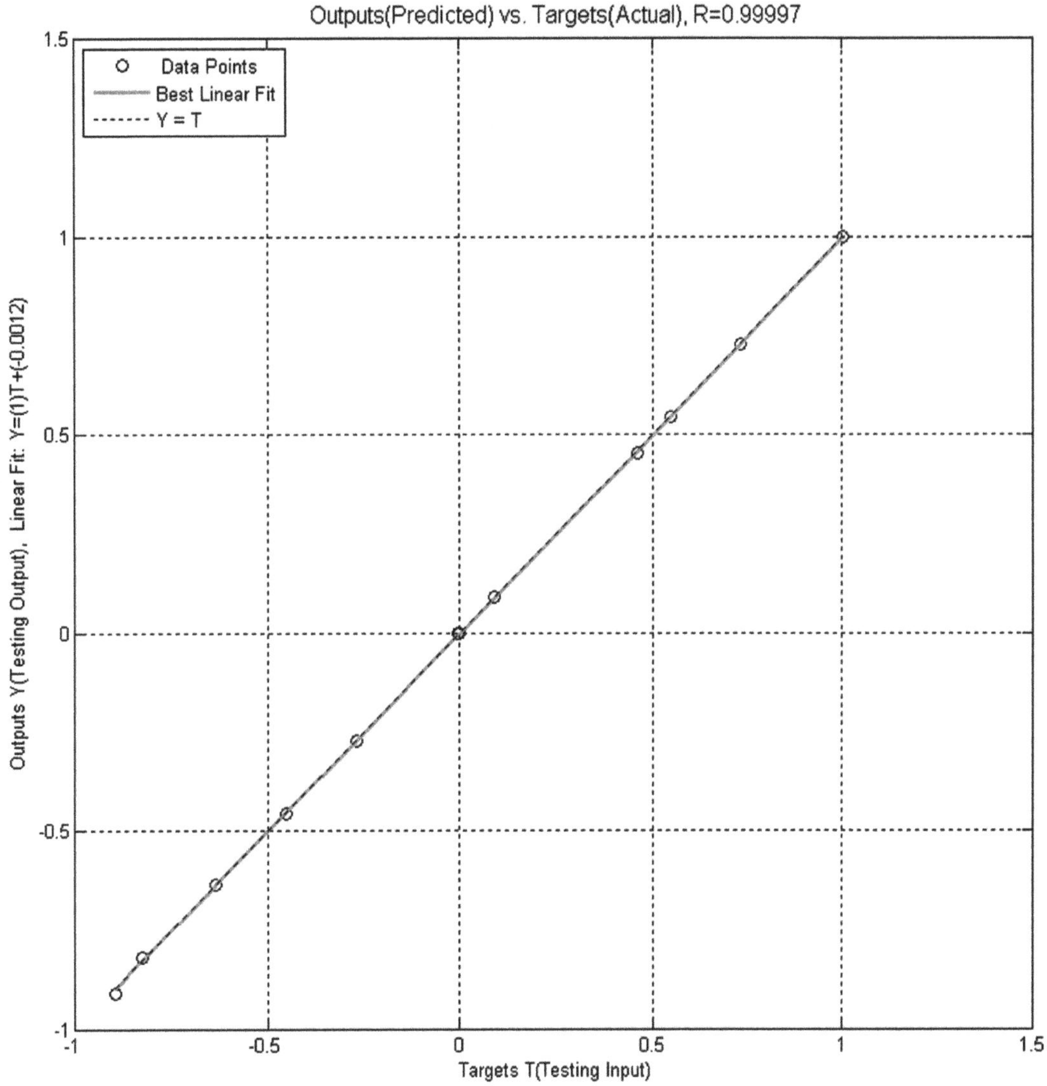

The performance of the GP model was depicted in Figures 9, 10, and 11.

The mean (μ) and standard deviation (σ) of Q_p/Q_m are the valuable indices of the exactness and meticulousness of the prediction method, where Q_p is the predicted parameter and Q_m is the measured parameter (Farsakh & Titi, 2004; Das & Basudhar, 2006). Under a perfect stipulation an exact and flawless method provides the mean value as 1.0 and the standard deviation to be 0. When the value is superior to 1 then it is said to be over prediction, if it is lesser than 1 it implies the under prediction. The best model is represented by value close to 1.0 and close to 0 for mean and standard deviation respectively. Different statistical criteria like Root Mean Square Error (RMSE), Mean Absolute Error (MAE) and Performance Index (ρ) have been used to compare the developed model. The above mentioned statistical criteria have been determined by using the following equations.

Figure 4. Regression plot of actual verses predicted for Cp4 by SVM

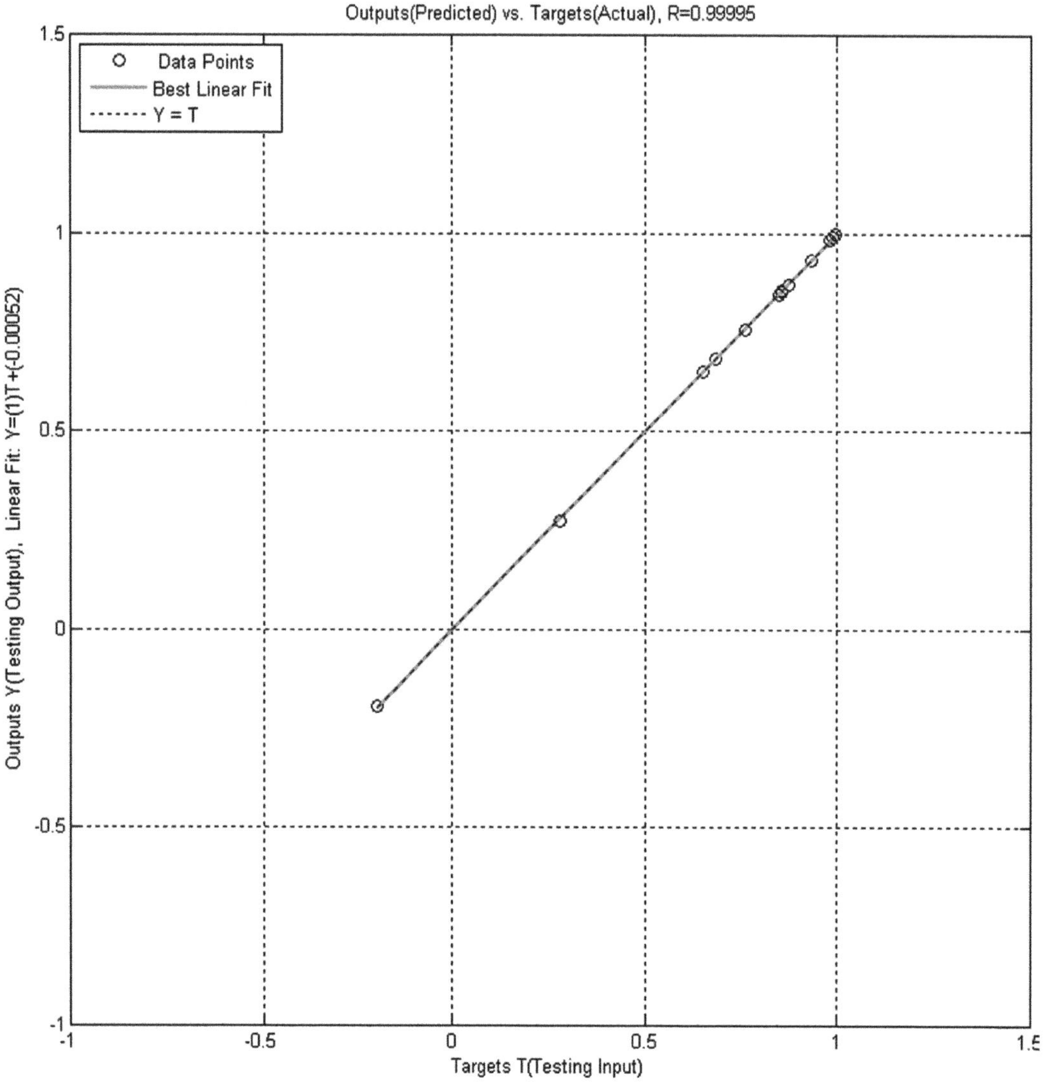

$$MAE = \frac{\sum_{i=1}^{n} \left| h_i - t_i \right|}{n} \tag{22}$$

$$RMSE = \sqrt{\frac{\sum_{i=1}^{n} \left(h_i - t_i \right)^2}{n}} \tag{23}$$

Figure 5. Regression plot of actual verses predicted for TSR by SVM

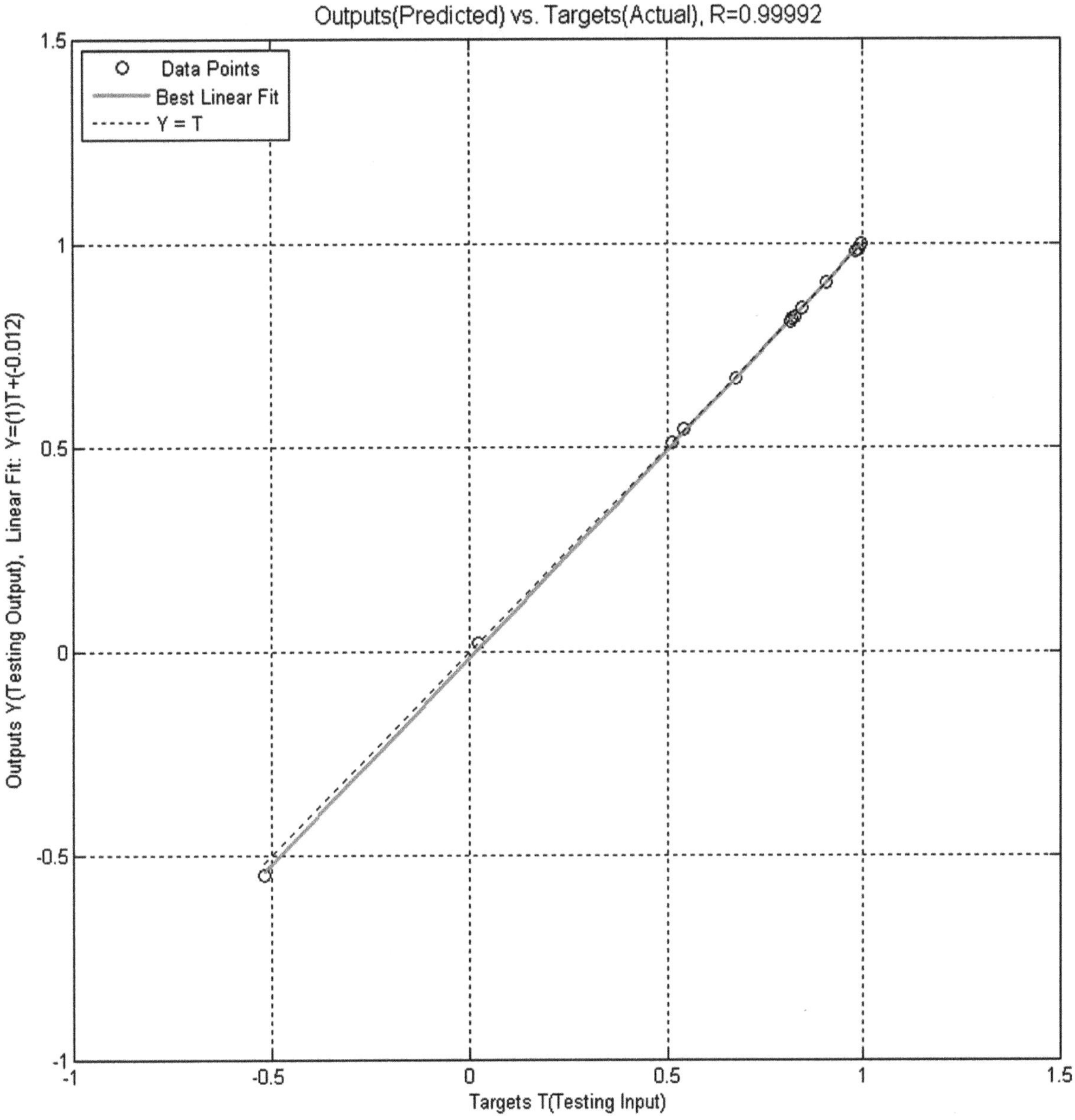

$$\rho = \frac{RMSE}{\bar{h_i}} \frac{1}{R+1} \tag{24}$$

where h_i is the measured value of the ith output

t_i is the predicted value for the ith output,

Figure 6. Regression plot of actual verses predicted for Cp3 by RVM

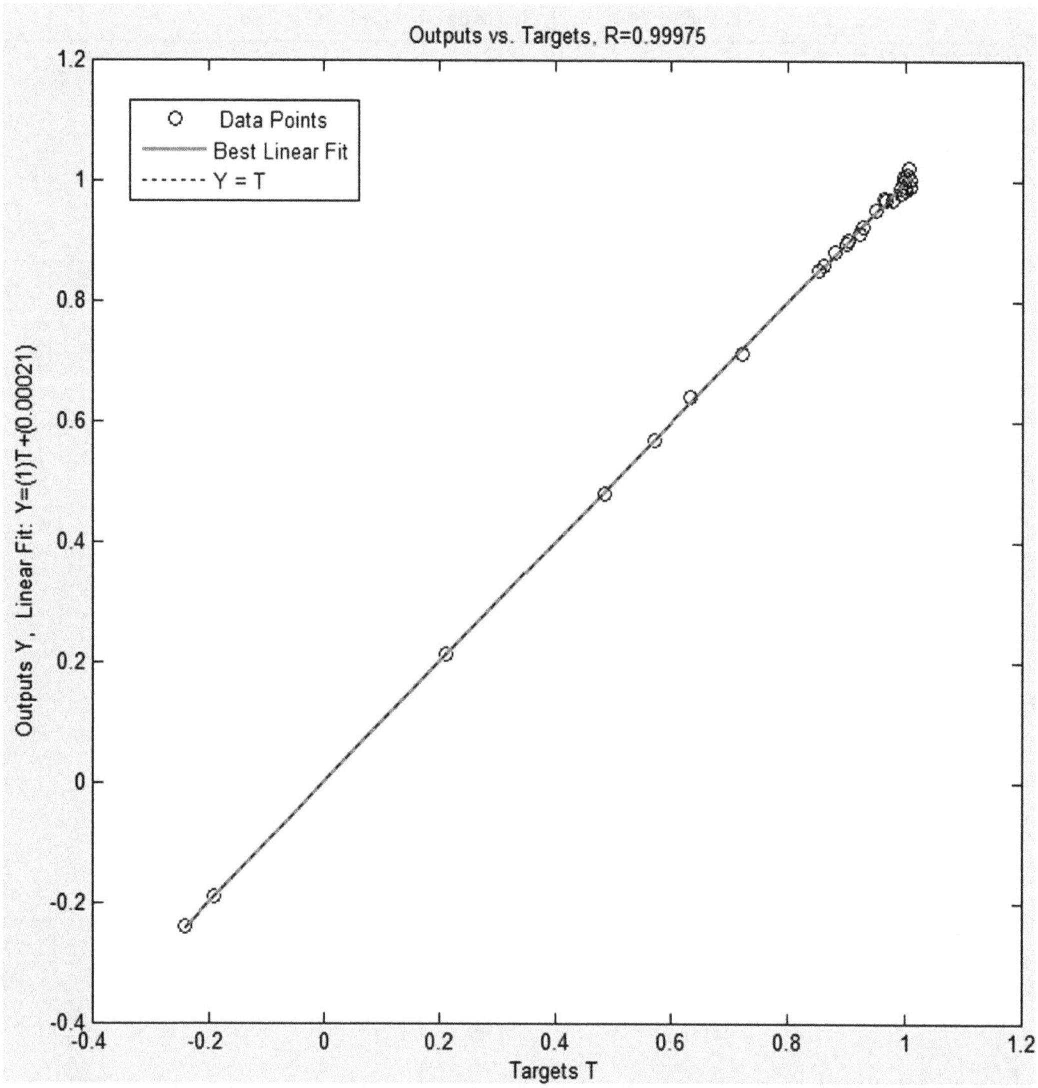

$\bar{h}_i$ is the average of the measured outputs

$\bar{t}_i$ is the average of the predicted outputs,
n is the total number of datasets.

The following table exposes the various statistical criteria's which delivers the potential of the different methods in designing the optimal wind turbine.

The other criterion like cumulative probability and normal distribution of the ratio, Q_p/Q_m, has also been considered for the evaluation of different models (Farsakh & Titi, 2004; Das & Basudhar, 2006).

Figure 7. Regression plot of actual verses predicted for Cp4 by RVM

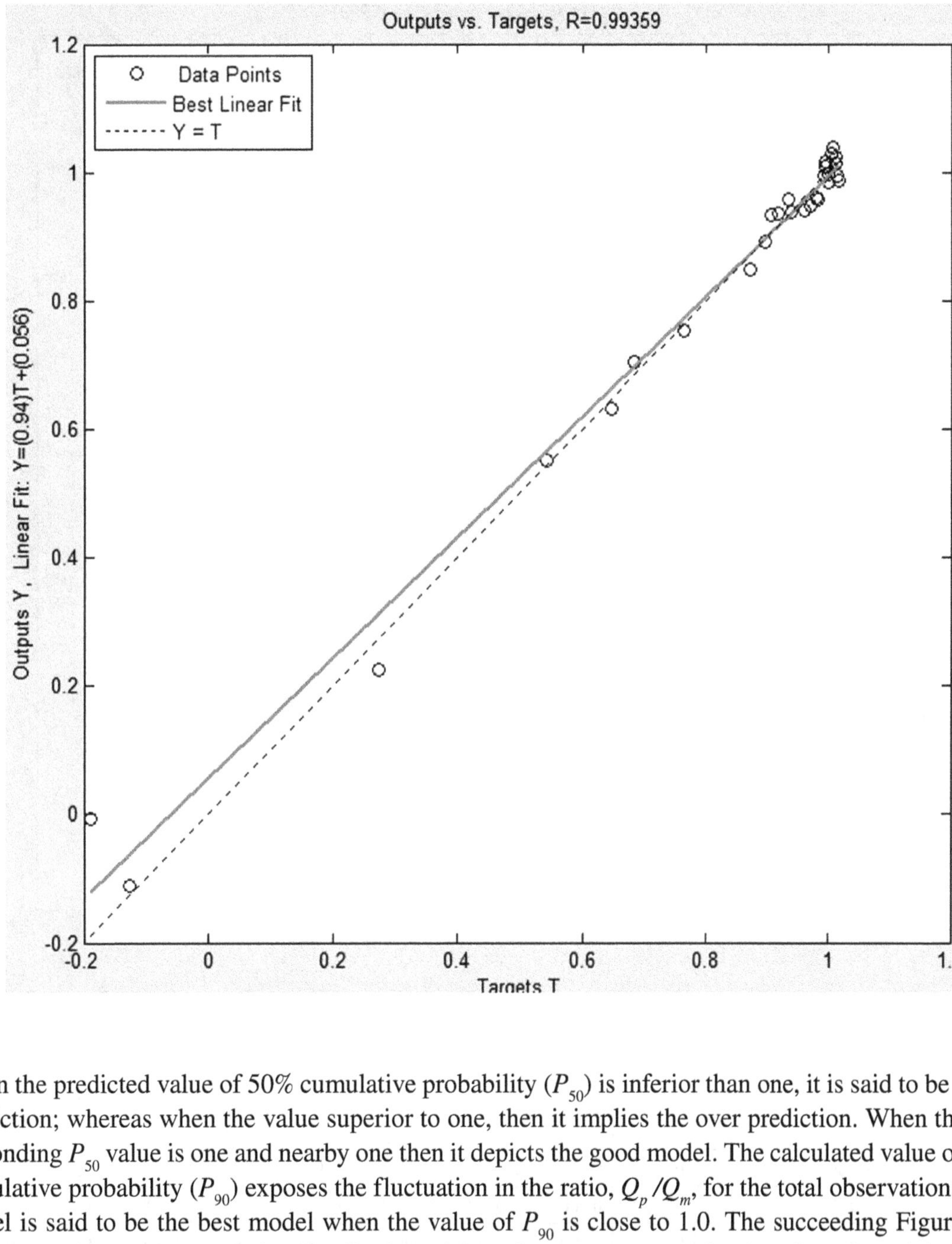

When the predicted value of 50% cumulative probability (P_{50}) is inferior than one, it is said to be under prediction; whereas when the value superior to one, then it implies the over prediction. When the corresponding P_{50} value is one and nearby one then it depicts the good model. The calculated value of 90% cumulative probability (P_{90}) exposes the fluctuation in the ratio, Q_p/Q_m, for the total observations. The model is said to be the best model when the value of P_{90} is close to 1.0. The succeeding Figures 12, 13, and 14 shows the cumulative distribution of the primary parameters by the adopted model and the ANN model.

The Normal distributions of the Q_p/Q_m primary parameters by the adopted model and the ANN model plotted in the following Figures 15, 16, and 17.

The adopted models depicted their capacity in determining the optimal design for the wind turbine.

Figure 8. Regression plot of actual verses predicted for TSR by RVM

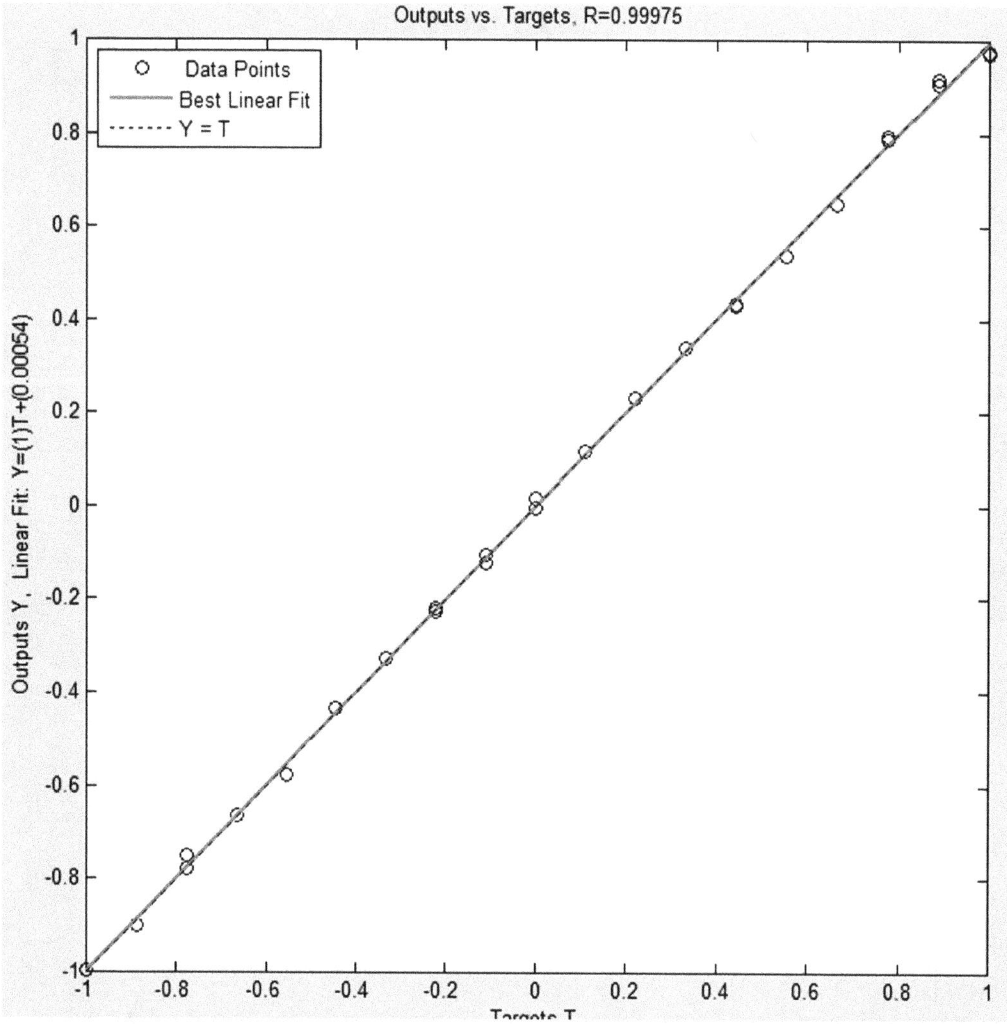

Table 2. Statistical criteria's for the developed models

	SVM			RVM			GP		
	C_{p3}	C_{p4}	TSR	C_{p3}	C_{p4}	TSR	C_{p3}	C_{p4}	TSR
MAE	0.0012	0.0006	0.0192	0.004105	0.0057	0.06912	0.0006	0.0006	0.008
RMSE	0.0022	0.0007	0.0187	0.00568	0.0138	0.0923	0.0008	0.0008	0.011
R	0.9999	0.999	0.999	0.99952	0.999	0.99978	0.9999	0.9999	0.9999
ρ	0.0027	0.0008	0.0192	0.00653	0.0158	0.0083	0.001	0.0009	0.0009
Mean (μ)	0.997	1	1.003	1.002	1.016	0.9974	0.999	0.999	1
Standard Deviation (σ)	0.014	0.002	0.0198	0.013	0.064	0.0283	0.0037	0.002	0.0044

Figure 9. Regression plot Cp3 by GP

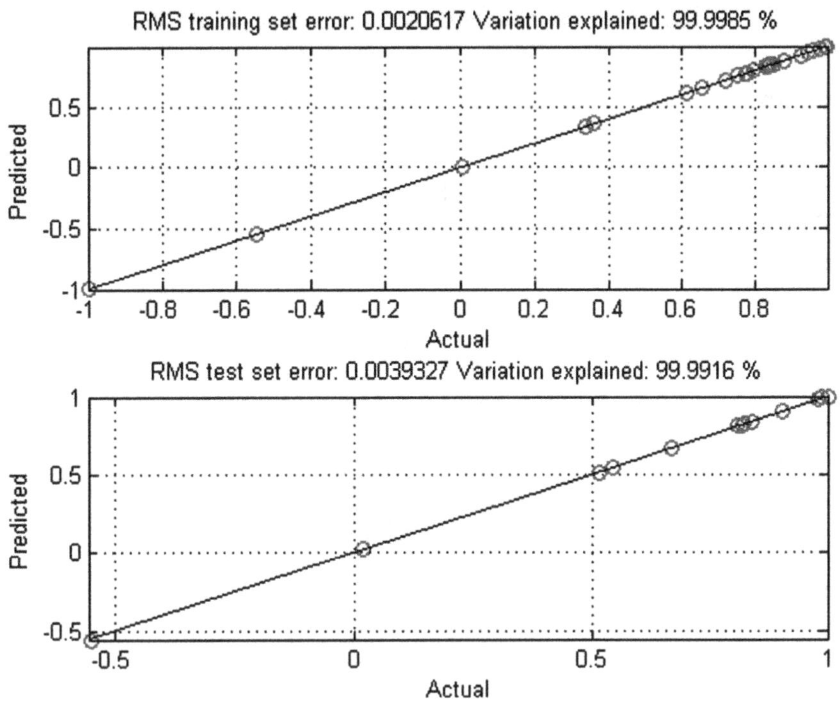

Figure 10. Regression plot Cp4 by GP

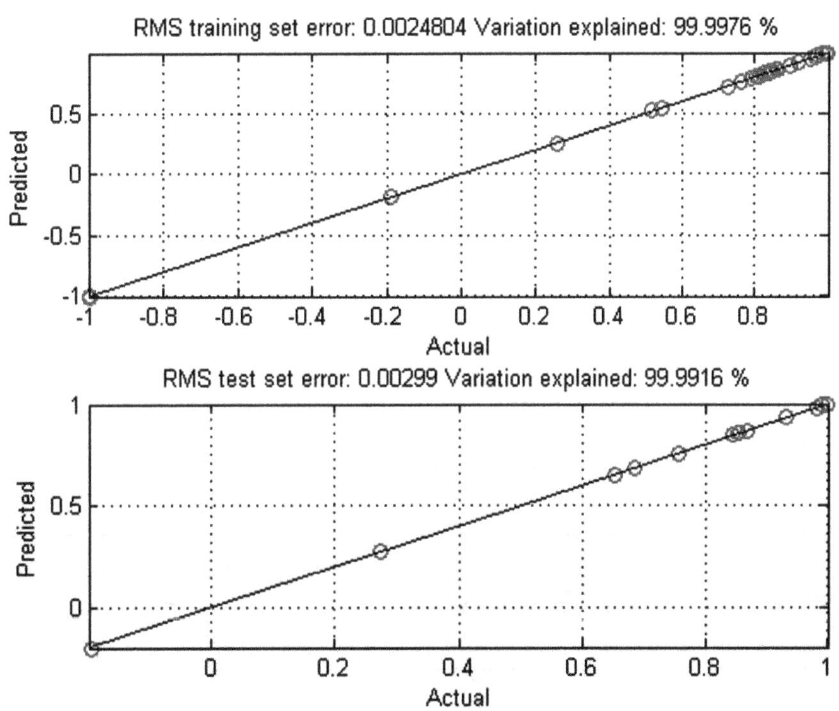

Figure 11. Regression plot TSR GP

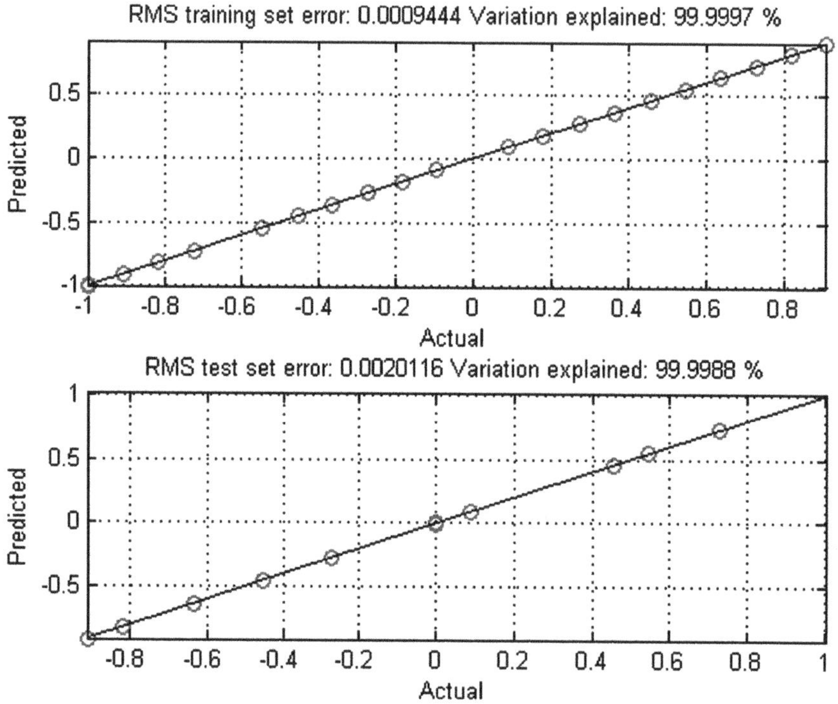

Figure 12. Cumulative probability plots of Cp3 for different models

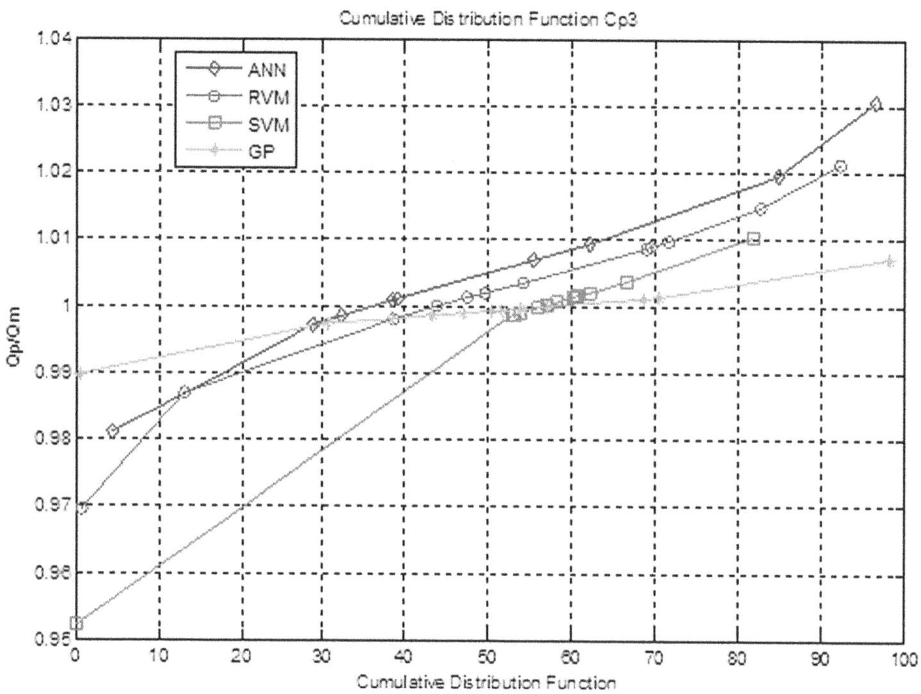

Figure 13. Cumulative probability plots of Cp4 for different models

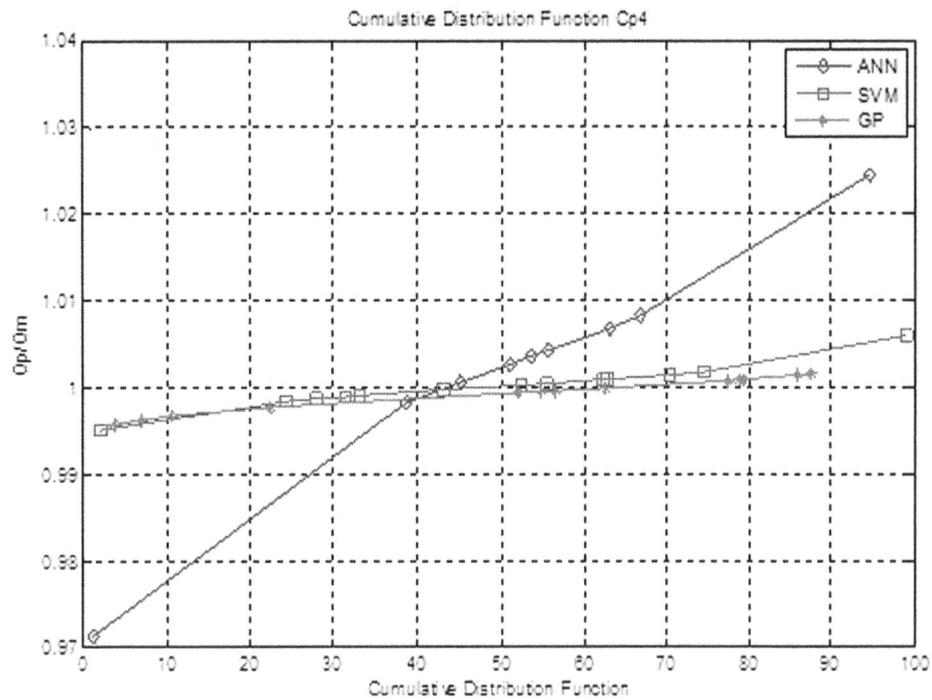

Figure 14. Cumulative probability plots of TSR for different models

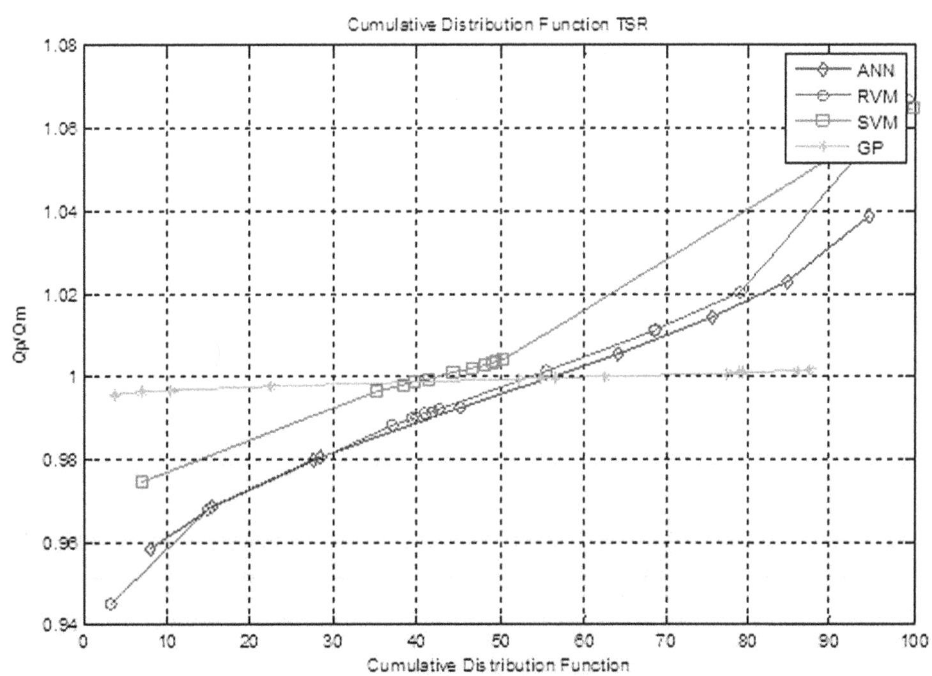

Figure 15. Normal distribution of Q_p/Q_m for C_{p3} of different models

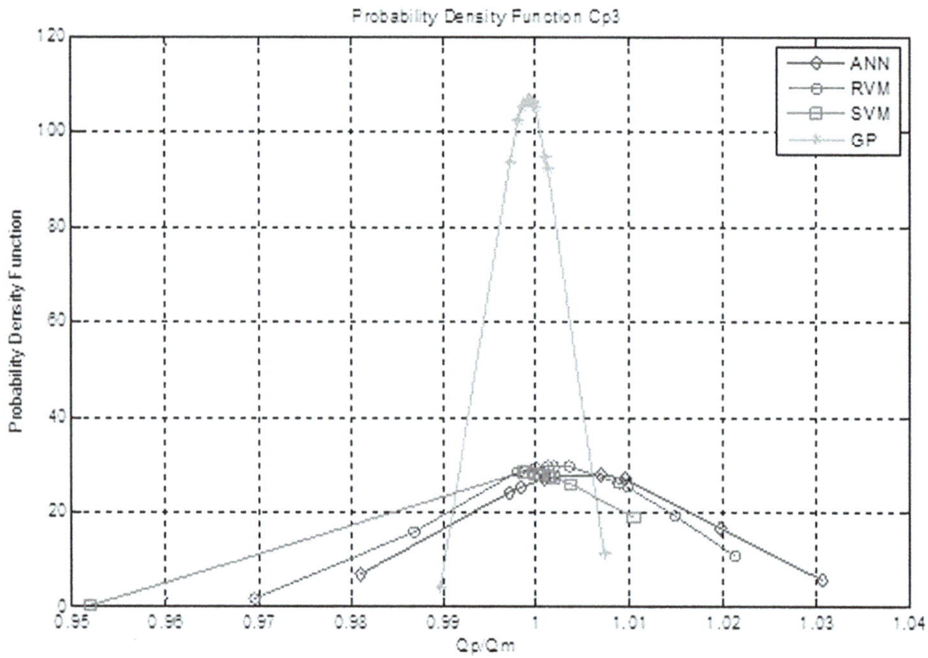

Figure 16. Normal distribution of Q_p/Q_m for C_{p4} of different models

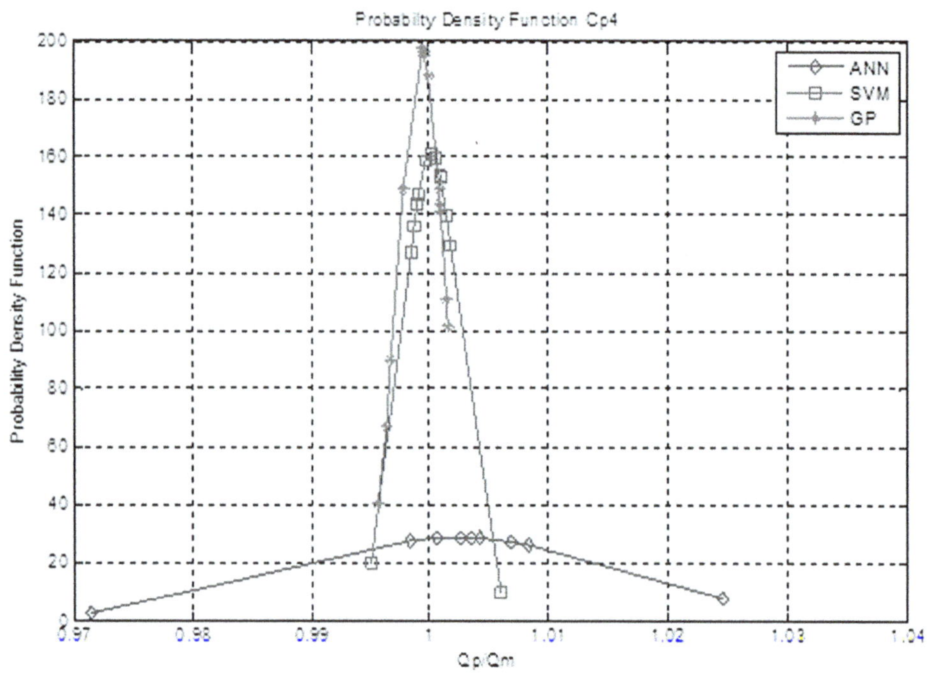

Figure 17. Normal distribution of Q_p/Q_m for TSR of different models

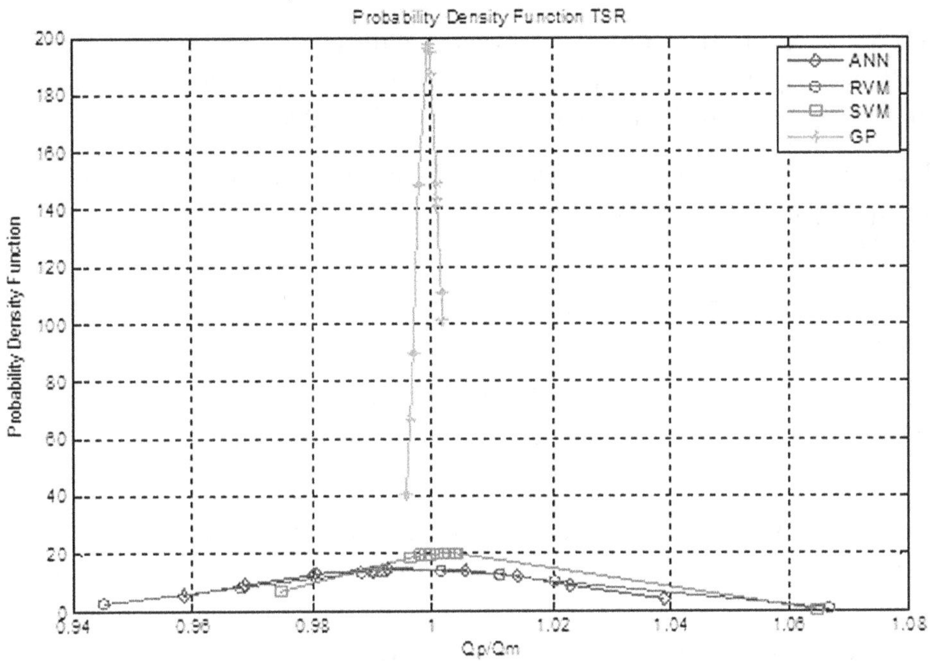

SOLUTIONS AND RECOMMENDATIONS

It depicts the path for easy and optimal design of the wind turbine. With any number of inputs, the predicted equation will provide the outcomes directly without any conditions.

FUTURE RESEARCH DIRECTIONS

The variation in the dataset ranges may lead to the inappropriate results. This can be resolved by the researchers in upcoming years. Various datasets can be collected and utilize the different procedures for foretelling optimal design for the wind turbine. It is also suggested for the scientists and researchers to adopt various techniques for determining the optimal designs for different turbines like steam turbine, Gas turbine, Transonic turbine, Water turbine, etc.

CONCLUSION

This chapter scrutinized the proficiency of SVM, RVM and GP for determining the best values for the parameters C_{p3}, C_{p4} and TSR which are responsible for the design of wind turbine. The developed SVM, RVM and GP give elegant and encouraging performances. The developed RVM only produces sparse solution. This article gives robust models based on SVM, RVM and GP for prediction of optimal design of wind turbine.

REFERENCES

Abu Farsakh, M. Y., & Titi, H. H. (2004). Assessment of direct cone penetration test methods for predicting the ultimate capacity of friction driven piles. *Journal of Geotechnical and Geoenvironmental Engineering*, *130*(9), 935–944. doi:10.1061/(ASCE)1090-0241(2004)130:9(935)

Abu Farsakh, M. Y., & Titi, H. H. (2004). Assessment of direct cone penetration test methods for predicting the ultimate capacity of friction driven piles. *J. Geotech. Geoenviron.*, *130*(90), 935–944. doi:10.1061/(ASCE)1090-0241(2004)130:9(935)

Adil, B., Güllüb, H., Çanakçıb, H., & Özbakır, L. (2008). Prediction of compressive and tensile strength of limestone via genetic programming. *Expert Systems with Applications*, *35*(1–2), 111–123.

Adil, B. L., Gullu, H., Canakc, H., & Ozbakır, L. (2008). Prediction of compressive and tensile strength of limestone via genetic programming. *Expert Systems with Applications*, *35*(1-2), 111–123. doi:10.1016/j.eswa.2007.06.006

Bao, Y., Wang, H., & Wang, B. (2014). Short-term wind power prediction using differential EMD and relevance vector machine. *Neural Computing & Applications*, *25*(2), 283–289. doi:10.1007/s00521-013-1482-z

Boser, B. E., Isabelle, M. G., & Vapnik, V. M. (1992). A Training Algorithm for Optimal Margin Classifiers. In *Proceedings of the 5th Annual Workshop on Computational Learning Theory (COLT'92)* (pp. 144-152). Pittsburgh, PA: ACM Press. doi:10.1145/130385.130401

Brezocnik, M., Kovacic, M., & Ficko, M. (2004). Prediction of surface roughness with genetic programming. *Journal of Materials Processing Technology*, *157–158*, 28–36. doi:10.1016/j.jmatprotec.2004.09.004

Ceryan, N., Okkan, U., Samui, P., & Ceryan, S. (2013). Modeling of tensile strength of rocks materials based on support vector machines approaches. *International Journal for Numerical and Analytical Methods in Geomechanics*, *37*(16), 2655–2670.

Cherkassky, V., & Mulier, F. (1998). *Learning from data: Concepts, theory, and methods*. New York: Wiley.

Coasta, A., Crespo, A., Navarro, J., Lizcano, G., Madsen, H., & Feitosa, E. (2008). A review on the young history of the wind power short-term prediction. *Renewable & Sustainable Energy Reviews*, *12*(6), 1725–1744. doi:10.1016/j.rser.2007.01.015

Das, S. K., & Basudhar, P. K. (2006). Undrained lateral load capacity of piles in clay using artificial neural network. *Computers and Geotechnics*, *33*(8), 454–459. doi:10.1016/j.compgeo.2006.08.006

Ernst, B., Oakleaf, B., Ahlstrom, M., Lange, M., Moehrlen, C., Lange, B., & Rohrig, K. et al. (2007). Predicting the wind. *Power and Energy Magazine*, *5*(6), 78–89. doi:10.1109/MPE.2007.906306

Ernst, B., Reyer, F., & Vanzetta, J. (2009). Integration of wide-scale renewable resources into the power delivery system. In *CIGRE/IEEE PES Joint Symposium* (pp.1–9).

Gandomi, M., Soltanpour, M., Zolfaghari, M. R., & Gandomi, A. H. (2016). Prediction of peak ground acceleration of Irans tectonic regions using a hybrid soft computing technique. *Geoscience Frontiers*, *7*(1), 75–82. doi:10.1016/j.gsf.2014.10.004

Goldberg, D. E. (1989). *Genetic Algorithms in Search*. Addison-Wesley Professional.

Hastie, T., Tibshirani, R., & Friedman, J. (2001). *The elements of statistical learning: Data mining, inference and prediction*. Berlin: Springer. doi:10.1007/978-0-387-21606-5

Koza, J. R. (1992). *Genetic Programming: On the Programming of Computers by Means of Natural Selection*. Cambridge, MA: MIT press.

Maria, G. D. G., Stefano, C., Antonio, F., & Paolo, M. C. (2014). Comparison between wind power prediction models based on wavelet decomposition with Least-Squares Support Vector Machine (LS-SVM) and Artificial Neural Network (ANN). *Energies*, *7*(8), 5251–5272. doi:10.3390/en7085251

Milligan, M., Porter, K., DeMeo, E., Denholm, P., Holttinen, H., Kirby, B., & Soder, L. et al. (2009). *Wind power myths debunked*. IEEE Power and Energy Society.

Muljadi, E., & Butterfield, C. P. (2001). Pitch-controlled variable-speed wind turbine generation. *IEEE Trans. Ind. Application*, *37*(1), 240–246. doi:10.1109/28.903156

Negnevitsky, M., Mandal, P., & Srivastava, A. (2009). Machine learning applications for load, price and wind power prediction in power systems. *Intelligent System Applications to Power Systems*, 1–6.

Ochieng, P. O. (2014). Mathematical analysis of tip speed ratio of a wind turbine and its effects on power coefficient. *International Journal of Mathematics and Soft Computing*, *4*, 61–66.

Pradyut, K. M., Manas, R. D., Samui, P., & Sarat, K. D. (2013). Uplift Capacity of Suction Caisson in Clay Using Artificial Intelligence Techniques. *Marine Georesources and Geotechnology*, *31*(4), 375–390. doi:10.1080/1064119X.2012.690827

Ragheb, M., & Adam, M. R. (2011). Wind turbines theory - The betz equation and optimal rotor tip speed ratio, fundamental and advanced topics in wind power. In C. Rupp (Ed.), *Fundamental and Advanced Topics in Wind Power*. doi:10.5772/21398

Santos, P., Luisa, F. V., Aníbal, R., Andres, B., & Maudes, J. (2015). An SVM-Based Solution for Fault Detection in Wind Turbines. *Sensors (Basel, Switzerland)*, *15*(3), 5627–5648. doi:10.3390/s150305627 PMID:25760051

Smola, A., & Scholkopf, B. (1998). *A tutorial on support vector regression*. NeuroCOLT Technical Report NC-TR-98-030, Royal Holloway College, University of London, UK.

Smola, A., & Scholkopf, B. (1998). On a kernel based method for pattern recognition, regression, approximation and operator inversion. *Algorithmica*, *22*(1-2), 211–231. doi:10.1007/PL00013831

Stamatios, V. K. (1995). *Understanding Neural Networks and Fuzzy Logic: Basic Concepts and Applications*. IEEE Press.

Tipping, M. E. (2000). *The relevance vector machine*. In S. A. Solla, T. K. Leen, & K. R. Müller (Eds.), Advances in Neural Information Processing Systems (Vol. 12, pp. 652–658). Cambridge, MA: MIT Press.

Tipping, M. E. (2001). Sparse bayesian learning and the relevance vector machine. *Journal of Machine Learning Research*, *1*, 211–244.

Tu, J. V. (1996). Advantages and disadvantages of using artificial neural networks versus logistic regression for predicting medical outcomes. *Journal of Clinical Epidemiology*, *49*(11), 1225–1231. doi:10.1016/S0895-4356(96)00002-9 PMID:8892489

Tzikas, D., Wei, L., Likas, A., Yang, Y., & Galatsanos, N. (2006). A tutorial on relevance vector machines for regression and classification with applications. *EURASIP Newsletter*, *17*(2), 1–25.

Vapnik, V. N. (1998). *Statistical learning theory*. New York: Wiley.

Vapnik, V. N. (1999). *The nature of statistical learning theory* (2nd ed.). Berlin: Springer.

Yan, J., Yongqian, L., Shuang, H., & Meng, Q. (2013). Wind power grouping forecasts and its uncertainty analysis using optimized relevance vector machine. *Renewable & Sustainable Energy Reviews*, *27*, 613–621. doi:10.1016/j.rser.2013.07.026

Yurdusev, M. A., Ata, R., & Cetin, N. S. (2006). Assessment of optimum tip speed ratio in wind turbines using artificial neural networks. *Energy*, *31*(12), 2153–2161. doi:10.1016/j.energy.2005.09.007

KEY TERMS AND DEFINITIONS

Artificial Intelligence: Artificial Intelligence is the field that studies the synthesis and analysis of computational agents that act intelligently.

Genetic Programming: Genetic Programming is a set of instructions and a fitness function to measure how well a computer has performed a task. It is a specialization of genetic algorithms (GA) where each individual is a computer program.

Power Coefficient: Power coefficient is a quantity that expresses what fraction of the power in the wind is being extracted by the wind turbine.

Prediction: The action of forecasting something is called as prediction.

Relevance Vector Machine: Relevance Vector Machine (RVM) uses Bayesian inference to get the solutions for classification and regression problems. It has same functional form of SVM, but provides the probabilistic classification.

Support Vector Machine: Support Vector Machine (SVM) is supervised learning algorithm used for analyzing data and recognizes designs, also used for classification and regression analysis.

Tip Speed Ratio: The ratio between the tangential speed of the tip of blade and the actual velocity of the wind is called tip speed ratio. The tip-speed ratio is related to efficiency, with the optimum varying with blade design.

Wind Turbine: A device which converts kinetic energy from wind into electric power is called as wind turbine.

Chapter 4
Optimization of Process Parameters Using Taguchi Coupled Genetic Algorithm:
Machining in CNC Lathe

Supriyo Roy
Birla Institute of Technology, India

J. Paulo Davim
University of Aveiro, Portugal

Kaushik Kumar
Birla Institute of Technology, India

ABSTRACT

In the era of mass manufacturing, Material removal rate and Surface Roughness are of primary concern even in manufacturing using contemporary CNC machines. In this work, L27 Orthogonal Array of Taguchi method is selected for three parameters (Depth of cut, Feed and Speed) with three different levels to optimize the turning parameters for Material Removal Rate and Surface Roughness on an EMCO Concept Turn 105 CNC lathe for machining SAE 1020 material with carbide tool. The MRR and SR are observed as the objective to develop the combination of optimum cutting parameters. The objectives were optimized using Taguchi, Grey Taguchi and NSGA-II. The result from these techniques was compared to identify the optimal values of cutting parameters for maximum MRR, minimum SR and best combination of both. This study also produced a predictive equation for determining MRR and SR for a given set of parameters outside the considered values. Thus, with the proposed optimal parameters it is possible to increase the efficiency of machining process and decrease production cost in CNC Lathe.

DOI: 10.4018/978-1-5225-1639-2.ch004

INTRODUCTION

Computer (Aided) Numerical Controlled (CNC) machining has grown to be an indispensible part of machining industry for fast machining, controlled time limits, etc. Traditional machines such as vertical millers, centre lathes, shaping machines, routers, etc. as operated by workers in the manufacturing shop floors been replaced nowadays by computer control machines. Pertinent parameters like accuracy and precision achieved through CNC could not be achieved by conventional manufacturing machines. These machines are programmed with a design which can then be manufactured hundreds or even thousands of times. Each manufactured product will be exactly the same shape and size. Any skilled engineer can make the same component many times. However, if each component is observed very carefully, each one may vary very slightly. A CNC machine will manufacture each component as an exact match. Moreover, very significantly, modern design software allows the designer to simulate the manufacture with his own idea. There is no need to make a prototype or a model; saves time and money.

CNC machining, though, is today's voice of workers in the shop floors; still there is a room for errors in it. CNC machining is basically depends on the skill and experience of the workers to get right dimension. Machine performance and product characteristics are not guaranteed to be acceptable always. In most cases, CNC machines are more expensive than manually operated machines. Out of the various reasons, which could be considered as manufacturing goal, academic researchers along with industry practitioners found Material Removal Rate (MRR) and Surface roughness (SR) that may be considered for the present work as the factors directly affects quality, cost of machining and the machining hour rate.

In order to find the optimized set of input parameters and also to identify the effect of each towards a particular output, researchers have been trying for years together. Meng (2000) tried to calculate optimum cutting condition for turning operation using a machining theory. Researchers (Aggarwal et al., 2008; Asilturk et al., 2016; Dutta & Majumdar, 2010; Liu et al., 2010; Singh & Kumar, 2006) have tried to optimize machining parameters using various methods like Genetic Algorithm, Simulated Annealing method, Multi-Objective Evolutionary Algorithm etc. Bharathi and Baskar (2010) reviewed the researches made on optimization techniques for machining using various mathematical models. Chen and Su (1998) tried for optimization of machining conditions for turning cylindrical stocks into continuous finished profiles. Negrete et. al. (2013) tried to optimize the cutting parameters for minimizing cutting power where as Cayda (2010) tried to vary the cutting tool to evaluate the machinability of AISI 4340 steel.

TERMINOLOGY: TURNING OPERATION

It is a machining process in which a rotating workpiece is machined using a non-rotary tool bit. The tool follows a helical path while moving linearly over the work piece. Usually the term 'turning' means generation of external surfaces by cutting action resulting from the shearing between the tool and the work piece. Turning can be done manually, in a traditional form of lathe, which frequently requires continuous supervision by the operator, or by using an automated lathe which does not require any supervision. In this work a computer numerical control lathe is used for turning operations. The turning processes can be of broadly classified into four different types straight turning, taper turning, profiling and grooving. These types of turning processes produce various shapes of materials such as straight, conical, curved or grooved workpiece. In this present study, work with straight turning was performed by using single-point cutting tool.

Necessity of Optimization in Turning Operation

Present age of globalization forced technocrats to develop technology of CNC turning machine in several directions. In order to meet the advance requirements in various manufacturing fields and also to satisfy the ever growing quality consciousness amongst the customers, manufacturers need optimized process parameters to save time and money. Researchers (Aloufi 2011; Korkut et al., 2004; Lan, 2010; Suhail et al., 2010; Tzeng et al., 2009; Kolahan & Abachizadeh, 2008; Mahapatra et al., 2006; Sankaya & Guilu, 2015) have tried to use various optimization methods to provide manufacturers a ready to use set of processing parameters. MRR and Ra, being the measurement of productivity and quality of machining component, are widely used in a variety of products / components by almost all the manufacturing industries. So in order to improve overall machining characteristics, minimized value of Ra and maximized value of MRR with appropriate values of machining process parameters like depth of cut, feed rate and cutting speed are required to be finalised.

In tune with the above the usage of the CNC by the industrial world falls under three groups:

1. Products under consideration are for mass production and hence the Material Removal Rate (MRR) is to be maximised to reduce the production time. In this the Surface Roughness (SR) does not play an important role. Usually the machine components for fast moving consumer goods (FMCG) fall under this category.
2. Products in this are usually tailor made and has specific application in which SR plays the vital role and MRR is unimportant. The components used in aviation e.g. space shuttle, aircraft etc. comes under this.
3. Machine component used in FMCG where SR is equally important comes in this category. In this both MRR and SR are considered simultaneously. The engine components in a car, for example piston, cylinder etc., are some examples of products in this section.

For an industrialist in any of the above cases, the 'best sets of values' of the operating parameters are always desired and a readily available data saves time and money otherwise the production has to go for a research to get the same.

There are many methods which scientists and mathematicians use for optimization of different processes and systems (Myers et al., 2009; Freund & Wilson, 1998; Latha & Senthilkumar, 2010; Li & Liang, 2007; Palanikumar, 2010; Sun & Lee, 2005; Yusoff et al., 2011).

OPTIMIZATION METHODS

In this work, authors have tried to provide best possible solution to the above problem using three methods namely, Taguchi, Grey Taguchi and Non-Dominated Sorting Genetic Algorithm (NSGA).

Taguchi Method

Taguchi method emphasizes over the selection of most optimal solution (i.e. MRR & SR) over the set of given inputs (i.e. depth of cut, feed rate and cutting speed) with a reduced cost and increased quality.

The optimal solution so obtained is least affected by any outside disturbances like the noise or any other environmental conditions (Montgomery, 2001; Park 1998). Thus, modern day approach to find the optimal output over a set of given input can be easily carried out by the use of Taguchi method (Heieh et al., 2005; Kamaruddin et al., 2010; Kathlee et al., 2004; Kaymaz & McMahon, 2005) rather than using any other conventional methods. This method has a wide scope of use varying from the agricultural field to medical field and various fields of engineering sciences. It has been used by many agricultural scientists to evaluate the production of their crops based on the different sets of inputs like the waterfall level, fertility of the land, types of seeds used and many other factors. It is also used by the weather department in forecasting of the various environmental conditions. In the field of Science and Engineering, it is used for obtaining optimal results based on the various engineering inputs. Thus, the Taguchi approach has a wide variety of applications in the different fields because of its easiness and optimized results (Fraley et al., 2006). However, even being having a lot of advantages, it suffers from a major disadvantage that the results obtained are only relative to the levels selected and do not exactly indicate the effect of entire range of each parameter on the performance characteristic value. They also do not test the output for all the combinations of input variables as the combination selected here is limited. Moreover, another major disadvantage of the Taguchi method is that it is not a dynamic process i.e. it is not concerned with those variables which discretely or continuously change their values from time to time. It deals with mostly the static variables i.e. variables whose values are not time dependent. Moreover, it deals with designing of quality rather than improving the quality. Because of this, they are set to be applied in the industries at the early stages only (Srinivas & Deb, 1995). However, despite of all these disadvantages, it is used widely in the different areas very effectively. To reduce the error while using an increased number of inputs, the orthogonal matrix of Taguchi can be used, which is selected as per the requirements of the person. In this work L27 (9^3) Orthogonal Array would be employed to obtain optimized economical results. Taguchi method for the Design of Experiments also emphasizes over the use of Loss function, which is the deviation from the desired value of the quality characteristics. Based on these Loss functions, S/N ratio for each experimental set would be evaluated and accordingly the optimal results will be derived. ANOVA is then generated for the results and finally, the confirmations results are obtained between the experimental values and the determined values.

In single objective problem, optimization problem deal with a global minimum or maximum, while in multiple objectives responses, there may not exist unique solution, best solution depend on the requirement of the product. For both MRR and SR influence of machining parameters are quite different. Hence, optimization of more than one functions needed one at a time to arrive at the optimal solutions. Various traditional and non-traditional methods are available for obtaining the solutions for multi-objective problems. Some examples are Distance Function methods, simulated annealing, Min–Max, grey-approach and Weighted Sum method. These techniques convert the multi-objective problem into a single objective problem, with the corresponding weights based on their relative importance. But these method need thorough knowledge, and accuracy of these methods reduced as the function become discontinuous. Genetic algorithm (GA) is free from such type of drawback; it does not require any knowledge of gradient and inherent parallelism for multi-objective problem. For multi-objective optimization problem, some alteration is needed to simple GA. Multi-Objective Genetic Algorithm (MOGA), Vector Evaluated Genetic Algorithm (VEGA) and Non-Dominated Sorting Genetic Algorithm (NSGA) are some examples of GA based multi-objective solution methods.

Non-Dominated Sorting Genetic (NSGA-II) Algorithm

Genetic algorithms (GA) mimic the principal of natural genetics and natural selection to constitute search and optimization procedure (Srinivas & Deb, 1995; Ganesan & Mohankumar, 2013). The non-dominated sorting genetic algorithm has been criticized for its high computational complexity, lack of elitism and its choice of the optimal parameter value for sharing parameter σ. The NSGA-II is a modified version, which has a better sorting algorithm, incorporates elitism and does not require the choosing of a sharing. There are two key concepts in NSGA-II: a fast non-dominated sorting of the population and a crowding distance (Deb et al., 2002; Deb, 2002). This algorithm uses the elite-preserving operator, which favours the elites of a population by giving them an opportunity to be directly carried over to the next generation (Deb et al., 2002). A typical algorithm for NSGA-II can be:

1. A fitness function is chosen to represent the problems, a selection operator, a crossover operator, and a mutation operator. Select population size, n, crossover probability, pc, and mutation probability, pm. initialize the random population of strings of size L. Assume a maximum allowable generation number tmax. Set t = 0.
2. Evaluate each string in the population.
3. If t > tmax or other termination criteria is satisfied, then Terminate.
4. Pre-form reproduction on the population.
5. Perform, crossover on random pairs of strings.
6. Perform mutation on every string.
7. Evaluate strings in the new population. Set t = t + 1 and go to Step 3.

EXPERIMENTATION

Input Process Parameters Considered

In the present work depth of cut, feed rate and cutting speed are taken into consideration as the input machining parameters. Three process parameters considered in the study with equally spaced three levels within the operating range for each of the process parameters as input parameters Table 1.

Table 1. Process parameters and their levels for turning

Process Parameter	Unit	Level 1	Level 2	Level 3
Depth of cut	mm	1	1.5	2
Feed rate	mm/rev	0.15	0.2	0.25
Cutting speed	m/min	50	100	150

Output Considered

MRR

It is the rate at which the material is removed from the work-piece. The weight of the work-piece material is recorded before carrying out the machining process and is compared with the weight after machining (using Equation 1).

$$MRR = \frac{w_i - w_f}{t} \quad (1)$$

Where,

MRR = Materials Removal Rate (gm/min)

W_i = Initial weight of work piece material (gm)

W_f = Final weight of work piece material (gm)

t = Time taken for machining in minutes

MRR is considered t be the most important parameters and for industrial purpose. Since larger MRR is always desired hence, it is optimized on the basis of Larger the Better condition.

Surface Roughness

Surface roughness is usually defined as the deviation of a surface from an ideal level and is defined according to international standard (ISO 4287:1997). The arithmetic mean surface roughness Ra is defined as deviation of a surface from a theoretical center line (Rmean) over a measuring length Lm Figure 1.

Since minimum SR is always desired hence, it is optimized on the basis of Smaller the Better condition.

Figure 1. Arithmetic mean surface roughness Ra

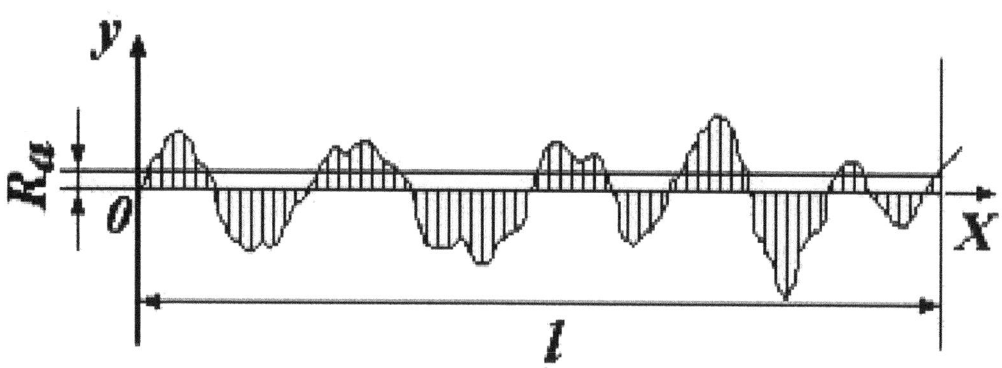

Material

Tungsten Carbide and mild steel (SAE 1020) are identified as tool and material respectively. Figure 2 shows the machined samples.

It can be observed from the figure that turning has been done in steps. For calculation of the MRR, the machining was done for a constant time (1 min.). Depending on the feed and speed the length has changed, moreover in a single sample machining with more than one set of parameters has been used to save material.

Machine and Instrument

CNC lathe utilised for the experimentation was EMCO Concept Turn 105 Figure 3. Although there are different methods of measurement of surface roughness (Correia & Davim, 2011; Wang & Meng, 2010) in this work a contact type Mitutoyo SJ201 Figure 4 instrument was used to measure surface roughness (R_a) and weight measurement was done using a Mettle Toddler electronic balance Figure 5.

DESIGN OF EXPERIMENT

Design of experiments (DOE) is a systematic process to determine the relationship between factors affecting a process and output of that process. This information is required to manage the process inputs in order to optimize the output. Most commonly used terms in design of experiments methodology include: controllable and uncontrollable input factors, responses, hypothesis testing, blocking, replication and interaction, etc. Controllable input factors are those input parameters that can be modified in an experiment or process. Uncontrollable input factors are those parameters that cannot be changed over time complexity. These factors need to be recognized to understand how they may affect the response. Responses or output measures are the elements of the process outcome that gage desired effect. Experi-

Figure 2. Machined Samples

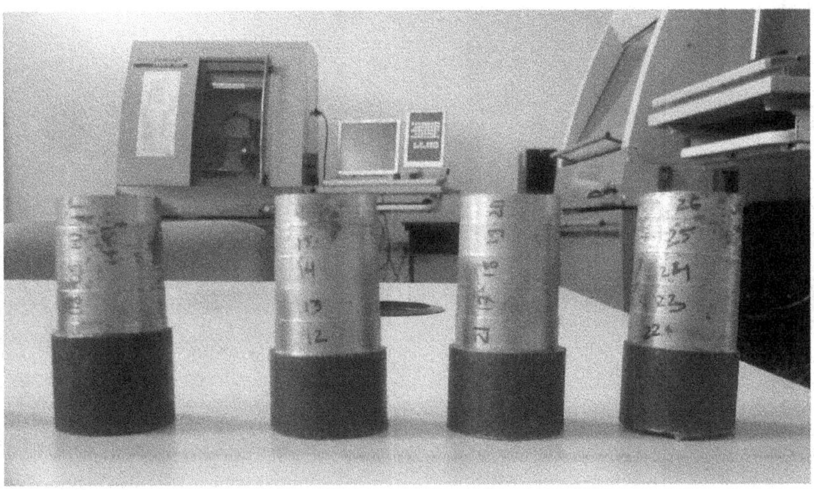

Figure 3. EMCO Concept Turn 105

Figure 4. Mitutoyo SJ201

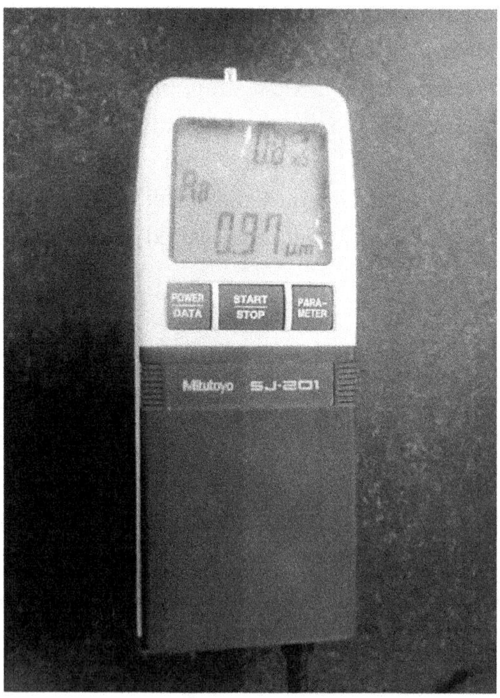

Figure 5. Mettle toddler electronic balance

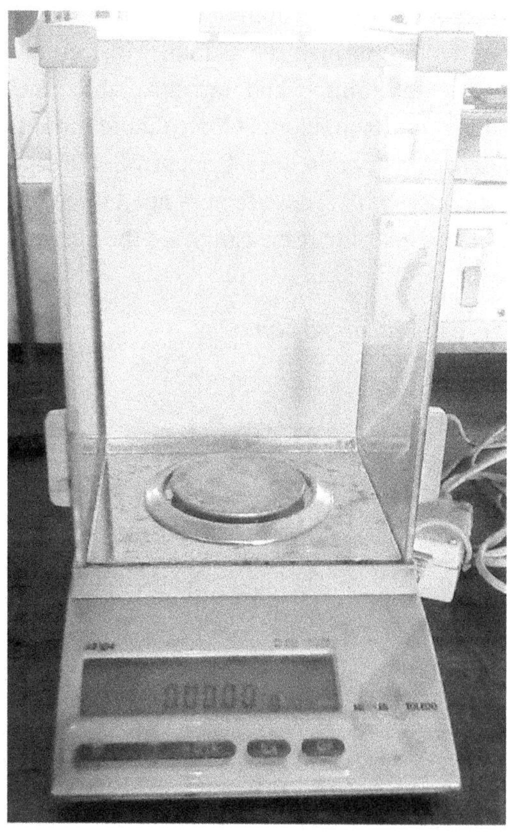

mental design can be used at point of greatest leverage to reduce design costs by speeding up design process, reducing late engineering design changes and reducing product material and labor complexity. Designed Experiments are also used as very important tool to achieve manufacturing cost savings by minimizing process variation and reducing rework, scrap and need for inspection.

Three process parameters, as already stated above, Depth of cut (A), Feed rate (B) and Cutting speed (C) were considered in the study with equally spaced three levels within the operating range for each of the process parameters as input parameters. In the present research work experiment was planned according to the Taguchi's design of experiment method. Taguchi (1990) suggested that, total Degree of freedom (DOF) of selected orthogonal array must be greater than or equal to the total DOF of standard Orthogonal Array (OA). Since total DOF of four factor and their interactions is 20 (4X2+3X4) and standard OA suggested for 4 factor with 3 level are L9 (DOF 8) and L27 (DOF 26), hence the L27 (DOF 26) OA has been selected for experimental work and tabulated as shown in Table 2.

OPTIMIZATION

Taguchi's Approach to Parameter Design

Designers can utilize standard and systematic approach for executing experimentation to obtain optimum settings of design parameters for quality and cost very efficiently. The method gives priority to move quality back to the design stage, pursuing to design a/process or product. Orthogonal arrays are utilized in course of Taguchi method, to analyse a large number of variables with a fewer number of experiments.

Table 2. Design of Experiments using L27 orthogonal array

Exp.No.	Control Factor Levels			Exp.No.	Control Factor Levels		
	Depth of Cut (A) mm	Feed Rate (B) mm/rev	Cutting Speed (C) m/min		Depth of Cut (A) mm	Feed Rate (B) mm/rev	Cutting Speed (C) m/min
1	1.0	0.15	50	15	2.0	0.20	100
2	1.5	0.15	50	16	1.0	0.25	100
3	2.0	0.15	50	17	1.5	0.25	100
4	1.0	0.20	50	18	2.0	0.25	100
5	1.5	0.20	50	19	1.0	0.15	150
6	2.0	0.20	50	20	1.5	0.15	150
7	1.0	0.25	50	21	2.0	0.15	150
8	1.5	0.25	50	22	1.0	0.20	150
9	2.0	0.25	50	23	1.5	0.20	150
10	1.0	0.15	100	24	2.0	0.20	150
11	1.5	0.15	100	25	1.0	0.25	150
12	2.0	0.15	100	26	1.5	0.25	150
13	1.0	0.20	100	27	2.0	0.25	150
14	1.5	0.20	100				

The conclusions made from small-scale experiments are genuine over the complete experimental domain consists of the control factors and their corresponding level settings. Taguchi method utilizes a statistical measure of performance called Signal-to-Noise (S/N) ratio, in order to examine the results. The S/N ratio takes into account both the mean and the variability of the response data. After performing the statistical analysis of S/N ratio, an ANOVA requires to be applied for estimating the relative importance of various factors and for computing error variance. The predicted optimum setting need not correspond to one of the rows of the matrix experiment in Taguchi method for parameter design. Therefore, an experimental confirmation is run utilizing the predicted optimum levels for the process parameters being investigated. The intention is to confirm that the optimum conditions suggested by the matrix experiments do indeed contribute the projected improvement. If the projected improvements and observed match, the suggested optimum conditions will be adopted.

Taguchi's S/N Ratio for Performance Evaluation

There is a loss function which describes the deviation from the target and further transformed into S/N ratio (Signal-to-Noise Ratio). The transformed S/N ratio is also defined as quality evaluation index. The least variation and the optimal design are obtained by analyzing S/N ratio. The higher the S/N ratio, the more stable the achievable quality. There are three Signal-to-Noise ratios of common interest for optimization of Static Problems:

1. **Nominal-the-Best (NB):** In this approach, the closer to the target value, the better and the deviation is quadratic. The formula (1) for these characteristics is:

$$\frac{S}{N}\,Ratio = -10\log\left(\frac{y}{S_y^{2}}\right) \tag{1}$$

2. **Lower-Is-Better (LB):** The Lower-Is-Better approach held when a company desires smaller values. The formula (2) for these characteristics is:

$$\frac{S}{N}\,Ratio = -10\log\left(\frac{1}{n}\sum_{i=1}^{n} y^{2}\right) \tag{2}$$

3. **Higher-Is-Better (HB):** Higher-Is-Better (HB) is required when a manufacturer desires higher values of a characteristic. The formula (3) for these characteristics is:

$$\frac{S}{N}\,Ratio = -10\log\left(\frac{1}{n}\sum_{i=1}^{n}\frac{1}{y^{2}}\right) \tag{3}$$

here,

y = Average of observed values;

s_y^2 = Variance of y;

N = Number of observations

However, Taguchi method is considered only for single objective optimization problems. It cannot be utilized for getting the single optimal setting of process parameters considering more than one performance parameter.

Optimization of Material Removal Rate Using Taguchi Method

The research is dedicated to find out the optimum combination of machining process parameters within a given range for better outputs in machining, with the application of taguchi approach. L27 orthogonal array (OA) design is utilised and data obtained from the experiments were analyzed to assess the effect of various controllable machining parameter variations on material removal rate (MRR). Material removal rate (MRR) taken as the response. Since for all practical purposes the MRR requires to be maximum hence for the analysis the maximization equation has been applied.

Analysis of variance (ANOVA) was also carried out to identify the level of significance of each factor and their interactions. Emphasis on the development of comprehensive mathematical models for material removal rate (MRR) describing the interactive and higher-order influences of the various machining process parameters. Finally, the three dimensional surface plot for MRR and machining parameter (i.e. Feed, Speed and depth of cut) was generated using statistical software Minitab 16 (Minitab, 2010). A regression equation was also formulated to speculate the values of MRR outside the considered level. Finally a confirmatory test was conducted to evaluate the betterment in the results.

Optimization of Surface Roughness Using Taguchi

The mode of operation was similar to the MRR as described earlier. The difference was selection of SR as the response. The surface roughness is required to be minimum hence the minimization equation was used for the calculations. In this also ANOVA table was generated along with regression equation and the confirmatory test results.

Multi Objective Optimization of Material Removal Rate and Surface Roughness Using Grey-Taguchi Approach

The earlier optimization technique predicts the optimal responses individually which necessarily does not optimize both i.e. with maximum MRR the SR may not be minimum and vice versa. So here SR and MRR characteristics were taken together as the responses and a grey relational grade was generated considering the effect of both the responses. This activity converted both MRR and SR into a grade, which in turn was optimised. This now got converted into a single response and was optimised in the same manner as done earlier for MRR and SR. In this also ANOVA table was generated along with confirmatory test results.

Multi Objective Optimization of Material Removal Rate and Surface Roughness Using Non-Dominated Sorting Genetic Algorithm

In the present work, objectives are minimization of average SR and simultaneously maximization of MRR. Since objective functions are of contradictory in nature, Genetic algorithm leads the minimizing behaviour therefore some modification are required to be done for MRR model and Objective function for MRR requires to be suitably minimized. The required constraints are formulated as shown below:

Objective 1 = 1/MRR

Objective 2 = SR

Subjected to:

$10 \leq A \leq 20$

$08 \leq B \leq 12$

$0.1 \leq C \leq 0.32$

$0.2 \leq C \leq 0.30$

RESULTS AND DISCUSSION

Optimization of Material Removal Rate Using Taguchi Method

The experiments are conducted according to the sequence of L27 OA and the experimental results of MRR are calculated. The analysis of experimental results is carried out by using Minitab 16 statistical software. The interaction between the parameters is also considered before analyzing the experimental data using this software. As per Taguchi method, the experimental results are converted into S/N ratio on the basis of larger-the-better criterion which is selected for oblivious reasons. The MRR value and the calculated S/N ratio are given in Table 3. The overall mean S/N ratio of friction coefficient is found to be 12.8143 dB. The mean response value for each level of design factors are computed by averaging the S/N ratio of their respective experiments (for example, factor A at level 1, 2, and 3 are calculated by averaging the S/N ratio for the experiment run 1-9, 10-18 and 19-27) and summarized in Table 4. Further, ranks are assigned to the factors based on Delta statistics i.e., rank 1 is assigned to the highest delta value, rank 2 to the second highest delta value and so on. The delta values are the indication of effectiveness of the parameters on the responses. From the mean response value of each level, the main effect and interaction plots between the design factors are graphically determined and shown in Figure 6 and Figure 7 (a, b, c) respectively. It is much clear from the main effect plot (Figure 6) that factor Depth of Cut (A) has the most significant effect as the line for this factor at level 1 has the highest inclination from the overall mean line while factor Feed (B) and Speed (C) has some influence on MRR. In case of interaction plots (Figure 7), it can be seen that there is no strong interaction between the parameters as lines are much parallel and not intersecting each other. The analysis of the results leads to the conclusion that optimal combination factors are found to be A1B3C3 which gives maximum MRR and factor A (Depth of Cut) has most effective on the response followed by B (Feed) and C (Speed). The phenomena is also supported by the general understanding that more depth of cut would surely result in more MRR.

Further, analysis of variance (ANOVA) is carried out and given in Table 5. It is seen that factor Depth of Cut (A) has the most significant influence at 95% level of confidence while Feed (B) has some contribution in controlling the minimum friction coefficient response within the specified test range. The interaction of parameters has no influence on the response for this study. It is well supported by the interaction plots shown in Figure 7.

Regression analysis is statistical tool for analysing the relation between variable. It is one of the commonly used techniques to predict the behaviour of response with respect to process variable. With the help of Minitab 16 regression equation (Equation 2) was generated for MRR and the various variables and is expressed as:

$$MRR = 0.124789 + 0.152033\ A - 0.0105444\ B - 0.0116667\ C - 0.032225\ A*B + 0.00893333\ B*C - 0.0121167\ C*A \quad (4)$$

A confirmation test is carried out to verify and validate the accuracy of analysis by comparing the initial condition with the optimal condition. Table 6 shows the comparison of initial response value (A2B2C2) with the estimated and actual response value using the optimal parameters. It is seen that the increase of response value from initial to optimal condition is found to be 4.6442 dB, which means an improvement of about 28% on MRR. This improvement confirms the design factors and analysis can be suitable for enhancing the MRR in turning operation.

Table 3. Design of experiments and results (MRR)

Exp. No.	A	B	C	MRR	S/N Ratio	Exp. No.	A	B	C	MRR	S/N Ratio
1	1.0	0.15	50	0.2820	10.9950	15	2.0	0.20	100	0.2625	11.6174
2	1.5	0.15	50	0.1539	16.2552	16	1.0	0.25	100	0.2505	12.0238
3	2.0	0.15	50	0.1813	14.8320	17	1.5	0.25	100	0.2398	12.4030
4	1.0	0.20	50	0.1172	18.6214	18	2.0	0.25	100	0.2066	13.6974
5	1.5	0.20	50	0.1556	16.1598	19	1.0	0.15	150	0.3835	8.3247
6	2.0	0.20	50	0.1504	16.4550	20	1.5	0.15	150	0.4520	6.8972
7	1.0	0.25	50	0.1298	17.7345	21	2.0	0.15	150	0.3416	9.3296
8	1.5	0.25	50	0.1587	15.9885	22	1.0	0.20	150	0.3353	9.4913
9	2.0	0.25	50	0.1491	16.5304	23	1.5	0.20	150	0.2654	11.5220
10	1.0	0.15	100	0.3299	9.6324	24	2.0	0.20	150	0.2297	12.7768
11	1.5	0.15	100	0.3548	9.0003	25	1.0	0.25	150	0.2141	13.3877
12	2.0	0.15	100	0.2946	10.6153	26	1.5	0.25	150	0.2287	12.8147
13	1.0	0.20	100	0.2661	11.4991	27	2.0	0.25	150	0.1680	15.4938
14	1.5	0.20	100	0.2545	11.8862						

Table 4. Response Table for Signal to Noise Ratios (Larger is better)

Average S/N Ratio for Each Factor Level (MRR)			
Level	**A**	**B**	**C**
1	15.95*	10.65	12.41
2	11.38	13.34	12.55
3	11.12	14.45*	13.48*
Delta	4.84	3.8	1.07
Rank	1	2	3

'*' indicates optimal process level

Figure 6. Main effect plot for MRR

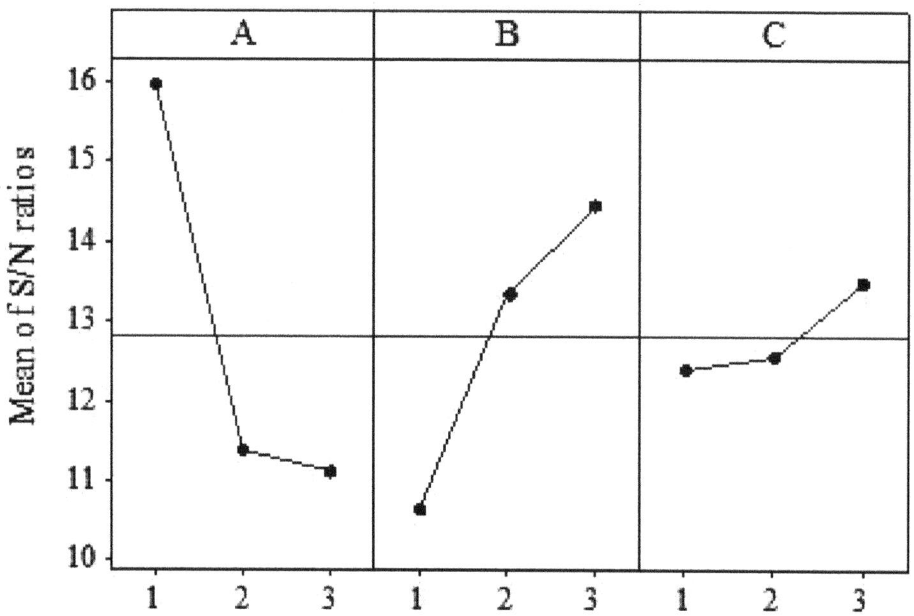

Figure 7. Interaction effect plot for mean S/N ratio of MRR (a) AxB (b) AxC (c) BxC

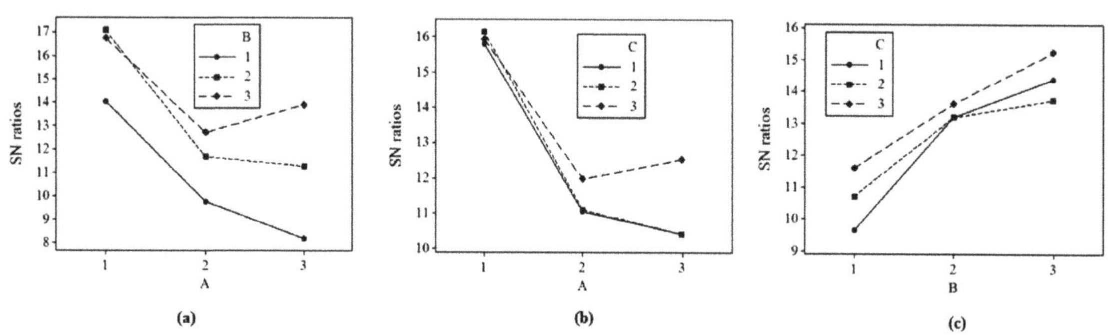

Table 5. ANOVA table for MRR

Source	DOF	Sum of Square	Mean Square	F-Ratio	% Contribution
A	2	133.26	66.627	25.10*	53.71
B	2	68.63	34.316	12.93*	27.66
C	2	6.12	3.061	1.15	2.47
AxB	4	10.81	2.702	1.02	4.36
AxC	4	4.75	1.186	0.45	1.91
BxC	4	3.33	0.831	0.31	1.34
Error	8	21.23	2.6540		8.56
Total	26	248.12			100

Significant at 95% confidence level (F0.05,2,8 = 4.46 & F0.05,4,8 = 3.84)
*" indicates most significant factor.

Table 6. Confirmation test table for MRR

	Initial Parameters	Optimal Parameters	
		Theoretical	Experimental
Level	A2B2C2	A1B3C3	A1B3C3
COF	0.2545		0.1491
S/N ratio	11.8862	17.5856	16.5304

Improvement in S/N ratio: 4.6442 dB

Optimization of Surface Roughness Using Taguchi Method

The present work deals with the optimization of surface roughness by varying process parameters. The R_a value is measured and tabulated. According to Taguchi method, the surface roughness is transformed into loss function (S/N ratio). Based on the objective of the study, smaller-the-better criterion is selected for the analysis. The calculated results of surface roughness along with their S/N ratio are given in Table 7. The mean S/N ratio for each level of factor A, B and C are summarized in Table 8 as response table for SR. The total mean S/N ratio is found to be 57.7728 dB. The parameter A ranked 1 as of delta statistics has the most effect on the response followed by parameter C and B. The corresponding main effects and interaction effects plots are shown in Figure 8 and Figure 9 (a, b, and c) respectively. It is found from the main effect plot (Figure 8) that the parameter A with level 1 has the most significant effect on SR among all three parameters considered in the present study. From interaction plot (Figure 9) that there is minimum interaction between the parameters (AxB, AxC and BxC) as some of the lines are intersecting each other. The optimal parameter combination for minimum SR is found to be A1B3C1. It is also seen from the ANOVA table (Table 9), that the parameter A has the most significant influence on SR at the confidence level of 95% within the specific test range. The parameter B and C along with all the interactions has no significant effect on the response according to Table 9.

With the help of Minitab 16 regression equation (Equation 5) was generated for SR and the various variables and was expressed as:

Table 7. Design of experiments and results (SR)

Exp. No.	A	B	C	SR	S/N Ratio	Exp. No.	A	B	C	SR	S/N Ratio
1	1.0	0.15	50	0.000190	74.4154	15	2.0	0.20	100	0.001917	54.3466
2	1.5	0.15	50	0.000990	60.0870	16	1.0	0.25	100	0.001295	57.7518
3	2.0	0.15	50	0.000443	67.0697	17	1.5	0.25	100	0.002807	51.0355
4	1.0	0.20	50	0.000456	66.8112	18	2.0	0.25	100	0.001420	56.9532
5	1.5	0.20	50	0.000503	65.9750	19	1.0	0.15	150	0.001920	54.3360
6	2.0	0.20	50	0.000532	65.4861	20	1.5	0.15	150	0.002339	52.6191
7	1.0	0.25	50	0.000408	67.7955	21	2.0	0.15	150	0.002722	51.3023
8	1.5	0.25	50	0.000522	65.6430	22	1.0	0.20	150	0.001903	54.4119
9	2.0	0.25	50	0.000597	64.4828	23	1.5	0.20	150	0.003088	50.2076
10	1.0	0.15	100	0.002400	52.3945	24	2.0	0.20	150	0.001633	55.7393
11	1.5	0.15	100	0.004129	47.6831	25	1.0	0.25	150	0.001610	55.8658
12	2.0	0.15	100	0.003195	49.9096	26	1.5	0.25	150	0.001690	55.4429
13	1.0	0.20	100	0.003521	49.0679	27	2.0	0.25	150	0.001024	59.7964
14	1.5	0.20	100	0.002178	53.2373						

Table 8. Response table for SR

	Average S/N Ratio for Each Factor Level		
Level	**A**	**B**	**C**
1	66.42*	56.65	59.21*
2	52.49	57.25	55.77
3	54.41	59.42*	58.34
Delta	13.93	2.77	3.44
Rank	1	3	2

'*' indicates optimal process level

$$SR = -0.000748222 + 0.00126039\,A + 0.000394111\,B + 0.000436778\,C - 0.000213417\,A*B - 4.76667e\text{-}005\,A*C - 0.000176833\,B*C \quad (5)$$

After finding out the optimal level of parameters, a verification test is carried in order to check the accuracy of analysis. Table 10 shows the initial (A2B2C2) condition as well as estimated and actual S/N ratio for the optimal parameters. It may be noted that the estimated and actual S/N ratios are in good agreement. The increase of S/N ratio from initial to the optimal condition is found to be 14.5582 dB, which means the SR decreased by about 22%. This improvement confirms that the main objective of the optimization study is obtained.

Figure 8. Main effect plot for SR

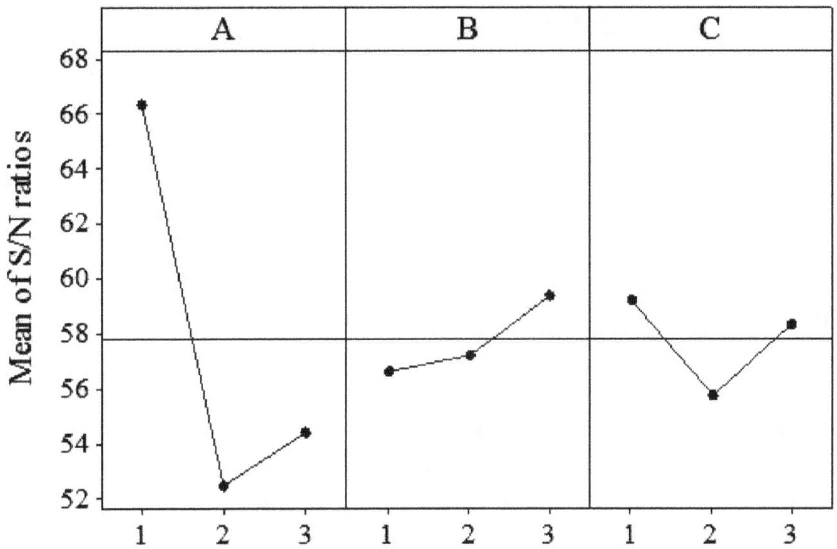

Figure 9. Interaction effect plot for mean S/N ratio of SR (a) AxB (b) AxC (c) BxC

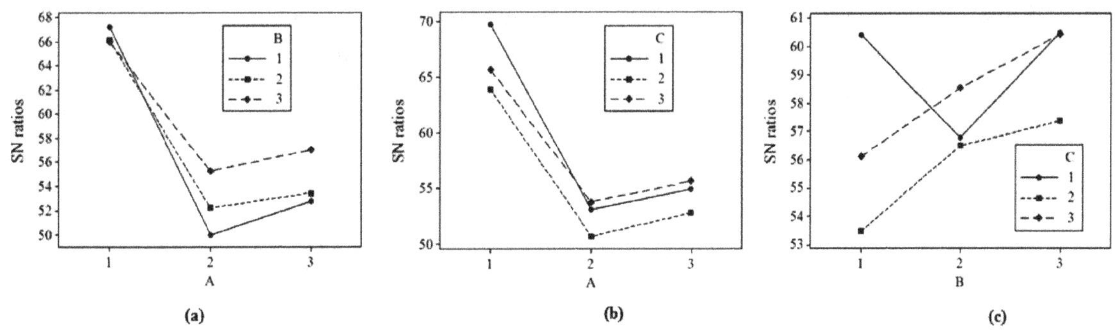

Table 9. ANOVA table for SR

Source	DOF	Sum of Square	Mean Square	F-Ratio	% Contribution
A	2	1025.78	512.891	56.78*	79.06
B	2	38.22	19.112	2.12	2.95
C	2	57.50	28.749	3.18	4.43
AxB	4	37.83	9.458	1.05	2.92
AxC	4	23.93	5.982	0.66	1.84
BxC	4	41.89	10.472	1.16	3.23
Error	8	72.27	9.033		5.57
Total	26	1297.42			100

Significant at 95% confidence level (F0.05,2,8 = 4.46 & F0.05,4,8 = 3.84)
"*" indicates most significant factor.

Table 10. Confirmation test table for SR

	Initial Parameters	Optimal Parameters	
		Theoretical	Experimental
Level	A2B2C2	A1B3C1	A1B3C1
SR	0.0022		0.0004
S/N ratio	53.2373	66.4200	67.7955

Improvement in S/N ratio: 14.5582 dB

Multi Objective Optimization of Material Removal Rate and Surface Roughness Using Grey-Taguchi Approach

The two responses MRR and SR are converted into a single response i.e. grey relational grade. The detailed process has already been discussed earlier. The experimental results of MRR and SR are given in Table 11. As per grey relational analysis, the multiple responses are initially normalized ranging from zero to one using lower-the-better criterion due to minimum values of responses are required for the study. It is done to avoid the problems of differences in units, scales and targets of the responses. The intermediate data of each stage of analysis are presented in Table 12. Finally by averaging the grey relational coefficient of the responses, grey relational grade can be obtained. The final grey relational grade is the combination result of the multiple responses. The grey relational grade and their orders of each experiment are shown in Table 13. The orders with numbers are given considering the highest to lowest grey relational grade. The optimum level of factors will be the level with highest grey relational grade, but the

Table 11. Experimental results for MRR and SR

Exp. No.	A	B	C	MRR	SR	Exp. No.	A	B	C	MRR	SR
1	1.0	0.15	50	0.2820	0.000190	**15**	2.0	0.20	100	0.2625	0.001917
2	1.5	0.15	50	0.1539	0.000990	**16**	1.0	0.25	100	0.2505	0.001295
3	2.0	0.15	50	0.1813	0.000443	**17**	1.5	0.25	100	0.2398	0.002807
4	1.0	0.20	50	0.1172	0.000456	**18**	2.0	0.25	100	0.2066	0.001420
5	1.5	0.20	50	0.1556	0.000503	**19**	1.0	0.15	150	0.3835	0.001920
6	2.0	0.20	50	0.1504	0.000532	**20**	1.5	0.15	150	0.4520	0.002339
7	1.0	0.25	50	0.1298	0.000408	**21**	2.0	0.15	150	0.3416	0.002722
8	1.5	0.25	50	0.1587	0.000522	**22**	1.0	0.20	150	0.3353	0.001903
9	2.0	0.25	50	0.1491	0.000597	**23**	1.5	0.20	150	0.2654	0.003088
10	1.0	0.15	100	0.3299	0.002400	**24**	2.0	0.20	150	0.2297	0.001633
11	1.5	0.15	100	0.3548	0.004129	**25**	1.0	0.25	150	0.2141	0.001610
12	2.0	0.15	100	0.2946	0.003195	**26**	1.5	0.25	150	0.2287	0.001690
13	1.0	0.20	100	0.2661	0.003521	**27**	2.0	0.25	150	0.1680	0.001024
14	1.5	0.20	100	0.2545	0.002178						

Table 12. Grey relational Analysis of MRR and SR

Exp No	Normalized Data		Values of Δ0i		Grey Relational Coefficient	
	MRR	SR	MRR	SR	MRR	SR
1	0.5078	0.9999	0.4922	0.0001	0.5039	0.9999
2	0.8904	0.7969	0.1096	0.2031	0.8202	0.7112
3	0.8085	0.9358	0.1915	0.0642	0.7231	0.8861
4	1.0000	0.9324	0.0000	0.0676	1.0000	0.8808
5	0.8853	0.9207	0.1147	0.0793	0.8134	0.8630
6	0.9008	0.9133	0.0992	0.0867	0.8345	0.8522
7	0.9624	0.9448	0.0376	0.0552	0.9300	0.9005
8	0.8760	0.9157	0.1240	0.0843	0.8013	0.8557
9	0.9047	0.8967	0.0953	0.1033	0.8399	0.8288
10	0.3647	0.4390	0.6353	0.5610	0.4404	0.4713
11	0.2903	0.0003	0.7097	0.9997	0.4133	0.3334
12	0.4701	0.2372	0.5299	0.7628	0.4855	0.3960
13	0.5553	0.1547	0.4447	0.8453	0.5292	0.3717
14	0.5899	0.4953	0.4101	0.5047	0.5494	0.4977
15	0.5660	0.5616	0.4340	0.4384	0.5353	0.5328
16	0.6019	0.7194	0.3981	0.2806	0.5567	0.6406
17	0.6338	0.3358	0.3662	0.6642	0.5772	0.4295
18	0.7330	0.6878	0.2670	0.3122	0.6519	0.6156
19	0.2046	0.5610	0.7954	0.4390	0.3860	0.5325
20	0.0000	0.4545	1.0000	0.5455	0.3333	0.4783
21	0.3297	0.3574	0.6703	0.6426	0.4273	0.4376
22	0.3486	0.5653	0.6514	0.4347	0.4342	0.5349
23	0.5573	0.2646	0.4427	0.7354	0.5304	0.4047
24	0.6640	0.6337	0.3360	0.3663	0.5981	0.5772
25	0.7106	0.6397	0.2894	0.3603	0.6334	0.5812
26	0.6670	0.6193	0.3330	0.3807	0.6002	0.5677
27	0.8483	0.7884	0.1517	0.2116	0.7672	0.7026

optimal parameters obtained from the highest grey relational grade are independent because design of experiments is orthogonal. Hence Taguchi method is used to find out the optimum process parameters.

The grey relational grade is used as the performance index for Taguchi method. The grey grade is then converted in to S/N. The mean of each factor with their levels are calculated and presented as response table for grey relational grade (Table 14). The overall mean grey relational grade is found to be 0.6222. It is seen from the response table (Table 14) that parameter A has the most influence on grey grade followed by B and C on the basis of delta statistics. The main effect plot is drawn from the mean values of each factor levels are shown in Figure 10. From the main effects plot, the optimal combination of process parameter is found to be A1B3C3. The interaction plots between the parameters are given in

Table 13. Grey relational grade and its order

Exp No	Grey Relational Grade	Grey Relational Order	Exp No	Grey Relational Grade	Grey Relational Order
1	0.7519	9	15	0.5341	16
2	0.7657	8	16	0.5986	13
3	0.8046	7	17	0.5034	18
4	0.9404	1	18	0.6337	11
5	0.8382	4	19	0.4592	21
6	0.8433	3	20	0.4058	26
7	0.9153	2	21	0.4324	25
8	0.8285	6	22	0.4846	19
9	0.8344	5	23	0.4676	20
10	0.4558	22	24	0.5876	14
11	0.3734	27	25	0.6073	12
12	0.4407	24	26	0.5840	15
13	0.4505	23	27	0.7349	10
14	0.5235	17			

Table 14. Response table for grey relational grade

	Average Grey Grade for Each Factor Level		
Level	A	B	C
1	0.8358*	0.5433	0.6293
2	0.5015	0.6300	0.5878
3	0.5293	0.6933*	0.6495*
Delta	0.3343	0.1501	0.0618
Rank	1	2	3

'*' indicates optimal process level

Figure 11. The interaction AxB and AxC has minor influence on grey relational grade as lines are slightly crossing each other. Further analysis of variance (ANOVA) is performed at 95% level of confidence for grey relational grade and shown in Table 15. With this level of confidence, the parameter A has the most significant effect on grey grade with 78% contribution on the analysis followed by factor B which has nearly 13% contribution. The parameter C and all the interaction sources are not significant for the analysis as the percentage contribution is less than the error percentage (2.24%).

A confirmation test is conducted for the mean grey relational grade and their values are shown in Table 16. It is found that the optimal parameter combination enhances the grey relational grade from 0.5235 to 0.8344 by about 38%. Thus both MRR and SR were optimized simultaneously from the present study.

Figure 10. Main effect plot for grey relational grade

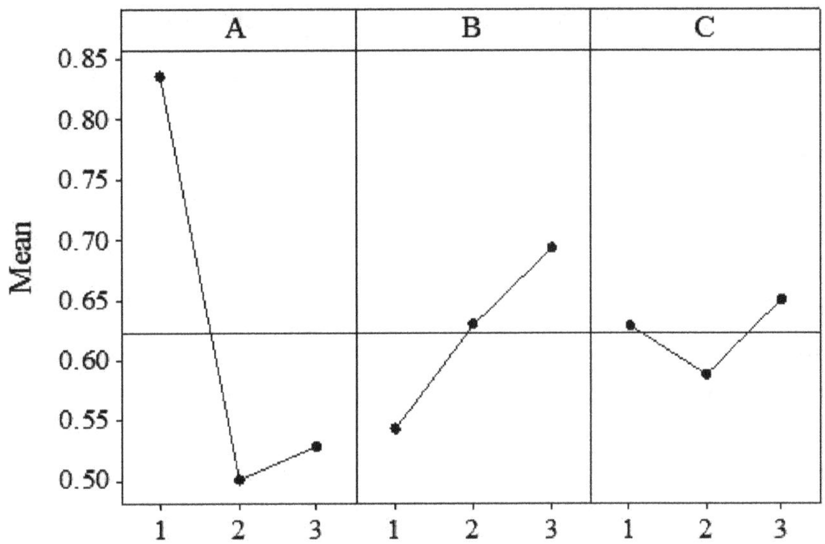

Figure 11. Interaction effect plot for grey relational grade (a) AxB (b) AxC (c) BxC

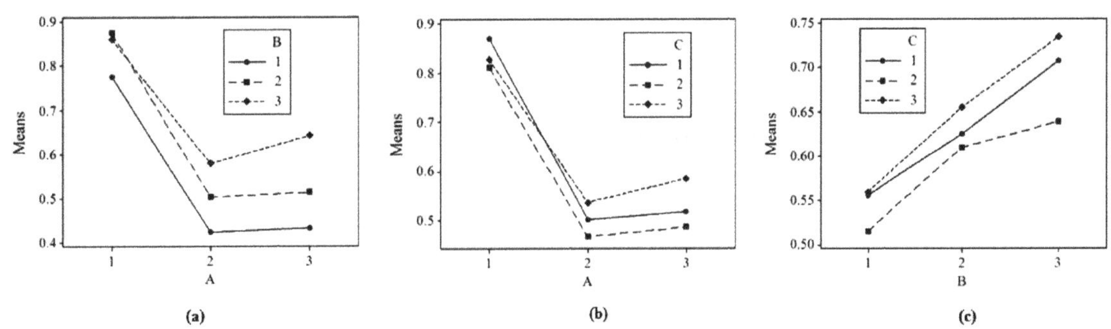

Table 15. ANOVA table for grey relational grade

Source	DOF	Sum of Square	Mean Square	F-Ratio	% Contribution
A	2	0.6195	0.61949	116.05*	78.1001
B	2	0.1021	0.1021	19.13*	12.8767
C	2	0.0178	0.0178	3.34	2.2492
AxB	4	0.0185	0.0185	1.74	2.3376
AxC	4	0.0103	0.0103	0.96	1.2927
BxC	4	0.0036	0.0036	0.33	0.4486
Error	8	0.0214	0.0214		2.6920
Total	26	0.7932			100

Significant at 95% confidence level (F0.05,2,8 = 4.46 & F0.05,4,8 = 3.84)
"*" indicates most significant factor.

Table 16. Confirmation test table for grey relational grade

	Initial Parameters	Optimal Parameters	
		Theoretical	Experimental
Level	A2B2C2	A1B3C3	A1B3C3
MRR	0.254500		0.149100
SR	0.002178		0.000597
Grey grade	0.523500	0.9069	0.834400

Improvement in grey relational grade: 0.3109

Multi Objective Optimization of Material Removal Rate and Surface Roughness Using Non-Dominated Sorting Genetic Algorithm

Using the values of MRR and SR (Table 3 and Table 7) and the regression equations (Equations 4 and 5), the Normal probability plots for MRR & SR are drawn and are depicted in Figure 12 and Figure 13 respectively. It can be seen that the residuals value fall close to the straight line, which signify that the errors are normally distributed and the regression model is well fitted. The value of R^2 is 95%, which indicates that the developed regression model is adequately significant.

ANOVA tests performed earlier for MRR and SR individually are tabulated together (Table 17) to evaluate the suitability of developed statistical models for the outputs and the results.

With the help of Objective function discussed earlier, non-dominated parameter setting obtained within the upper and lower bound for MRR and SR and the variation of the same was plotted as shown in Figure 14. This figure shows the Pareto front, for the optimal response at different optimal parameter setting.

Figure 12. Normal probability plot for MRR

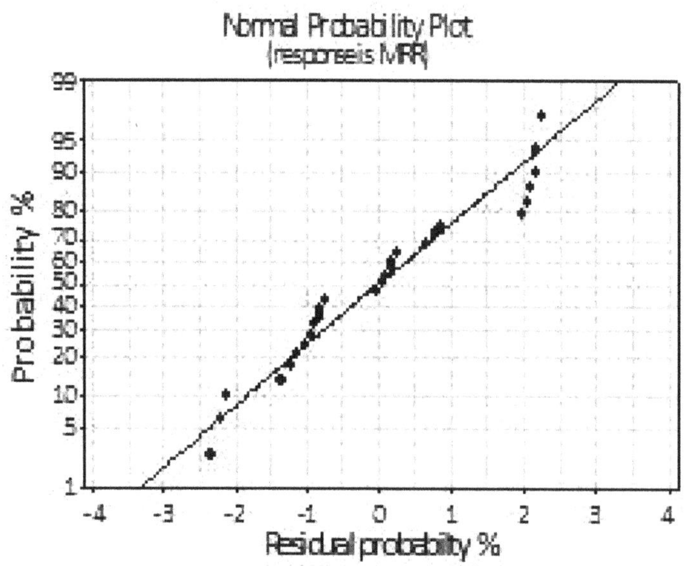

Figure 13. Normal probability plot for Ra

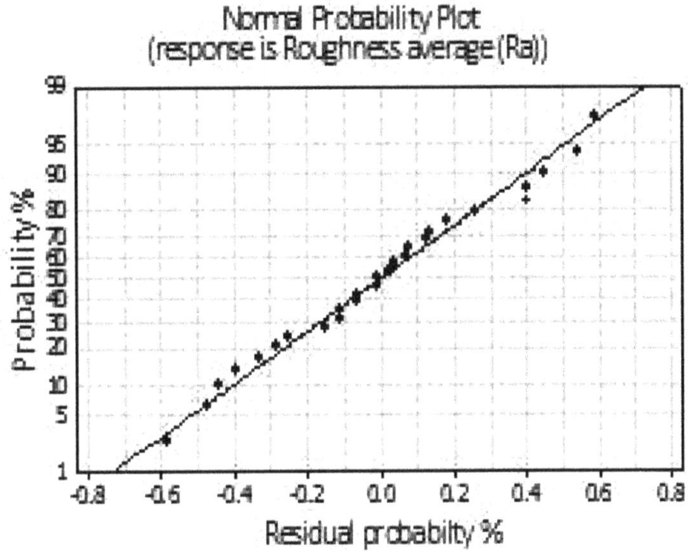

Figure 14. Optimal chart obtained through NSGA-II

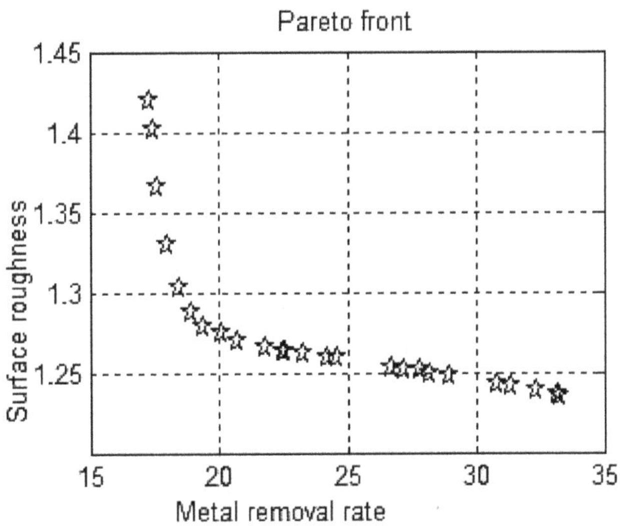

From the optimality Table 17 and the Optimal Chart it can be observed that the solution set obtained from the NSGA-II are in close agreement with the experiment result. The confirmation test was conducted, for the selected optimum conditions for the cases of MRR and Ra are presented in Table 18. The performance for the optimum machining condition and the actual machining condition is measured. From the verification Table, it can be observed that the calculated percentage errors are within acceptable limit. The percentage error between the experimental and optimum values for MRR and SR are 7.4% and 5.2%, respectively. Clearly, this confirms excellent reproducibility of the experimental conclusions.

Table 17. Analysis of Variance Table

Source	DF	Sum of Square		Mean Sum of Square		F- Value		P-Value	
		MRR	Ra	MRR	Ra	MRR	Ra	MRR	Ra
A	2	133.26	1025.78	66.627	512.891	25.10	56.78	0.5371	0.7906
B	2	68.63	38.22	34.316	19.112	12.93	2.12	0.2766	0.295
C	2	6.12	57.50	3.061	28.749	1.15	3.18	0.247	0.443
AB	4	10.81	37.83	2.702	9.458	1.02	1.05	0.436	0.292
AC	4	4.75	23.93	1.186	5.982	0.45	0.66	0.191	0.184
BC	4	3.33	41.89	0.831	10.472	0.31	1.16	0.134	0.323
Error	8	21.23	72.27	2.6540	9.033			0.856	0.557
Total	26	248.12	1297.42					1	1

Table 18. Validation test results.

MRR			Surface Roughness		
Predicted	Experimental	Error %	Predicted	Experimental	Error %
27.149	25. 269	7.4	1.251	1.32	5.2

CONCLUSION

The main aim of the investigation was to optimize the straight turning of a work piece on CNC lathe using Taguchi, Grey Taguchi and Non-dominated Sorting Genetic Algorithm-II (NSGA-II) techniques for maximizing the Metal removal rate and minimizing the Surface roughness. Depth of cut, Feed rate (B) and Cutting speed (C) were considered in the study with equally spaced three levels within the operating range for each of the process parameters as input parameters. From the results obtained conclusion can be drawn as:

1. Industrialist / Manufacturer undertaking mass production of machine components for fast moving consumer goods can use the optimised values of parameters obtained from maximized MRR i.e. A1B3C3. In case the levels indicated in this work do not suit their requirement, regression equation of MRR can be used to predict the output for levels outside the considered ones.
2. For manufacturing of products in which SR plays a vital role with minimum significance of MRR, optimal combination of parameters resulting in minimized SR (i.e. A1B3C1) can be employed whereas regression equation can be used for prediction of SR with levels different from those used here.
3. In products were both MRR and SR are to be given equal consideration, optimized results obtained in Grey or NSGA-II (A1B3C3) can be used anytime.

Future direction of this present study may further be explored in several ways. The same work may be further enriched by taking more input parameters like tool wear, cooling medium etc. and responses like modality, developed stress etc. For solution platform, other parallel methods like Simulated annealing may work handy. Synchronization or hybridization of two or more methods like Simulated annealing in Genetic algorithm (SAGA) may work even better and optimized in a more realistic manner ...further work may be envisaged.

REFERENCES

Aggarwal, A., Singh, H., Kumar, P., & Singh, M. (2008). Optimizing power consumption for CNC turned parts using response surface methodology, Taguchis technique- a comparative analysis. *Journal of Materials Processing Technology*, *200*(1-3), 373–384. doi:10.1016/j.jmatprotec.2007.09.041

Aloufi, M., and Kazmierski, T. J. (2011). A response surface modelling approach to performance optimisation of kinetic energy harvesters IJRRCS Simulation. *Benchmarking and Modeling of Systems and Communication Networks,* 1-8.

Asilturk, I., Neseli, S., & Ince, M. A. (2016). Optimisation of parameters affecting surface roughness of Co28Cr6Mo medical material during CNC lathe machining by using the Taguchi and RSM methods. *Measurement*, *78*, 120–128. doi:10.1016/j.measurement.2015.09.052

Bharathi, R. S., & Baskar, N. (2010). Optimization techniques for machining operations: A retrospective research based on various mathematical models. *International Journal of Advanced Manufacturing Technology*, *48*(9-12), 1075–1090. doi:10.1007/s00170-009-2351-x

Cayda U. (2010). Machinability evaluation in hard turning of AISI 4340 steel with different cutting tools using statistical techniques. *Proc. Inst. Mech. Eng. B-J Eng. Ma., 224*(7), 1043-1055. DOI: 10.1243/09544054JEM1822

Chen, M. C., & Su, C. T. (1998). Optimization of machining conditions for turning cylindrical stocks into continuous finished profiles. *International Journal of Production Research*, *36*(8), 2115–2130. doi:10.1080/002075498192805

Correia, A. E., & Davim, J. P. (2011). Surface roughness measurement in turning carbon steel AISI 1045 using wiper inserts. *Measurement*, *44*(5), 1000–1005. doi:10.1016/j.measurement.2011.01.018

Datta, R., & Majumder, A. (2010). *Optimization of Turning Process parameters Using Multi-Objective Evolutionary Algorithm*. IEEE; doi:10.1109/CEC.2010.5586296

Deb, K. (2002). *Multi-Objective Optimization using Evolutionary Algorithms*. Singapore: John Wiley & Sons.

Deb, K., Pratap, A., Agarwal, S., & Meyarivan, T. A. (2002). Fast and elitist multi-objective genetic algorithm: NSGA-II. *IEEE Transactions on Evolutionary Computation*, *6*(2), 182–197. doi:10.1109/4235.996017

Fraley, S., Oom, M., Terrien, B., & Date, J. Z. (2006). *Design of Experiments via Taguchi Methods: Orthogonal Arrays*. The Michigan Chemical Process Dynamic and Controls Open Text Book.

Freund, R. J., & Wilson, W. J. (1998). *Regression Analysis*. San Diego, CA: Academic Press.

Ganesan, H., & Mohankumar, G. (2013). Optimization of Machining Techniques in CNC Turning Centre using Genetic Algorithm. *Arab. J. Sci. Engg., 38*(6), 1529–1538. doi:10.1007/s13369-013-0539-8

Hsieh, K. L., Tong, L. I., Chiu, H. P., & Yeh, H. Y. (2005). Optimization of a multi-response problem in Taguchis dynamic system. *Computers & Industrial Engineering, 49*(4), 556–571. doi:10.1016/j.cie.2005.08.002

Kamaruddin, S., Khan, Z. A., & Foong, S. H. (2010). Application of Taguchi method in the optimization of injection moulding parameters for manufacturing products from plastic blend. *IACSIT International Journal of Engineering and Technology, 2*(6), 574–580. doi:10.7763/IJET.2010.V2.184

Kathleen, M.C., Natalia, Y.K., & Jeff, R. (2004). *Response surface methodology.* Center for computational analysis of social and organizational systems (CASOS), Technical Report.

Kaymaz, I., & McMahon, C. A. (2005). A response surface method based on weighted regression for structural reliability analysis. *Probabilistic Engineering Mechanics, 20*(1), 11–17. doi:10.1016/j.probengmech.2004.05.005

Kolahan, F., & Abachizadeh, M. (2008). Optimizing turning parameters for cylindrical parts using simulated annealing method. *Proceedings of World Academy of Science, Engineering and Technology., 36*, 436–439.

Korkut, I., Kasap, M., Çiftci, I., & Seker, U. (2004). Determination of optimum cutting parameters during machining of AISI 304 austenitic stainless steel. *Materials & Design, 25*(4), 303–305. doi:10.1016/j.matdes.2003.10.011

Lan, T. S. (2010). Fuzzy Deduction Material Removal Rate Optimization for Computer Numerical Control Turning. *American Journal of Applied Sciences, 7*(7), 1026–1031. doi:10.3844/ajassp.2010.1026.1031

Latha, B., & Senthilkumar, V. S. (2010). Modeling and analysis of surface roughness parameters in drilling GFRP composites using fuzzy logic. *Materials and Manufacturing Processes, 25*(8), 817–827. doi:10.1080/10426910903447261

Li, Z., & Liang, X. (2007). Vibro-acoustic analysis and optimization of damping structure with response surface method. *Materials & Design, 28*(7), 1999–2007. doi:10.1016/j.matdes.2006.07.006

Liu, Y. T., Chang, W. C., & Yamagata, Y. (2010). A study on optimal compensation cutting for an aspheric surface using the Taguchi method. *Journal of Manufacturing Science and Technology., 3*(1), 40–48. doi:10.1016/j.cirpj.2010.03.001

Mahapatra, S. S., Patnaik, A., & Patnaik, P. (2006). Parametric analysis and optimization of cutting parameters for turning operations based on Taguchi method. *Proceedings of the Int. Conference on Global Manufacturing and Innovation.*

Meng, Q., Arsecularatne, J. A., & Mathew, P. (2000). Calculation of optimum cutting condition for turning operation using a machining theory. *International Journal of Machine Tools & Manufacture, 40*(12), 1709–1733. doi:10.1016/S0890-6955(00)00026-2

Minitab. (2010). *Minitab user manual release 16.* MINITAB Inc.

Montgomery, D. C. (2001). *Design and analysis of experiments*. New York: John Wiley.

Myers, R. H., Douglas, C. M., & Anderson, C. M. (2009). *Process and Product Optimization Using Designed Experiments*. John Wiley & Sons.

Negrete, C.C, Nájera, C.J. & Valenzuela, M.J.C. (2013). *Optimization of Cutting Parameters in Turning of AISI 1018 Steel With Constant Material Removal Rate Using Robust Design for Minimizing Cutting Power*. Paper No. IMECE2013-62520.

Palanikumar, K. (2010). Modeling and analysis of delamination factor and surface roughness in drilling GFRP composites. *Materials and Manufacturing Processes*, *25*(10), 1059–1067. doi:10.1080/10426910903575830

Park, S. H. (1998). *Robust Design and Analysis for Quality Engineering*. London: Chapman & Hall.

Sankaya, M., & Guilu, A. (2015). Multi-response optimization of minimum quantity lubrication parameters using Taguchi-based grey relational analysis in turning of difficult-to-cut alloy Haynes 25. *Journal of Cleaner Production*, *91*, 347–357. doi:10.1016/j.jclepro.2014.12.020

Singh, H., & Kumar, P. (2006). Optimizing multi-machining characteristics through Taguchis approach and utility concept. *Journal of Manufacturing Technology Management*, *17*(2), 255–274. doi:10.1108/17410380610642304

Srinivas, N., & Deb, K. (1995). Multi-objective optimization using non dominated sorting in genetic algorithms. *Evolutionary Computation*, *3*, 221–248.

Suhail, A. H., Ismail, N., Wong, S. V., & Abdul, N. A. (2010). Optimization of cutting parameters based on surface roughness and assistance of workpiece surface temperature in turning process. *American Journal of Engineering and Applied Sciences.*, *3*(1), 102–108. doi:10.3844/ajeassp.2010.102.108

Sun, H., & Lee, S. (2005). Response surface approach to aerodynamic optimization design of helicopter rotor blade. *International Journal for Numerical Methods in Engineering*, *64*(1), 125–142. doi:10.1002/nme.1391

Taguchi, G. (1990). *Introduction to Quality Engineering*. Tokyo: Asian Productivity Organization.

Tzeng, C. J., Lin, Y. H., Yang, Y. K., & Jeng, M. C. (2009). Optimization of turning operations with multiple performance characteristics using the Taguchi method and Grey relational analysis. *Journal of Materials Processing Technology*, *209*(6), 2753–2759. doi:10.1016/j.jmatprotec.2008.06.046

Wang, Z. H., Meng, J., & Fu, J. (2010). Novel method for evaluating surface roughness by grey dynamic filtering. *Measurement.*, *43*(1), 78–82. doi:10.1016/j.measurement.2009.06.008

Yusoff, N., Ramasamy, M., & Yusup, S. (2011). Taguchis parametric design approach for the selection of optimization variables in a refrigerated gas plant. *Chemical Engineering Research & Design*, *89*(6), 665–675. doi:10.1016/j.cherd.2010.09.021

Chapter 5
Reliability Optimization of Complex Systems Using Cuckoo Search Algorithm

Anuj Kumar
University of Petroleum and Energy Studies, India

Sangeeta Pant
University of Petroleum and Energy Studies, India

S. B. Singh
G. B. Pant University of Agriculture and Technology, India

ABSTRACT

In this chapter, authors briefly discussed about the classification of reliability optimization problems and their nature. Background of reliability and optimization has also been provided separately so that one can clearly understand the basic terminology used in the field of reliability optimization. Classification of various optimization techniques have also been discussed by the authors. Few metaheuristic techniques related to reliability optimization problems like Genetic Algorithm (GA), Ant Colony Optimization (ACO) and Particle Swarm Optimization (PSO) have been discussed in brief. Thereafter, authors have discussed about Cuckoo Search Algorithm (CSA) which is the main focus of this chapter. Finally, Cuckoo Search Algorithm has been applied for solving reliability optimization problems of two complex systems namely complex bridge system and life support system in space capsule. Simulation results and conclusion have been presented in the last followed by the references.

INTRODUCTION

Almost every one of us is acquainted with the term reliability in day-to-day life. When someone assigns attribute 'reliable' to a component or a system (a system may be consist of collection of several components) it precisely mean to say that the same will render service for a good or at least reasonable period of time. Overall corporate success depends on the reliability of a company's products process

DOI: 10.4018/978-1-5225-1639-2.ch005

and services. Reliability is always a top customer concern and is increasingly vocalized by customers as a major factor in purchasing decisions. All these factors and many more, demanded high reliability in the design and operations of components/systems/equipments in various reliability models of practical utility, at the same time researchers often across with the situations of minimizing the cost. Due to those facts the problem of reliability optimization has gained importance in the last few decades (Kumar & Singh, 2008), (Ram et al., 2013) and various metaheuristic algorithms like Particle Swarm Optimization (PSO), Ant Colony Optimization (ACO), Cuckoo Search Algorithm (CSA) and Genetic Algorithm (GA) etc. have become popular over the years (Geem et al., 2001), (Qin et al., 2006).

Reliability optimization problems can be classified; according to the types of their decision variables; into three typical problems:

1. Reliability allocation problems
2. Redundancy allocation problems
3. Reliability-redundancy allocation problems.

Reliability allocation is a continuous nonlinear programming problem while redundancy allocation problem can be viewed as a pure integer nonlinear programming problem, and reliability-redundancy allocation is a mixed integer programming problem which is nonlinear in nature (Pant et al., 2015). In general, obtaining optimal reliability design is a tedious task because of the NP-hard nature of reliability optimization problems. Further, it is more tedious and time consuming to find the optimal solutions of those problems through exact/heuristics algorithms because these optimization problems generate a very large search space. Therefore, metaheuristic algorithms are more suitable for solving reliability optimization problems. Recently, many metaheuristics have been employed to solve reliability optimization problems (Kuo & Wan, 2007), (Pant & Singh, 2011), (Pant et al., 2015), (Ravi et al., 1997). The main reason behind the popularity of metaheuristic is that they are inspired by very simple physical phenomena, animals' behaviors, or evolutionary concepts. Most of them have derivation free mechanism and finally, they make a difference with conventional optimization techniques due to their superior ability of avoiding local optima. Cuckoo search Algorithm is an evolutionary algorithm inspired by the life of a bird family, called Cuckoos and the main inspiration behind this algorithm is the obligate brood parasitism of some cuckoos species in combination with L´evy Flights (Yang & Deb, 2010). In this chapter, authors have applied Cuckoo search algorithm for reliability optimization of complex bridge system and life support system in space capsule.

BACKGROUND

Reliability

In the modern age of sciences, a high degree of reliability is being demanded from all kind of users whether it is in general, public sectors, industries, defence and space research programs. There is too much at stake in term of cost, human life, and national security to take any risks with equipment which might not function properly when required. Moreover the present day weapons used for military purposes consist of thousands of small parts, each interwoven into a complex web which constitutes the weapons. The failure of any one of these could adversely affect the operation of the weapon. Therefore, it becomes more important that each part of the complex equipment must be highly reliable so the equipment as a whole must be reliable. The concept of high reliability is equally important outside the Military field. Computers or any other engineering system which are complex as well as expensive, play a major role in industrial and scientific activities.

The theory of reliability is not very old; usually world war second in 1939 is regarded as the starting point of reliability discipline. Before world war second, in the first quarter of 20th century a team of workers in 'Bell Telephone Laboratories' developed statistical methods for solving there quality control problems which is strongly linked with quality control. They provided the basis for development of statistical quality control. The American Society for Testing and Materials, The American Standard Association and The American Society for Mechanical Engineers also worked for the quality control techniques. But these technique were widely used till world war second 1939. Complexity and automation of equipments used in the war resulted in severe problems of maintenance and repairs. The equipment/component failed beyond the expectation. During this war army and navy in U.S.A set up a joint committee known as Vacuum Tube Development Committee for the study of failure in vacuum tube which is considered to be one of the root causes of the trouble. The major committee on reliability was set up by U.S defence department in 1950. This was latter called the Advisory Group on Reliability of Electronic Equipment (AGREE). During 1950's Germany, Japan, and Britain also took interest in such type of study. The last 20 years have seen remarkable progress in the application of reliability in industries and in other departments of all the developed and developing countries.

A high degree of reliability is also desirable from the economic point of view so as to reduce the overall costs. Sometimes the annual maintaining cost of some system in operable state is much higher than its original cost. Insufficient reliability of units engenders great loss in servicing, partial stoppage of equipment, and there may be accidents with considerable damage to the equipment and even the cost may be more serious in term of human life, national prestige and security.

Mathematically Reliability can be defined as the ability of an equipment to preserve its output characteristic (parameters) within established limits under given operating conditions."

From all these definitions, quantitatively, reliability is the probability that a device will operate as anticipated. Quantitatively, Reliability is given by

$$R(t) = P(T > t) \qquad 0 \le t < \infty \ldots \tag{1}$$

Where T is the time of failure of the device (T itself is a variable) and t is any time.

Generally the device is assumed to be working normally initially and every device definitely suffers failures. Therefore symbolically

$$R(0) = 1 \quad and \quad \lim_{t\to\infty} R(t) = 0 \ldots \tag{2}$$

Furthermore R (t) is non-increasing (at most constant, generally decreasing) function for $[0 \le t < \infty]$.

The theory of reliability is a scientific disciplined that studies the general regularity that must be maintained under design, experimentation, manufacture, acceptance, and use of units/components in order to maximal effectiveness from their use. The need of obtaining highly reliable systems and components has acquired special importance with the development of the present day technology and the problem of optimizing reliability along with the cost of equipments/components becomes more important and urgent in connection with the complex mechanization, modern sophistication and automation of industrial Process in many fields of industry, transportation, communication and space technology etc. The complex system, equipments, machines etc. are not of much use if they cannot perform the work adequately for which they are intended.

A system is a combination of elements forming a planetary whole i.e. there is a functional relationship between its components. The properties and behavior of each component ultimately affects the properties of the system. Any system has a hierarchy of components that pass through the different stages of operations which can be operational, failure, degraded or in repair. Failure doesn't mean that it will always be complete; it can be partial as well. But both these types affect the performance of system and hence the reliability. Majority of the systems in the industries are repairable. The performance of these systems can influence the quality of product, the cost of business, the service to the customers, and thereby the profit of enterprises directly. Modern repairable systems tend to be highly complex due to increase in convolution and automation of systems. During the last years reliability concepts have been applied in various manufacturing and technological fields. Earlier researcher discussed reliability and steady state analysis of some realistic engineering systems by using different approaches. Reliability techniques have also been applied to a number of industrial and transportation problems including automobile industry.

Overall corporate success depends on the reliability of a company's products process and services. Reliability is always a top customer concern and is increasingly vocalized by customers as a major factor in purchasing decisions. Reliability plays a role in many aspects of business. A few examples of the impact of reliability on key corporate objectives are listed below:

- **Reputation:** A company's reputation is very closely related to the reliability of its products. The more reliable a product is, the more likely the company is to have a favorable reputation.
- **Warranty Costs:** If a product fails to perform its function within the warranty period, its repair or replacement negatively affects profits. Additionally, if the failure is either large in magnitude or causes injuries or death, the company may become the focus of unwanted negative attention. Introducing reliability analyses in the early design stage is an important step in taking preventive actions, ultimately leading to a product that is more reliable.
- **Future Business:** A concentrated effort towards improved reliability shows existing customers that the manufacturer is serious about its product and committed to customer satisfaction. This type of attitude has a positive impact on future business. Many companies publish their predicted or field measured reliability metrics to help gain an advantage over competitors who either do not publish their metrics or have poor reliability metrics.
- **Contract Requirements:** Many customers in today's market demand that their suppliers have an effective reliability program. These customers have learned the benefits of reliability analysis from experience.
- **Cost of Ownership:** Manufacturers may take reliability data and combine it with other cost information to illustrate the cost-effectiveness of their products. This life cycle cost analysis can prove that although the initial cost of their product might be higher, the overall lifetime cost is lower than a competing product because their product requires fewer repairs or less maintenance.
- **Custom Satisfaction:** While a reliable product may not dramatically affect customer satisfaction in a positive manner, an unreliable product will severely affect customer satisfaction in a negative way. Thus, high reliability is ultimately a mandatory requirement for customer satisfaction.

- **Safety Regulations:** Manufacturers of highly critical devices must design their products to comply with current safety regulations. Failure to meet standards for protecting the health and safety of users and/or the general public can result in the breach of contract and even the loss of life.

Reliability is used to describe the overall consistency of a measure. This is the introduction of terminology and models that will be used to describe quantity product reliability. The terminology probability distributions and models used for reliability analysis differ in many cases from those used in other statistical applications. In today's technological world nearly everyone depends upon the continued functioning of wide array of complex machinery and equipment for their everyday health safety, mobility and economic welfare. We expect our car, computers, electrical appliances, lights etc. to function whenever we need them day after day, year after year.

A measure is said to have a high reliability if it produces similar results under consistent conditions. For example, measurements of people's height and weight are often extremely reliable.

There are some general classes of reliability estimates:

- **Inter-Rater Reliability:** Assesses the degree of agreement between two or more raters in their appraisals.
- **Test-Retest Reliability:** Assesses the degree to which test scores are consistent from one test administration to the next. Measurements are gathered from a single rater who uses the same methods or instruments and the same testing conditions. This includes *intra-rater reliability.*
- **Inter-Method Reliability:** Assesses the degree to which test scores are consistent when there is a variation in the methods or instruments used. This allows inter-rater reliability to be ruled out. When dealing with forms, it may be termed *parallel-forms reliability.*
- **Internal Consistency Reliability:** Assesses the consistency of results across items within a test.

Reliability does not imply validity. That is, a reliable measure that is measuring something consistently, may not be measuring what you want to be measuring. For example, while there are many reliable tests of specific abilities, not all of them would be valid for predicting, say, job performance. In terms of accuracy and precision, Reliability is a more accurate way of describing precision, while validity is a more precise way of describing accuracy.

While reliability does not imply validity, a lack of reliability does place a limit on the overall validity of a test. A test that is not perfectly reliable cannot be perfectly valid, either as a means of measuring attributes of a person or as a means of predicting scores on a criterion. While a reliable test may provide useful valid information, a test that is not reliable cannot possibly be valid.

An example often used to illustrate the difference between reliability and validity in the experimental sciences involves a common bathroom scale. If someone who is 200 pounds steps on a scale 10 times and gets readings of 15, 250, 95, 140, etc., the scale is not reliable. If the scale consistently reads "150", then it is reliable, but not valid. If it reads "200" each time, then the measurement is both reliable and valid.

Optimization

Optimization is the act of determining the optimal value of cost function in a domain of definition, subject to various constraints on the variable values.

Optimization methods are applicable to almost every activity of engineering, science, commerce, management, economics etc. in which numerical information is processed. For example, an architect is interested in design of structures such as towers, dam of minimum weight subjected to some constrains like reliability, cost and other specifications.

Variables, constraints and objective functions are three essential ingredients for an optimization problem. Variables represent components of the optimization model that can be changed to create different possibilities and constraints represent limitations on the variables and objective function. Objective functions assign a value to each different possibility. The general optimization problem is mathematically defined as:

Given $\mathrm{f} : \mathrm{D}^{\mathrm{N}} \rightarrow \mathrm{R}$, $f : D^N \rightarrow R$

Find $\vec{X}^* = \left[X_1^*, X_2^*, \ldots\ldots\ldots, X_N^*\right] \in D^N = D_1 \cap D_2 \cap \ldots\ldots \cap D_N$

which will satisfy

$$c_i^{=}(\vec{X}) = 0,\ 1 \le i \le N_{c^=}\ \mathrm{c}_{\mathrm{i}}^{=}\left(\vec{\mathrm{x}}\right) = 0, 1 \le \mathrm{i} \le \mathrm{N}_{\mathrm{c}^=} \tag{3}$$

$$c_i^{+}(\vec{X}) \ge 0,\ 1 \le i \le N_{c^+}\ \mathrm{c}_{\mathrm{i}}^{+}\left(\vec{\mathrm{x}}\right) \ge 0,\ 1 \le \mathrm{i} \le \mathrm{N}_{\mathrm{c}^+} \ldots \tag{4}$$

$$c_i^{-}(\vec{X}) \le 0,\ 1 \le i \le N_{c^-}\ \mathrm{c}_{\mathrm{i}}^{-}\left(\vec{\mathrm{x}}\right) \le 0,\ 1 \le \mathrm{i} \le \mathrm{N}_{\mathrm{c}^-} \ldots \tag{5}$$

and will optimize (minimize or maximize) the function $f(\vec{x})\ \mathrm{f}\left(\vec{\mathrm{x}}\right)$.

Optimal solutions can be classified as global optimal solution and local optimal solution. Lowest/highest output value among all inputs is known as global optimal solution, whereas lowest/highest output value among all nearby inputs is known as local optimal solution.

Classification of optimization problems can be done on the basis of number of characteristics. It can be classified as nonlinear, linear, quadratic, geometric programming problems on the basis of the nature of expression for objective function and constraints. It can also be classified as convex optimization problems and nonconvex problems based on convex/concave Nature of Equations Involved. In Convex optimization problems any local optimum is necessarily a global optimum so it is more tractable than nonconvex optimization problems. Thus, if the problem is convex this is enough to ensure that the solution is globally optimal. In contrast, nonconvex problems may have many different local optima and it is tedious to choose the best one among them.

If the classification has been done on the basis of number of objective function it can be named as multi-objective optimization problem (MOOP) and single objective optimization problem (SOOP).

MOOP involves more than one objective functions which have to be optimize simultaneously while SOOP has only one objective function for optimization.

Most of the real world problems have multi criteria nature so multi-objective optimization problems (MOOP) are very common, especially in engineering applications. Atomic structure determination of proteins (Bush et al., 1995), x- ray diffraction pattern recognition (Paszkowicz, 1996), curve fitting (Ahonen et al., 1997) and production scheduling (Swinehart et al., 1996) are such applications in which different objectives cannot be arbitrarily improved without decreasing the value of another. While SOOPs may have a unique optimal solution, MOOPs have a possible set of uncountable set of solutions from which a decision maker can choose an acceptable solution (or solutions).

For example, imagine a situation of summer with different cooling devices available in market (Figure 1). The cooling device may be an expensive air conditioner or a cheapest handmade fan. Now, let two objectives, say cost and comfort level. Obviously for more comfort level one have to pay more cost. One may choose other cooling system between these two extreme situations so that he/she can trade off between cost and comfort level. Now it depends upon the budget of the customer which device he/she wants to buy according to his expected comfort level.

Solutions of an optimization problem can be obtained by different optimization methods. Optimization methods can be categorized into three classes:

1. Exact methods,
2. Heuristics and
3. Meta Heuristics,

which can be further sub classified as shown in Figure 2.

Figure 1. Hypothetical trade off solutions for cooling system buying decision making problem

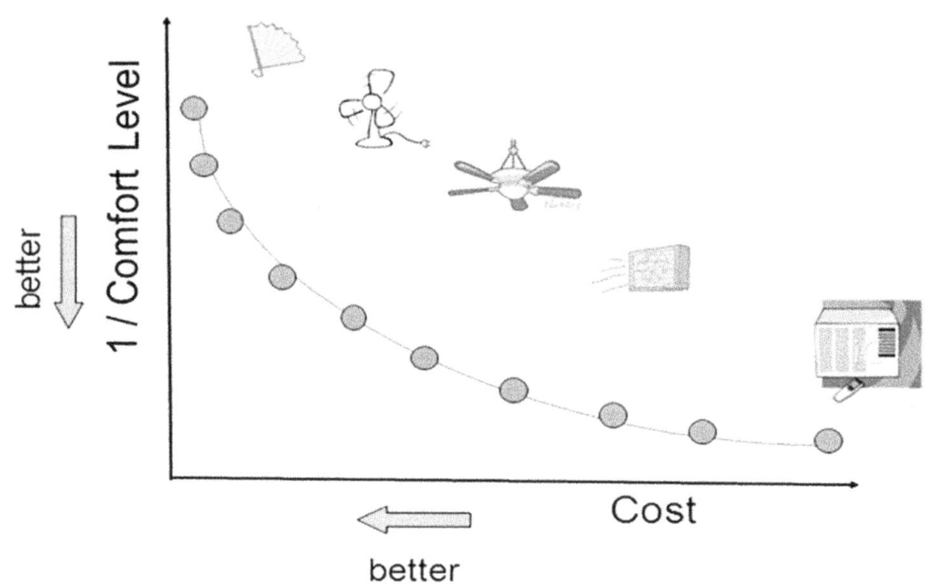

Figure 2. Optimization techniques classification

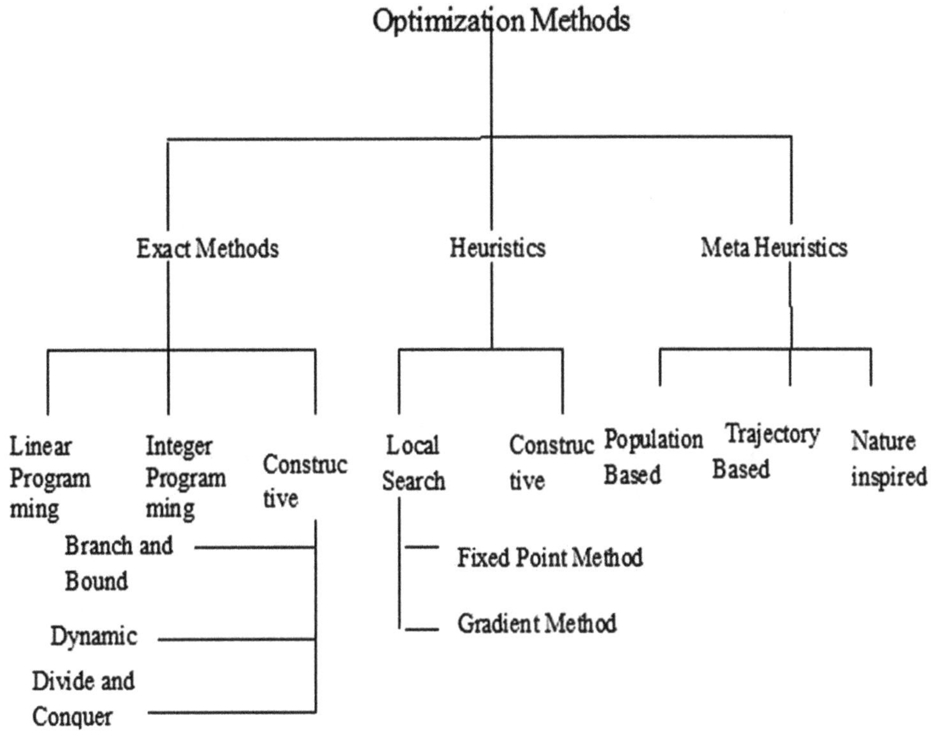

Exact Methods

Exact methods have been used in traditional optimization algorithms for finding the best solution. By the mean of those methods one can find an optimal solution to a given optimization problem and solve it by searching the entire solution space exhaustively but it takes a lot of time. Linear Programming, Integer programming and Dynamic Programming are few example of exact methods. More details about exact methods can be found in Michalewicz and Fogel (2004). But most of the exact methods are not appropriate for NP-hard problems. For most of the real- world problem the size of the search space is very large and exact method require enormous computational time to check every possible solution. So one can find exact optimal solution by exact methods but he/she has to put more computational efforts. Approximate methods viz. Heuristics and metaheuristics could be useful in case of failure of exact methods.

Heuristics

A heuristic is used to solve those optimization problems which can not be solve by exact method. In spite of focusing on optimal solution, the heuristics solution approaches focus on approximate solution that can be computed in an acceptable amount of time. This is the only way to find the solutions of very difficult optimization problems for which the exact algorithms are incapable of providing an optimal solution in reasonable times.

A heuristic approach used some piece of knowledge about the structure of specific problem and used it to devise a strategy for solving the same. One can device a good heuristic if it has the knowledge of the problem. It is best to use more general heuristics, called meta heuristics, when someone does not have strong insight into how is this specific problem might be solved.

Metaheuristics

Metaheuristic optimization techniques have been gaining popularity from last two decades. Some of them are Particle Swarm Optimization (PSO), Ant Colony Optimization (ACO), Cuckoo Search Algorithm (CSA) and Genetic Algorithm (GA). Main Reasons behind the popularity of metaheuristic algorithms are their simplicity, derivation-free mechanism, flexibility and local optima avoidance nature. They also known as black box optimization algorithms. Classification of the metaheuristics optimization techniques have been shown in Figure 3 (Larrosa, 2008). It has been done by the following two main criteria:

- Metaheuristics which are based on a population (e.g., particle swarm optimization, genetic algorithms, scatter search, ant colony optimization, estimation of distribution algorithms etc.) or based on a trajectory (e.g., hill climbing, tabu search, simulated annealing, iterated local search etc.).
- Metaheuristics which are inspired by nature (e.g., genetic algorithms, evolution strategies, ant colony optimization, particle swarm optimization, simulated annealing etc.) or not (e.g., iterated local search, GRASP, scatter search, tabu search, hill climbing, estimation of distribution algorithms etc.).

Some popular metaheuristic algorithms are briefly discussed below:

Figure 3. Classifications of metaheuristics

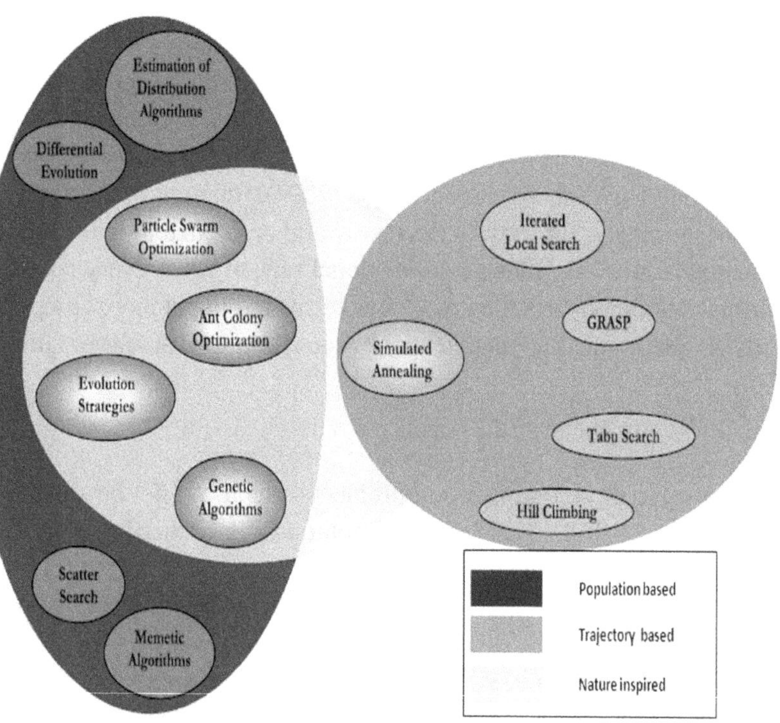

Genetic Algorithm (GA)

Holland (1962) proposed Genetic algorithm. It modeled Genetic evolution. GA was inspired by Darwinian evolution and Mendelian genetics. Individuals in GAs are called chromosome and characteristics of individual are expressed using genotypes (Engelbrecht, 2007). String of cells attached with each chromosome is known as genes and allele denoted the value of each gene. GA is a popular evolutionary algorithm which is used to solve complex optimization problems arise in the field of pattern recognition, image processing, machine learning, etc.

Ant Colony Optimization (ACO)

Dorrigo (1992) proposed Ant Colony Optimization (ACO). Inspiration behind this optimization technique is the foraging behavior of some ant species. Nature blessed ants with a unique quality of finding the shortest path between their nest and food source only if they are in groups. A single ant can't do that means ants living in groups are well cooperated. As far as searching for food is concerned, ants initially explore randomly the area near their nest. As a result of this random searching, when an ant find a food source it evaluates the quantity and the quality of the food and carries some of food back to the nest. During its return journey ant deposits a chemical pheromone trail on the ground. The other ants can sense this chemical and follow the path to the food source. Ants tend to move to the way having higher pheromone resulting this chemical accumulates faster in the shortest path and all the ants converges to it.

Particle Swarm Optimization (PSO)

Particle swarm optimization (PSO) was first proposed by R. Eberhart and J. Kennedy in 1995 (Eberhart & Kennedy, 1995), (Kennedy & Eberhart, 1995). It is a population-based search algorithm which based on the simulation of the animal's societies that don't have any leader in their group or swarm. Bird flocking

Figure 4. The behavior of real ant movements

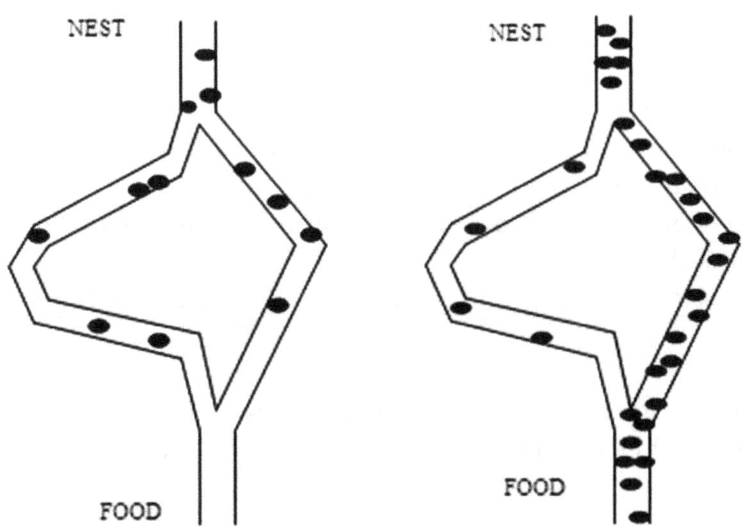

and fish schooling are the perfect example of it. These groups find food by random by following one of its members who has the closest position to the food source. Animal which has a better condition will inform it to its flocks and the others will move simultaneously to that place. This happened repeatedly until the food source discovered. So the PSO algorithm uses the food searching behavior of such animal society in finding optimal solution of a complex optimization problem.

MAIN FOCUS OF THE CHAPTER

Cuckoo Search Algorithm (CSA)

Yang and Deb (2009) proposed Cuckoo search Algorithm. It is an evolutionary algorithm inspired by the life of a bird family, called Cuckoo.

In order to understand CSA more clearly let us have a quick look over the interesting breeding behavior of certain cuckoo species.

Breeding behavior of Cuckoo: Cuckoos are medium sized slender birds known for their beautiful sound. Some of the species among them are brood parasites and follow very aggressive reproduction strategy. Some species lay their eggs in communal net and remove others eggs to increase the hatching probability of their own eggs (Payne et al., 2005). Hosts are manipulated by broad parasites to raise its young as if it were its own. The host may belong to brood parasites own species or of another species. Nest takeover, cooperative breeding and intraspecific brood parasitism, are three basic types of brood parasitism. Some host birds identified the cuckoo's eggs and they throw these eggs out of their net or simply leave their net and built a new nest somewhere else. Some host birds resist the intruding cuckoos and engage in direct conflict with them. During this evolution of brood parasitism among cuckoos, some cuckoo species enable them to mimic their eggs in same pattern and colour as their few chosen host species. By doing so, they reduce the probability of their eggs being abandoned. Even cuckoo chick can also mimic the call of host chick for increasing its feeding opportunity.

L´evy Flights

Benoit Mandelbrot coined the term L´evy flight. It is inspired by the food searching pattern of animals and insects. They search for food in a random or quasi-random manner (Brown et al., 2007). A recent study by Reynolds and Frye (2007) shows that fruit flies explore their landscape using a series of straight flight paths punctuated by a sudden turn of ninety degree, leading to a L´evy-flight-style intermittent scale-free. So L´evy flight can be defined as a random walk and the steps in this random walk are defined in terms of the step-lengths, which follow a certain probability distribution, with the directions of the steps being isotropic and random.

For modelling the brood parasitism among cuckoos, Yang and Deb (2009) considered the following three rules:

1. A cuckoo can dump its only egg in a randomly chose nest.
2. Only best nests, which have the high quality of eggs (solutions), will carry over to the next generation.

Figure 5. L´evy flights in two dimensions

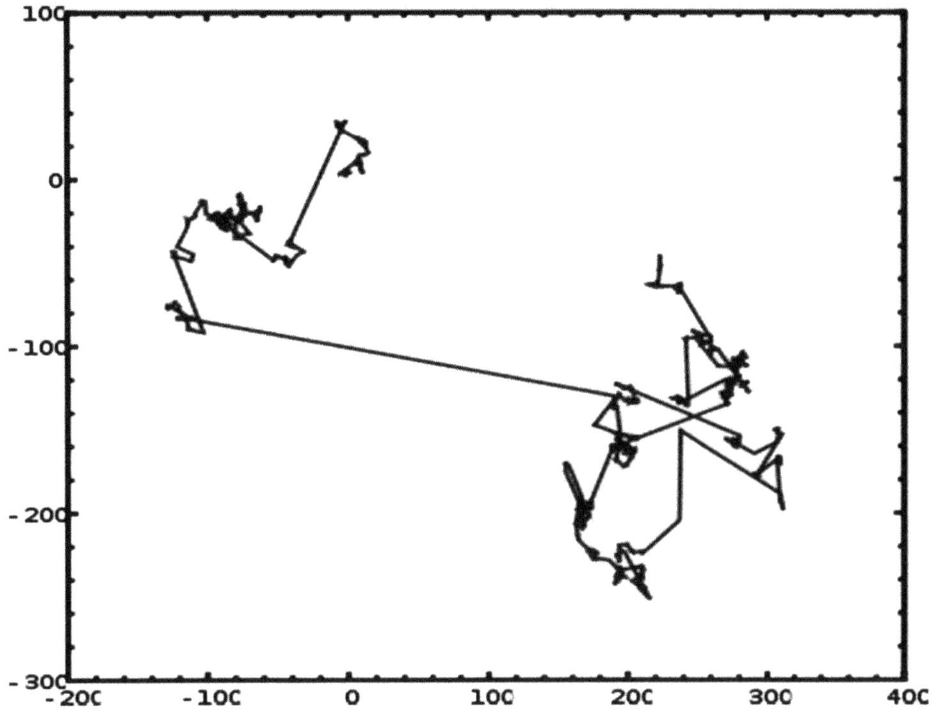

The number of available host nests is fixed. A host can identify an alien egg with a probability p_a lying between 0 and 1. In this case, either the host will throw the egg or abandon the nest and built a new nest in a new location.

When generating new solution $x^{(t+1)}$ for, say cuckoo i, a L´evy Flight is performed,

$$x_i^{(t+1)} = x_i^{(t)} + \alpha \oplus Levy(\lambda) \tag{6}$$

where α is greater than zero and it is denoting the step size which is depending upon the scale of the problem of interest. The consecutive steps of cuckoo form a random walk which follows a power-law step-length distribution with a heavy tail. So the Cuckoo search algorithm can be summarized as the pseudo code given below (Yang & Deb, 2009):

Pseudo Code

Objective function $f(x)$, $\quad x = (x_1, x_2, ..., x_d)^T$;
Initial a population of n host nests x_i $(i = 1, 2, ..., n)$;
while (t <MaxGeneration) or (stop criterion);
Get a cuckoo (say i) randomly by L´evy flights;
Evaluate its quality/fitness F_i;

Choose a nest among n (say j) randomly;
if $(F_i > F_j)$,
Replace j by the new solution;
end
Abandon a fraction (p_a) of worse nests
[and build new ones at new locations via L´evy flights];
Keep the best solutions (or nests with quality solutions);
Rank the solutions and find the current best;
end while
Postprocess results and visualisation;

Reliability Allocation Problems

In this chapter, two case studies are considered, they are:

- Complex bridge system and
- Life support system in space capsule.

Complex Bridge System

Figure 6 represents the block diagram of the complex bridge system (mohan & Shanker, 1988).

The system reliability (R_s) and system cost (C_s) are given by

$$R_s = r_1 r_4 + r_2 r_5 + r_2 r_3 r_4 + r_1 r_3 r_5 + 2 r_1 r_2 r_3 r_4 r_5 - r_1 r_2 r_4 r_5 - r_1 r_2 r_3 r_4 - r_2 r_3 r_4 r_5 - r_1 r_2 r_3 r_5 - r_1 r_3 r_4 r_5 \tag{7}$$

Figure 6. Complex bridge system

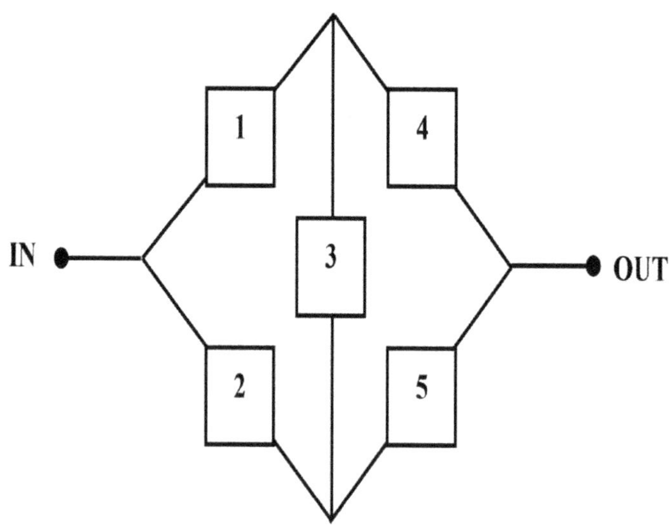

$$C_S = \sum_{i=1}^{5} a_i \exp\left[\frac{b_i}{(1-r_i)}\right] \dots \tag{8}$$

Here authors have considered the optimization of this system with nonlinear constraint as per the mathematical expressions given below:

$$Minimize\ C_S$$

subjected to

$$0 \le r_i \le 1, \quad i = 1,2,3,4,5$$

$$0.99 \le R_S \le 1$$

$$a_i = 1\, and\, b_i = 0.0003, for\, i = 1,2,3,4,5$$

Life Support System in Space Capsule

Figure 7 shows the block diagram of life support system in space capsule (Tillman et al., 1980).

The system reliability R_s and system cost C_s are given by:

$$R_S = 1 - r_3\left[(1-r_1)(1-r_4)\right]^2 - (1-r_3)\left[1 - r_2\left\{1-(1-r_1)(1-r_4)\right\}\right]^2 \tag{9}$$

Figure 7. Life-support system in a space capsule

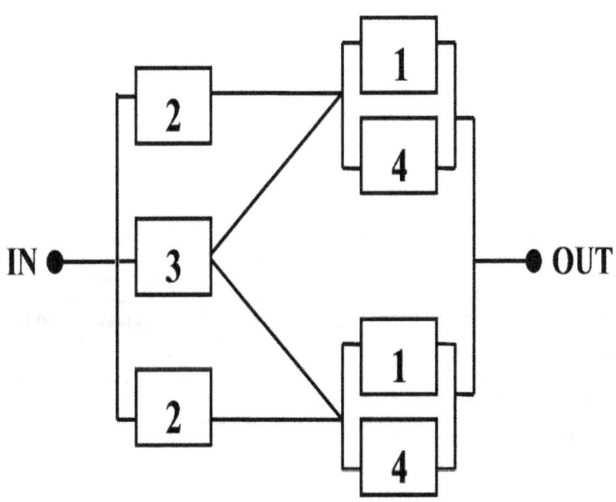

$$C_S = 2K_1 r_1^{\alpha_1} + 2K_2 r_2^{\alpha_2} + K_3 r_3^{\alpha_3} + 2K_4 r_4^{\alpha_4} \quad (10)$$

where, K_1=100, K_2=100, K_3=200, K_4=150and $\alpha_i = 0.6, i = 1,2,3,4.$.

This problem is continuous nonlinear in nature, where objective is to find the minimum cost of this complex system subject to the constraints on reliability. The problem formulation is as follows:

$$Minimize\, C_S$$

subjected to

$$0.5 \le r_i \le 1 \qquad i = 1,2,3,4$$

$$0.9 \le R_S \le 1$$

where, R_i is i^{th} component's reliability.

CONCLUSION

Number of nests (or different solutions) 'n' has been fixed at 30 while discovery rate of alien eggs/solutions (p_a) has been set at 0.25 for both the problems. Total 2000 and 500 iterations have been performed for complex bridge system and life support system in space capsule respectively and table 1 and table 2 represent the results obtained by CSA.

For the large search space L´evy flights are usually more efficient while most metaheuristic algorithms use either Gaussian or uniform distributions to generate new explorative moves. Apart from this, In

Table 1. Result for complex bridge system

	CSA
r_1	0.934899908273690
r_2	0.935545755733345
r_3	0.788533770251255
r_4	0.941231205173011
r_5	0.927707487562083
R_S	0.990000
C_S	5.019980140492710
FE	60000

Table 2. Result for life support system in space capsule

	CSA
r_1	0.500000000000000
r_2	0.838920107107387
r_3	0.500000000000000
r_4	0.500000000000000
R_S	0.900000000000000
C_S	641.8235631303452
FE	15000

Cuckoo Search Algorithm there are fewer parameters (number of nests 'n' and discovery rate of alien eggs $\left(p_a\right)$) which have to fine-tune so Cuckoo Search in combination with L´evy flights is very effective for most of the complex reliability optimization problem.

REFERENCES

Ahonen, H., de Souza Júnior, P. A., & Garg, V. K. (1997). A genetic algorithm for fitting Lorentzian line shapes in Mössbauer spectra. *Nuclear Instruments & Methods in Physics Research. Section B, Beam Interactions with Materials and Atoms*, *124*(4), 633–638. doi:10.1016/S0168-583X(97)00107-9

Brown, C., Liebovitch, L. S., & Glendon, R. (2007). Levy flights in Dobe Ju/'hoansi foraging patterns. *Human Ecology*, *35*(1), 129–138. doi:10.1007/s10745-006-9083-4

Bush, T. S., Catlow, C. R. A., & Battle, P. D. (1995). Evolutionary programming techniques for predicting inorganic crystal structures. *Journal of Materials Chemistry*, *5*(8), 1269–1272. doi:10.1039/jm9950501269

Dorigo, M. (1992). *Optimization, learning and natural algorithms. Ph.D. book.* Politecnico di Milano.

Eberhart, R., & Kennedy, J. (1995). A new optimizer using particle swarm theory. In *Proceeding of International Symposium on Micro Machine and Human Science*, (pp. 39-43).

Engelbrecht, A. P. (2007). *Computational intelligence: An introduction.* John Wiley & Sons Ltd. doi:10.1002/9780470512517

Geem, Z. W., Kim, J. H., & Loganathan, G. V. (2001). A new heuristic optimization algorithm: Harmony search. *Simulation*, *76*(2), 60–68. doi:10.1177/003754970107600201

Holland, J. H. (1962). Outline for a logical theory of adaptive systems. *Journal of the Association for Computing Machinery*, *9*(3), 297–314. doi:10.1145/321127.321128

Kennedy, J., & Eberhart, R. (1995). Particle swarm optimization. In *Proceeding of IEEE International Conference on Neural Networks,* (vol. 4, pp. 1942-1948). IEEE.

Kumar, A., & Singh, S. B. (2008). Reliability analysis of an n-unit parallel standby system under imperfect switching using copula. *Computer Modeling and New Technologies*, *12*(1), 47–55.

Kuo, W., & Wan, R. (2007). Recent Advances in Optimal Reliability Allocation. *IEEE Transactions on Systems, Man, and Cybernetics. Part A, Systems and Humans*, *37*(2), 143–156. doi:10.1109/TSMCA.2006.889476

Larrosa, J. A. E. (2008). *New heuristics for global optimization of complex bioprocesses.* Universidade de Vigo.

Michalewicz, Z., & Fogel, D. B. (2004). *How to solve it: Modern heuristics.* Springer-Verlag. doi:10.1007/978-3-662-07807-5

Mohan, C., & Shanker, K. (1988). Reliability Optimization of Complex Systems using Random Search Technique. *Microelectronics and Reliability*, *28*(4), 513–518. doi:10.1016/0026-2714(88)90133-3

Pant, S., Anand, D., Kishor, A., & Singh, S. B. (2015). A Particle Swarm Algorithm for Optimization of Complex System Reliability. *International Journal of Performability Engineering*, *11*(1), 33–42.

Pant, S., Kumar, A., Kishor, A., Anand, D., & Singh, S. B. (2015). Application of a Multi-Objective Particle Swarm optimization Technique to Solve Reliability Optimization Problem. In *Proceeding of IEEE Int. Conf. on Next Generation Computing Technologies*, (pp. 1004-1007). Doi:10.1109/NGCT.2015.7375272

Pant, S., & Singh, S. B. (2011). Particle Swarm Optimization to Reliability Optimization in Complex System. In *Proceeding of IEEE Int.Conf. on Quality and Reliability*. Doi:10.1109/ICQR.2011.6031711

Paszkowicz, W. (1996). Application of the Smooth Genetic Algorithm for Indexing Powder Patterns–Tests for the Orthorhombic System. *Materials Science Forum*, *228*, 19–24. doi:10.4028/www.scientific.net/MSF.228-231.19

Payne, R. B., Sorenson, M. D., & Klitz, K. (2005). *The Cuckoos*. Oxford University Press.

Qin, Z., Yu, F., Shi, Z., & Wang, Y. (2006). Adaptive inertia weight particle swarm optimization. In *Proceeding of International conference on Artificial Intelligence and Soft Computing,* (pp. 450-459).

Ram, M., Singh, S. B., & Singh, V. V. (2013) Stochastic Analysis of a Standby System with Waiting Repair Strategy. *IEEE Transactions on Systems, Man, and Cybernetics, 43*(3), 698-707.

Ravi, V., Murty, B. S. N., & Reddy, J. (1997). Nonequilibrium Simulated-Annealing Algorithm Applied to Reliability Optimization of Complex Systems. *IEEE Transactions on Reliability*, *46*(2), 233–239. doi:10.1109/24.589951

Reynolds, A. M., & Frye, M. A. (2007). Free-flight odor tracking in Drosophila is consistent with an optimal intermittent scale-free search. *PLoS ONE*, *2*(4), e354. doi:10.1371/journal.pone.0000354 PMID:17406678

Swinehart, K., Yasin, M., & Guimaraes, E. (1996). Applying an analytical approach to shop-floor scheduling: A case study. *International Journal of Materials and Product Technology*, *1*, 98–107.

Tillman, F. A., Hwang, C. L., & Kuo, W. (1980). *Optimization of Systems Reliability*. New York: Marcel Dekker Inc.

Yang, X. S., & Deb, S. (2009). Cuckoo search via Levy flights. In *Proceeding of World Congress on Nature & Biologically Inspired Computing (NaBIC),* (pp. 210-214).

Yang, X. S., & Deb, S. (2010). Engineering optimization by cuckoo search. *Int. J. Mathematical Modelling and Numerical Optimization*, *1*(4), 330–343. doi:10.1504/IJMMNO.2010.035430

Chapter 6
Multi-Objective Optimization of Die-Sinking EDM Process on AISI P-20 Tool Steel Using Cuckoo Search and Genetic Algorithm

Goutam Kumar Bose
Haldia Institute of Technology, India & West Bengal University of Technology, India

Pritam Pain
Haldia Institute of Technology, India

Sayak Mukhopadhyay
Haldia Institute of Technology, India

ABSTRACT

Electrical Discharge Machining (EDM) is nontraditional machining processes applied for precise machining and developing intricate geometries on work materials which are difficult to machine by conventional process. The present research work emphases on the die sinking EDM of AISI P20 tool steel, to study the effect of machining parameters such as pulse on time (POT), pulse off time (POF), discharge current (GI) and spark gap (SG) on performance response like Material removal rate (MRR), Surface Roughness (Ra) and Overcut (OC) using square-shaped Cu tool with Lateral flushing. The experimentation is performed using L27 orthogonal array and significant process parameters are ascertained using Regression analysis. The influence of the important process parameters on individual responses are detected using Cuckoo search algorithm. The present chapter is aimed at multi-response optimization i.e. higher MRR, lower Ra and minimum OC, which is conceded out using Genetic Algorithm.

DOI: 10.4018/978-1-5225-1639-2.ch006

INTRODUCTION

The intention of secondary manufacturing processes or material removal processes is to impart desired shape and size to the work material with acceptable tolerances, surface finish and surface integrity. Electric Discharge Machining (EDM) is anextensively used non-conventional machining process in the manufacturing of complex shaped dies, molds and critical parts of plastic molding, automobile, aerospace and other industrial applications. The process uses thermal energy of the sparks to machine electrically conductive parts regardless of the hardness of the work materials. This unique feature of EDM provides a noticeable advantage in the manufacture of complex shaped die and molds made up of hard materials which are complicated to machine by traditional machining processes (El-Hofy, 2005). The process involves removal of material through action of series of repeated electrical discharges of short duration and high current density between the tool–electrode and the workpiece under the effect of die–electric medium. The electrode never touches the workpiece but spark gets discharge through the dielectric across a very small gap. Thermal energy generates a plasma channel between the electrode and workpiece at a temperature range between 8000^{0}C to 12000^{0}C. EDM has proven to be applicable to machine electrically conductive materials such as stainless steels, tool steels, carbides, super alloys, ceramics etc., regardless of their other physical and metallurgical properties. The major factors influence in EDM machining are:

- Spark gap,
- Electrical parameters (pulse frequency, duty cycle, current and voltage) and
- Physical properties of electrode, workpiece and dielectric fluid.

AISI P-20 tool steel is being employed for making plastic injection molding dies for various polymer products. The plastic injection molding process requires hardened and precision dies to meet stringent surface quality. This material is difficult to cut by conventional process hence EDM technique is adapted to process the material. The knowledge of cutting parameters and their responses is essential to manufacture precision components. Hence an investigation has been focused on EDM machining of AISI P-20 Tool steel with squared copper electrode.

Presently, problem solving optimization algorithms are mostly metaheuristics and all of them are nature motivated. The moment they originate, they are acknowledged by researchers and applied widely. The potency of all modern metaheuristic algorithms is the fact that they imitate the paramount properties in nature, particularly biological systems materialize from natural selection over millions of years. Cuckoo Search (CS) is a novel metaheuristic algorithm. It is being used for solving optimization problem. It was exploited in 2009 by Xin- She Yang and Suash Deb. Exceptionality of this algorithm is the compulsory brood parasitism behavior of several cuckoo species besides the Levy Flight behavior of some birds and fruit flies. A Lévy flight is a haphazard move in which the movement-lengths are spread according to a heavy-tailed probability distribution. After a considerable amount of steps, the distance from the origin of the random walk has a tendency to a stable distribution.

Genetic Algorithm (GA) is a search heuristic that emulate the process of natural selection. This heuristic process is used to generate useful solutions to optimization problems. GA belong to the larger class of Evolutionary Algorithms (EA), which yield solutions to optimization problems by means of techniques inspired by the process of natural evolution, such as birthright, metamorphosis, selection, and crossover (Rao, 2011). It simulates the survival of the fittest amongst individuals over successive generation for elucidating a problem. Each generation comprises of a population of character strings that are similar to the chromosome that we perceive in our DNA. Each individual represents a point in a search space and a probable solution. The entities in the population are then made to proceed through a system of evolution.

A brief literature of the past research work is presented here. Bose and Mahapatra (2015) studied the important process parameters of die sinking EDM process on P-20 tool steel by applying ANOVA method. These experimental data are further investigated using Response Surface Methodology (RSM) where response surfaces and contour plots are plotted. Finally, multi-response optimization is conceded out by means of overlaid contour plots and desirability functions. Bose and Mahapatra (2014) experimented the significant process parameters of die sinking EDM process on AISI H-13 tool steel by applying Taguchi methodology. These test data are additionally studied by applying Response Surface Methodology (RSM). Here multi-response optimization is applied using overlaid contour plots and desirability functions. Bose and Mahapatra (2015) experimented the significant process parameters of die sinking EDM process by applying ANOVA. It is find that MRR is affected by gap current & Ra is affected by pulse on time. These experimental data are further investigated using Grey Relational Analysis to optimize multiple performances in which different levels combination of the factors are ranked based on grey relational grade. Kuriakose and Shunmugam (2005) studied on a multiple regression model to represent a relationship between input and output variables and a multi-objective optimization method based on a Non- Dominated Sorting Genetic Algorithm is used to optimize wire-EDM process and a non-domination solution set has been obtained from this experimentation. Somashekhar et al. (2010) developed a model to optimize the process parameters for micro-electric discharge machining process by using Artificial Neural Network (ANN) and Genetic Algorithm (GA). Zhang et al. (2014) designed an experimental plan for central composite design of processing tool steel according to the response surface methodology for medium speed wire electric discharge machining process. Pavlyukevich (2007) Solved a problem of non-convex stochastic optimization with help of simulated annealing of Lévy flights of a variable stable index and find out the convergence to the ground state is fast due to a polynomial decrease rate of the temperature. Abu-Srhan and Daoud (2013) studied on the traveling salesman problem and use Genetic Algorithms (GA) so solve the problem and also used cuckoo search optimizer to avoid the local minima problem from GA. Fister et al. (2014) summarized and categorized the majority of the literature about cuckoo search in pre-reviewed journals and conferences. Goswami and Chakraborty (2015) applied firework algorithm and cuckoo search (CS) algorithm for single as well as multi-response optimization of two laser machining process and find out CS algorithm outperforms firework algorithm with respect to average computation time. Kanagaraj et al. (2014) applied a hybrid algorithm cuckoo search with genetic algorithm (hybrid-CSGA) to solve the path optimization problem for printed circuit board holes drilling process and find out that the hybrid-CSGA reaches the near-optimal solution much earlier than the CS and GA approach for both small and large problems. Kanagaraj et al. (2013) studied on meta-heuristic optimization algorithm cuckoo search, hybridized with genetic algorithm to solve the reliability and redundancy allocation problems and find out that the proposed CS-GA algorithm proves to be extremely effective and efficient al local optimal solution. Madić and Radovanović (2013) studied on empirical modeling of surface roughness in CO2 laser cutting of stainless steel using feed forward ANN method along with cuckoo search algorithm and find out the percentage errors on training and testing data. Civicioglu and Besdok (2013) used cuckoo search, particle swarm optimization (PSO), differential evolution (DE) and artificial bee colony (ABC) to compare statically by testing over 50 different benchmark functions and reveal that:

1. Problem solving effectiveness of CS algorithm is very close to the DE algorithm,
2. The performance of the CK and PSO algorithms are statistically very similar to the performance of the DE algorithm than the ABC algorithm,
3. The CS and DE algorithm supply more robust and precise result than the PSO and ABC algorithms.

Rani et al. (2012) experimented a newly modified cuckoo search algorithm integrated with the roulette wheel selection operator and the inertia weight controlling the search ability toward synthesizing symmetric linear array geometry with minimum side lobe level and/or nulls control and finally they compared this result with evolutionary algorithm techniques particle swarm optimization and genetic algorithms. Roy and Chaudhuri (2013) studied Cuckoo Hashing and Modified CS algorithm then test results compared with GA and PSO and find out CS is superior than other two algorithms. Valian et al. (2011, a) proposed an improved cuckoo search algorithm to enhance the accuracy and revel that the proposed algorithm is capable of finding better solution in comparison with the solution obtain by the CS. Valian et al. (2011, b) employed CS algorithm for training feedforward neural network for two benchmark classification problems and find out simulated result has more effectiveness than normal CS algorithm. Yang and Deb (2010) applied CS algorithm to solve engineering design optimization problems, including the design of springs and welded beam structures and reveal that CS is far better than an efficient PSO. Zhou et al. (2014) studied CS algorithm to solve travelling salesman problem and find out the numerical experimental result from the proposed algorithm can find the global optimization solution with rapid convergences and stable generation.

The objective of the work is to study the characteristic features of the EDM process while machining AISI P-20 Tool Steel as reflected through the experimental studies with several process parametric combinations like pulse-on-time (POT), pulse-off-time (POF), gap current (GI) and spark gap (SG) on material removal rate (MRR), surface roughness (Ra) &overcut (OC). These experimental data are investigated using Regression analysis. The effect significant process parameters on individual responses are identified using Cuckoo search algorithm. The present chapter is aimed at fulfillment of the basic but conflicting objectives concurrently during machining – higher material removal rate (MRR), minimum surface roughness (Ra) and lower overcut (OC) by employing a single set of optimal or near optimal process variables. The multi-response optimization is carried out using Genetic Algorithm.

EXPERIMENTAL DESIGN

The present experimental study is conducted on Electric Discharge Machine (ACTSPARK SP1, China) die-sinking type with servo-head having constant gap and positive polarity for electrode is used for experimentation. Commercial grade EDM-30 oil (specific gravity of 0.80 @ 25°C, Viscosity of 3.11 CSt. @ 38°C) is used as dielectric fluid. A square-shaped Cu tool (12x12 mm) having external lateral flushing with pressure 0.2 kgf/cm^2 is used. AISI P -20 tool steel work piece material is selected for the experiment. P20 tool steels are either nitride or carburized and are pre-hardened to (29-33) HRC. These steels are capable of being machined into complex and large dies and molds. The presence of chromium and nickel improves the toughness and hardness of P20 steels. The pulsed discharge current is applied in various steps in positive mode. The EDM setup comprises of dielectric reservoir, pump and flow system, power supply and control unit, operational tank with work holding device, X-Y table compliant to the working table, tool holder and the servo system to feed the tool part as shown in Figure 1.

Figure 1. (a) Working setup (b) servo feed system (c) control panel (d) tool holder (e) machined work surface

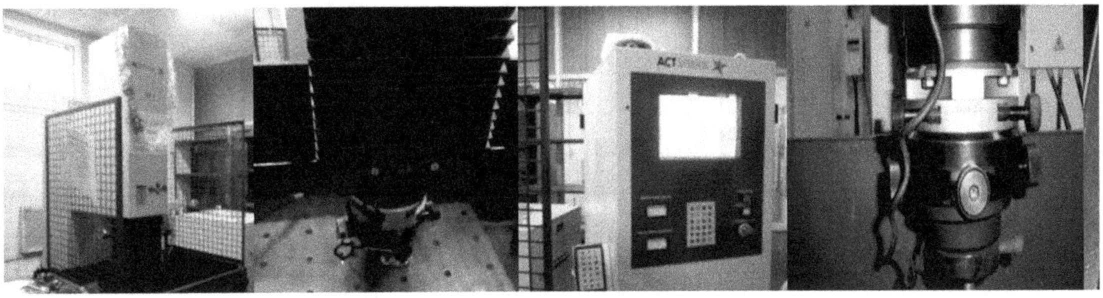

The servo control unit is applied to maintain the pre-determined gap. It senses the gap voltage and compares it with the existing value and the change in voltage is then used to control the movement of servo motor to adjust the gap.

The MRR is articulated as the ratio of the volume of the work piece material removed during machining the cavity to the machining time. Surface finish of the machined surface is expressed as Ra expressed in µm, is measured using stylus type profilometer named Perthometer-M1, of MahrGmbh make. Overcut (OC) is expressed as half the difference of area of the cavity produced to the tool frontal area that is shown in these equations. Area of Cavity & frontal area of electrode can be calculated by measuring the respective length & width using Toolmaker's microscope.

While performing an experiment, varying the levels of the factors in tandem rather than one at a time is proficient in terms of time and cost and also permits for the study of interactions between the factors. Based on previous research works and preliminary investigation, four parameters are selected as input. Four input parameters are varied with three levels in twenty-seven experimental run. The control parameters and their levels are shown in Table 1 below. There are other factors which may influence the measured performance like Duty cycle, Flushing pressure, Lift time etc., however, are kept constant during experimentation. The combination for factors and the three responses are shown in Table 2 below.

CUCKOO SEARCH ALGORITHM FOLLOWING LEVY FLIGHTS

Cuckoo Search (CS) is a newly developed metaheuristic nature inspired search algorithm. It is characterized by three basic laws (Yang & Deb, 2009):

Table 1. Control parameters and levels

Cutting Parameters	Symbol	Units	Levels		
			L1	L2	L3
Pulse on time	POT	(µSec)	16	20	24
Pulse off time	POF	(µSec)	12	16	20
Gap current	GI	(Amp)	7	9	11
Spark gap	SG	(mm)	0.16	0.18	0.20

Table 2. Combination of factors and responses

Exp No.	POT (μSec)	POF (μSec)	GI (Amp)	SG (mm)	MRR (mm^3/min)	Ra (μmm)	Overcut (mm^2)
1	20	12	11	0.18	0.0071	2.772	5.7312
2	24	20	9	0.18	0.0893	7.502	4.4166
3	24	16	11	0.18	0.1414	5.846	2.8788
4	20	20	7	0.18	0.1721	10.738	5.1352
5	16	16	11	0.18	0.0787	7.148	3.6724
6	20	16	7	0.16	0.0096	4.52	3.824
7	16	20	9	0.18	0.1186	7.34	6.6948
8	20	20	11	0.18	0.0098	5.262	3.4664
9	20	16	11	0.2	0.0229	2.626	3.6001
10	24	16	9	0.16	0.1311	6.243	7.6843
11	20	16	9	0.18	0.2157	10.462	8.8853
12	16	16	7	0.18	0.0297	2.217	6.084
13	20	12	9	0.2	0.2880	10.345	5.8502
14	16	16	9	0.16	0.0225	2.742	5.2415
15	20	16	9	0.18	0.0907	7.512	5.484
16	20	16	11	0.16	0.0210	2.356	7.6843
17	20	20	9	0.16	0.2251	6.817	6.847
18	20	16	9	0.18	0.0048	4.977	7.1376
19	20	20	9	0.2	0.0933	7.279	4.7739
20	20	12	9	0.16	0.0366	4.453	1.9083
21	20	16	7	0.2	0.1140	5.9776	3.6768
22	16	16	9	0.2	0.1213	7.944	3.4358
23	16	12	9	0.18	0.0557	6.343	7.441
24	24	12	9	0.18	0.0886	8.08	7.0674
25	24	16	9	0.2	0.0123	2.353	3.4818
26	24	16	7	0.18	0.2342	9.307	5.624
27	20	12	7	0.18	0.0737	6.611	6.198

1. First each cuckoo lays one egg at a time and disposes the egg at a randomly chosen host nest,
2. The finest nest with outstanding quality of eggs will convey over to the next generation, and

Lastly the number of host nest is constant and the probability of parasitic egg-detection by the host bird is $p_a \in [0,1]$. The host bird can either throw out the sponging egg or forgo that nest for a better new nest. For practicality, in the later hypothesis it can be assumed that fraction p_a of the nests is substituted by new nests (with new random solutions).In one form of fitness the quality or fitness of a solution is suggested to be proportional to the value of the objective function in case of a maximization problem.

In a simple manner of representation, each egg in a nest is represented by a solution and each cuckoo egg implies a new solution. The aim is to utilize the new and probably finer/ superior solutions (cuckoos) to substitute an inferior solution in the nests. More complex cases like multiple eggs in a nest, symbolizing a set of solution, can be incorporated in this algorithm. Guided by the above three rules, the CS is abridged by the pseudo code.

Algorithm 1: Cuckoo Search via Le'vy Flights

```
begin
        Objective function f(x), x=(x1, x2,.........xd)T
        Generate initial population of
        n host nests xi (i=1, 2, 3......, n)
while
        (t<Maxgeneration) or (stop criterion)
        Get a cuckoo randomly by Levy Flights
        evaluate its quality/fitness Fi
        Choose a nest among n (say, j) randomly
if
        (Fi > F j),
        Replace j by the new solution;
end
        A function (pa) of worse nests
                are abandoned and new ones are built;
        Keep the best solutions
                (or nests with quality solutions);
        Rank the solutions and find the current best
end while
        Postprocess results and visualization
end
```

When making new solutions x_i (t +1) for the ith cuckoo, the following Lévy flight is performed

$$x_i\left(t\ +1\right)=\ x_i\left(t\right)+\alpha\otimes L\acute{e}vy\ \left(\lambda\right) \tag{1}$$

where a (>0) is the stage size, which should be correlated to the scale of the problem of interest. The product ⊗ means entry-wise multiplications. In this experimental investigation, a Lévy flight is considered, in which the step-lengths are distributed according to the following probabilitydistribution

$$L\acute{e}vy\ u\ =\ t^{-\lambda},1<\ \lambda\ \leq\ 3 \tag{2}$$

which has an infinite variance. Here, the successive jumps/steps of a cuckoo essentially outline arandom walk process which conform a power-law step-length distribution to a heavy tail. It is significant tomention that, in the practical world, if a cuckoo's egg is very similar to a host's eggs, then this cuckoo's egg is most unlikely to be discovered, thus the fitness should be related to thedifference in solutions. Therefore, it is a noble idea to do a random walk in a prejudiced way withsome random step sizes.

GENETIC ALGORITHM

Genetic Algorithm (GA) is established on a resemblance with the genetic structure and behavior of chromosomes within a population of individuals by applying the following fundamentals (Goldberg, 1989):

- Living beings in a population compete for resources and mates.
- Those individuals most victorious in each 'competition' will yield more offspring than those individuals that perform poorly.
- Genes from 'good' individuals reproduce throughout the population so that two good parents will sometimes produce offspring that are superior than either parent. Thus each consecutive generation will become more suited to their environment.

A population of individuals is sustained within search space for a GA, each representing a possible solution to a given problem. Every individual is coded as a finite length vector of components, or variables, in terms of some alphabet, usually the binary alphabet (i.e. 0, 1). To continue the genetic analogy these individuals are likened to chromosomes and the variables are similar to genes. Therefore, a chromosome (solution) is composed of several genes (variables). A fitness score is allocated to each solution signifying the abilities of an individual to 'compete'. The individual with the optimal (or preferably near optimal) fitness score is pursued. The GA aims to use selective `breeding' of the solutions to yield 'offspring' better than the parents by uniting information from the chromosomes. The steps for GA are shown in Figure 2.

Figure 2. Normal Plot of Residuals for MRR

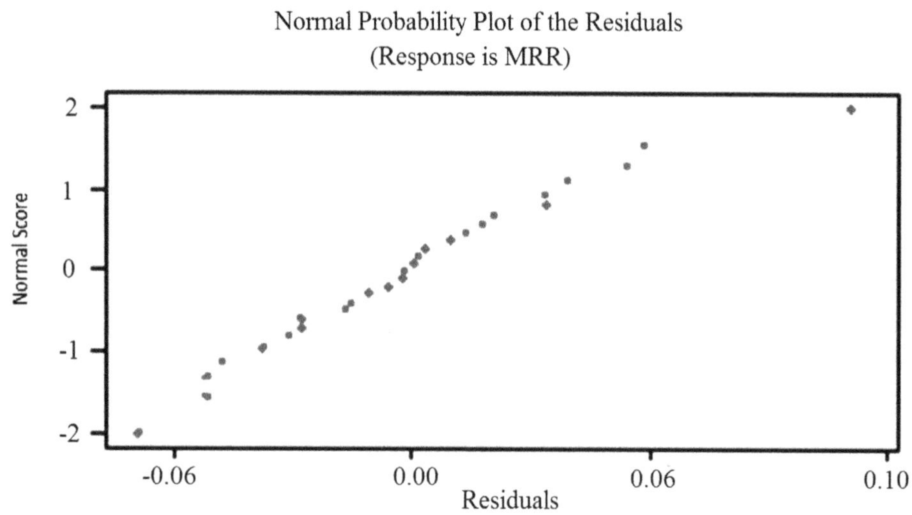

The genetic algorithm conserves a population of 'n' chromosomes (possible solutions) with associated fitness values. Parents are selected to mate, based on their fitness, producing offspring via a reproductive plan (Deb, 2001). Consequently, highly fit solutions are assumed to offer more opportunities to reproduce, so that offspring inherit characteristics from each parent. As parents mate and make offspring, scope must be created for the new arrivals since the population is kept at a static size. The individuals in the population expire and are substituted by the new solutions. This eventually creates a new generation once all mating prospects in the old population have been expended. In this way it is hoped that over successive generations better solutions will flourish while the least fit solutions perish out.

Algorithm 2: Genetic Algorithm

```
Begin
        choose initial population.
Initialize max_genration
        evaluate each individual's fitness.
        determine population's average fitness.
While (i<max_genration)
        select best-ranking individuals to reproduce.
        mate pairs at random.
        apply crossover operator.
        apply mutation operator.
        evaluate each individual's fitness.
        determine population's average fitness.
        i=i+1.
        End
End
```

New generations of solutions are produced containing, on average, better genes than a typical solution in a previous generation. Each successive generation will contain better 'partial solutions' than previous generations. Eventually, once the population has converged and is not producing offspring noticeably different from those in previous generations, the algorithm itself is said to have converged to a set of solutions to the problem at hand.

EXPERIMENTAL RESULTS

Developing a regression model that includes only a subset of the available regression involves two confliction objectives such as (Montgomery et al., 2003):

1. Regression model has to include as many regressors as possible in order to make the resultant equation useful for predictive purpose.
2. Regression model has to maintain as few regressors as possible due to the cost involved in obtaining and maintaining information on a large scale.

In this present experimentation backward elimination method is applied.

Regression Analysis for MRR

The empirical model for MRR is shown in Equation 7 below:
The estimated regression equation for MRR is:

$$MRR = -0.159 - 0.00862POT - 0.00361POF + 0.0494GI + 0.206SG \quad (7)$$

The empirical model has R-square value of 94.9% and it provides good explanation of the relationship between parameters and responses. The details of the regression analysis result and subsequent ANOVA are presented in Table 3 and Table 4 respectively.

R-square as well as R-square (adjusted) assumes a value of 94.9% and 89.0% respectively, that implies the model is poised to explain 94.9% variability with process variable POT, POF, GI & SG. From the T values of the process variables, it can be concluded that GI is the most significant process variables followed by SG, POF and POT.

The normal plot of residuals for MRR is shown in Figure 2. It can be observed that the residuals are located on a straight line, which indicates that the errors are normally distributed and the non-linear models are fairly well fitted with the experimental values.

Table 3. Estimated regression coefficients for MRR

Term	Coef	SE Coef	T	P
Constant	0.10449	0.015112	6.914	0.000
POT	-0.03448	0.007556	-4.563	0.001
POF	-0.01443	0.007556	-1.909	0.080
GI	0.09884	0.007556	13.082	0.000
SG	0.00413	0.007556	0.546	0.595
POT*POT	-0.03394	0.011334	-2.995	0.011
POF*POF	-0.01603	0.011334	-1.414	0.183
GI*GI	0.02464	0.011334	2.174	0.050
SG*SG	-0.00074	0.011334	-0.066	0.949
POT*POF	0.01355	0.013087	1.036	0.321
POT*GI	-0.01553	0.013087	-1.187	0.258
POT*SG	0.00810	0.013087	0.619	0.547
POF*GI	-0.01613	0.013087	-1.232	0.241
POF*SG	-0.00972	0.013087	-0.743	0.472
GI*SG	0.01745	0.013087	1.333	0.207

Notes: S = 0.02617 R-Sq = 94.9% R-Sq(adj) = 89.0%.

Table 4. ANOVA for MRR

Source	DF	Seq SS	Adj SS	Adj MS	F - Value	P - Value
Regression	14	0.154026	0.154026	0.011002	16.06	0.000
Linear	4	0.134203	0.134203	0.033551	48.97	0.000
Square	4	0.015224	0.015224	0.003806	5.56	0.009
Interaction	6	0.004599	0.004599	0.000767	1.12	0.407
ResidualError	12	0.008221	0.008221	0.000685		
Lack-of-Fit	10	0.006066	0.006066	0.000607	0.56	0.781
Pure Error	2	0.002155	0.002155	0.001077		
Total	26	0.162247				

Regression Analysis for Surface Roughness (Ra)

The empirical model for Ra is shown in Equation 8 below:

The estimated regression surface equation for Rais:

$$Ra = 4.28 - 0.413POT - 0.036POF + 1.23GI - 2.3SG \quad (8)$$

The empirical model has R-square value of 78.8% and it present satisfactory details of the relationship between parameters and responses. The particulars of the regression analysis outcome and subsequent ANOVA are presented in Table 5 and Table 6 respectively. R-square as well as R-square (adjusted) furnishes a value of 78.8% and 54.1% respectively that implies the model is balanced to explain 78.8% variability with process variable POT, POF, GI & SG. From the T values of the process variables, it can be concluded that GI is the most significant process variables followed by SG, POF and POT.

The normal plot of residuals for Ra is shown in Figure 3. It is pragmatic that the residuals are positioned on a straight line, which signifies that the errors are normally distributed and the non-linear models are moderately healthy fitted with the experimental values.

Regression Analysis for Overcut (OC)

The empirical model for Ra is shown in Equation 9 below:

The estimated regression surface equation for Overcut (OC)is:

$$OC = -3.12 + 0.084POT + 0.069POF + 0.603GI + 1.3SG \quad (9)$$

The empirical model has R-square value of 51.2% and it signifies details of the relationship between parameters and responses. The particulars of the regression analysis and subsequent ANOVA details are presented in Table 7 and Table 8 respectively. R-square furnishes a value of 51.2% that implies the model is balanced to explain 51% variability with process variable POT, POF, GI & SG. From the T values of the process variables, it is observed that GI relatively affects the process.

Table 5. Estimated regression coefficients for Ra

Term	Coef	SE Coef	T	P
Constant	6.758	1.0000	6.757	0.000
POT	-1.651	0.5000	-3.303	0.006
POF	-0.143	0.5000	-0.285	0.780
GI	2.465	0.5000	4.929	0.000
SG	-0.045	0.5000	-0.090	0.929
POT*POT	-1.674	0.7500	-2.232	0.045
POF*POF	-0.251	0.7500	-0.335	0.743
GI*GI	0.340	0.7500	0.453	0.659
SG*SG	0.195	0.7500	0.260	0.799
POT*POF	0.284	0.8661	0.328	0.749
POT*GI	-0.274	0.8661	-0.317	0.258
POT*SG	-0.005	0.8661	-0.006	0.995
POF*GI	1.085	0.8661	1.253	0.234
POF*SG	0.161	0.8661	0.186	0.856
GI*SG	-0.466	0.8661	-0.538	0.600

Notes: S = 1.732 R-Sq = 78.8% R-Sq(adj) = 54.1%.

Table 6. ANOVA for Ra

Source	DF	Seq SS	Adj SS	Adj MS	F – Value	P – Value
Regression	14	133.766	133.766	9.5547	3.18	0.026
Linear	4	105.882	105.882	26.4705	8.82	0.001
Square	4	21.578	21.578	5.3944	1.80	0.194
Interaction	6	6.306	6.306	1.0510	0.35	0.896
ResidualError	12	36.003	36.003	3.0003		
Lack-of-Fit	10	34.748	34.748	3.4748	5.54	0.163
Pure Error	2	1.255	1.255	0.6276		
Total	26	169.769				

The normal plot of residuals for OC is shown in Figure 4. It is revealed that the residuals are positioned on a straight line, which signifies that the errors are normally distributed and the non-linear models are moderately healthy fitted with the experimental values.

SIMULATION RESULTS

In the preliminary study, four parameters such as Pulse on Time (POT), Pulse off Time (POF), Gap Current (GI) and Spark Gap (SG)in standard CS optimizers is analyzed specifically on their implicit

Figure 3. Normal plot of residuals for Ra

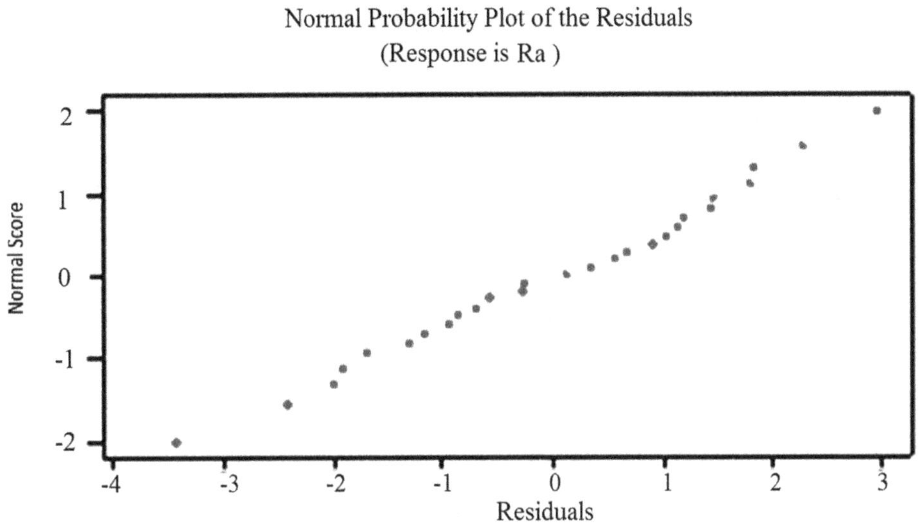

Table 7. Estimated regression coefficients for OC

Term	Coef	SE Coef	T	P
Constant	3.775	1.0452	3.612	0.004
POT	0.336	0.5226	0.644	0.532
POF	0.277	0.5226	0.530	0.605
GI	1.206	0.5226	2.308	0.040
SG	0.025	0.5226	0.049	0.962
POT*POT	1.441	0.7839	1.838	0.091
POF*POF	0.761	0.7839	0.971	0.351
GI*GI	0.784	0.7839	1.001	0.337
SG*SG	0.514	0.7839	0.656	0.525
POT*POF	0.214	0.9052	0.236	0.817
POT*GI	-1.098	0.9052	-1.213	0.248
POT*SG	-0.498	0.9052	-0.550	0.593
POF*GI	-0.126	0.9052	-0.139	0.892
POF*SG	0.897	0.9052	0.991	0.341
GI*SG	-0.282	0.9052	-0.312	0.761

Notes: S = 1.810 R-Sq= 51.2% R-Sq(adj) = 0.0%.

effects on Material Removal Rate (MRR), Surface Roughness (Ra) & Overcut (OC). To optimize the given test functions, a MATLAB code, by using MATLAB Version 7.10, is established based on the CS. In this version of MATLAB, the numbers smaller than $4.9407e^{-324}$ are considered as zero. The PC used is an Intel(R) Core i5 3230 M CPU @ 2.60GHz having 4GB of internal memory, and 64 Bit OS.

Table 8. ANOVA for OC

Source	DF	Seq SS	Adj SS	Adj MS	F Value	P Value
Regression	14	41.289	41.289	2.9492	0.90	0.580
Linear	4	19.753	19.753	4.9383	1.51	0.262
Square	4	11.940	11.940	2.9850	0.91	0.489
Interaction	6	9.596	9.596	1.5994	0.49	0.805
Residual Error	12	39.330	39.330	3.2775		
Lack-of-Fit	10	37.519	37.519	3.7519	4.14	0.210
Pure Error	2	1.812	1.812	0.9058		
Total	26	80.619				

Figure 4. Normal Plot of Residuals for OC

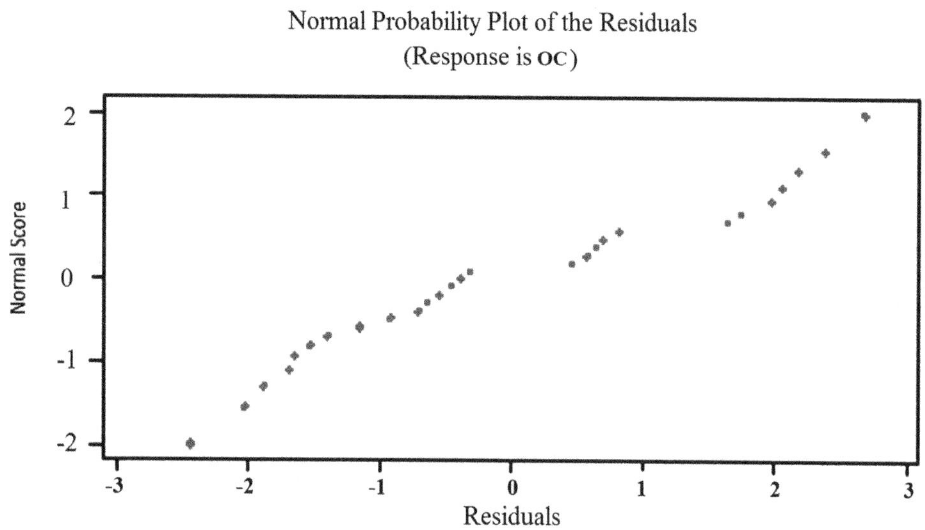

Optimization problem can be formulated as follows:

- *Find* POT, POF, GI, SG
- *To Maximize:* MRR and
- *To Minimize:* Ra, OC

Within the cutting parameter ranges (constraints):

- $16 \leq POT \leq 24$ µSec
- $12 \leq POF \leq 20$ µSec
- $7 \leq GI \leq 11 amp$
- $0.16 \leq SG \leq 0.20\ mm$

Results Applying Cuckoo Search Algorithm Via Levy Flights

Initially the individual responses are optimized applying Cuckoo Search algorithm via Levy flights. The parameters were α is distributiontype number of host nest (size of population), and fractionprobability, Pa (discovery rate). Firstly, thecharacteristic of α is analyzed. In this case, the CS-optimizer withnest = 25, Pa = 0.25 and step size factor = L/100 are examined. Total number of iterations carried out is 50,000. Maximum MRR is obtained at 0.2432 mm^3/min for parametric combination of 16μSec (POT), 12μSec (POF), 11amp (GI) and 0.20 mm (SG). Whereas minimum Ra is obtained at 1.799μ mm with the parametric combination of 24 μSec (POT), 20 μSec (POF), 7 amp (GI) and 0.20 mm (SG).Finally, minimum OC is obtained at 21.82 mm^2 having parametric combination of 16 μSec (POT), 12 μSec (POF), 7 amp (GI) and 0.16 mm (SG).

To confirm the proposed results, acquired from the algorithm, verification experiments have been performed and the results are listed in the Table 9 below.

From the Table 9, it is observed that the predicted values through additive model are almost similar to the experimental observations. Therefore, this model is plausible for prediction.

Results Applying Genetic Algorithm

Based on the conflicting nature of the objectives, multi response optimization is carried out in order to achieve the goal by a single parametric combination. Here Genetic Algorithm is applied for the three major attributes high MRR and low Ra and Overcut.

The conditions considered for GA are as follows:

1. **Population**
    a. **Population Type:** Double Vector
    b. **Population Size:** 15 x number of variables
2. **Selection Function:** Tournament
    a. **Tournament Size:** 2
3. Crossover Fraction = 0.8
4. **Mutation Function:** Gaussian if there are no constraints and Adaptive feasible otherwise
5. **Crossover Function:** Intermediate

Table 9. Verification experiments

Response	Parametric Combination				Predicted Results	Validated Results	Error %
	POT (μSec)	POF (μSec)	GI (amp)	SG (mm)			
MRR	16	12	11	0.20	0.2432	0.24	3.33
OC	16	12	7	0.16	21.820	21.40	1.74
Ra	24	20	7	0.20	1.799	1.68	6.56

6. **Migrationa**
   a. **Direction:** forward
   b. **Fraction:** 0.2
   c. **Interval:** 20
7. **Stopping Criteria**
   a. **Generation:** 10000
   b. **Time Limit:** Infinite
   c. **Fitness Limit:** infinite
   d. **Stall Generation:** 100
   e. **Fitness Tolerance:** 1e-5

Based on the conflicting nature of the objectives, multi response optimization is carried out in order to achieve the goal by a single parametric combination. Here Genetic Algorithm is applied for the three major attributes high MRR and low Ra and Overcut. EDM primarily being a complex operation, there can be two situations during commercial machining. One in case of rough cutting where more emphasis is given on MRR rather than on Ra leading to an assignment of biased weights amongst the attributes. Secondly during finish cutting where importance is given on Ra rather than on MRR. The different parametric combinations with respective responses at rough cut as well as at finish cut are shown in Table 10 below.

CONCLUSION

The experimental investigation for evaluating the optimal parametric combination and the subsequent effect of the parameters over the responses can act as an efficient & useful guideline for machining and manufacturing various metallic products. From commercial point of view, the parametric combinations

Table 10. Combination of factors and responses

Rough Cut						
POT (μSec)	POF (μSec)	GI (amp)	SG (mm)	MRR (mm^3/min)	Ra (μ mm)	OC (mm^2)
16	12.0	11	0.2	0.24436	10.31	24.281
16	12.0	11	0.16	0.23585	10.37	24.223
16	13.6	11	0.18	0.22968	10.12	24.506
18	12.6	10	0.2	0.21451	9.18	26.064
16	13.7	11	0.17	0.21389	9.83	24.302
Finish Cut						
16	12.0	7	0.16	0.03853	5.482	21.817
16	12.0	8	0.19	0.09165	6.565	22.426
16	12.5	9	0.18	0.11952	7.313	23.038
16	14.6	9	0.19	0.11737	7.288	23.510
16	14.4	7	0.19	0.03416	5.247	22.279

obtained during rough cut and finish cut machining will help to a great extent as well as it will act as a guideline for the operator / machinist working in this area. The future work in this emerging area can be considered with other parameters and different responses such as cutting force, tool life etc. to capture the process in full perspective. This set of inputs can be further used to optimize other functions like machining cost and can form the backbone of adaptive control strategies (adaptive control with optimization and geometric adaptive control).

REFERENCES

Abu-Srhan, A., & Daoud, E. A. (2013). A Hybrid Algorithm Using a Genetic Algorithm and Cuckoo Search Algorithm to Solve the Traveling Salesman Problem and its Application to Multiple Sequence Alignment. *International Journal of Advanced Science and Technology*, *61*, 29–38. doi:10.14257/ijast.2013.61.04

Bose, G. K., & Mahapatra, K. K. (2014). Parametric study of Die Sinking EDM process on AISI H13 tool steel using statistical techniques. *Journal of Advances in Production Engineering and Management*, *8*(4), 168–180. doi:10.14743/apem2014.4.185

Bose, G. K., & Mahapatra, K. K. (2015). Experimental analysis of Die Sinking EDM Process on AISI P20 tool steel using response surface methodology. *Manufacturing Engineering*, 163–188.

Bose, G. K., & Mahapatra, K. K. (2015). Multi-criteria decision making of machining parameters for Die Sinking EDM process. *International Journal of Industrial Engineering Computations*, *6*(2), 241–252. doi:10.5267/j.ijiec.2014.10.005

Civicioglu, P., & Besdok, E. (2013). A conceptual comparison of the Cuckoo-search, particle swarm optimization, differential evolution and artificial bee colony algorithms. *Artificial Intelligence Review*, *39*(4), 315–346. doi:10.1007/s10462-011-9276-0

Deb, K. (2001). *Multi-objective optimization using evolutionary algorithms*. Hoboken, NJ: Wiley.

El-Hofy, H. (2005). *Advance Machining Process*. New York: McGraw Hill.

Fister, I. J., Xin-She, Y., Fister, D., & Fister, I. (2014). Cuckoo Search: A Brief Literature Review. *Studies in Computational Intelligence*, *516*, 49–62.

Goldberg, D. E. (1989). *Genetic algorithms in search, optimization and machine learning*. Boston: Addison Wesley.

Goswami, D. & Chakraborty, S. (2015). A Study on the Optimization Performance of Fireworks and Cuckoo Search Algorithms in Laser Machining Processes. *Journal of The Institution of Engineers*, *96*(3), 215-229.

Kanagaraj, G., Ponnambalam, S. G., & Jawahar, N. (2013). A hybrid cuckoo search and genetic algorithm for reliability–redundancy allocation problems. *Computers & Industrial Engineering*, *66*(4), 1115–1124. doi:10.1016/j.cie.2013.08.003

Kanagaraj, G., Ponnambalam, S. G., & Lim, W. C. E. (2014). Application of a hybridized cuckoo search-genetic algorithm to path optimization for PCB holes drilling process. *2014 IEEE International Conference on Automation Science and Engineering (CASE)*. doi:10.1109/CoASE.2014.6899353

Kuriakose, S., & Shunmugam, M. S. (2005). Multi-objective optimization of wire EDM process by non-dominated sorting genetic algorithms. *Journal of Materials Processing Technology*, *170*, 133–141. doi:10.1016/j.jmatprotec.2005.04.105

Madić, M., & Radovanović, M. (2013). Application of Cuckoo Search Algorithm for Surface Roughness Optimization in CO2 Laser Cutting, Annals of Faculty Engineering Hunedoara. *International Journal of Engineering*.

Montgomery, D. C., Peck, E. A., & Vining, G. G. (2003). *Introduction to linear regression analysis* (3rd ed.). New York: Wiley.

Pavlyukevich, I. (2007). Lévy flights, non-local search and simulated annealing. *Journal of Computational Physics*, *226*(2), 1830–1844. doi:10.1016/j.jcp.2007.06.008

Pavlyukevich, I. (2007). Lévy flights, non-local search and simulated annealing. *Journal of Computational Physics*, *226*(2), 1830–1844. doi:10.1016/j.jcp.2007.06.008

Rani, K. A. N., Malek, M. F. A. & Siew-Chin, N. (2012). Nature-inspired Cuckoo Search Algorithm for Side Lobe Suppression in a Symmetric Linear Antenna Array. *Radio Engineering*, *21*(3), 865 – 874.

Rao, R. V. (2011). *Advanced modeling and optimization of manufacturing processes international research and development*. London: Springer. doi:10.1007/978-0-85729-015-1

Roy, S., & Chaudhuri, S. S. (2013). Cuckoo Search Algorithm using Lèvy Flight: A Review. *International Journal of Modern Education and Computer Science*, *12*(12), 10–15. doi:10.5815/ijmecs.2013.12.02

Somashekhar, K. P., Ramachandran, N., & Mathew, J. (2010). Optimization of Material Removal Rate in Micro-EDM Using Artificial Neural Network and Genetic Algorithms. *Materials and Manufacturing Processes*, *25*(6), 467–475. doi:10.1080/10426910903365760

Valian, E., Mohanna, S., & Tavakoli, S. (2011). a). Improved Cuckoo Search Algorithm for Global Optimization. *International Journal of Communications and Information Technology*, *1*(1), 31–44.

Valian, E., Mohanna, S., & Tavakoli, S. (2011). b). Improved Cuckoo Search Algorithm For Feedforward Neural Network Training. *International Journal of Artificial Intelligence & Applications*, *2*(3), 36–43. doi:10.5121/ijaia.2011.2304

Yang, X. S., & Deb, S. (2009). Cuckoo Search via Lévy Flights. *World Congress on Nature & Biologically Inspired Computing (NaBIC 2009)*, (pp. 210-214). IEEE Publications. doi:10.1109/NABIC.2009.5393690

Yang, X. S., & Deb, S. (2010). Engineering optimization by cuckoo search. *International Journal of Mathematical Modelling and Numerical Optimization*, *1*(4), 330–343. doi:10.1504/IJMMNO.2010.035430

Zhang, G., Zhang, Z., Ming, W., Guo, J., Huang, Y., & Shao, X. (2014). The multi-objective optimization of medium-speed WEDM process parameters for machining SKD11 steel by the hybrid method of RSM and NSGA-II. *International Journal of Advanced Manufacturing Technology, 70*(9-12), 2097–2109. doi:10.1007/s00170-013-5427-6

Zhou, Y., Ouyang, X., & Xie, J. (2014). *A discrete cuckoo search algorithm for travelling salesman problem*. Inderscience Enterprises Ltd.

Chapter 7
Introduction to System Reliability Evaluation through Bayesian Approach

M. Kumar
National Institute of Technology Calicut, India

P. N. Bajeel
National Institute of Technology Calicut, India

ABSTRACT

Consider a parallel system with n independent components. Assume that lifetime of i-th component follows exponential distribution with unknown parameter λ_i, $1 \leq i \leq n$. We assume each λ_i, $1 \leq i \leq n$, is distinct and the priori information can be modeled by quasi-density function given by $g(\lambda_i) = \frac{1}{\lambda_i^k}$, $k \geq 1$, $\lambda_i \leq u_i$, where u_i is a predefined upper bound. A Bayesian estimator for λ_i, $1 \leq i \leq n$, based on type-I censoring is used to get an estimate of system reliability. Optimal reliability test plan is designed, and an optimization problem is formulated satisfying usual probability requirements. Several numerical examples are considered to illustrate the Bayesian approach of obtaining optimal reliability test plan for a parallel system.

INTRODUCTION TO LIFE TESTING

The expected lifetime and failure rate of units or products are of great significant for any manufacturing industry. If a statistically significant sample is selected randomly from a lot of similar units and if the selected units are tested till all units fail, then enough data will get generated for confidently predicting the life expectancy of any unit contained in the lot. Life testing is carried out for estimating the mean time to failure and the expected life of units.

When units are produced in the manufacturing industry, the manufacturer would like to study the exact behavior of the unit under the normal working environment. For that he needs the data and evaluates

DOI: 10.4018/978-1-5225-1639-2.ch007

the life and reliability of the units produced. By studying this, he can improve the design, manufacturing, and other related goals. This will improve the confidence of the manufacturer and thus to provide higher guarantee periods and better service facilities. In addition, he can plan maintenance schedules, replacements of parts, warranty policies, etc.

Life Testing Methods

Some of the important life testing methods are briefly discussed below:

Life Test with Censoring

The procedure adopted for terminating a life test is called censoring. Censoring of data arises when exact lifetimes are known only for a portion of the individuals under study. A life test will get terminated when all units undergoing a test fail. However, it might take lot of time and immense effort for this to happen in practice, and sometimes it is even not possible to attain failures of all units under the specified test. Therefore censoring methods are developed for terminating life tests, and thereby to study the lifetime characteristics of units.

Life Test with Replacement

In tests with replacement, life testing is started with n units. Whenever a unit fails, it is replaced with a new unit and the testing is continued without any halt. The test is terminated either after the pre-decided time or after the occurrence of the *r-th* failure. Test with replacement will generate more data over a period of time as compared to the data generated by tests without replacement over the same time period.

Simulated Life Tests

In this procedure units are tested in simulated operating conditions; equipment such as test chambers are used for obtaining the simulated operating conditions. Some of the factors that significantly influence the operating conditions are temperature, humidity, pressure, etc. The actual tests are carried out with various combinations of these factors. Best results are achieved through the proper control of simulated test conditions.

Accelerated Life Testing

Most of the highly reliable products manufactured in industry do not fail very easily. To obtain failure data of such products further techniques are required. Accelerated life testing is one such technique to obtain faster failures. A variety of procedures is used to accelerate failures due to the difficulties encountered in performing life tests with time deadlines. Accelerated tests are designed and carried out for accelerating failures so that more number of units fail in a short test duration. Compressed-time test and advanced stress tests are two different types of accelerated life tests. In compressed-time testing, loads and environmental stresses on the product are maintained at the same level as in normal use but the product is used more frequently in the life test than it would normally be used. In advanced stress test, life testing is conducted at higher stress levels to obtain more data within a short time. Some of the common accelerated tests are design qualification test, environmental stress screening, etc. (Naikan, 2010)

In this chapter, the authors consider a censored life test. To do a censored life test, either the test time or the number of failures should be pre-fixed. For this, some reliability test plan is essential.

THE NEED OF RELIABILITY TEST PLANS

A dedicated program of testing is an important ingredient in the development of a complex system. The usual objective is to evaluate the reliability of a system and demonstrate that it will perform satisfactorily, prior to actual deployment in the field. To derive optimum component test plans, the problem can be formulated based on the notion of producers' and consumers' risks in acceptance sampling plans. Formally, producers' risk is defined as the probability that an acceptable system may be erroneously rejected by the test plan, whereas consumers' risk is defined as the probability that an unacceptable system may be erroneously accepted by the plan. The errors associated with producers' and consumers' risks are the type-I and type-II errors, respectively. The main step of development of highly reliable system is the testing and analysis of the system. The purpose of this testing and analyzing is to demonstrate that the system under consideration will perform at a level of reliability that is acceptable with respect to the mission for which it is designed. Before the actual production, put the system for testing and evaluate the reliability. The reason for doing this is to ensure that the system will achieve some predefined reliability requirements. Analogous to standard acceptance sampling plans, reliability test plans can be expressed in terms of maximum acceptable values for producer's risk (type-I error) and consumer's risk (type-II error) . System level tests tend to be very highly scrutinized because of the length of time and the cost that they add to the overall design and development schedule; however, they are often unavoidable. On the other hand, there are situations where a better approach might be to use component test plans for system reliability demonstration. There are numerous advantages to this approach: it is less expensive and more time taking, testing can proceed at different times and locations, test facilities for components tend to be simpler and less expensive, and it is more cost-effective since the entire system is not assembled until it is guaranteed to be reliable.

Several variations of the two basic options (component level and system level) are possible. For instance, one may conduct tests at the level of subassemblies (or subsystems) that are composed of individual components, and that are themselves part of the entire system. Alternatively, one could adopt some combination of system, subsystem and component testing. The choice of an appropriate test plan depends on the characteristics of the specific system under consideration, as well as cost and feasibility. System level tests are typically more expensive as well as more inconvenient than component tests. There are several reasons for this.

1. First, the instrumentation, test fixtures and facilities, materials and the actual test units all usually cost more for systems than for components.
2. Second, and perhaps most importantly, with component tests the entire system need not be assembled prior to testing. This results in significant savings both with respect to costs as well as the times associated with system development.

3. Third, testing may proceed independently at different times and locations. This is an important issue since components (or subsystems) of many large systems are typically contracted to different organizations who develop these components independently.
4. Fourth, component tests tend to be more informative in that the behavior of individual components can often be understood better than in the case where they are assembled into a larger system.
5. Finally, component testing goes well with the notions of total quality since the final system is assembled only after a performance guarantee is obtained.

Conversely, system testing is preferable when the component failures are highly dependent on each other since component testing assumes independence of component failure times, and the availability of a mathematical model that expresses system reliability in terms of component reliabilities. Another situation where system testing may be preferable is when the interfaces between components are inherently unreliable; however, in such situations it may still be more economical to use a combination of system and component testing. (Rajgopal & Mazumdar, 1997, 1998).

Introduction to System Based Component Reliability Test Plans

As with any test plan, the objective is to minimize the total cost of the test program while ensuring that they offer sufficient protection from errors that result from making wrong decisions. Specific examples of reliability test plans can be found in the Department of the Navy document MIL-HDBK-781, which offers protection from the probabilities of type-I error α and type-II error β. In this document values are specified for *a* and *b* (with $\alpha + \beta < 1$), and for R_0 and R_1 (with $0 < R_0 < R_1 < 1$) where a system with reliability less than R_0 over its mission time would be considered definitely unacceptable and a system with reliability higher than R_1 over its mission time would be considered definitely acceptable. Suppose R_S denotes the reliability of the system over its mission time. Then the two constraints to be met by any test plan are:

- $P(Plan\, rejects\, the\, system \mid System\, is\, good) \leq \alpha,$
- $P(Plan\, accepts\, the\, system \mid System\, is\, bad) \leq \beta.$

In general, there may be many different test plans that meet the above constraints. Each such plan could have its own format (test procedure, acceptance rule, etc.) and associated total test cost, and ideally, one would like to select the plan that is the least expensive. Unfortunately, it is not possible to exhaustively list all procedures that will lead to feasible test plans. Therefore, one can take the approach of deciding in advance upon a plan format that is plausible for the specific system under consideration, and then searching for plan parameters which minimize total test costs. The test programs in general, can be conducted at various levels; for example, component-level, subsystem-level, and system level. The advantages of component-level tests are described in Rajgopal and Mazumdar (1993) and Rajgopal and Mazumdar (1995). System-level tests are generally more expensive and time consuming. Thus most often, component-level testing provides a superior alternative, as long as the system-level objectives are followed. More discussions on component-level test plans can be seen in (Easterling, Mazumdar, Spencer, & Diegert, 1991).

DIFFERENT TYPES OF CENSORING SCHEMES FOR LIFE TESTING

Observation will need to be censored if they are above certain pre-defined bounds. Different types of censoring schemes are available in literature. Some of the methods used for censoring in reliability test plans are discussed below:

Type-I Censoring

In type-I censoring, components (units) are tested for a predetermined duration of time. When a component fails before this fixed time it is replaced with an identical component and testing is continued. The total number of failures during this test period is noted. Let $T_1, T_2, \ldots, T_n$ be independent, identically distributed random variables each with cumulative distribution function *F*. Assume also that t_c is a fixed censoring time. Instead of observing $T_1, T_2, \ldots, T_n$, observe $Y_1, Y_2, \ldots, Y_n$, where

$$Y_i = \begin{cases} T_i, & if \quad T_i \le t_c \\ t_c, & if \quad T_i > t_c \end{cases}.$$

Type-II Censoring

In this procedure, the number of failures is predetermined and components (units) are tested till the required number of failures is achieved. The time duration between each failure is observed and noted. Let $r < n$ be fixed, and let $T_{(1)} < T_{(2)} < \ldots < T_{(n)}$ be the order statistics of $T_1, T_2, \ldots, T_n$. Observation ceases after *r-th* failure, so one can only observe $T_{(1)}, T_{(2)}, \ldots, T_{(n)}$. The full ordered observed sample can be written as:

$$Y_{(1)} = T_{(1)}$$

$$Y_{(2)} = T_{(2)}$$

$$\vdots$$

$$Y_{(r)} = T_{(r)}$$

$$Y_{(r+1)} = T_{(r)}$$

$$\vdots$$

$Y_{(n)} = T_{(r)}.$

(Miller, 2011).

Mixed Censoring

In mixed censoring, a predetermined number of units of a component are tested for a predetermined unit time. If a component survives the test for this unit time, then the component is classified as a successful unit; if a component fails the test during this unit time, then it is declared as a failed unit. In either case, a new component is immediately put on test and this is continued until the predetermined number of components is tested.

Progressive Type-II Censoring

In life testing experiments, data is often censored. Among the different censoring schemes, type-I and type-II are the most popular. Unfortunately, in these censoring schemes, it is not possible to withdraw live units during the experiment. A generalization of the classical type-II censoring scheme allows for the withdrawal of live units during experiment, and it is called the progressive type-II censoring scheme. In this scheme, *n* units are placed on a life test for observing a predetermined *m* number of failures. At the time of the first failure, R_1 of the remaining $n-1$ surviving units are randomly removed from the experiment. At the time of the second failure, R_2 of the remaining $n-2-R_1$ units are randomly removed from the experiment. Finally, at the time of the *m-th* failure, all the remaining units $R_m = n - m - R_1 - \ldots - R_{m-1}$ are removed from the experiment (Dey, Dey, & Kundu, 2016).

BRIEF REVIEW OF LITERATURE

The basic problem was first considered by Gal (1994). Gal studied a situation where a certain unacceptable reliability level, R_0, needs to be demonstrated at a specified confidence $1-\alpha$. He assumed exponential life distributions for components and derived a general solution procedure to compute the optimum component test times which minimize the total test cost while guaranteeing the probability requirement

$$P(accept\,the\,system\;|System\,is\,bad) \leq \alpha.$$

The system is accepted if and only if there is no component failure during the test.

Mazumdar (1997) extended Gal (1994) model to the situation where in addition to an unacceptable system reliability level, certain acceptable reliability level needs to be demonstrated at a specific confidence $1-\beta$. He added the constraint

$$P(reject\,the\,system\;|System\,is\,good) \leq \beta.$$

In the context of quality control, the significance levels α and β are also known respectively as consumer's and producer's risks. Mazumdar (1997) used the following decision rule to accept a system: "Test (with replacement) each component j for a total of t_j time units. Observe the number of failures for each component. If the total number of failures is less than a pre-assigned number, accept the system; otherwise reject it". This rule for accepting a system is referred to as the sum rule in the sequel. Using the sum rule, Mazumdar (1997) gave a solution for the optimum component test times for a series system. Within the framework of their formulations, both Gal (1994) and Mazumdar (1997) showed that for a series system, the optimum component test times are independent of component test costs, and are in fact identical. In all this work, it was assumed that no prior information is available about component reliabilities.

Altinel (1992) studied this problem first for a series system of k components when upper bounds on individual component failures are given as the prior information; he has shown that optimum component test times are not identical, and the use of such prior information also leads to reduced total test costs. He also developed a procedure to compute optimum component test times.

Altinel (1994) extended these results for a problem which is more general in two aspects. First of all, the only assumption he made on the system reliability function is that it is an increasing function of component reliabilities, or equivalently a decreasing function of component failure rates, which is true for all coherent systems. In addition, the deterministic a priori information on component failure rates can restrict their selection to any compact and non-empty subset of non-negative real numbers. This subset does not necessarily have to be in the form of simple upper bounds on the failure rates, which limits each of them with a closed real interval; it can be a polyhedral set, or a hyper-sphere, etc.

Different aspects of the test design problem for a series system have been addressed in Rajgopal and Mazumdar (1995), Mazumdar (1980), and Altinel (1992). Rajgopal and Mazumdar (1995) considered a series system consisting of components with unequal test costs and unknown constant failure rates with pre-assigned bounds on the failure rate. They derived minimum cost component test plans that guarantee specified values of producers' and consumers' risks on the system reliability. These plans are derived for the case in which the component failure times are exponentially distributed, as well as for the case in which they follow a Gamma distribution. The optimization problems are formulated and solved via standard mathematical programming techniques. Mazumdar (1980) considered the problem of optimizing component testing for a series system with redundant subsystems so as to minimize the total cost of testing with given constraints as: the probability of accepting the system must be less than α if the system is not good (that is, the reliability of the system is less than a pre-assigned unacceptable level), and the probability of rejecting the system must be less than β if the system is good (that is, the reliability of the system is greater than a pre-assigned acceptable level). This problem was solved previously for a series system under the sum rule acceptance criterion. This result is extended in Mazumdar (1980) to cover the more general case of a series system that has built in redundancy in that it consists of subsystems each of which contains several identical components in parallel. The optimum testing time obtained is shown to be independent of the individual component testing costs. Altinel (1992) developed an acceptance procedure based on component tests that guarantee certain type-I and type-II errors for a series system. Methods for computing the individual component test times which minimize total test cost and guarantee the probability requirements for this acceptance procedure have also been given. In the paper Altinel (1992), a priori deterministic information on individual component failure rates is available. Results have shown that it is profitable to use the given information, and optimum component test times are not necessarily equal: they depend on individual component test costs. They provide a procedure to compute optimum component test times. This procedure is based on the well known cutting plane idea and column generation technique. They also discuss the conditions for the existence of an optimum solution. Rajgopal, Mazumdar, and Savits (1994) derive some of the properties of the Poisson distribution that are not commonly known or used to construct certain class of component tests for verification of series system reliability. Further, Rajgopal, Mazumdar, and Majety (1999) address the need of development of test plans that explicitly consider the possibility of interface failures. This paper analyzes a series system to determine when testing should be performed on the system alone, on the components only, and on both, depending on test costs and interface reliabilities. Optimum test plans are derived by solving a two-stage mathematical program.

Parallel systems have also been addressed in Yan and Mazumdar (1987) and Rajgopal and Mazumdar (1988). Yan and Mazumdar (1987) considered a parallel system of n components with constant failure rates and the component testing procedure guarantee that the given consumer and producer risks are not exceeded. They give certain restrictions on the magnitude of the unknown failure rates for guaranteeing the producer risk. The component test procedures use type-I censoring and use decision rules based on the total number of component failures during the testing periods, the number of failures for each individual component, and the maximum likelihood estimate of system reliability. Rajgopal and Mazumdar (1988) considered the problem of acceptance testing for a parallel system of different components with constant failure rates. The components are individually tested and the tests are terminated as soon as a pre-assigned number of each component fails. This paper provides a criterion for accepting or rejecting the system based on the sum of the logarithms of the total times on test for each component. The critical level for the test statistic is chosen so as to guarantee that the specified consumer and producer risks on the system reliability are not exceeded. The use of this statistic makes the computation of these critical values much simpler as compared with that of a previously used statistic based on the product of the total times on test for each component. Several approximate procedures are considered for deriving these critical values. The paper also formulates the optimization problem for deriving the minimum cost component testing plans when a type-II censored component-test procedure is used for a parallel system.

Rajgopal and Mazumdar (1996) considered the problem of acceptance testing for a series system of n different components, each having an unknown, constant failure rate. Components are individually tested and the tests are terminated when a pre-assigned number of failures are observed for each component. The total time on test for each component is noted and a statistic is constructed by using the observed test times and the number of failures of the different components; the statistic is based on the MLE of system reliability. This statistic based on component test times is then used in specifying a decision rule for accepting or rejecting the entire system. The design of the test plan is stated as an optimization problem which minimizes test costs while ensuring that specified consumer and producer risks on the system reliability are not exceeded. Numerical examples are provided and implications of the test plan are discussed.

Altinel, Özekici, and Feyzioğlu (2001) considered the component testing problem of a series system with redundant subsystems, where all components fail exponentially. In this work, the component failure rates are not constant parameters, but they change in a dynamic fashion with respect to time. In this paper, the optimal component testing problem is formulated as a semi-infinite linear programming problem.

They also presented an algorithmic procedure to compute optimal test times based on the column generation technique. Altinel, Özekici, and Feyzioğlu (2002) considered the component testing of a series system in a random mission. In this paper, they have shown that the realistic model can be handled using available results in semi-infinite linear programming and difference of convex functions programming.

Vellaisamy and Kumar (2008) constructed a reliability test plan for parallel system with n independent components under type-II censoring with the assumption that the lifetime of each component follows an exponential distribution with a constant but unknown failure rate parameter. They derived optimal reliability test plans which ensure the usual probability requirements on system reliability. Further, they solved the associated nonlinear integer programming problem by a simple enumeration of integers over the feasible range and developed an algorithm to obtain integer solutions with minimum cost. Vellaisamy and Kumar (2010) constructed a reliability test plan for series system with n components under mixed censoring, a combination of both type-I and type-II censoring, with the assumption that the lifetime of each component follows an exponential distribution with a constant but unknown failure rate parameter. In their work, they have showed that optimal sample size is same for all components.

Most of the component level reliability test plans that are available in literature have been developed for series and parallel systems under the assumption that components have constant failure rates. Some works in which failure rate is not constant are Sabnis and Agnihothram (2006), Sabnis and Agnihothram (2007), and Bajeel and Kumar (2015). Sabnis and Agnihothram (2006) construct a test plan for a parallel system with the failure rates of the corresponding exponential distributions depending upon covariates such as room temperature, pressure etc. Unlike in the works existing in literature, component failure rates are explicitly expressed in terms of covariates via exponential relationship. They use the arithmetic-geometric mean inequality to develop the design. With the motivation of developing more realistic test plans Sabnis and Agnihothram (2007) construct reliability test plans for the components of series systems under the assumptions that the component lifetimes are independent and identically distributed exponential random variables. The failure rates of these exponential random variables depend on covariates, such as room temperature, humidity etc. through linear and exponential relationships. The test plans differ from the classical test plans in the sense that the former assume dependence on covariates through explicit relationships. It should be noted that as far as the construction of reliability test plans on component level is concerned, this is the first instance of the use of the penalty function methodology. Bajeel and Kumar (2015) considered a parallel system with n different components by assuming that lifetime of each component follows exponential distribution with failure rate as a parameter and depends upon k covariates such as temperature, pressure, humidity etc. through exponential relationships. Using data obtained through type-II censoring, they derived an optimal reliability test plan. In most of the reliability studies, acceptable reliability level (ARL) and unacceptable reliability level (URL) are considered as constants. Bajeel and Kumar (2015) defined a new strategy replacing ARL and URL, as acceptable reliability interval (ARI) and unacceptable reliability interval (URI) respectively. This will enhance the possibility of reducing the burden of rejection cost, and also enhance the possibility of accepting the good system (rejecting the bad system) when the estimated reliability belongs to half open intervals ARI (URI). They also use prior information available in the form of upper bounds on failure rates in their design and an integer optimization problem is formulated satisfying the usual probability requirements.

A brief summary of literature that is used for authors work is presented in the following table.

Table 1. Summary of literature

Authors	Contributions	Drawbacks
Gal (1974)	• Optimal reliability test plan is introduced.	• System is accepted if and only if there is no component failure during the test. • Only consumers risk is considered as a constraint.
Rajgopal and Mazumdar (1988)	• Considered a parallel system with components having constant failure rate. • Type-II censoring is used to generate data. • Acceptance rule is based on the sum of logarithms of the total times on test for each component. • Consumers' and producers' risks are considered as constraints.	• Failure rate is considered as constant. • No prior information is used to develop the design.
Rajgopal and Mazumdar (1997)	• Considered a parallel system with components having constant failure rate. • Type-I censoring is used to generate data. • Acceptance rule is based on the total number of failures. • Upper bounds on failure rates are used as priori information.	• Failure rate is considered as constant.
Sabnis and Agnihothram (2006)	• Considered a parallel system • Failure rate of the component depends on covariates. • Acceptance rule is based on the total number of failures. • Type-I censoring is used to generate data. • Used arithmetic-geometric mean inequality to develop the design.	• No prior information is used other than covariates.
Vellaisamy and Kumar (2008)	• Considered a parallel system with components having constant failure rate. • Type-II censoring is used to generate data.	• Failure rate is considered as constant.
Bajeel and Kumar (2015)	• Considered a parallel system • Failure rate of the component depends on covariates. • Type-II censoring is used to generate data. • Acceptable reliability interval and Unacceptable reliability interval are introduced.	• No prior information is used other than covariates and upper bounds on failure rates.

WHAT IS LACKING IN LITERATURE?

Nair and Sabnis (2002) developed a test plan for a series system by considering that the failure rate of the component follows Uniform distribution over an interval, while the underlying lifetime distribution is Exponential. All problems addressed so far in literature are classical in nature; in some cases the priori information in the form of upper bound is used and test plan is derived satisfying the usual probability requirements. In the literature, a pure Bayesian approach is not seen to be adopted to design optimal component reliability test plans for parallel systems under type-I censoring.

INTRODUCTION TO BAYESIAN CONCEPTS

There are broadly two approaches in statistics to make inferences by quantifying the uncertainty of the data generating process; they are the Bayesian Statistical Methods and Frequentist Statistical Methods. Both approaches strive to characterize the degree of uncertainty arising from the underlying physical processes and the data collection processes. The Frequentist approach to statistics considers the parameters to be fixed but unknown. Procedures are judged by how they perform in an infinite number of hypothetical repetitions of the experiments. The Bayesian approach to statistics allows the parameter to be considered as a random variable and the probabilities can be calculated for parameters as well as observations. Probabilities calculated for parameters are interpreted as "degree of belief" and can be subjective and the rules of probability are used to revise 'degree of beliefs' about the parameters based on the observed data.

To infer possible values of the parameter vector θ (or its function $g(\theta)$), assume that the uncertainty about θ before observing data can be expressed by a probability density $\pi\left(\theta\right)$. This density is called prior density. To learn about θ, a sample of observations $X = x_1,\ x_2,\ \ldots,\ x_n$ is collected assuming that the sample is generated by a conditional probability density $f(x \mid \theta)$. The goal is to update the uncertainty about θ in the light of the data X. That is, to make an attempt to find the conditional distribution of θ given X. The solution is provided by the Bayes' theorem. The conditional density of θ given x is

$$P\left(\theta|x\right) = \frac{f\left(x|\theta\right)\pi\left(\theta\right)}{\int f\left(x|\theta\right)\pi\left(\theta\right)d\theta} = \frac{f\left(x|\theta\right)\pi\left(\theta\right)}{m\left(x\right)}.$$

This density is called the posterior density. The expression that appears in the denominator $m\left(x\right)$ is often called the marginal density of the observation x. The posterior density is defined only for those x which satisfy $m\left(x\right)>0$. If a vector of response $X = (x_1, x_2, \ldots, x_n)$ is observed and if $f(X \mid \theta)$ denotes the joint sampling density of X given θ, then $f(x \mid \theta)$ can be simply replaced by $f(X \mid \theta)$. Then $p(\theta \mid X)$ can be written as

$$P\left(\theta|X\right) = \frac{f\left(X|\theta\right)\pi\left(\theta\right)}{m\left(X\right)},$$

where $m\left(X\right)$ denotes the joint marginal density of X. Suppose $f(X \mid \theta) = C\left(X\right)L\left(\theta;X\right)$ where $C\left(X\right)>0$ is a function of data X only, then $L\left(\theta;X\right)$ is called the likelihood function of θ. From Bayes theorem it is clear that the posterior density is determined completely by the posterior kernel function: $K\left(\theta;X\right)=L\left(\theta;X\right)k\left(\theta\right)$. The posterior density now becomes

$$P\left(\theta|X\right) = \frac{K\left(\theta;X\right)}{\int K\left(\theta;X\right)d\theta} = \frac{L\left(\theta;X\right)k\left(\theta\right)}{\int L\left(\theta;X\right)k\left(\theta\right)d\theta}.$$

Thus, the posterior inference can be made only by obtaining the prior kernel and the likelihood function (normalizing constants can safely be ignored). As the denominator of $\mathrm{P}\left(\theta|\mathrm{x}\right)$ does not depend on θ (but it does depend on X), the above equation is often expressed as the posterior, which is proportional to the likelihood times the prior kernel, $P\left(\theta|X\right) \propto L\left(\theta;X\right) \times k\left(\theta\right)$ (Bansal, 2007; Ross 2014; Rohatgi & Saleh, 2011).

HOW BAYESIAN CONCEPT IS USEFUL FOR LIFE TESTING?

The totality of evidence is made up of two components; data and prior information. The prior information may come from the data obtained from the past experiment or from the theoretical background of the decision maker about the data generating mechanism. It is important to consider the order in which they are to be incorporated in calculations to reach some inference or decision. The prior does not necessarily mean earlier in time. The prior information may be considered as any information other than the current data. The information which is updated in the light of information obtained by the current data is called the posterior information. Ample amount of failure data, which can be incorporated in to the design of optimal reliability test plans, can easily be obtained from industries and hence Bayesian approach can be adopted. Such an approach will bring out good estimates and cost effective optimal design plans. We introduce the following notations and assumptions:

NOTATIONS

n Number of components in the parallel system

λ_i Failure rate of component i

u_i Upper bound for λ_i

R Parallel system reliability for unit time period

k Quasi density parameter

L_i Total testing time for *i-th* component

C Total testing cost of the system

c_i Cost of testing the *i-th* component per unit time

α Pre-assigned producer risk

β Pre-assigned producer risk

T_{ij} Lifetime of *j-th* component of type *i*

L Likelihood function

X_i Total number of failures occurred before total testing time L_i

$E(X)$ Expected value of the random variable X

$\hat{}$ Estimate of a random variable; $\hat{R}$ is the estimate of R

d A constant to chosen in $(0,1)$

φ Natural logarithm of $\left(1-d\right)$

R_1 Acceptable reliability level; if the system reliability is larger than or equal to R_1, the system is considered satisfactory

R_0 Unacceptable reliability level; if the system reliability is less than or equal to R_0, the system is considered unsatisfactory

φ_1 Natural logarithm of $\left(1-R_1\right)$

φ_2 Natural logarithm of $\left(1-R_0\right)$

ASSUMPTIONS

1. The parallel system consists of n independent components.
2. The lifetime of *i-th* component follows exponential distribution with parameter λ_i.
3. Each λ_i has an upper bound u_i, which is known.
4. In Bayesian paradigm, each λ_i is treated as a random variable.
5. Density of λ_i is assumed as Quasi density given by $g(\lambda_i) = \frac{1}{\lambda_i^k},\ k \geq 1$.

PROBLEM FORMULATION

If a manufacturer produces a new product, then he should test the reliability of the product for a particular duration of time to make sure that the product will perform to the best as expected by the consumer. Fixing of the testing time based on prior information is one of the difficult tasks. The incorporation of priori information about the failure rate and its upper bounds, take a significant role for fixing the duration of testing time in the design of reliability test plans. In this chapter, the authors are considering the problem of testing the reliability of a highly reliable parallel system with n independent components under type-I censoring, where the *i-th* component has exponential lifetime with unknown parameter $\lambda_i, \lambda_i \leq u_i\ \forall\ i = 1, 2, \ldots, n$. Then the parallel system reliability for unit time period is given by

$$R = 1 - \prod_{i=1}^{n} \left(1 - e^{-\lambda_i}\right).$$

Since the system is highly reliable, the system reliability can be approximated as

$$R = 1 - \prod_{i=1}^{n} \left(1 - e^{-\lambda_i}\right) \simeq 1 - \prod_{i=1}^{n} \lambda_i.$$

Since the quasi-prior with $k>1$ give rise to intractable optimization problem which is difficult to solve in general set up. We consider the case $g(\lambda_i) = 1/\lambda_i$, a simple prior with $k = 1$ as the non-informative quasi-prior for λ_i. Let $L_i, 1 \leq i \leq n$ be the total testing time for the *i-th* component. Let c_i denote the cost of testing the *i-th* component per unit time. Then the aim is to find the time periods $L_i, 1 \leq i \leq n$ that minimize the total testing cost subjected to type-I and type-II error constraints. That is, the problem is to determine the optimum values of L_i, $1 \leq i \leq n$ by formulating the following optimization problem:

$$\text{Minimize } C = \sum_{i=1}^{n} c_i L_i$$

such that

$$P(\text{Accept the system} \mid \text{System is good}) \geq 1-\alpha,$$

$$P(\text{Accept the system} \mid \text{System is bad}) \leq \beta,$$

where $0 < \beta < 1-\alpha < 1$. Here α is usually referred to as producer's risk while the β is the consumer's risk.

Likelihood Function of Exponential Distribution Based on Type-I Censoring

Since T_{ij} follows Exponential distribution with failure rate λ_i, the p.d.f of T_{ij} is given by

$$f(t_{ij}) = \lambda_i e^{-\lambda_i t_{ij}}.$$

Then the likelihood function based on type-I censoring is given by

$$L = \prod_{j\in N} \left(\lambda_i e^{-\lambda_i t_{ij}}\right)^{\delta_j} \left(e^{-\lambda_i t_{ij}}\right)^{1-\delta_j} = \lambda_i^{X_i} e^{-\lambda_i L_i}, \text{ where } X_i = \sum_{j\in N} \delta_j \text{ and } L_i = \sum_{j\in N} t_{ij}. \text{ (Lawless, 1982).}$$

Posterior Distribution

Let $f(t_{ij}, \lambda_i)$, $g(\lambda_i|t_{ij})$ and $f(t_{ij} \mid \lambda_i)$ be the joint density of t_{ij} and λ_i, the conditional density of λ_i, given t_{ij}, and the conditional density of t_{ij}, given λ_i, respectively. Under the assumption that the marginal densities $m(t_{ij})$ and $g(\lambda_i)$ of t_{ij} and λ_i, respectively, satisfy the conditions required for the existence of conditional densities, we have $g(\lambda_i|t_{ij}) = \frac{f(t_{ij}, \lambda_i)}{m(t_{ij})}$ and $f(t_{ij}|\lambda_i) = \frac{f(t_{ij}, \lambda_i)}{g(\lambda_i)}$. Thus the posterior distribution of λ_i is given by

$$g(\lambda_i|t_{ij}) = \frac{g(\lambda_i)f(t_{ij} \mid \lambda_i)}{m(t_{ij})} = \frac{g(\lambda_i)f(t_{ij} \mid \lambda_i)}{\int_{\lambda_i=0}^{\infty} g(\lambda_i)f(t_{ij} \mid \lambda_i)d\lambda_i} = \frac{\lambda_i^{X_i} e^{-\lambda_i L_i} \frac{1}{\lambda_i}}{\int_0^{\infty} \lambda_i^{X_i} e^{-\lambda_i L_i} \frac{1}{\lambda_i} d\lambda_i} = \frac{L_i^{X_i}}{\Gamma X_i} \lambda_i^{X_i-1} e^{-\lambda_i L_i}.$$

(Bansal, 2007).

Bayesian Estimate of Failure Rate Using Squared Error Loss Function

$$E\left[\left(\lambda_i - \widehat{\lambda_i}\right)^2\right] = \int \left(\lambda_i - \widehat{\lambda_i}\right)^2 f\left(\lambda_i\right) d\lambda_i$$

Differentiate on $\widehat{\lambda_i}$ and equate to zero implies we have, $\widehat{\lambda_i} = E\left(\lambda_i\right)$. Now $E\left(\lambda_i\right)$ from posterior distribution is nothing but the mean of gamma random variable λ_i. That is, $\widehat{\lambda_i} = E\left(\lambda_i\right) = \frac{X_i}{L_i}$. (Balakrishnan, 1996).

Reliability Estimate of the System at Unit Period

The reliability estimate of the system reliability is obtained using the Bayesian estimator of failure rates.

$$\hat{R} = 1 - \prod_{i=1}^{n} \widehat{\lambda_i} = 1 - \prod_{i=1}^{n} \frac{X_i}{L_i}.$$

Acceptance Rule Based on Reliability Estimate

Accept the system if the estimate of the system reliability based on Bayesian estimator of λ_i, given by

$$\hat{R} = 1 - \prod_{i=1}^{n} \frac{X_i}{L_i},$$

is greater than or equal to some number d, where $d \in (0, 1)$. Then

$$\hat{R} \geq d \Rightarrow 1 - \prod_{i=1}^{n} \frac{X_i}{L_i} \geq d \Rightarrow \sum_{i=1}^{n} \ln\left(\frac{X_i}{L_i}\right) \leq \ln(1 - d).$$

Let $\varphi = \ln(1 - d)$. Then the acceptance rule can be written as

$$\sum_{i=1}^{n} \ln\left(\frac{X_i}{L_i}\right) \leq \varphi.$$

Acceptable and Unacceptable Reliability Levels

A system is said to be satisfactory for unit time if R, the survival probability, is greater than or equal to R_1, the acceptable reliability level (ARL) and, it is said to be unsatisfactory if R is less than or equal to R_0, the unacceptable reliability level (URL), where R_0 and R_1 are constants such that $0 < R_0 < R_1 < 1$. Let $\varphi_1 = \ln(1 - R_1)$ and $\varphi_2 = \ln(1 - R_0)$, then we have the following relations:

$$R \geq R_1 \Rightarrow \prod_{i=1}^{n} \lambda_i \leq 1 - R_1 \Rightarrow \sum_{i=1}^{n} \ln \lambda_i \leq \varphi_1$$

$$R \leq R_0 \Rightarrow \prod_{i=1}^{n} \lambda_i \geq 1 - R_0 \Rightarrow \sum_{i=1}^{n} \ln \lambda_i \geq \varphi_2.$$

Normal Approximation of the Test Statistic Using Delta Method

Since the lifetime is exponentially distributed, the number of failures follows Poisson distribution with mean and variance $\lambda_i L_i$. Define $g(X_i) = \ln\left(\frac{X_i}{L_i}\right)$, then $g'(X_i) = \frac{1}{X_i}$. Then by Delta method $g(X_i)$ follows Normal distribution with mean $g(\mu) = \ln(\lambda_i)$ and variance $(g'(\mu))^2 \sigma^2(X_i) = \frac{1}{\lambda_i L_i}$. That is, $g(X_i) \sim N\left[\ln(\lambda_i), \frac{1}{\lambda_i L_i}\right]$. Now $\sum_{i=1}^{n} g(X_i) = \sum_{i=1}^{n} \ln\left(\frac{X_i}{L_i}\right)$, then by Lindeberg central limit theorem $\sum_{i=1}^{n} g(X_i)$ follows Normal distribution with mean $\sum_{i=1}^{n} \ln(\lambda_i)$ and variance $\sum_{i=1}^{n} \frac{1}{\lambda_i L_i}$. (Papanicolaou A, 2009).

OPTIMAL DESIGN OF THE PROBLEM

Let the cost of testing the *i-th* component per unit time be denoted by c_i. Then based on type-I censoring scheme, the total cost of testing is

$$C = \sum_{i=1}^{n} c_i L_i$$

Then the problem is to minimize C subjected to producer's risk and consumer's risk. That is, the aim is to find the optimum values of L_i from the optimization problem

$$\text{Minimize } C = \sum_{i=1}^{n} c_i L_i$$

such that

$$\mathrm{P}(\text{Accept the system} \mid \text{System is good}) \geq 1-\alpha, \tag{1}$$

$$\mathrm{P}(\text{Accept the system} \mid \text{System is bad}) \leq \beta. \tag{2}$$

Using the acceptance rule defined in the previous section, the constraints (1) and (2) can be written as

$$\mathrm{P}\left(\sum_{i=1}^{n} \ln\left(\frac{X_i}{L_i}\right) \leq \varphi \mid \sum_{i=1}^{n} \ln \lambda_i \leq \varphi_1,\ \lambda_i \leq u_i\ \forall\ i\right) \geq 1-\alpha, \tag{3}$$

$$\mathrm{P}\left(\sum_{i=1}^{n} \ln\left(\frac{X_i}{L_i}\right) \leq \varphi \mid \sum_{i=1}^{n} \ln \lambda_i \geq \varphi_2,\ \lambda_i \leq u_i\ \forall\ i\right) \leq \beta. \tag{4}$$

Then in terms of probability of acceptance, constraint (3) states that the probability of acceptance should be at least $1-\alpha$ for all combinations of λ_i values that satisfy the conditions

$$\sum_{i=1}^{n} \ln \lambda_i \leq \varphi_1, \lambda_i \leq u_i\ \forall i = 1,2,\ldots,n.$$

That is, the minimum probability of acceptance over all such λ_i should exceed $1-\alpha$. The constraint (4) states that the probability of acceptance should be at most β for all combinations of λ_i values that satisfy

$$\sum_{i=1}^{n} \ln \lambda_i \geq \varphi_2, \lambda_i \leq u_i\ \forall i = 1,2,\ldots,n.$$

That is, the maximum probability of acceptance over all such λ_i should not exceed β. Therefore constraints (3) and (4) can be rewritten as

$$\min_{\lambda_i} \mathrm{P}\left(\sum_{i=1}^{n} \ln\left(\frac{X_i}{L_i}\right) \leq \varphi \mid \sum_{i=1}^{n} \ln \lambda_i \leq \varphi_1,\ \lambda_i \leq u_i\ \forall\ i\right) \geq 1-\alpha, \tag{5}$$

$$\max_{\lambda_i} \mathrm{P}\left(\sum_{i=1}^{n} \ln\left(\frac{X_i}{L_i}\right) \leq \varphi \mid \sum_{i=1}^{n} \ln \lambda_i \geq \varphi_2,\ \lambda_i \leq u_i\ \forall\ i\right) \leq \beta. \tag{6}$$

The exact distribution of $\sum_{i=1}^{n} \ln\left(\frac{X_i}{L_i}\right)$ is not easy to obtain, and in order to obtain the tractable problem, we therefore need to approximate the distribution of $\sum_{i=1}^{n} \ln\left(\frac{X_i}{L_i}\right)$. Recall that $\sum_{i=1}^{n} \ln\left(\frac{X_i}{L_i}\right)$ has mean $\sum_{i=1}^{n} \ln\left(\lambda_i\right)$ and variance $\sum_{i=1}^{n} \frac{1}{\lambda_i L_i}$. Then constraints (5) and (6) become

$$\min_{\lambda_i} P\left(Z \leq \frac{\varphi - \sum_{i=1}^{n} \ln\left(\lambda_i\right)}{\sqrt{\sum_{i=1}^{n} \frac{1}{\lambda_i L_i}}} \middle| \sum_{i=1}^{n} \ln \lambda_i \leq \varphi_1,\ \lambda_i \leq u_i\ \forall\ i\right) \geq 1 - \alpha, \tag{7}$$

$$\max_{\lambda_i} P\left(Z \leq \frac{\varphi - \sum_{i=1}^{n} \ln\left(\lambda_i\right)}{\sqrt{\sum_{i=1}^{n} \frac{1}{\lambda_i L_i}}} \middle| \sum_{i=1}^{n} \ln \lambda_i \geq \varphi_2,\ \lambda_i \leq u_i\ \forall\ i\right) \leq \beta. \tag{8}$$

Since the cumulative distribution function of standard normal random variable is strictly increasing function in its arguments, the constraints (7) and (8) can be written as

$$\min_{\lambda_i} \left(\frac{\varphi - \sum_{i=1}^{n} \ln\left(\lambda_i\right)}{\sqrt{\sum_{i=1}^{n} \frac{1}{\lambda_i L_i}}} \middle| \sum_{i=1}^{n} \ln \lambda_i \leq \varphi_1,\ \lambda_i \leq u_i\ \forall\ i\right) \geq Z_{1-\alpha}, \tag{9}$$

$$\max_{\lambda_i} \left(\frac{\varphi - \sum_{i=1}^{n} \ln\left(\lambda_i\right)}{\sqrt{\sum_{i=1}^{n} \frac{1}{\lambda_i L_i}}} \middle| \sum_{i=1}^{n} \ln \lambda_i \geq \varphi_2,\ \lambda_i \leq u_i\ \forall\ i\right) \leq Z_{\beta}. \tag{10}$$

$Z_{1-\alpha}$ and Z_β are strictly positive and negative respectively for all values of $\alpha,\ \beta < 0.5$, hence $\varphi \in \left[\ln R_0, \ln R_1\right]$. Now consider the left hand side of the constraint (9)

$$\min_{\lambda_i} \frac{\varphi - \sum_{i=1}^{n} \ln(\lambda_i)}{\sqrt{\sum_{i=1}^{n} \frac{1}{\lambda_i L_i}}}$$

such that

$$\sum_{i=1}^{n} \ln \lambda_i \le \varphi_1$$

$$\lambda_i \le u_i \ \forall \ i = 1, 2, \ldots, n.$$

Clearly, this will attain the optimum when $\sum_{i=1}^{n} \ln \lambda_i = \varphi_1$, then this optimization problem can be rewritten as

$$\min_{\lambda_i} \frac{\varphi - \varphi_1}{\sqrt{\sum_{i=1}^{n} \frac{1}{\lambda_i L_i}}}$$

such that

$$\sum_{i=1}^{n} \ln \lambda_i = \varphi_1$$

$$\lambda_i \le u_i \ \forall \ i = 1, 2, \ldots, n.$$

Since the numerator is a positive and independent of λ_i, to minimize the objective function, it is enough to maximize the denominator. That is,

$$\max_{\lambda_i} \sum_{i=1}^{n} \frac{1}{\lambda_i L_i}$$

such that

$$\sum_{i=1}^{n} \ln \lambda_i = \varphi_1$$

$\lambda_i \le u_i \ \forall \ i = 1, 2, \ldots, n.$

This optimization problem can be rewritten as

$$\max_{\lambda_i} \sum_{i=1}^{n} \frac{1}{e^{\ln \lambda_i} L_i}$$

such that

$$\sum_{i=1}^{n} \ln \lambda_i = \varphi_1$$

$\ln \lambda_i \le \ln u_i \ \forall \ i = 1, 2, \ldots, n.$

This is an optimization problem in $\ln \lambda_i$. Define

$$\vartheta 1_i = \begin{cases} \ln u_i & if\, i \ne j \\ \varphi_1 - \sum_{i \ne j} \ln u_i & if\, i = j \end{cases} \quad \text{for } j = 1, 2, \ldots, n.$$

Then it is clear that by assuming feasibility, the optimum solution to above two maximization problem will be at any one of these $\vartheta 1_i$'s, let it be $\vartheta 1_i^*$. Then the constraint (9) can be written as

$$\frac{\varphi - \varphi_1}{\sqrt{\sum_{i=1}^{n} \frac{1}{e^{\vartheta 1_i^*} L_i}}} \ge Z_{1-\alpha}$$

That is,

$$\sum_{i=1}^{n} \frac{1}{e^{\vartheta 1_i^*} L_i} \le \left(\frac{\varphi - \varphi_1}{Z_{1-\alpha}} \right)^2 \tag{11}$$

Similarly by defining

$$\vartheta 2_i = \begin{cases} \ln u_i & if\, i \ne j \\ \varphi_2 - \sum_{i \ne j} \ln u_i & if\, i = j \end{cases} \quad \text{for } j = 1, 2, \ldots, n.$$

Then it is clear that by assuming feasibility, the optimum solution to above two maximization problem will be at any one of these $\vartheta 2_i$'s, let it be $\vartheta 2_i^*$, and then the constraint (10) can be rewritten as

$$\sum_{i=1}^{n} \frac{1}{e^{\vartheta 2_i^*} L_i} \leq \left(\frac{\varphi - \varphi_2}{Z_\beta} \right)^2 \tag{12}$$

Now for $\varphi \in \left[\ln R_0, \ln R_1\right]$, the optimal design can be written as follows:

$$\text{Minimize } C = \sum_{i=1}^{n} c_i L_i$$

such that

$$\sum_{i=1}^{n} \frac{1}{e^{\vartheta 1_i^*} L_i} \leq \left(\frac{\varphi - \varphi_1}{Z_{1-\alpha}} \right)^2$$

$$\sum_{i=1}^{n} \frac{1}{e^{\vartheta 2_i^*} L_i} \leq \left(\frac{\varphi - \varphi_2}{Z_\beta} \right)^2$$

This is a convex programming problem, and can be solved easily.

NUMERICAL RESULTS

The method presented in the previous section is illustrated below with the help of numerical computation. Let the number of components in a parallel system be 3 and the cost vector $c = \left(1,\ 1.5, 0.75\right)$. Then the following Table 2 gives the optimal test times and total testing cost for different inputs.

Now, let us consider a two component parallel system with cost vector denoted by $c = \left(c_1, c_2\right)$ and upper bound vector of failure rate denoted by $u = \left(u_1,\ u_2\right)$. Corresponding to the inputs $c = \left(1,\ 1.5\right)$, $u = \left(0.2, 0.15\right)$, $R_0 = 0.97$, $R_1 = 0.999$, $\alpha = 0.1$ and $\beta = 0.1$, the optimum test times are $L_1 = 16.0918$ and $L_2 = 1.68004$. The total testing cost C and optimum value of φ respectively are 18.6119 units and -0.0304592.

Rajgopal and Mazumdar (1997) obtained the component reliability test plan for a highly reliable parallel system under type-I censoring with optimum test times 77.12,102.82 and corresponding testing cost was 231.35 units for the data $n = 2$, $c = \left(1,\ 1.5\right)$, $u = \left(0.2, 0.15\right)$, $R_0 = 0.98$, $R_1 = 0.999$, $\alpha = 0.05$ and $\beta = 0.05$. For the same data our Bayesian plan gives the optimum test times $L_1 = 26.5852$ units

Table 2. Numerical examples for a three component parallel system

α	β	R_0	R_1	u_1	u_2	u_3	L_1	L_2	L_3	C	$-\varphi$
0.1	0.1	0.95	0.9999	0.2	0.12	0.15	1.27057	1.26737	15.6436	14.9044	0.0512933
0.1	0.1	0.99	0.9999	0.2	0.12	0.15	1.20486	1.27003	15.8918	15.0288	0.0100503
0.1	0.1	0.95	0.999	0.12	0.9	0.15	2.6345	2.03911	2.87663	7.85064	0.0512933
0.1	0.1	0.95	0.999	0.1	0.2	0.3	1.92282	1.90522	2.51101	6.66391	0.0512933
0.001	0.001	0.95	0.999	0.1	0.2	0.3	11.1802	11.0778	14.6002	38.747	0.0512933
0.001	0.001	0.95	0.9999	0.1	0.2	0.3	9.26874	5.35131	78.9137	76.481	0.0100503

and $L_2 = 2.7756$ units. The total testing cost C is 30.7486 units and optimum value of $\varphi = -0.0202027$. On Comparison of these two results, the proposed method has about 86% savings in total testing costs.

The results are generated by running Visual C++ and LINGO11 in tandem. The programming is done in Visual C++, within which LINGO11 is called whenever an optimization required.

FUTURE RESEARCH DIRECTIONS

This chapter uses the non-informative quasi- prior ($k = 1$). For values of $k > 1$, solving the optimization problem considered in this chapter is an open problem. Also one can use different priors, and lifetime distributions for arriving at a more complicated optimization problem, and the same is not addressed in literature.

CONCLUSION

In this chapter, the designing of an optimal reliability test plan for a parallel system with a failure rate as random variable having quasi-density is discussed in detail. The data are obtained through type-I censoring scheme, and the reliability estimator is obtained by estimating a Bayesian estimator of failure rate. Some numerical examples are also computed to illustrate the Bayesian approach of estimating system reliability and thereby to test the system reliability.

REFERENCES

Altinel, I. K. (1992). The design of optimum component test plans in the demonstration of a series system reliability. *Computational Statistics & Data Analysis*, *14*(3), 281–292. doi:10.1016/0167-9473(92)90039-I

Altinel, I. K. (1994). The design of optimum component test plans in the demonstration of system reliability. *European Journal of Operational Research*, *78*(3), 318–333. doi:10.1016/0377-2217(94)90043-4

Altinel, İ. K., & Özekici, S. (1998). Optimum component test plans for systems with dependent components. *European Journal of Operational Research*, *111*(1), 175–186. doi:10.1016/S0377-2217(97)00355-X

Altinel, İ. K., Özekici, S., & Feyzioğlu, O. (2001). Dynamic component testing of a series system with redundant subsystems. *IIE Transactions*, *33*(12), 1093–1108. doi:10.1080/07408170108936899

Altinel, I. K., Özekici, S., & Feyzioğlu, O. (2002). Component testing of a series system in a random mission. *Reliability Engineering & System Safety*, *78*(1), 33–43. doi:10.1016/S0951-8320(02)00047-9

Bajeel, P. N., & Kumar, M. (2015). A component reliability test plan for a parallel system with failure rate as the exponential function of covariates. *Journal MESA*, *6*(2), 177-190.

Balakrishnan, K. (1996). *Exponential distribution: theory, methods and applications*. CRC Press.

Bansal, A. K. (2007). *Bayesian parametric inference*. Narosa Publishing House.

Dey, T., Dey, S., & Kundu, D. (2016). On progressively type-II censored two-parameter rayleigh distribution. *Communications in Statistics. Simulation and Computation*, *45*(2), 438–455. doi:10.1080/03610918.2013.856921

Easterling, R. G., Mazumdar, M., Spencer, F. W., & Diegert, K. V. (1991). System-based component-test plans and operating characteristics: Binomial data. *Technometrics*, *33*(3), 287–298. doi:10.1080/00401706.1991.10484835

Gal, S. (1974). Optimal test design for reliability demonstration. *Operations Research*, *22*(6), 1236–1242. doi:10.1287/opre.22.6.1236

Handbook, M. (1987). Reliability Test Methods, Plans, and Environments for Engineering Development, Qualification, and Production. MIL-HDBK-781.

Lawless, J. F. (2011). *Statistical models and methods for lifetime data* (Vol. 362). John Wiley & Sons.

Mazumdar, M. (1977). An optimum procedure for component testing in the demonstration of series system reliability. *Reliability. IEEE Transactions on*, *26*(5), 342–345.

Mazumdar, M. (1980). An optimum component testing procedure for a series system with redundant subsystems. *Technometrics*, *22*(1), 23–27. doi:10.2307/1268379

Miller, R. G. Jr. (2011). *Survival analysis* (Vol. 66). John Wiley & Sons.

Naikan, V. A. (2008). *Reliability Engineering and life testing*. PHI Learning Pvt. Ltd.

Nair, J. H., & Sabnis, S. V. (2002). A reliability test-plan for series systems with components having stochastic failure rates. *Reliability. IEEE Transactions on*, *51*(1), 17–22.

Papanicolaou, A. (2009). *Taylor approximation and the delta method*. Retrieved May 3, 2015, from http://web.stanford.edu/class/cme308/OldWebsite/notes/TaylorAppDeltaMethod.pdf

Rajgopal, J., & Mazumdar, M. (1988). A type-II censored, log test time based, component-testing procedure for a parallel system. *Reliability. IEEE Transactions on*, *37*(4), 406–412.

Rajgopal, J., & Mazumdar, M. (1993). Design of system based component test plans.*Proceedings of 2nd Industrial Engineering Research Conference.*

Rajgopal, J., & Mazumdar, M. (1995). Designing component test plans for series system reliability via mathematical programming. *Technometrics*, *37*(2), 195–212. doi:10.1080/00401706.1995.10484304

Rajgopal, J., & Mazumdar, M. (1996). A system-based component test plan for a series system, with type-II censoring. *Reliability. IEEE Transactions on*, *45*(3), 375–378.

Rajgopal, J., & Mazumdar, M. (1997). Minimum cost component test plans for evaluating reliability of a highly reliable parallel system. *Naval Research Logistics*, *44*(5), 401–418. doi:10.1002/(SICI)1520-6750(199708)44:5<401::AID-NAV1>3.0.CO;2-8

Rajgopal, J., & Mazumdar, M. (1997). System-based component test plans for reliability inferences. *Frontiers in Reliability*, *1998*, 295–302.

Rajgopal, J., Mazumdar, M., & Majety, S. R. V. (1999). Optimum combined test plans for systems and components. *IIE Transactions*, *31*(6), 481–490. doi:10.1080/07408179908969851

Rajgopal, J., Mazumdar, M., & Savits, T. H. (1994). Some properties of the Poisson distribution with an application to reliability testing. *Probability in the Engineering and Informational Sciences*, *8*(03), 345–354. doi:10.1017/S0269964800003454

Rohatgi, V. K., & Saleh, A. M. E. (2011). *An introduction to probability and statistics* (Vol. 910). John Wiley & Sons.

Ross, S. M. (2014). *Introduction to probability and statistics for engineers and scientists*. Academic Press.

Sabnis, S. V., & Agnihothram, G. (2006). Application of arithmetic-geometric mean inequality for construction of reliability test plan for parallel systems in the presence of covariates. *Economic Quality Control*, *21*(2), 219–230. doi:10.1515/EQC.2006.219

Sabnis, S. V., & Agnihothram, G. (2007). Reliability test plans for series systems in the presence of covariates. *Economic Quality Control*, *22*(2), 197–209. doi:10.1515/EQC.2007.197

Vellaisamy, P., & Kumar, M. (2008). Optimal component test plans for a parallel system based on type-II censoring. *Statistical Methodology*, *5*(5), 454–461. doi:10.1016/j.stamet.2007.11.001

Vellaisamy, P., & Kumar, M. (2010). Design of optimal reliability test plan for a series system based on mixed-censoring. *Journal of Applied Statistical Science*, *17*, 1–10.

Yan, J. H., & Mazumdar, M. (1986). A comparison of several component-testing plans for a series system. *Reliability. IEEE Transactions on*, *35*(4), 437–443.

Yan, J. H., & Mazumdar, M. (1987). A comparison of several component-testing plans for a parallel system. *Reliability. IEEE Transactions on*, *36*(4), 419–424.

Chapter 8
Reliability Analysis of Engineering Systems:
An Accelerated Life Testing for Boiler Tubes

Taha-Hossein Hejazi
Sadjad University of Technology, Iran

Hossein Poursabbagh
Ferdowsi University of Mashhad, Iran

ABSTRACT

Nowadays, reliability and reliability-based design in any system has become very important. There are some parts in many systems that designing their test process take too much time and money in order to be analyzed for their reliability. Boilers, as the hearts of plants, contain various components. One of the most important components is Super-heater tubes. Creep failure is the most common failure mode in these tubes which grows faster by increasing the temperature. Accelerated Life Testing (ALT) is used to study the performance of a component under high stress in a short period of time and use of the test result in order to assess the performance of the component under design stress in a desired lifetime. In this chapter, the reliability of Super-heater boiler tubes in plants was examined by applying ALTs. The obtained results indicated that considering the life of components at the beginning of the test, reusing them under the same stress did not exhibit desirable reliability. Furthermore, some recommendations have been suggested to improve the reliability.

INTRODUCTION

The reliability of a plant is the most basic concern in the power industry. The reliability of the boilers is usually calculated by using functional data analysis in the long term or by utilizing short-term testing and extrapolating them for long term functions (Suik & Pihu, 2009).

DOI: 10.4018/978-1-5225-1639-2.ch008

Capital cost, power consumption, and availability are all necessary considerations in the economic optimization of power-generation equipment design. The capital and running costs of the boiler are relatively small compared to the plant for which it is an auxiliary. However, such plant depends implicitly for its revenue-earning capacity on the reliability of the boiler. Awareness of these overall economic considerations has led to the maximum availability design concept for large boiler (Weldon, 1964-1970).

Boilers are essentially shell-and-tube heat exchangers, with water being boiled to steam or transformed to high pressure hot water. Heat is generated by burning a fossil fuel; the combustion gas can be ducted around tubes containing the water (water-tube boiler), or the hot gas can pass inside the tubes with the water being in the shell (fire tube boiler) (Hall, 2012).

Boilers are designed to produce the steam for plants. They act as a heart of the plants and are composed of many components and parts. Therefore, in case of its failure for whatever reason, the plant would completely shut down. Steam is produced in some components and in the other components, the temperature are increased for further use.

The components studied in this chapter were the tubes, that the produced steam was directed through them. Their temperature was increased under extreme heat. The section containing Super-heater and tubes are called Super-heater tubes.

These tubes were made of seamless Molybdenum Chrome by the method of forging[1]. The purpose of this project was to design various ALTs and analyze and assess the reliability of boiler tubes. The chapter was organized in a way that the principles and generalities of ALT will be discussed in the next the section. In Section 3, the particular issue will be discussed along with the scope of the research, the system performance and the work condition. Method for designing the ALTs as well as the environment requirements for the tests were included in Section 4. Finally, analysis of the exam data, assessment, and reliability analysis will be presented in the last section.

A plant in Australia is discussed in this study to illustrate the economic value of boiler tubes. The program to reduce the failures in boiler tubes in NSW was arranged in 1988 and continued to work ceaselessly. It was a huge success and reduced the unavailability from %15 to less than %1.

When you discuss availability requirements with a user or project leader, he will invariably tell you that 100 percent availability is required: “Our project is so important that we can’t have any downtime at all.” But the tune usually changes when the project leader finds out how much 100 percent availability would cost. Then the discussion becomes a matter of money, and more of a negotiation process (Marcus & Stern, 2003).

At its simplest level, availability, whether high, low, or in between, is a measure of the time that a server is functioning normally. In other word, Availability is the probability of finding system on a required condition for future use (Roy & Ronald, 1983).

$$A = \left(T_a - T_{po-}T_{uo}\right) / \left(T_a - T_{po}\right) \quad (1)$$

where

T_a is the total time period of availability.
T_{po} is the total break time when the plant is out of service due to expected maintenance.
T_{uo} is the total break time when the plant is out of service due to unexpected maintenance.

The condition of boiler tubes in the plant was similar to the one in this study. The nominal condition of steam for stations was at the temperature of 540°C and the pressure of 16MPa. The cost of failure in boiler tubes includes three parts: the cost of maintenance, the cost of applying oil to prepare the unit for operation, and the cost of the reduced production. Depending on the extent of failure and the difficulty of access to the failed section, the cost of maintenance usually ranges from $10,000 to $100,000. The unavailability costs $26,000 per each boiler and the price of utilizing operating oil costs about $200,000. However, estimating the damage of reduced production is almost impossible.

In the literature, extensive research has been conducted on the optimal design of ALT plans by nelson (2005) following the work done by Chernoff (1962). Nelson and Kielpinski obtained the optimal plans and the best traditional plans for estimating the medians of the normal and lognormal distributions respectively under constant-stress ALT (1976). Meeker and Hahn considered the optimal allocation of test units to accelerated stress conditions with the objective of minimizing the estimate of the product reliability under the design stress condition (1977). Nelson and Meeker provided the optimal test plans to estimate the percentiles of the Weibull and extreme-value distributions at a specified design stress (1978). Meeker compared the statistically optimal test plans to a more practical test plan involving three stress levels (1984). Nelson reviewed statistically optimal and compromise plans for the single stress ALT planning problem (1990). Meeter and Meeker extended the maximum likelihood theory for planning ALT to the nonconstant scale parameter model and presented test plans for a large range of practical tests (1994). Yang proposed an optimal design of 4-level constant-stress ALT plans considering different censoring times (1994). Regarding the step-stress ALT, Miller and Nelson presented a simple SSALT plans assuming exponential life distribution (1983). Khamis proposed an optimal m-step SSALT design with k stress variables, assuming complete knowledge of the stress–life relation with multiple stress variables (1997). Bai and Chun obtained the optimal simple step-stress ALT with competing causes of failure (1991). Chung and Bai considered the optimal design of step-stress ALT in which a cumulative exposure model for the effect of changing stress is assumed (1998). Bai and Kim presented an optimal simple step-stress accelerated life test for the Weibull distribution under Type I censoring. A log-linear relationship and the cumulative exposure (CE) model were assumed, and nomographs were used to find the optimal plan (1993). Khamis and Higgins considered quadratic stress–life relation and derived the optimum three-step SSALT for the exponentially distributed Type I censored data (1996). Xiong addressed the statistical inferences on the parameters of a simple step-stress ALT model with Type II censoring (1998). Yeo and Tang derived a simple SSALT for an optimum hold time under low stress and an optimum low stress level by taking the target acceleration factor into consideration. These studies were based on the assumption that the failure time follows exponential distribution because of its simplicity. However, the assumption of exponential time to failure may not be applicable in many cases. Due to the flexibility of the Weibull distribution, it is more reasonable to assume that the failure time distribution of the test units is Weibull (1999). Khamis and Higgins proposed a time transformation of the exponential CE model for analyzing Weibull SSALT data, which made the computation of Weibull model less complicated (1998). Li and Fard developed a SSALT for two stress variables considering censored Weibull distributed failure data (2007).

Xiong and Milliken studied the statistical models in step-stress ALT when the stress change times are random (1999). Xiong and Ji studied the optimal design of a simple step-stress test plan involving grouped and censored data (2004).

Reliability prediction in a product design and during the developmental testing process is an important factor. In order to accurately estimate the product reliability, the test design criterion is defined to minimize the AV of the reliability estimate at a time under normal operating condition. A few studies have focused on the reliability estimates as a test criterion (2009). Elsayed and Zhang determined the optimum simple step-stress ALT plans based on Cox's proportional hazards model with a linear baseline hazard function. The optimization criterion was used to minimize the AV of the reliability function over a pre-specified period of time at a design stress level (2004).

Accelerated Life Testing

One of the three basic methods of accelerated testing is the Accelerated Life Testing (ALT). The other tastings include Reliability Enhancement Testing (RET) and Environmental Stress Screening (ESS). Accelerated testing consists of a variety of test methods for shortening the life of products or hastening the degradation of their performance. The aim of such testing is to quickly obtain data which, properly modeled and analyzed, yield desired information on product life or performance under normal use. Such testing saves much time and money(Nelson, 2008). ALTs are used to determine the service life of the components, substance, and the processes used in the production. Alts are not applied to detect errors and failures but to identify and reduce failures and failure mechanisms that lead to permanent breakdowns at the end of their service life (Condra, 1994). Carrying out ALTs requires spending a long time in order to reveal the service life of a product. Tastings may sometimes take forever, therefore accelerated stress are being applied to decrease the required time. The ALTs are the most expensive and time-consuming methods among the above methods. The tests should be designed meticulously so that the results of the tests would be used generally for any products.

Accelerated tests have a number of intrinsic problems which have to be considered both when designing the trial and its analysis and when estimating the significance and limitations of the results. There is need for a proven procedure to relate the degree of degradation with the level of the degrading agent and time. This is more critical with tests for performance prediction or design data but in all cases some understanding of the degradation process in relation to degree of acceleration is necessary. Any relation between degradation, acceleration and time may only be valid for a limited range of acceleration i.e. the degradation mechanisms may change with level of agent.

It is not generally easy to simulate the service environment accurately as there may be a number of agents acting simultaneously, the level of agent may vary with time and its exact nature may be difficult to match.

If the accelerated tests do not include all the degradation agents present in service, procedures are needed to model any synergistic or simply superimposed effects.

These problems form the basic limitation of accelerated tests simply because the requirements of relationships between degree of acceleration and time and between degree of acceleration and degradation mechanisms, the interaction of degradation agents and the validity of test conditions in relation to service are not generally known with sufficient accuracy or certainty.

They are hence the reason why the vast majority of accelerated durability trials are really only comparative and cannot be used for prediction with confidence.

A further difficulty is that the degree of effect of a degradation agent will be different for different properties or parameters measured. Hence it is necessary to know the critical parameter for any given application in advance of designing the trial (Brown et al., 1995).

Also there are some other limitations of accelerated life tests:

- Determination of acceleration factors is very complex and cost and time prohibitive.
- Test equipment which includes automated test monitoring may be too complex to be affordable or manageable.
- Some means of test acceleration may not be attainable because of large thermal masses of the tested items or because of limited stress rating.
- In components testing, usually the curves are constructed based on times to failures, and those are used for determination of test acceleration and for information on components reliability. When the components are small and have failed catastrophically (burned or greatly changed physical properties), it often is not possible to determine in which failure mode they did fail, therefore, the results may be fitted with the wrong distributions, resulting in the wrong reliability information.
- The results of quantifying reliability through acceleration may not always be predictive on an individual product since it may operate at different stress levels than was tested (BS EN 62506:2013).

Accelerated test data can be divided into two types. Namely, the product characteristic of interest is life or some other measure of performance, such as tensile, strength or ductility. Such data are described below:

1. **Performance Data:** One may be interested in how product performance degrades with age. In such performance testing, specimens are aged under high stress, and their performance measured at different ages. Such performance data are analyzed by fitting a degradation model to the data to estimate the relationship between performance, age, and stress.
2. **Life Data:** The proper analysis of life data depends on the type of data.
3. **Complete Data:** consist of the exact life (failure age) of each sample unit. Figure 1 depicts a complete sample from a single test condition.
4. **Censored Data:** Often when life data are analyzed, some units are unfailed, and their failure times are known only to be beyond their present running times as shown in Figure 2. Such data are said to be censored on the right. Similarly, a failure time known only to be before a certain time is said to be censored on the left.

Figure 1. Complete data

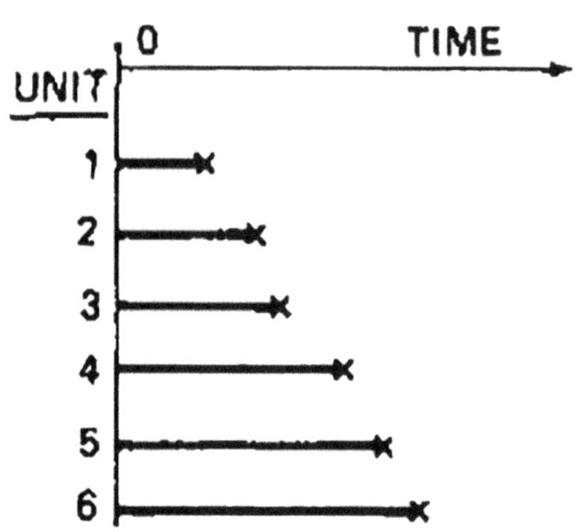

Figure 2. Censored data

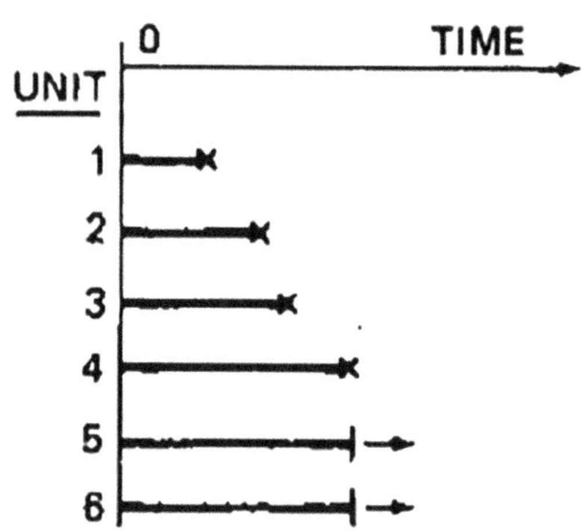

5. **Fully Censored Data:** A fully censored data is the type of data that is most often seen in "standardized" testing schemes for validating designs and processes. A set number of parts exposed to a defined environment for a set period of time. If no parts fail, then the parts pass. If any parts fail before the end of the defined time, the parts fail (Porter, 2003).

The stress loading in an accelerated test can be applied various ways. They include constant, cyclic, step, progressive, and random stress loading. Descriptions of some loadings follow.

- **Constant Stress:** The most common stress loading is constant stress. Each specimen is run at a constant stress level. Figure 3 depicts a constant stress test with three stress levels. At the highest level, all four specimens ran to failure. At the middle level, four ran to failure, and one was unfailed. At the lowest level, four ran to failure, and four were unfailed.
- **Step Stress:** In step-stress loading, a specimen is subjected to successively higher levels of stress. A specimen is first subjected to a specified constant stress for a specified length of time. If it does not fail, it is subjected to a higher stress level for a specified time. The stress on a specimen is thus in- creased step by step until it fails. Figure 4 depicts two such patterns. Such data may be

Figure 3. Constant stress test

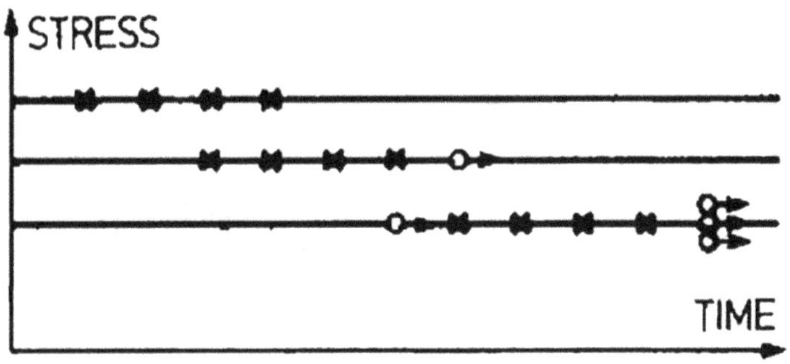

Figure 4. Step-stress test

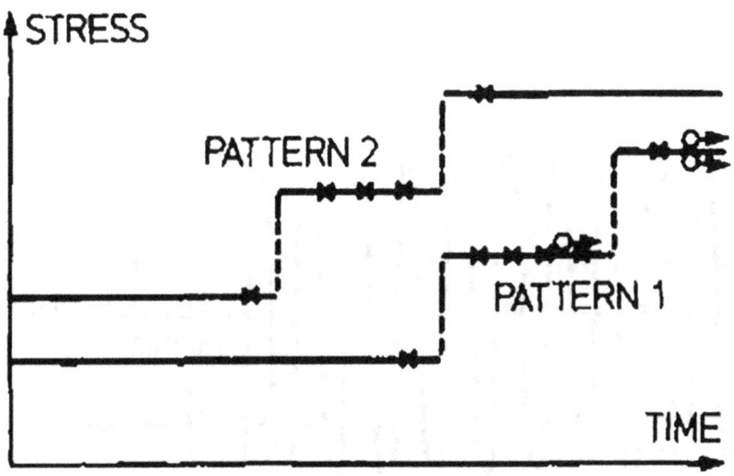

censored. Pattern 1 has six failures and three runouts. The step stress test is an accelerated test, not because it gets the same information in a shorter period of time, but because it gets several times the information in only twice the time (Porter, 2003).

- **Progressive Stress:** In progressive stress loading, a specimen undergoes a continuously increasing level of stress. Different groups of specimens may undergo different progressive stress patterns. Figure 5 depicts such a test with three patterns each a linearly increasing stress.
- **Cyclic Stress:** In use, some products repeatedly undergo a cyclic stress loading. For example, insulation under ac voltage sees a sinusoidal stress. A cyclic stress test for such a product repeatedly loads a specimen with the same stress pattern at high stress levels. Figure 6 depicts a cyclic stress test.
- **Random Stress:** Some products in use undergo randomly changing levels of stress, as depicted in Figure 7.

Figure 5. Progressive stress test

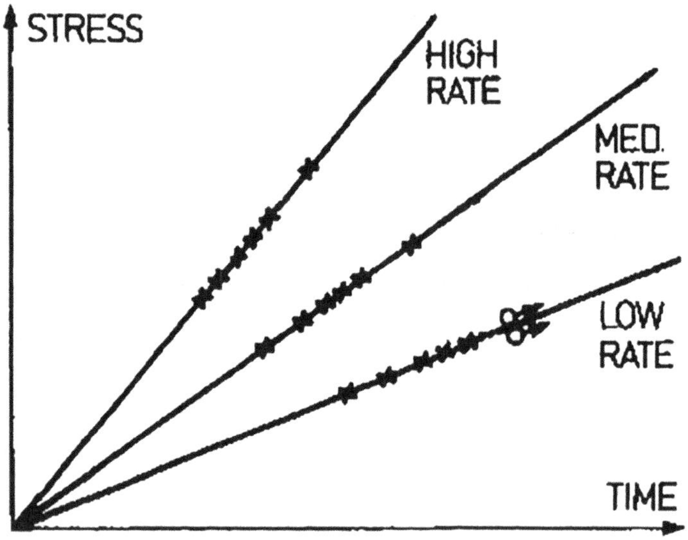

Figure 6. Cyclic-stress loading

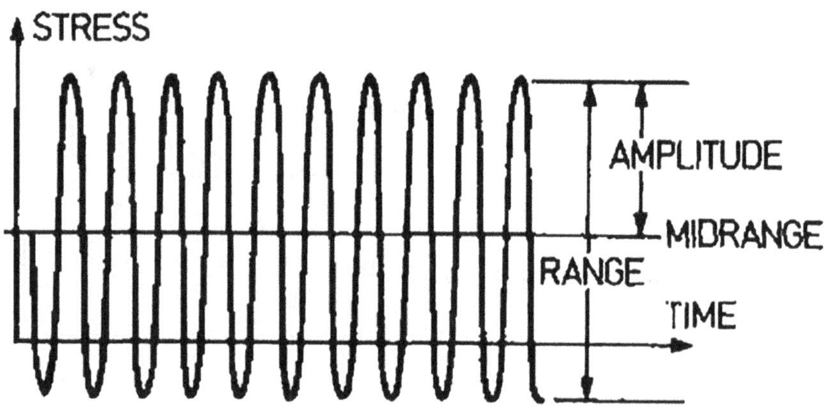

Figure 7. Random stress loading

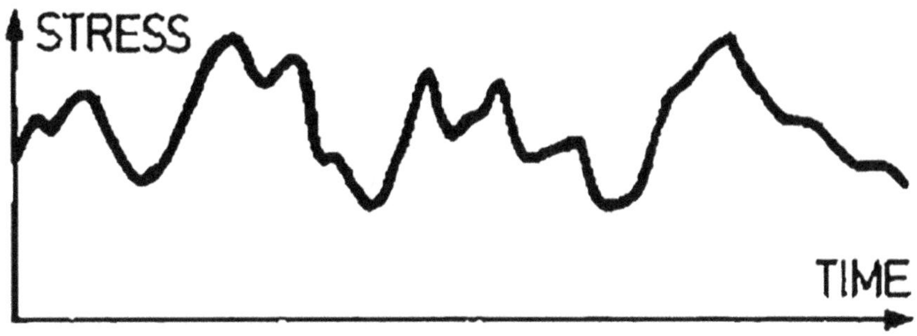

This method is not suitable for all types of failure mechanisms. Dasgupta and Petch classified the failure mechanisms into four types:

1. **Stress Strength:** When the stress of the component is more than the yield stress.
2. **Damage Endurance:** Although the stress exerted upon a component is smaller than the yield stress, recurring this type of stress and focus of breakdown in a particular spot leads to failure.
3. **Challenge Response:** Some defects remain to dwell in the component for a long time without being identified. They are only revealed under certain circumstances.
4. **Tolerance Requirements:** While the above failures occur suddenly, this type of failure occurs because of the gradual erosion during performance (Dasgupta & Pecht, 1991).

The principle for applying AlTs is that the behaviour of a component under short time and high stress experiment must be similar to its behaviour in a long term with low stress testing. Most of the failures occur under a time distribution upon a stress level where the most important types of distributions are Weibull, Log Normal, Rayleigh, and Extreme Value.

Description of the Sample System

Super-heater Tubes are a critical part of plants boilers. The produced steam is directed into super-heater tubes in boilers. Then, when it is exposed to extreme heat, its temperature gets increased. The components and parts of a boiler are illustrated in Figure 8. The area numbered 10 shows the super-heater tubes.

Affective Factors on the drop in performance or the existing stress include:

- External corrosion caused by being exposed to smoke and burning products
- Internal corrosion caused by being exposed to steam or additive substances
- Abrasion
- High Temperature
- The Steam Pressure in the Tube (Ameri at al., 2000)

The Affective Stress Mechanism on the Performance of Boiler Tube

As it was mentioned earlier, the most important role of super-heater boiler tubes is to increase the temperature of the steam. The tubes must show suitable strength and diameter and have to be well sealed.

Figure 8. Components of a plant boiler

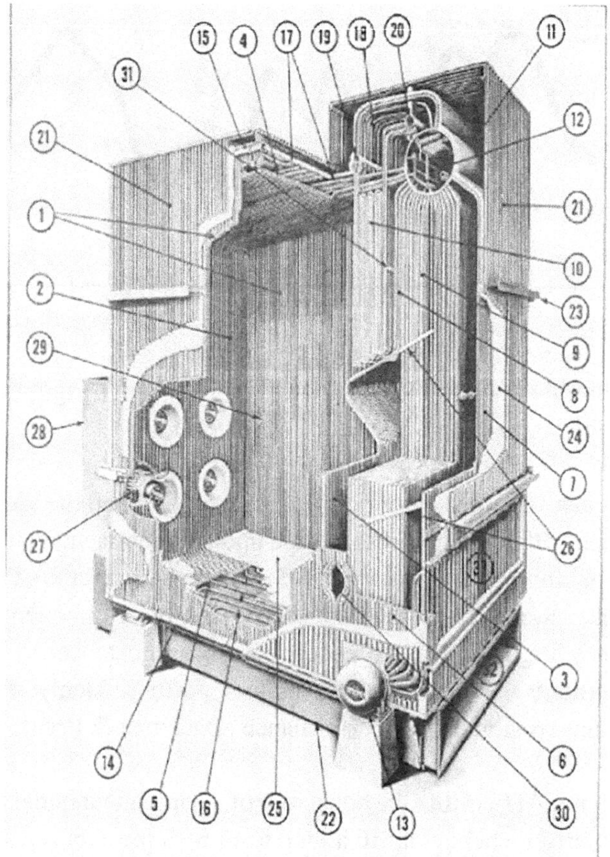

Each of the above factors can make changes in the mechanical or metallurgic properties or the micro structure of the substance and lead to rupture or puncture in the tube. As a result, the pressure and the temperature of the steam would not reach to the desired levels and a drop in performance would be observed which is called Tube Failure[5].

- Erosion, external and internal corrosion decrease the diameter and the strength of the tubes which at first leads to performance drop and gradually deviates the tubes from their functionality.
- If the tubes will be exposed to high temperature constantly, the strength of the tubes diminishes as well as their sealing.
- Moreover, if the steam pressure in the tubes would get larger than the assigned level, it may dominate the strength of the tubes and deform or puncture them.

The only stress examined in this test, among all the types of stress described earlier, was the temperature stress variable and the rest of them remained steady. During normal performance, the temperature of boiler tubes varies from 300°C to 600°C. Two types of stress are exerted upon this part: the pressure stress caused by the steam inside the tube and the temperature which is resulted from the heating outside the tube. In ALTs, the steam pressure is held steady while the temperature is incremented gradually.

The increment action is designed according to the defined allowable working temperature for each type of tubes. If the tube is made of Molybdenum Chrome, for example, which its phase changes at the temperature of 720°C so its allowable working temperature is 720°C, then the maximum temperature is assigned as 700 °C and it is incremented by 10. (The temperature is increased 10 by 10 from the normal working temperature until it reaches the maximum working temperature such as 700°C for thee Molybdenum Chrome tubes.)

Condition Leading to a Component Failure

Because reliability deals with the probability that a given item—a component, a subsystem, a system, etc.—will fulfill its intended function over time, we must be clear about what a product's intended function is and when a failure occurs: A failure is said to have occurred when one or more intended functions of a product are no longer fulfilled to the customer's satisfaction.

The definition of a failure is obvious when there is a total loss of product function. If something breaks during product usage, and this was not intended, then we all agree that it has failed. However, this will not be the case if only a partial loss of function is involved. In such instances the very definition of the term ''failure'' may not be precise when one observes a gradual or intermittent loss of performance over time (Wasserman, 2002).

Where a specific tube suffers from leakage under any circumstance, it is called a broken part. As it was mentioned earlier, it is important to note that various factors may lead to failure in super-heater tubes but the only failure mode for them is the tube leakage. Therefore, there is one failure mode in the test and the result of all the types of affective stress in performance drop impedes the transmission of steam in the tubes and decreases the boiler efficiency. If a large number of tubes suffer from the same problem, boiler goes out of service and eventually the plant shuts down.

The normal condition for the performance of boiler tubes ranges from 300°C to 600°C. Since these tubes were not exposed to the temperature of 600°C which was their maximum working temperature, it was only suitable to exert stress with a temperature of a little higher than 600°C for the ALTs. Thus, the upper and lower bound were determined as 700°C and 630°C respectively.

Designing Accelerated Life Tests

As it was mentioned, the most critical failure mode in boiler tubes is creep in high temperature. Furthermore, the type of the stress in this failure mode was the temperature and the inner pressure so the temperature was assumed to be the accelerated stress in the test. The pressure was assumed to remain steady and it was determined according to each type of the tubes.

Regarding the working allowable temperature of the tubes which was 300°C - 600°C, the temperature for the beginning of the accelerated test was assigned 630°C for each of the for tubes.

Table 1 shows the data of accelerated testing of the component A, Table 2 shows data of component B, Table 3 shows data of component C and Table 4 shows data of component D.

As it is illustrated, all the data for the test is comprehensive and the entire components encountered failure and breakdown.

Table 1. The data of accelerated testing of the component A

Stress Level	Stress (MPa)	Temperature (Kelvin)	Rupture Time (Hour)
Upper	40	923	196
Medium	40	943	134
Medium	40	953	82
lower	40	963	31

Table 2. The data of accelerated testing of the component B

Stress Level	Temperature (Kelvin)	Stress (MPa)	Rupture Time under Oxidation Condition (Hour)	Rupture Time under Corrected Condition (Hour)
Upper	963	21	50	120
Medium	953	21	95	205
Medium	943	21	160	300
Medium	923	21	420	749
Medium	913	21	730	1300
lower	903	21	1150	1900

Table 3. The data of accelerated testing of the component C

Stress Level	Temperature (Kelvin)	Stress (MPa)	Rupture Time under Oxidation Condition (Hour)	Rupture Time under Corrected Condition (Hour)
Upper	963	29	51	111
Medium	943	29	94	202
Medium	923	29	340	631
Medium	913	29	725	1297
lower	903	29	912	1630

Table 4. The data of accelerated testing of the component D

Rupture Time under Oxidation Condition (Hour)	Rupture Time under Corrected Condition (Hour)	Temperature (Kelvin)	Stress (MPa)	Stress Level
28	12	963	25	Medium
61	30	943	25	Medium
149	72	923	25	Medium
451	225	913	25	lower
780	449	903	25	Upper

Experimental Conditions

The life service of every parts and components utilized in this test was recorded in the table prior to the test. It is obvious the older the components, the sooner they approach to their corrosion and wear period.

The test also took place under the condition of 510°C and pressure of 8.62 MPa. Table 5 shows the service life of components at the beginning of the test.

Figure 9 shows Dimension of the Sample Component.

Ideal Life Model and Lifetime-Stress Model for Testing

A statistical model for an ALT is composed of a distribution for lifetime and a lifetime-stress relationship. Lifetime data related to steady stress is represented as variables dependent on stress level. However, taking use of scatter diagrams does not present a thorough view of the effect of stress on lifetime. Therefore, we'd better take use of the lifetime distribution at a stress level. As a result, we look up for models that not only reveal the relationship between lifetime and stress but also explain the lifetime distribution. The followings are examples of the lifetime-stress models: Arrhenius, Inverse Power-law, exponential, power-exponential, polynomial, Eyring, Cox, and etc.

Table 5. The service life of components at the beginning of the test

Component Title	Service Life Before the Beginning of the Test (Hour)
Tube A	166000
Tube B	166000
Tube C	166000
Tube D	166000

Figure 9. A profile of the dimensions of the component

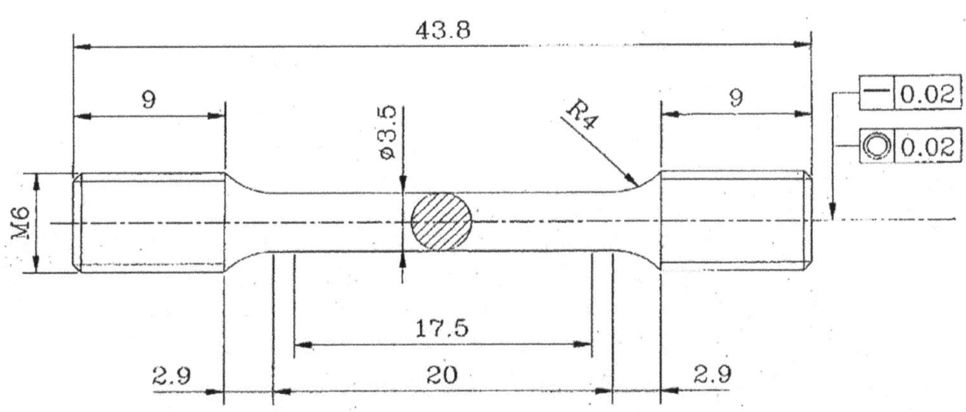

INVERSE POWER-LAW MODELS

The Coffin-Manson model described by Caruso and Dasgupta (1998) is a representative example of Power-Law Models for accelerated testing. The required number of cycles at acceleration level g_2, $N(g_2)$, that corresponds to an equivalent amount of damage to $N(g_1)$ cycles at acceleration level g_1 is

$$N\left(g_2\right) \;=\; N\left(g_1\right)\;\left(g_1 \,/\, g_2\right)^{\beta} \tag{2}$$

where

β= exponent based on material and damage mechanism.

The exponent, beta, represents a fatigue characteristic of a particular material, as derived from the slope of the S-N curve for that material. As noted by Caruso and Dasgupta, beta is a function of both the material and the damage mechanism (Caruso & Dasgupta, 1998).

Arrhenius Equation

We would expect that as temperature is increased, a reaction that is affected by temperature, such as a chemical reaction, will occur more rapidly. The Arrhenius equation embodies this idea and is named after S. A. Arrhenius (1859–1927). The equation when applied to accelerated testing is given by

$$\mu\left(s\right) = \log(\mathrm{R0}) + \frac{Ea}{k\left(S+273\right)} \tag{3}$$

with R_0 denoting the rate constant, E_a is the activation energy, k is Boltzmann's constant, S is the stress (temperature in degrees Celsius in the Arrhenius equation), and μ(S) is a function of the time to failure at stress S. Equation 3 does not by itself give the acceleration rate, which is obtained by using two temperatures, S_1 and S_2. That rate must be solved for by using Equation 2, and doing so produces

$$a\left(S2,S1\right) = \exp\left[\left(\frac{Ea}{k}\right)\left(\frac{1}{(S1+273} - \frac{1}{(S2+273)}\right)\right] \tag{4}$$

Regarding computation, Section 8.4.2.2 of the NIST/SEMATECH e-Handbook of Statistical Methods shows how to use JMP for Arrhenius parameter estimation (Ryan, 2007).

The Cox Model

Let

$$x\;\left(0\right) = \;\left(x_1\left(0\right),\ldots,\; x_m\left(0\right)\right)^T \tag{5}$$

be a possibly time-varying and multidimensional explanatory variable; here $x_1(0),\ldots, x_m(0)$ are one-variate explanatory variables.

Under the Cox model the hazard rate under x(0) is

$$\alpha_{x(0)}\left(t\right) = r\left\{x\left(t\right)\right\} \alpha\left(t\right) \tag{6}$$

The function r is parametrized in the following form:

$$r(x) = e^{(\beta^T z)} \tag{7}$$

where $\beta = (\beta_{1,\ldots,} \beta_m)^T$ is a vector of unknown parameters and

$$z = \left(z_{1,\ldots,} z_m\right)^T = \left(\varphi_1\left(x\right)_{,\ldots,} \varphi_m\left(x\right)\right)^T \tag{8}$$

is a vector of specified functions φ_i . In what follows we use the same notation x_j for $z_j = \varphi_j(x)$.

The parameterized Cox model has the form

$$\alpha_{x(0)}\left(t\right) = e^{\beta^T x(t)} \pm \left(\mathrm{t}\right) \tag{9}$$

The baseline hazard rate function α(t) is supposed to be unknown.

Eyring Model

The Eyring model comprises a general model for incorporating both temperature and one or more additional stress variables. It is based on a theoretical derivation on chemical reaction rate theory (kinetics) and quantum mechanics. In this derivation the term DH is used in chemical kinetics to denote the amount of energy needed to start a diffusion or migration process. The Eyring model equation is of the form (Wasserman, 2002).

$$t_{0.50} = A.T^{\alpha} e^{\Delta H/kT}_{temp.term} . e^{B+(C/T)S_1}_{stress\ term, S_1} .. e^{B+(E/T)S_2}_{stress\ term, S_2} \cdots \tag{10}$$

By Barriga et al. a Bayesian method based on MCMC for ALT under EW distribution (for life time) and Arrhenius models (relating the stress variable and parameters) was proposed. By the predictive density of a future observation, a procedure was developed to plan ALT and also to verify if the conformance fraction of the manufactured process reaches some desired level of quality (Barriga et al., 2008).

In order to acquire the best lifetime distribution as well as the best model for life-stress distribution, it was necessary to consider various modes for lifetime relationship and lifetime-stress relationship in Minitab software and draw the diagrams for each of them. The software presented an Anderson-Darling (AD) statistic for each of Lifetime-Stress and Lifetime distributions. The Anderson–Darling test is a statistical test of whether a given sample of data is drawn from a given probability distribution. In its basic form, the test assumes that there are no parameters to be estimated in the distribution being tested, in which case the test and its set of critical values is distribution-free. However, the test is most often used in contexts where a family of distributions is being tested, in which case the parameters of that family need to be estimated and account must be taken of this in adjusting either the test-statistic or its critical values. An average of them was calculated by taking into account the different stress levels. The smallest AD value indicated the best lifetime distribution and the best lifetime-stress relationships. As it is illustrated in the following table, it was deduced that Log logistics and Weibull distribution had the smallest AD value by comparing the values with linear lifetime-stress and lifetime distribution. In the following, the critical lifetime percentiles were calculated for the linear lifetime-stress model at first by using log logistics lifetime distribution and then by Weibull lifetime distribution. Finally, the value of reliability was obtained. Weibull lifetime distribution is more general and common in comparison to the other distributions according the recent studies. The AD value obtained by Minitab software also indicated that the Weibull distribution can be applied for lifetime distribution of the boiler tubes. As a result, Weibull lifetime distribution was selected and the rest of calculation was done by that. Table 6 shows AD Values for Lifetime Distribution and Lifetime-Stress Relationship.

Considering that simple models like Arrhenius and linear assume the slope of stress lines to be similar, each of the stress levels of the slopes were assimilated by using weighted mean. It is presented in the Figure 10.

The red points in the Figure 11 illustrate the life failure of the stress level of the sample. At this time, the percentiles of the lifetime for the design stress including 10, 20, 50, 63, and 90 were calculated. Design stress for boiler super-heater tubes was determined as about 560°C (833 Kelvin). Table 7 shows the critical lifetime percentile calculated by minitab software.

Assessment and Analysis of Reliability and the Other Factors

Since vital maintenance of boilers is usually taken care of every three years, the reliability is calculated for three years or 26780 hours of a desired lifetime at the design stress of 833Kelvin . The output of the software is presented in the table 8.

By placing values of Weibull parameters for an estimated desired lifetime of three years (26780 hours), the reliability was obtained by manual calculation:

$$R\left(t\right) = exp(- \ (t \ / \ \alpha)^{\beta}) \tag{11}$$

where

β was the shape parameter
α was the criterion parameter

For the sample, the above parameters was:

$\alpha = 40997$

Table 6. AD values for lifetime distribution and lifetime-stress relationship

		AD for Stress Level						AD* Average
Lifetime-Stress Relationship	**Lifetime Distribution**	**963**	**953**	**943**	**923**	**913**	**903**	
Arrhenius	*Smallest Extreme Value*	5.548	3.739	3.856	2.801	3.134	5.458	4.089
	Exponential	3.875	4.181	2.769	3.893	2.946	3.008	3.445
	Weibul	3.608	4.184	2.796	4.595	2.609	2.696	3.415
	Normal	6.517	4.554	3.419	2.736	3.046	4.003	4.0458
	Lognormal	3.602	4.148	3.048	4.063	2.747	2.895	3.417
	Logistic	6.584	4.808	3.326	2.725	2.979	3.88	4.050
	Loglogistic	6.584	4.808	3.326	2.725	2.979	3.88	4.050
Linear	*Smallest Extreme Value*	5.631	3.726	3.863	2.777	3.128	5.512	4.106
	Weibul	3.642	4.174	2.797	4.539	2.61	2.721	3.413
	Exponential	3.903	4.176	2.758	3.852	2.946	3.037	3.445
	Normal	6.593	4.545	3.433	2.724	3.043	4.032	4.062
	Lognormal	3.645	4.145	3.054	3.997	2.743	2.926	3.418
	Logistic	6.694	4.829	3.321	2.717	2.978	3.9	4.073
	Loglogistic	3.544	4.028	3.01	3.725	2.727	2.925	3.326
Inverse Temp	*Smallest Extreme Value*	5.548	3.739	3.856	2.801	3.134	5.458	4.089
	Exponential	3.875	4.181	2.769	3.893	2.946	3.008	3.445
	Weibul	3.608	4.184	2.796	4.595	2.609	2.696	3.415
	Normal	6.517	4.554	3.419	2.736	3.046	4.003	4.046
	Lognormal	3.602	4.148	3.048	4.063	2.747	2.895	3.417
	logistic	6.584	4.808	3.326	2.725	2.979	3.88	4.050
	loglogistic	3.512	4.038	3.006	3.781	2.731	2.893	3.327
ln(power)	*Smallest Extreme Value*	5.578	3.734	3.859	2.792	3.132	5.478	4.095
	Exponential	3.885	4.179	2.765	3.878	2.946	3.018	3.445
	Weibul	3.62	4.18	2.796	4.576	2.609	2.704	3.414
	Normal	6.543	4.551	3.424	2.731	3.045	4.013	4.051
	Lognormal	3.616	4.147	3.05	4.04	2.746	2.906	3.417
	logistic	6.623	4.816	3.324	2.722	2.979	3.887	4.058
	loglogistic	3.523	4.035	3.007	3.761	2.73	2.904	3.327

β=1.80283

$$R\left(t\right) = \ exp(-\ (26780\ /\ 40997)^{1.80283}) = 0.6287206 \quad (12)$$

which was exactly the same as the output of the software with precision of five decimals (Table 9).

Figure 10. Weibull lifetime distribution diagram and linear relation with stress after the slope assimilation

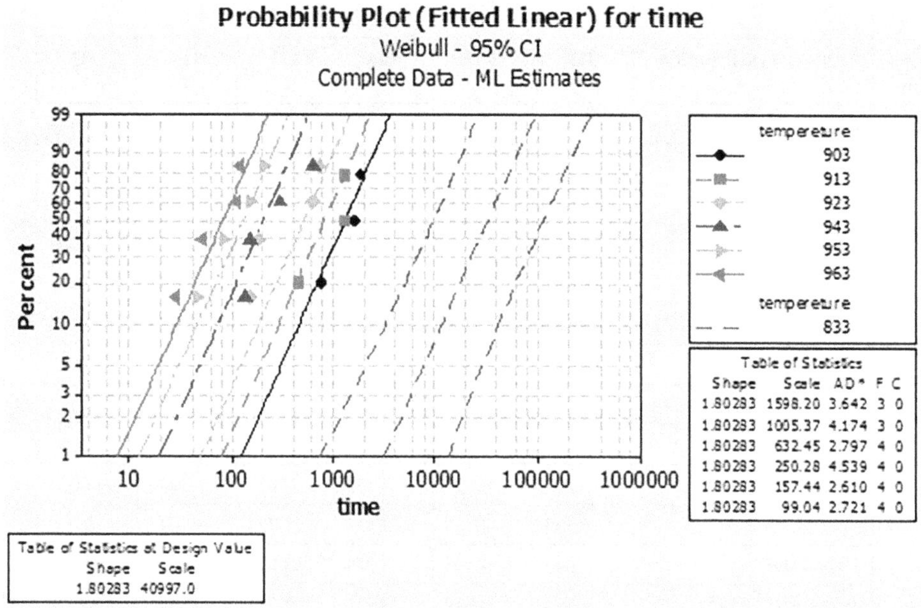

Figure 11. Linear lifetime-stress model diagram with Weibull lifetime distribution

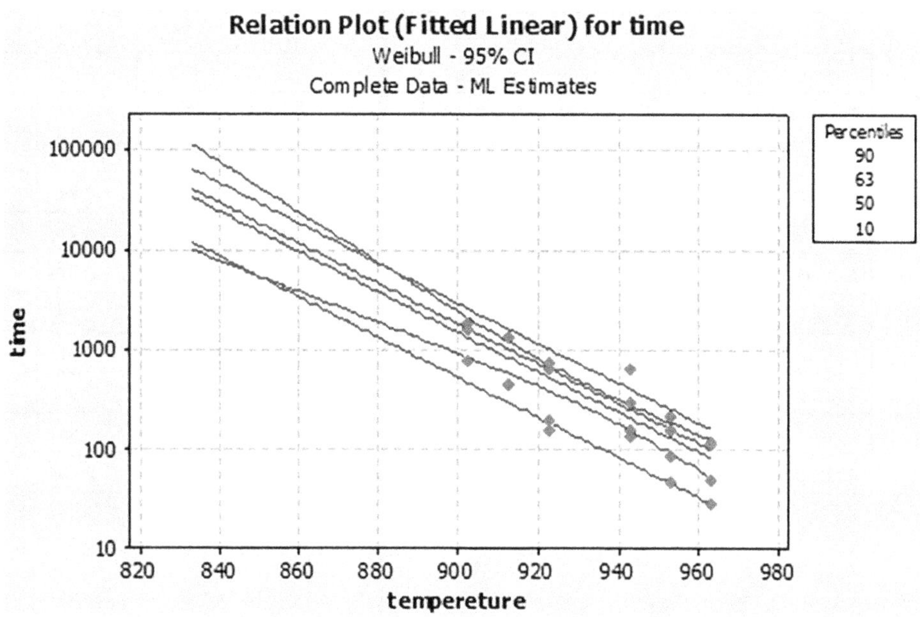

Table 7. Calculating the critical lifetime percentile by minitab software

Percent	Temperature	Per Each Percent	Error	Minimum	Maximum
10	833	11766.5	7699.53	3263.28	42426.8
20	833	17841.1	11373.7	5114.26	62238.9
50	833	33454.9	20909.9	9827.52	113888
63	833	40866.1	25519.5	12017.5	138967
90	833	65112.8	40998	18954.5	223677

Table 8. The obtained probability for the design stress in a desired lifetime

70.00%	Normal		CI	
Time	Temperature	Probability	Minimum	Maximum
26780	833	0.62872	0.22331	0.86619

Table 9. Verification of the direct calculations

0.6287200	Reliability calculated by Minitab software
0.6287206	Reliability calculated by manual calculation

Calculation of Reliability by Posterior Probability

Considering that the tubes had been used for 166000 hours before the beginning of the test, and the fact that no background of the tubes was presented to Minitab software, zero life in Minitab (or a trivial lifetime) was assumed to be equal to the preliminary function lifetime of the component. Therefore, the value of T was assigned zero and the rest of calculation was done.

If the component remains unbroken till T time, the success probability for a t time is equal to success in mission time divided to cumulative probability of a success from zero time to T lifetime.

$$R_W(T,t) = \frac{R(T+t)}{R(T)} \tag{13}$$

It was assumed that lifetime at the zero time in Minitab and in real time was not similar in this condition (Real time condition = 166000 hours and Minitab = 0). If the lifetime at the zero time would be regarded as trivial in comparison to the real preliminary lifetime (for example: 100 hours), then there is the following calculation:

R(T+t)=R(100+26870)=R(26970)= 0.624989

R(T)=R(100)= 0.99981

Rw(T,t)=0.625108

In order to improve reliability, the following approaches can be applied:

1. Using Parallel System

In parallel system, the probability of flawless performance of the system is easily gained by using the failure probability of each of the members, because the system will only fail when all the members stop working at the same time.

$$R_{sp}(t) = 1 - Q_{sp}(t) = 1 - [Q_1(t)Q_2(t)...Q_{n-1}(t)Q_n(t)] = \\ 1 - \{[1 - \mathrm{R}_1(t)][1 - \mathrm{R}_2(t)]...[1 - \mathrm{R}_n(t)]\} = 1 - \prod_{i=1}^{n} Q_i(t) = \\ 1 - \prod_{i=1}^{n} [1 - \mathrm{R}_i(t)] \quad (14)$$

Where the reliability is equal to:

R_{sp}(26780)=1-(1-0.62872)^2=0.862151

As it is shown, the reliability was improved from 0.62872 to 0.862151 by placing a parallel system with the super-heater tubes.

2. Avoid Using System at the Maximum Stress (Temperature) Constantly

To reach this goal, the stress level, for example, can be decreased by %10 (749 Kelvin would be considered instead of 833 Kelvin) and the reliability would be recalculated (Table 10).

As it is indicated, the reliability would be improved by %58 in this mode.

Table 10. Confidence interval for the reliability measurement

Reliability Probability Table				
95.0% Normal CI				
Time	Temperature	Probability	Minimum	Maximum
26780	749	0.999585	0.959916	1

Table 11. Reliability values vs. time

Time	1000	5000	10000	15000	20000	25000	30000	35000
Reliability	0.999	0.978	0.924	0.849	0.760	0.664	0.566	0.471
Time	40000	45000	50000	60000	70000	80000	90000	100000
Reliability	0.384	0.306	0.239	0.137	0.073	0.036	0.016	0.007

Figure 12. Reliability diagram of boiler super-heater tubes

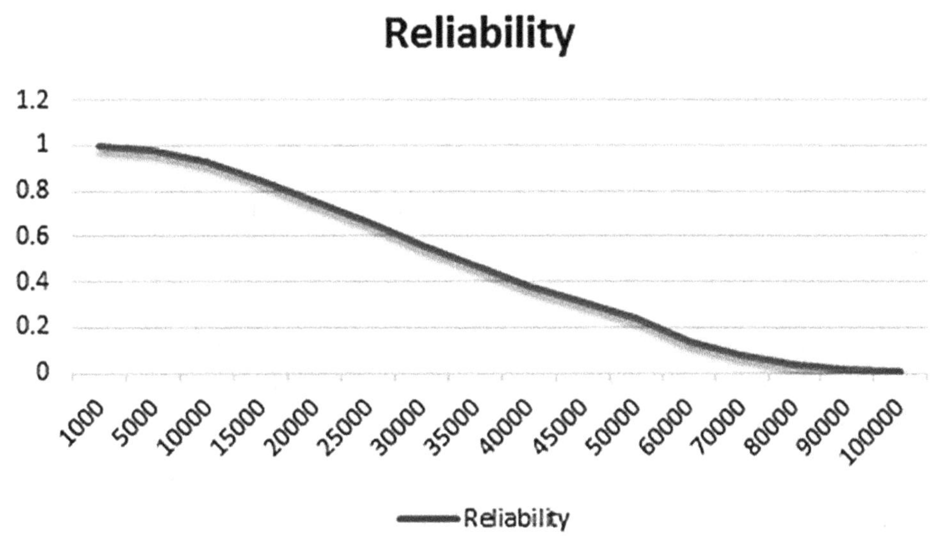

The reliability for different times can be obtained by the acquired relation (Table 11). Figure 12 shows the reliability diagram for boiler tubes.

CONCLUSION

In this chapter, the result of Accelerated Lifetime Testing for four boiler tubes along was studied considering temperature stress and creep failure mode. The AD value for various distributions was calculated by regarding the available temperature values in Minitab software. Since the AD values were so close and that Weibull distribution is accepted as general distribution, this type of distribution model was decided to be suitable for the lifetime factor. The Lifetime-Stress model was observed to be linear. As the boiler tubes are maintained every three years, the desired lifetime was estimated to be 26000 hours and the reliability was calculated for this lifetime at the working temperature of 833 Kelvin. Since the tubes had been working for 166000 hours before the testing which was far longer than the desired life of a boiler tube, their obtained reliability was at % 62.87 which indicated the critical condition of the tubes. Considering the great variation of performance with their desired lifetime, and in order to assess the wellbeing situation of the tube until the end of the test, using posterior probability did not reveal any particular change in the reliability.

Regarding the several number of tubes in plants, it is recommended to use parallel system to improve reliability. Calculation revealed that using a particular type of parallel system improved their reliability up to % 86.

Furthermore, if % 10 of the Maximum stress amount would be lessened and they will be used at the temperature of 749 Kelvin, the reliability would be increased by % 99.9585.

REFERENCES

Ameri, T. P. M., & Yousefi, M. Y. (2000). *Analysis of Affective Factors on Crisis Situation in Boilers of Steam Plants.* 15th International Conference Power Engineering Conference.

Bai, D. S., & Chun, Y. R. (1991). Optimum Simple Step-stress Accelerated Life Tests with Competing Causes of Failure. *IEEE Transactions on Reliability*, *40*(5), 622–627. doi:10.1109/24.106787

Bai, D. S., & Kim, M. S. (1993). Optimum simple step-stress accelerated life tests for Weibull distribution & Type-I censoring. *Naval Res. Logist.*, *40*(2), 193–210. doi:10.1002/1520-6750(199303)40:2<193::AID-NAV3220400205>3.0.CO;2-J

Barriga, G. D. C., Lee Ho, L., & Cancho, V. G. (2008). Planning accelerated life tests under Exponentiated-Weibull-Arrhenius model. *International Journal of Quality & Reliability Management*, *25*(6), 636–653. doi:10.1108/02656710810881926

Billinton, R., & Allan, R. N. (1983). *Reliability Evaluation of Engineering Systems.* New York: Springer. doi:10.1007/978-1-4615-7728-7

Brown, R. P., Kockott, D., Trubiroha, P., Ketola, W., & Shorthouse, J. (1995). *A Review of Accelerated Durability Tests.* VAMAS Technical Report No 18.

BSI Standards Publication. (2013). *Methods for product accelerated testing.* BS EN 62506.

Caruso, H., & Dasgupta, H. (1998). A Fundamental Overview of Accelerated Testing Analytical Models. *Journal of the IEST*, *41*(1), 16–20. doi:10.17764/jiet.41.1.3626j43p305q1978

Chernoff, H. (1962). Optimal Accelerated Life Designs for Estimation. *Technometrics*, *4*(4), 381–408. doi:10.1080/00401706.1962.10490020

Chung, S. W., & Bai, D. S. (1998). Optimal Designs of Simple Step-stress Accelerated Life Tests for Lognormal Lifetime Distributions. *International Journal of Reliability Quality and Safety Engineering*, *5*(04), 315–336. doi:10.1142/S0218539398000285

Condra, L. W. (1994). *Accelerated Testing for Product Reliability Assurance. Hanse Environmental Test Report HETR 94-001.* Hans Environmental. Inc. Hopkins.

Dasgupta, A., & Pecht, M. (1995). Material Failure Mechanisms & Damage Models. *IEEE Transactions on Reliability, 40*(5), 531-536.

Elsayed, E. A., & Zhang, H. (2004). Design of optimum simple step-stress accelerated life testing plans. In *Proceedings of the Second IIEC.*

Fard, N., & Li, C. (2009). Optimal simple step stress accelerated life test design for reliability prediction. *Journal of Statistical Planning and Inference*, *139*(5), 1799–1808. doi:10.1016/j.jspi.2008.05.046

Hall, S. (2012). *Rules of Thumb for Chemical Engineers. Elsevier Science.*

Khamis, I. H. (1997). Optimum M-step step-stress design with k stress variables. *Communications in Statistics. Simulation and Computation*, *26*(4), 1301–1313. doi:10.1080/03610919708813441

Khamis, I. H., & Higgins, J. J. (1996). Optimum 3-step Step-stress Tests. *IEEE Transactions on Reliability*, *45*(2), 341–345. doi:10.1109/24.510823

Khamis, I. H., & Higgins, J. J. (1998). A new model for step-stress testing. *IEEE Transactions on Reliability*, *47*(2), 131–134. doi:10.1109/24.722275

Li, C., & Fard, N. (2007). Optimum bivariate step-stress accelerated life test for censored data. *IEEE Transactions on Reliability*, *56*(1), 77–84. doi:10.1109/TR.2006.890897

Marcus, E., & Stern, H. (2003). *Blueprints for high availability*. John Wiley & Sons.

Meeker, W. Q. (1984). A Comparison of Accelerated Life Test Plans for Weibull & Lognormal Distributions & Type I Censoring. *Technometrics*, *26*(2), 157–171. doi:10.1080/00401706.1984.10487941

Meeker, W. Q., & Hahn, G. J. (1977). Asymptotically Optimum Over-stress tests to Estimate the Survival Probability at a Condition with a Low Expected Failure Probability. *Technometrics*, *19*(2), 381–399. doi:10.1080/00401706.1977.10489577

Meeter, C. A., & Meeker, W. Q. (1994). Optimum Accelerated Life Tests with a Nonconstant Sigma. *Technometrics*, *36*(27), 71–83.

Miller, R., & Nelson, W. B. (1983). Optimum Simple Step-stress Plans for Accelerated Life Testing. *IEEE Transactions on Reliability*, *32*(1), 59–65. doi:10.1109/TR.1983.5221475

Nelson, W. B. (2005). A Bibliography of Accelerated Test Plans. *IEEE Transactions on Reliability*, *54*(2), 194–197. doi:10.1109/TR.2005.847247

Nelson, W.B. (2008). *Accelerated Testing: Statistical Models, Test Plans, & Data Analysis*. John Wiley & Sons, Inc.

Nelson, W. B., & Kielpinski, T. J. (1976). Theory for Optimum Accelerated Life Tests for Normal & Lognormal Life Distributions. *Technometrics*, *18*(1), 105–114. doi:10.2307/1267923

Nelson, W. B., & Meeker, W. Q. (1978). Theory for Optimum Accelerated Censored Life Tests for Weibull & Extreme Value Distributions. *Technometrics*, *20*(2), 105–114. doi:10.1080/00401706.1978.10489643

Porter, A. (2004). *Accelerated Testing & Validation: Testing, Engineering, & Management Tools for Lean Development*. Elsevier.

Ryan, T. P. (2007). *Reliability Analysis & Life Testing, Modern Engineering Statistics*. John Wiley & Sons, Inc.

Small, R. F., & Wagner, C. (2003). Economic Analysis of Boiler Tube Failures in NSW. *OMMI*, *2*(2), 1–20.

Suik, H., & Pihu, T. (2009). *Warranty reliability of CFB boiler burning oil shale*. Academic Press.

Wasserman, G. S. (2002). *Reliability Verification, Testing, & Analysis in Engineering Design*. Marcel Dekker, Inc. doi:10.1201/9780203910443

Weldon, R. (1969, September). Paper 3: Boiler Feed Pump Design for Maximum Availability. In *Proceedings of the Institution of Mechanical Engineers,Conference Proceedings* (*Vol. 184*, No. 14, pp. 18-35). SAGE Publications. doi:10.1243/PIME_CONF_1969_184_421_02

Xiong, C. (1998). Inference on a Simple Step-stress Model with Type-II Censored Exponential Data. *IEEE Transactions on Reliability*, *47*(2), 142–146. doi:10.1109/24.722278

Xiong, C., & Ji, M. (2004). Analysis of Grouped & Censored Data from Step-stress Life Testing. *IEEE Transactions on Reliability*, *53*(1), 22–28. doi:10.1109/TR.2004.824832

Xiong, C., & Milliken, G. A. (1999). Step-stress Life-testing with R&om Stress-change Times for Exponential Data. *IEEE Transactions on Reliability*, *48*(2), 141–148. doi:10.1109/24.784272

Yang, G. B. (1994). Optimum Constant-stress Accelerated Life Test Plans. *IEEE Transactions on Reliability*, *43*(4), 575–581. doi:10.1109/24.370223

Yeo, K. P., & Tang, L. C. (1999). Planning step-stress life-test with a target acceleration-factor. *IEEE Transactions on Reliability*, *48*(1), 61–67. doi:10.1109/24.765928

Chapter 9
Monte Carlo Simulation for Reliability-Based Design of Automotive Complex Subsystems

Taha-Hossein Hejazi
Sadjad University of Technology, Iran

Amirmohsen Hejazi
Sahand University of Technology, Iran

ABSTRACT

Design and production of high reliable and safer systems with longer life has been a challenge because of high competitive market and recent safety issues of reputable car manufacturers. In this chapter, a methodology is introduced for reliability based design of automotive system. FMEA results are used in the process of failure rate estimation. The basic failure data are adjusted by multiplicative corrective factors to account for the design and environment impacts on system failure characteristics. The system is modeled by reliability block diagram (RBD) method, simulated by an efficient Monte Carlo method. According to the results of FMEA and reliability evaluation, the structure of system is improved by reducing the components failure rates and potential change of system configuration. The components' reliability is improved by increasing the quality of components by utilization of high quality materials and modern manufacturing techniques. The results showed the failure rate improvement for friction lining component in dry friction clutch system.

INTRODUCTION

Reliability is considered as an important characteristic of the products life performance. There are several purposes demanding on reliability improvement. This includes

1. Products become more complex; unless reliability target is not satisfied, the performance may become inadequate.

DOI: 10.4018/978-1-5225-1639-2.ch009

2. There exists a tough competition between the industrialized nations. Products with high reliability will finally survive.
3. There is a pressure to minimize waste as the world population grows as the resources keep on shrinking.
4. Product's high durability implies less environmental impacts (Misra, 2008).
5. Considering the warranty and callback costs, manufacturers motivates to invest more time and money into the designing phase which includes the reliability design of the product (Teixiera, 2007).

Design for reliability of automotive is an important research area, specifically in the early design phase. This requires proper knowledge about the potential failure modes and reliability characteristics of all relevant automotive parts and subsystems. The main objective of automotive design for reliability is to maximize the reliability and life time quality of products.

It is reported (Bhote & Bhote, 2004) that 80% of poor quality is caused by design. Over 90% of field failures are the result of poor design, and a high percentage of product recalls have their origin in design. Most law suits are filed on account of improper design and 70–75% of product costs are functions of design (Bhote & Bhote, 2004). Therefore, if there is any phase in the entire life cycle of a product that has maximum impact on reliability, it should be the design phase. It should be noted that reliability has found as a key product attribute in the costumer's point of view (see Figure 1). Nowadays there are an increasing number of callback reports in automotive industry due to lack of proper design which illustrates the importance of developing a reliability integration plan during the product designing phase.

In fact, reliability should be designed and built into products and systems at the earliest possible stages of product development. It is the most economical approach to minimize the lifecycle costs of the product or system. Better system reliability is achieved at much lower costs, otherwise because the majority of life-cycle costs are locked in phases other than design and development; one pays later on in the product life for poor reliability effort at the design stage. Therefore, a descent effort at the design stage offers the following advantages (Misra, 2008) of

Figure 1. Car purchase criteria (DAT-Report)
Source: Bretsche, 2008

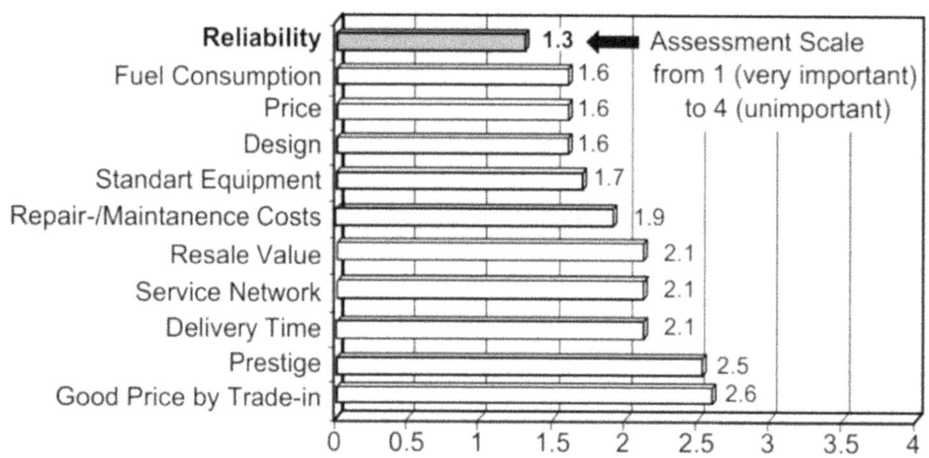

1. Ensures that the product meets the performance goals
2. Identifies potential failure mechanisms during product design
3. Provides an estimation of the product warranty costs and allows life-cycle and warranty cost analyses
4. Optimizes benefits accruing from design alternatives
5. Finds the best reliability allocation to meet system reliability goals; and
6. Permits prediction of product reliability prior to finalization of changes.

An efficient way to implement reliability methods is through a reliability program for a design group. Specific characteristics that are traditionally implemented in a reliability program include goal setting, prediction, statistics, and examination process (Schenkelberg, 2012). These traits are intended to be a framework to assess design reliability through the entire design process. Goal setting provides the context to the reliability program and enables predictions to be made. Common prediction techniques include Failure Modes and Effects Analysis (FMEA), Highly Accelerated Life Testing (HALT), Highly Accelerated Stress Screening (HASS), Reliability Block Diagrams (RDB) (Blischk,2000). These reliability evaluation methods are used at different stages throughout the design process. The design examination is the process of the estimation validation through experimental and field operation data.

Brief Review of Literature

In the recent decades, the requirement of modern technology leads to a growth in the amount of researches about design for reliability. This field of research has been developed especially for complex systems in highly important industries such as automotive and aerospace industry. Popovic (2005, 2007, 2011) illustrated a complete design for reliability procedure for a vehicle's power transmission system. That is used to designing for targeted reliability levels in order to reduce risk levels of critical failures during system useful life. Teixeira (2007) presented some tools for quality assurance. The proposed methodology is a combination of target cost, KANO method and value analysis with respect to the assessment of client requirement compliance levels and also the determination of function alternatives. Rao (1994) demonstrated a reliability-based approach for designing a triple speeded automotive transmission system. Procedure can be also applied to other mechanical transmission systems such as automatic and belt drives. Avontuur (2000, 2001) emphasized the importance of reliability analysis in the conceptual design phase. The feasibility has been evaluated to improve a design by applying reliability analysis techniques in the conceptual design phase. The aim is to quantify the cost of failure, unavailability and comparing them by investment cost. Al-kheer (2011) developed a design for reliability approach by integrating the randomness of tillage forces into the design analysis of tillage machines, aiming at achieving reliable machines. The presented approach was based on the uncertainty analysis of the failure probability of tillage machines and basic random variables. O'Halloran (2012) presented a case study for the early design reliability prediction method (EDRPM) to calculate function and component failure rate distributions during the design process such that the components and design alternatives can be selectively eliminated for search of optimum design. The output is a set of design alternatives with reliability value equal or greater than a preset reliability goal. Comprehensive design for reliability process is proposed for electromechanical complex systems in several research works (Soleimani & Pourgol-Mohammad, 2014; Soleimani et.al., 2014). This method is applicable in early design phase where only limited failure data is available. As a case study horizontal drilling equipment is used for assessment of the proposed method.

Most of recent reliability based design researches for automotive systems have some main infirmity:

1. Improper consideration of all systems components;
2. System reliability evaluation only base on customer demand and components' price;
3. Consideration of system failure modes only in some qualitative methods like FMEA.

This chapter proposes a new framework of design for reliability of automotive systems. This method addresses the deficiency of existing design for reliability approaches which is not comprehensive and suitable for automotive or its complex sub-systems. The advantages of the developed method include:

1. Consideration of all system's components in modeling and simulation;
2. Paying attention to customer demands and optimal balance between cost and benefit concurrently;
3. Considering almost all failure modes of system in FMEA;
4. Design verification with accelerated life tests for reliability optimization. Other advantage of the proposed method is to consider significant factors for correction of collected failure rates for different components in addition to considering the merits of the available methods.

The chapter is organized as follows: Section 2 describes the proposed methodology and illustrates its steps. Section 3 discusses the case study and demonstrates the implementation of proposed design for reliability results. And the final section presents some concluding remarks.

Methodology

There are several methods of designing based on reliability. Most of them are customized for designing a specific system (e.g. Electronic, Mechanical, Nuclear systems etc.). Figure 2 shows the general approach of the design for reliability of an automotive system. This flowchart includes six main phases.

The phases and steps of the methodology are briefly discussed in following sections:

Determining Reliability Targets

There are several methods used for determining costumer's satisfaction factors. The KANO model translates the clients attribute to the importance of product features, placing them in three distinct categories: basic needs, performance and aesthetics (Terninko, 1997). Quality Function Diagram (QFD) is another alternative to correlate the customer requirement to the design specifications.

One of the main manufacturing management policies is to make a balace between cost and benefits of items. Target reliability of an item is determined (step 1 of the flow chart) with warranty data, customer needs and other management policies such as high quality products. This determination lets one understand whether the system design should improve or not.

System Identification

Identification and familiarization of subsystems and components is the next step (step 2 of the flow chart) in RBD process of an automotive system. The functional relationships, items' operating conditions and their missions from the reliability point of view are the important technical information to be gathered. There are some logical structures for making an arrangement of system components in order to structure the reliability block diagram, including series, parallel, series-parallel, standby, load-sharing, and

Figure 2. Flow chart of design for reliability in this research

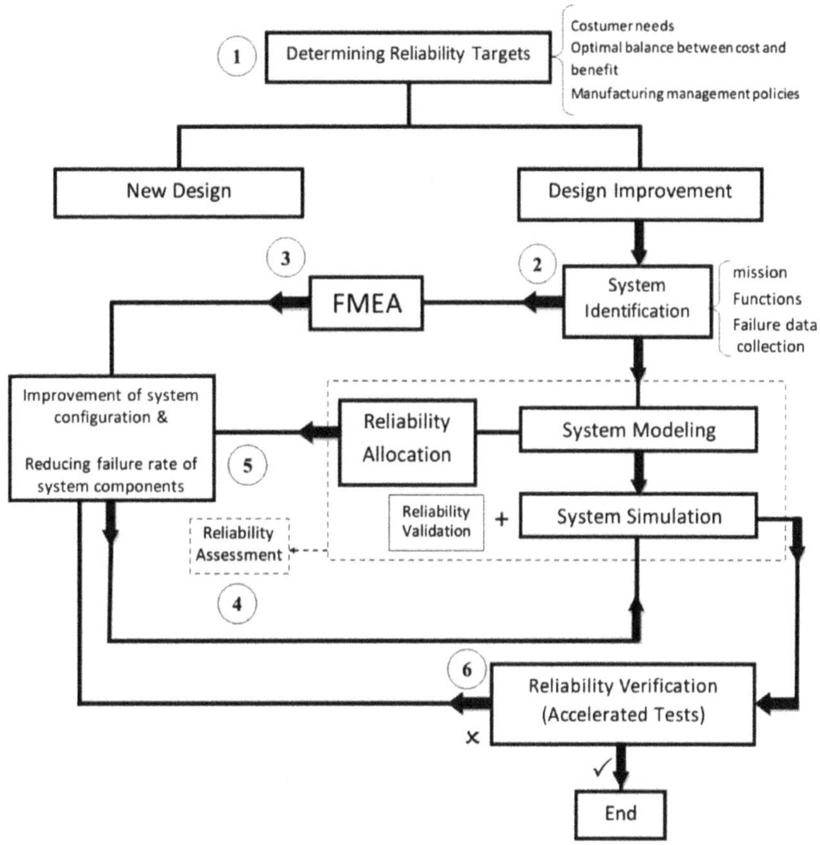

complex system (Modarres, 2010). Each of these structures needs their own formulations for estimating the reliability and failure probabilities. Furthermore, collecting components' failure and repair data is an essential task of this step.

System reliability assessment requires the failure, repair and maintenance data analysis. Classified generic reliability data is a suitable source in case there is only limited field data for the system components (see Soleimani (2014) for more details). Usually these reliability data are needed in systems' early design phase. Depending on the nature and kind of components (mechanical, electrical or electronic), some valuable databases are available for reliability prediction method. MIL-HDBK-217F) 1991), OREDA (2002), NPRD-95 (Denson, Crowell, Clark, & Jaworski, 1994) and NSWC (Jones & Tyrone, 2010) are some of these reliability data banks, utilized here when reliability data is needed. Generally, these data sources include base failure rate and maintenance for different components. In addition to the base failure rate λ_b, correcting factors are used here in determining a part's failure rate including the temperature factor π_T, the power factor π_p, the power stress factor π_s, the quality factor π_Q, and the environmental factor π_E. In this research, these factors are considered in obtaining corrected reliability values for the specified system.

As an example, the base failure rate of a compression spring is adjusted as follows (Jones & Tyrone, 2010).

$$\lambda_{SP} = \lambda_{SP,B} \cdot \pi_G \cdot \pi_{DW} \cdot \pi_{DC} \cdot \pi_N \cdot \pi_Y \cdot \pi_L \cdot \pi_K \cdot \pi_{CS} \cdot \pi_R \cdot \pi_M \quad (1)$$

where the multiplying factors are defined as:

λ_{SP} =failure rate of spring,
$\lambda_{SP,B}$ = base failure rate for spring (it is 23.8 failures per million hours),
π_G = effect of the material rigidity modulus.
π_{DW} =effect of the wire diameter,
π_{DC} = effect of the coil diameter,
π_N = effect of the number of active coils
π_Y = effect of material tensile strength,
π_L =effect of spring deflection,
π_K = spring concentration factor,
π_{CS} = effect of spring cycle rate,
π_R = effect of a corrosive environment,
π_M = effect of manufacturing process.

Failure Mode and Effect Analysis

Failure mode and effects analysis (FMEA) is a well deserved qualitative reliability procedure in automotive industry. It is a dynamic preventive method used in the modification of systems and accompanies the design cycle for components improvements. The overall aim is to analyze and modify components in the light of experience to achieve an optimum criterion of reliability assessment. FMEA is utilized here (step 3 of the flow chart) to examine probable failure modes and omit (or control) potential failures throughout system design (Xiao et al., 2011).

Reliability Assessment

Since system reliability design is essentially an iterative procedure, calculation of system reliability is needed at each cycle of iteration in the design process (step 4 of the flow chart). This necessitates the development of effective and rapidly methods for system reliability assessment in order to meet the relatively short period of time required for designing of system (Xie & Zhou, 2005; Cho & Lee, 2011).

System Modeling

Modeling is the first step in the system reliability evaluation with several approaches available. Reliability block diagram (RBD) is the method chosen here for reliability assessment. RBD is often used to model the influence of item failures on system performance. It mostly corresponds to the physical configuration of items in the system (Modarres, 2010). Calculating reliability is useful by RBD when requirements specify the level of reliability which a design can practically achieve or during component selection with each component's different reliability requirement. For complex systems, these diagrams are used as a visual tool to find out the occurrence of the failures (O'Halloran, 2012).

System Simulating

Monte Carlo analysis is utilized here for simulation of systems for reliability estimation. The Monte Carlo simulation is a numerical method based on a probabilistic interpretation of quantities obtained from generated random variables schemes. MC simulation technique uses statistics to mathematically model a real-life process and then estimating the likelihood of possible outcomes (Adjallah et al., 2007). Random sampling is designed for variables with considering the dependencies if these correlations significantly exist. This process is repeated for sufficient iterations to estimate availability values. Typical sampling is given for k elements in n iterations for estimating of availability function by following equations (Wang & Pham, 2006; Modarres, 2006):

$$\begin{cases} \boldsymbol{x}_1^1, \boldsymbol{x}_2^1, \ldots, \boldsymbol{x}_k^1 = \boldsymbol{A}(\boldsymbol{t})^1 \\ \boldsymbol{x}_1^2, \boldsymbol{x}_2^2, \ldots, \boldsymbol{x}_k^2 = \boldsymbol{A}(\boldsymbol{t})^2 \\ \vdots \\ \boldsymbol{x}_1^n, \boldsymbol{x}_2^n, \ldots, \boldsymbol{x}_k^n = \boldsymbol{A}(\boldsymbol{t})^n \end{cases} \tag{2}$$

where $\boldsymbol{x}_k^n$ is *k-th* parameter in *n-th* iteration, and $\boldsymbol{A}(\boldsymbol{t})$ is the availability value.

Results Validation

Simulating results are compared with warranty and reliability testing data for validation of the methodology. This provides a measure for verifying the results obtained from modeling and simulating of system. This helps the designer understands efficiency assessment of the design procedure and determines the confidence for continuing the other design steps. In this step, sub-systems and components with higher probability of failure, is identified and marked for improvement in the next step. The importance measure is a practical mean for identification of the most critical items. By ranking the items, prioritizing policy is planned in a way that the weakest items are identified and improved (Leemis, 1995). Although identification of weak components is easy in simple series systems, this becomes a difficult task in more complex systems. The value of the reliability importance depends on both the reliability of a component and its position in the system.

Importance measure $\boldsymbol{IR}_j$ is defined as the change in component $\boldsymbol{j}$ reliability effecting upon system failure and is calculated by (Sharirli, 1985):

$$\boldsymbol{IR}_j = \frac{\partial \boldsymbol{R}(\boldsymbol{t})}{\partial \boldsymbol{R}_j(\boldsymbol{t})} \tag{3}$$

where $\boldsymbol{R}(\boldsymbol{t})$ is reliability of the system, $\boldsymbol{R}_j(\boldsymbol{t})$ is reliability of the component j.

Reliability Allocation

Reliability allocation is the key step in design for reliability process which assures optimizing of system from life perspective. The reliability allocation is the process that the failure allowance is allocated for a system in some logical manner to its sub-systems and elements. The purpose of reliability allocation is to establish a target for the reliability of each component so that the manufacturers can have an idea of the life performance required of the product. An overview is recently published of the methods developed during the past 3 decades for solving various reliability optimization problems (Kuo & Prasad, 2000; Kuo 2001). The interested reader can consult references (Yalaoui et. al., 2005; Allella et. al., 2005; Salazar et. al., 2006) for further details on reliability allocation. The reliability allocation is a nonlinear programming problem from mathematical point of view. It can be shown as follows (Misra, 2008):

$$\text{Maximize}\, R_s = f\left(R_1, R_2, \dots, R_n\right) \tag{4}$$

Subject to

$$g_i\left(R_1, R_2, \dots, R_n\right) \le b_i;\; i = 1, 2, \dots, n. \tag{5}$$

$$R_{j\,min} \le R_j \le R_{j\,max};\; j = 1, 2, \dots, n \tag{6}$$

For separable constraints,

$$g_i\left(R_1, R_2, \dots, R_n\right) = \sum_{j=1}^{n} g_{ij}\left(R_j\right) \tag{7}$$

where:

R_s System reliability, $0 \le R_s \le 1$
F_s Unreliability of system
R_j Component reliability of stage *j*
$R_{j\,min}$ Lower bound on R_s
$R_{j\,max}$ Upper bound on R_s
b_i Resources allocated to i-*th* type of constraint
$f(.)$ The system reliability function
$g_i(.)$ The i-*th* constraint function
n Number of subsystems in the system
m Number of resources

In this optimization scheme, reliability of components takes any value in range between 0 and 1. Suppose that there are u_j separate choices for reliability of component at step j for $j = 1,\ldots,k(\leq n)$, and the selection for component reliability at step $k+1,\ldots,n$ is on a continuous scale. Let $R_j(1), R_j(2),\ldots,R_j(u_j)$ denote the component reliability choices at stage j for $j = 1,\ldots,k$. Then the problem of selecting optimal component reliabilities maximizing system reliability is written as (Kuo, 2001):

Maximize

$$R_j = h[R_1(x_1),\ldots R_k(x_k), R_{k+1},\ldots,R_n] \tag{8}$$

subject to

$$g_j = [R_1(x_1),\ldots R_k(x_k), R_{k+1},\ldots,R_n] \leq b_i,\ \mathrm{i} = 1,2,\ldots,\mathrm{m},\ x_j \in \{1,2,\ldots,u_j\}, j = 1,2,\ldots,k.$$

$$R_{j\,min} \leq R_j \leq R_{j\,max}; j = k+1, k+2,\ldots,n. \tag{9}$$

Aeronautical Radio Incorporated (ARINC) technique is one of the well-known reliability allocation based on subsystems weighting factors. This method supposes series subsystems with constant failure rates, such that any subsystem failure results to system failure with subsystem mission time equal to system mission time. This division method requires expression of reliability requirements in terms of failure rate. Equation 10 to 12 shows the mathematical formulation of this technique (Dodson & Noland, 1999).

$$\sum_{i=1}^{n} \lambda^*_i \leq \lambda^* \tag{10}$$

$$w_i = \frac{\lambda_i}{\sum_{i=1}^{n} \lambda_i} \tag{11}$$

$$\lambda^*_i = w_i \lambda^* \tag{12}$$

where:

n Number of sub-systems,
λ_i Failure rate of *i-th* sub-systems,
λ^* Required failure rate for system,

λ^*_i Allocated Failure rate for *i-th* sub-system,

w_i Weighting factors,

Design Improvement

The next step (Step 5) of proposed methodology is reliability improvement of under-performing sub-systems. These components are recognized with reliability assessment, especially with importance measure. There are several approaches for reliability optimization of low reliable items i.e.:

1. Using high quality materials in manufacturing,
2. Improvement of manufacturing procedure for increasing the quality of products,
3. Trade-off the characteristics of sub-systems' design like quality of performance and size of components,
4. Utilizing de-rated materials for prevention of critical failure occurrence e.g., utilization of rustproof and anti-abrasion. Additionally, changing system configuration is another well-known procedure, for example adding redundancy, parallel or standby sub-systems would be helpful for system reliability improvement.

Design Verification

Design verification is the final phase of the proposed method. In the reliability verification phase testing is the main part and its aimed especially at design reliability verification. In this phase of design, designers examine the reliability of improved systems with various reliability tests in order to realize if the improved design fulfills the requirements of the target system reliability. If a design fails to meet the reliability requirements during the test, it must be revised following careful failure analysis. Then the redesigned product is re-subjected to verification testing.

There are typically four types of reliability verification tests:

1. Bogey test,
2. Sequential life test,
3. Test-to-failure test and,
4. Degradation test. Bogey test is utilized here for the design verification however other test methods can be used as well.

Bogey test is used in industry to demonstrate reliability at a high confidence level. This method is simple to apply. For some products, a failure is defined in terms of a performance characteristic exceeding a specified threshold. For these products, it is possible to measure the performance characteristic at different times during the test. The measurement data can be employed to predict whether or not a test unit will fail by the end of the test. When there are sufficient data to make such a prediction with a high degree of confidence, the test can be terminated for the unit.

This method deals with the Weibull and binomial distributions. If the product life has a Weibull distribution with shape parameter β and scale parameter α, assuming the known $\beta and \alpha$, the time that test should be continued is defined by (Yang, 2007):

$$t_0 = t_l \left[\frac{\ln(1-CL)}{n_0 \ln(R_L)} \right]^{1/\beta} \quad (15)$$

where:

t_0 is test time,
t_l is time of system reliability evaluation,
CL is confidence level,
n_0 is sample size,
R_L is system reliability.

This test is sometimes called WeiBays if the parameter ² is assumed and the test's life characteristic is determined.

Design for Reliability of Dry Friction Clutch System: Case Study

In this chapter an automotive dry friction clutch has been selected as a case-study. Clutch is a critical mission part of automotive transmission system. This is due to its harsh operating condition and its safety level. Failure of clutch causes the failure of transmission system and results in automotive function failure. In the following sections, the steps of presented method are applied for reliability based design of a dry friction clutch.

Determining Reliability Targets for Clutch System

The analysis of products field data, e.g., warranty data, is the key factor to improve the product's quality. These represent the real operation. The information offered by the returns from warranty lies in the time of use and distance driven up to the occurrence of the failure. From KANO's survey, the users expect to drive about 72500 km without observing any lost in operation of clutch. Also the probability of driving of 72500 km during a year is approximately 1.52% (Teixiera, 2007). According to this information, if a reliability of 0.99 is targeted for clutch at 72500 km, failure of about 150 clutches would only anticipate for a year of warranty in one million produced clutches.

Clutch System Identification

Clutch is designed to engage and disengage for power transmission from the engine. This is the main mission of clutch systems.

Figure 3 shows the simple schematic of automotive dry friction clutch. The main functions of clutch system are (Teixiera, 2007):

- Moving vehicle,
- Allows changing gears,

Figure 3. Simple schematic of automotive dry friction clutch
Source: Heisler, 2002

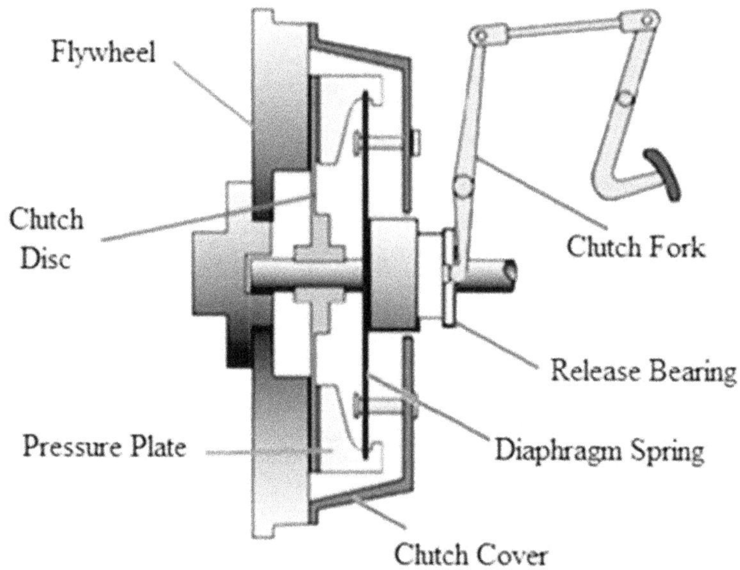

- Automotive transfer torque,
- Provide comfort to the user when engaging,
- Provide comfort to the user when disengaging,
- Clutch eliminates engine noises of transmission system.

From reliability point of view, the main sub-systems of dry friction clutch include: hydraulic system, fork, release bearing, diaphragm spring, pressure plate and disc. There are other decomposition structures for the system depending on the objectives, the resources and the required detail.

There are some important factors interfered in correcting reliability data estimated for dry friction clutch which is related to environmental and functional conditions. These factors are engagement rate, actuation time, engagement time and motor torque where all of these factors can be extracted from European standards such as "Worldwide Emission Standards and Related Regulations" and "ECE No.15". For example, According to the Figure 4, quantities of 44 clutch engagements are recorded in UDC (Urban Driving Cycle) during 780 seconds. Also in this time, the automotive is moved 1017 meter (ECE No.15). In addition, the mean revolution of automotive motor and clutch in UDC (Urban Driving Cycle) is about 1730 RPM (EURO Parliament, 2004).

Figure 5 shows the free dimensional alteration of revolution and torque of gasoline engine in UDC. According to the results of quality control laboratory of Paya Clutch Company (PAYACLUTCH Corp. 2013), these mentioned factors are applied as correction factors in reliability data estimation. Expert judgment is also used for some especial components and factors (Hejazi, 2013).

Table 1 shows the MTTF and failure rates of clutch sub-systems considering operational hours of life. These results are calculated using several correction factors such as environmental, temperature effect, motor torque and the engagement rate. As an example, the return spring's failure rate obtained from the following equation (extracted from Equation 1). (See details in (Hejazi, 2013) for different components' failure rate correction):

Figure 4. Urban driving cycle from 0-780 second and extra- urban driving cycle from 780-1180 second (ECE No.15).

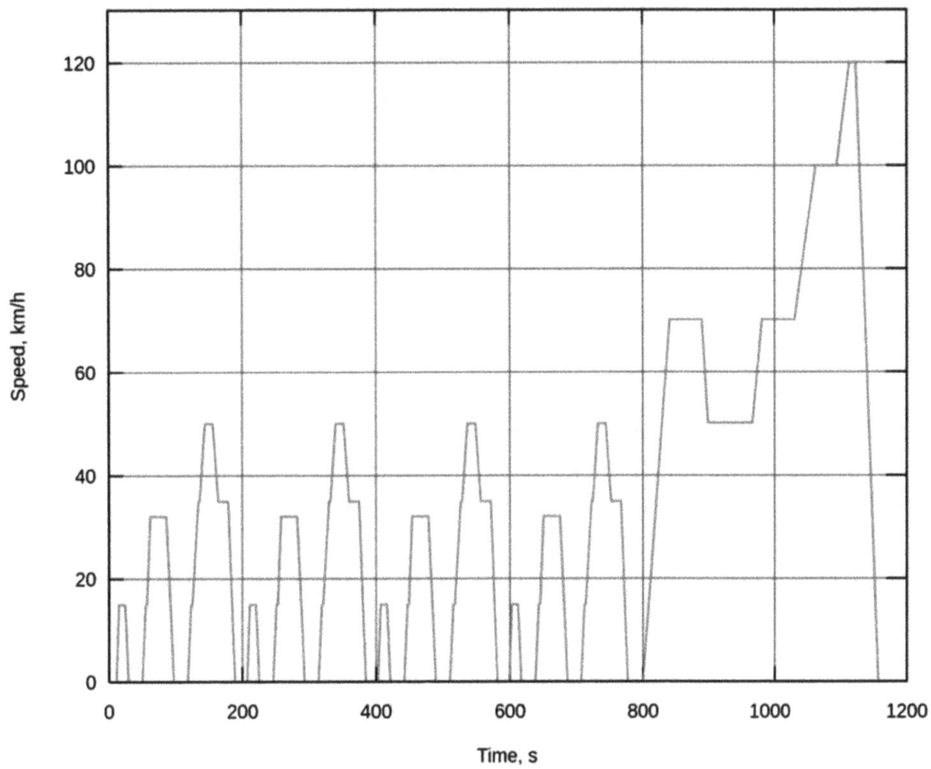

Figure 5. ETC dynamometer schedule
Source: EURO Parliament, 2004

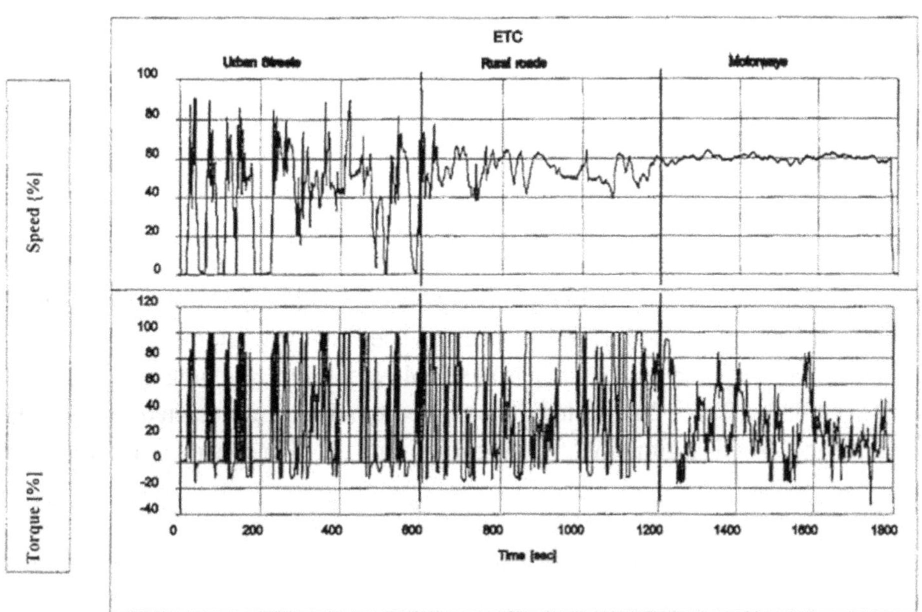

Table 1. MTTF and failure rate of dry friction clutch sub-systems

$t'\left(\text{hours of operation}\right) = 0.25t\left(\text{hours of life}\right)$		
$\lambda\left(10^6\,\text{hrs}^{-1}\right)$	MTTF (hrs)	Sub-system
0.6	1.67E+06	Pedal
0.08	1.25E+07	Return Spring
3.96	2.53E+05	Master Hydraulic Cylinder
0.2	5.00E+06	Hydraulic Line
3.92	2.55E+05	Slave Hydraulic Cylinder
1.6	6.25E+05	Fork
6.72	1.49E+05	Release Bearing
7.6	1.32E+05	Diaphragm Spring
1.44	6.94E+05	Pressure Plate
1.74	5.75E+05	Spline Hub
2.8	3.57E+05	Pre-Damper Spring
3.56	2.81E+05	Main-Damper Spring
22.80	4.39E+04	Friction Lining
76.1	1.31E+04	Clutch System

$$\lambda_{SP} = \lambda_{\mathrm{SP,B}}.\pi_{\mathrm{G}}.\pi_{\mathrm{DW}}.\pi_{\mathrm{DC}}.\pi_{\mathrm{N}}.\pi_{\mathrm{Y}}.\pi_{\mathrm{L}}.\pi_{\mathrm{K}}.\pi_{\mathrm{CS}}.\pi_{\mathrm{R}}.\pi_{\mathrm{M}}.\pi_{\mathrm{C}} \tag{16}$$

where π_C = the effects of automotive operational condition. So:

$$\lambda_{\mathrm{sp}} = 23.85\left(0.97\right)\left(2.2\right)\left(0.038\right)\left(0.16\right)\left(0.47\right)\left(0.8\right)\left(1.14\right)\left(0.1\right)\left(1\right)\left(1\right)\left(1.5\right)$$
$$= 0.02 \text{ Failure per } 10^6 \text{hours}$$

The estimated failure rate is close to the value in reliability data reported by NPRD for dry friction clutch. Availability assessment is a key design measure used for evaluation of repairable systems' performance. In this research, clutch repair data has been collected from 10 clutch specialists (repair shops) in a period of two months. Table 2 indicates the obtained repair data associated to a popular sedan car used for urban driving. In this estimation, the diagnosis time are also considered as the repair time.

Table 2. Repair parameter estimation for dry friction clutch components

Sub-System	Total Repair Time (Hours)	
	Average	Standard Deviation
Pedal	5.25	1.375
Return Spring	5.25	1.375
Master Hydraulic Cylinder	10.75	2.5
Hydraulic Line	10.75	2.5
Slave Hydraulic Cylinder	11.25	2.5
Fork	6.75	1.375
Release Bearing	10.75	2.5
Diaphragm Spring	8.75	1.75
Pressure Plate	11.75	2.25
Spline Hub	8.75	1.75
Pre-Damper Spring	8.75	1.75
Main-Damper Spring	8.75	1.75

Failure Mode and Effect Analysis for Clutch System

FMEA is executed for all components of clutch. As an example, Table 3 shows a summarized worksheet for FMEA of Spline Hub. (See detailed complete FMEA for dry friction clutch in (Hejazi, 2013))

Reliability Assessment of Clutch System

Clutch System Modeling

RBD modeling method utilized here for reliability modeling of the clutch system. The dry friction clutch has a series configuration in the reliability point of view. Figure 6 shows this modeling created by ReliasoftBlocksim8 software (ReliaSoft Corp, 2012). Hydraulic system and disc blocks contain several components and both have series configuration.

Table 3. FMEA worksheet of Spline Hub

Item: Spline Hub **Function: Transferring Received Torque from Pressure Plate to Gearbox Axle**		
Potential Failure Mode	Effects of Failure	Failure reasons / Mechanism
Erosion	• Clutch is not released • Wrap in disc	• Unbalance power transmission • Shaft erosion
Corrosion	• Clutch is not released	• Improper greasing of gearbox • Shaft corrosion
Grease on spline	• Clutch is not released	• Utilize metal with grease • leakage

Clutch is considered as a repairable system which most of its failures occurring in wear out phase of its life cycle. Weibull distribution (O'Connor, 2011) is determined as the best fit and suitable distribution (goodness of fit tests are utilized for this selection). This distribution has the capability of reliability modeling in different phases of life-cycle (especially in wear-out phase).

Clutch System Simulating

Clutch system has been simulated for 1 million iterations in 53k driving hours (10^6 km). Then The reliability parameters are calculated and has been analyzed.

Reliability

Figure 7 shows the estimated reliability diagram of the dry friction. According to this figure, the reliability of the dry friction clutch is almost 100% for 4500 kilometers of driving (24000 driving hours) and tends to 0% for 6500 kilometers of driving. It should be pointed out that the reliability of the clutch system is 0.77 in 72500 kilometers of driving with 95% confidence level, seems it must get improved to the higher values.

Availability Estimation

In designing repairable systems, availability is used as a combination of reliability and maintainability parameters (Dodson, Noland, 1999). Figure 8 shows the mean availability diagram of the dry friction clutch at 53277 driving hours (10^6 km).

Figure 6. RBD of dry friction clutch system

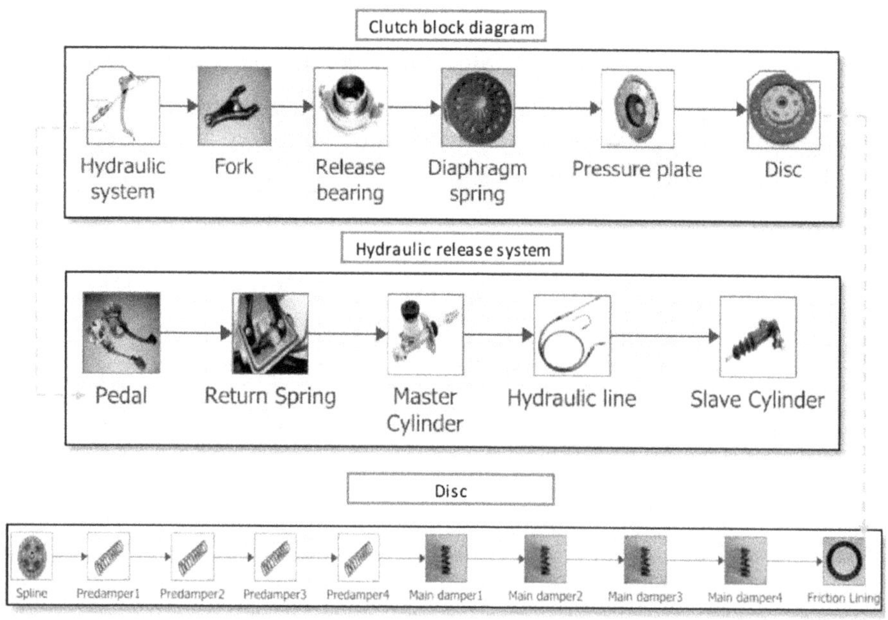

Figure 7. Reliability diagram for the dry friction clutch

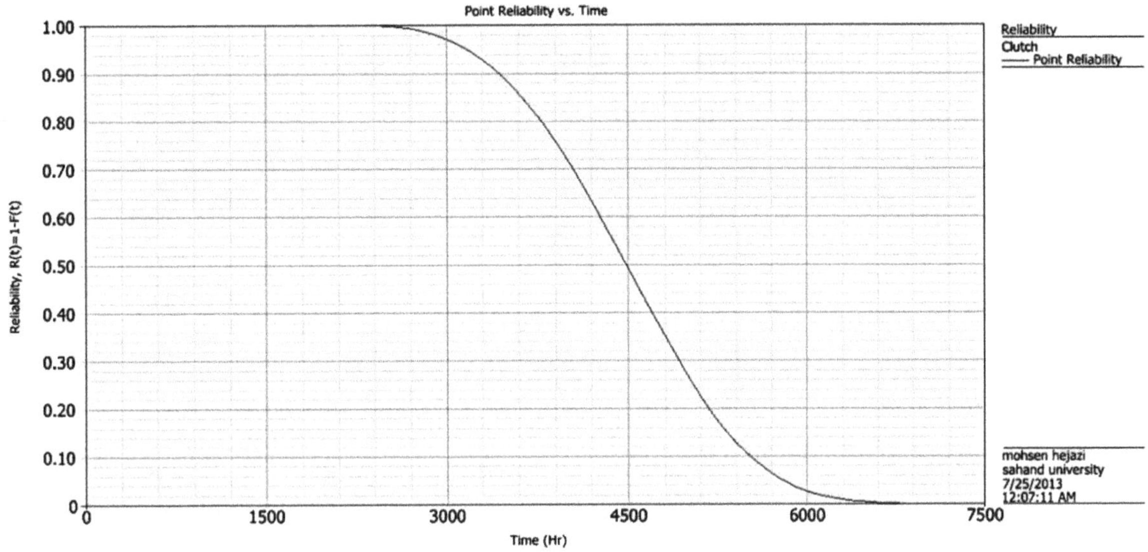

Figure 8. Mean availability of dry friction clutch system at 53277 driving hours (10^6 km)

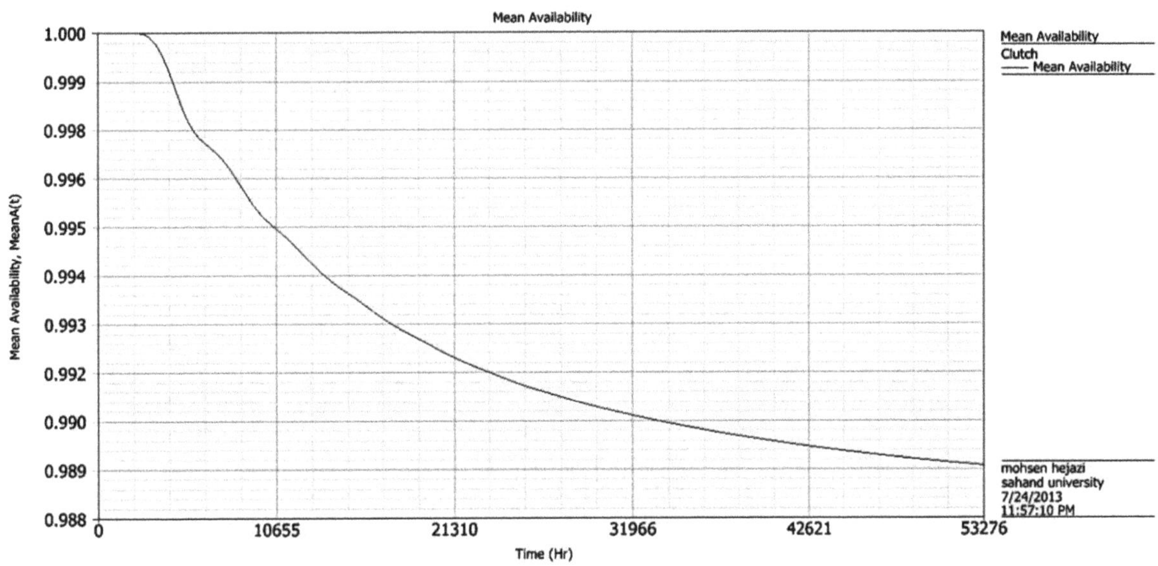

According to the field survival rates and the manufacturing company's reliability targets the value 0.989 is acceptable for mean availability of a dry friction clutch system in 10^6 kilometers of driving. Table 4 shows the simulation results.

Reliability Allocation

In this case, ARINC technique is used to determine reliability allocation. Table 5 shows the results of reliability allocation for the dry friction clutch subsystems. Reliability of 0.99 is considered as a target

Table 4. Simulation results for availability estimation of the clutch system at 53277 driving hours

Feature	Value
Mean Availability (All Events)	0.989
Std Deviation (Mean Availability)	7E-006
Point Availability (All Events)	0.987
Expected Number of Failures	63.9
MTTFF (hr)	4473.2
Uptime (hr)	52694.1
Total Downtime (hr)	581.9

Table 5. Initial and target reliability (for 3862 driving hours) of clutch system

Block Name	Reliability Importance	Reliability	Weight	Target Reliability
Pedal	0.770	0.998	0.000019	0.999
Return Spring	0.769	0.999	3.00E-07	0.999
Master Cylinder	0.775	0.992	0.000059	0.999
Hydraulic Line	0.769	0.999	0.000008	0.999
Slave Cylinder	0.775	0.992	0.000060	0.999
Fork	0.772	0.996	0.000034	0.999
Release bearing	0.780	0.986	0.000078	0.999
Diaphragm Spring	0.786	0.978	0.000097	0.999
Pressure plate	0.771	0.997	0.000034	0.999
Spline	0.772	0.996	0.000038	0.999
Maindamper1	0.780	0.985	0.000080	0.999
Maindamper2	0.780	0.985	0.000080	0.999
Maindamper3	0.780	0.985	0.000080	0.999
Maindamper4	0.780	0.985	0.000080	0.999
Friction lining	0.867	0.887	0.000240	0.997
Predamper1	0.773	0.994	0.000050	0.999
Predamper2	0.773	0.994	0.000050	0.999
Predamper3	0.773	0.994	0.000050	0.999
Predamper4	0.773	0.994	0.000050	0.999
Clutch System	-	0.769	-	0.990

for 3862 driving hours (72500 km). This is an optimal target in a trade-off between costs and benefits. It should be noted that these results are obtained in 95% confidence level. In addition, the reliability importance of components is calculated for critical items' identification.

Design Improvement

Considering the importance values, friction lining has the maximum reliability importance and the diaphragm spring has the second highest reliability importance value. Therefore failure occurrence is more susceptible in these components. At the lower importance level, main damper 1-4 and the release bearing have the higher values of reliability importance.

With regard to the reliability importance values, reliability optimization can be planned for the dry friction system. For instance, increasing the quality of critical components (especially friction lining) could be done by selecting the material with higher quality, improving the manufacturing procedure and or using anti-erosion and anti-corrosion material. Tests would be planned to examine the amount of reliability growth. For example, if the reliability value for the friction lining can be improved to 0.95 (from 0.887332) in 72500 kilometers of driving, the clutch reliability can be optimized from 0.76959 to 0.82394.

Design Verification

The final step is examining the reliability of optimized system in the proposed design for reliability method. This could be executed by performing different reliability tests in order to verify the target reliability of the system and it's components.

In this study, Reliasoft ALTA (Accelerated Life Testing Analysis) software (ReliaSoft Corp., 2012) is used for design of reliability tests. Bogey testing method is selected for reliability verification. Table 6 demonstrates the characteristics of designed test for some of the clutch components at 80% confidence level (see more details in (Hejazi, 2013)). This means that, if the tested component can survive in the test time (cycle), the probability that this component can guarantee the allocated reliability is 80%. Acceleration factor is an important concept in ALTs, defined as the ratio of a life percentile at a use stress level to the percentile observed at an elevated stress level.

According to Table 6, only one sample is examined in friction lining and pressure plate reliability tests. This is planned for managing the limited financial issues or samples available. Acceleration factor is estimated for different components of clutch system according to the effects of mechanical and thermal

Table 6. characteristics of the zero failure reliability test for clutch components

Zero Failure Test Time or Cycle	Acceleration Factor	Test stress Coefficient	Test Speed	Sample Size	Allocated Reliability	Components
1.85 hours of continuous slip	60	1	1730 rpm	1	0.9979	Friction lining
983284 Cycle	4	2	60 cpm	10	0.9995	Master Cylinder
983284 Cycle	4	2	60 cpm	10	0.9995	Slave Cylinder
695321 Cycle	8	4	60 rpm	10	0.9997	Fork
28 hours of continuous slip	60	1	1730 rpm	1	0.9997	Pressure plate
983358 Cycle	8	4	60 cpm	10	0.9998	Pedal
983358 Cycle	8	4	60 cpm	10	0.9999	Hydraulic Line

stress enhancements. Higher acceleration factor is calculated for the friction lining due to its higher rate of skidding during the test in comparison to the normal functioning condition.

CONCLUSION

In this chapter, a new design for reliability framework is developed for automotive systems. The advantages of this method in comparison to the other processes are the comprehensive failure mode and effect analysis and the consideration of almost all environmental effects and operating conditions in components failure rates estimation. The main steps of this approach were presented and a case study is demonstrated for a dry friction clutch.

According to the reliability importance analysis, friction lining and diaphragm spring are the critical elements in the reliability point of view. So the higher quality of components of the subsystems and the better reliability designing (e.g., using some redundancies), the higher reliability of the system. In the final step of the presented method, a reliability verification testing utilized for realizing that if the improved design fulfills the requirments of the target system reliability. Results illustrates that the proposed design for reliability methodology is practical for reliability improvement of automotive systems and its main subsystems. It is worthy to note that the following aspects could be considered for further studies.

- Design for reliability of other types of clutches
- Optimization of clutches' reliability with respect to clutch operating conditions or design parameters
- Design improvement of clutch components in order to achieve the target allocated reliability

REFERENCES

Adjallah, K. H., Crespo Marquez, A., & Iung, B. (2007). A structured approach for the assessment of system availability and reliability using Monte Carlo simulation. *Journal of Quality in Maintenance Engineering*, *13*(2), 125–136. doi:10.1108/13552510710753032

Al-Kheer, A. A., El-Hami, A., Kharmanda, M. G., & Mouazen, A. M. (2011). Reliability-based design for soil tillage machines. *Journal of Terramechanics*, *48*(1), 57–64. doi:10.1016/j.jterra.2010.06.001

Allella, F., Chiodo, E., & Lauria, D. (2005). Optimal reliability allocation under uncertain conditions, with application to hybrid electric vehicle design. *International Journal of Quality & Reliability Management*, *22*(6), 626–641. doi:10.1108/02656710510604926

Avontuur, G. C. (2000). *Reliability analysis in mechanical engineering design: TU Delft*. Delft University of Technology.

Avontuur, G. C., & van der Werff, K. (2001). An implementation of reliability analysis in the conceptual design phase of drive trains. *Reliability Engineering & System Safety*, *73*(2), 155–165. doi:10.1016/S0951-8320(01)00041-2

Bertsche, B. (2008). *Reliability in automotive and mechanical engineering: determination of component and system reliability*. Springer Science & Business Media.

Bhote, K. R., & Bhote, A. K. (2004). *World class reliability: Using multiple environment overstress tests to make it happen*. AMACOM Div American Management Association.

Blischke, W. R., & Murthy, D. P. (2011). *Reliability: modeling, prediction, and optimization* (Vol. 767). John Wiley & Sons.

Cho, T. M., & Lee, B. C. (2011). Reliability-based design optimization using convex linearization and sequential optimization and reliability assessment method. *Structural Safety*, *33*(1), 42–50. doi:10.1016/j.strusafe.2010.05.003

Denson, W., Chandler, G., Crowell, W., Clark, A., & Jaworski, P. (1994). *Nonelectronic Parts Reliability Data 1995*. DTIC Document.

Dodson, B., & Nolan, D. (1999). *Reliability engineering handbook*. Marcel Dekker New York.

EURO Parliament. (2004), *EURO Emission Regulations*. Author.

Heisler, H. (2002). *Advanced vehicle technology*. Elsevier.

Hejazi, A. M. (2013). *Reliability Based Design of an Automotive Dry Friction Clutch* (Master of science thesis). Sahand University of Technology, Tabriz, Iran.

Japan Automobile Standards Internationalization Center. ECE No.15. (2012). Exhaust Emission Standard.

Jones, T. L. (2010). *Handbook of reliability prediction procedures for mechanical equipment. Naval Surface Warfare Center*. West Bethesda, MD: Carderock Division.

Kuo, W. (2001). *Optimal reliability design: Fundamentals and applications*. Cambridge University Press.

Kuo, W., & Prasad, V. R. (2000). An annotated overview of system-reliability optimization. *IEEE Transactions on Reliability*, *49*(2), 176–187. doi:10.1109/24.877336

Leemis, L. M. (1995). *Reliability: Probabilistic models and statistical methods*. Prentice-Hall, Inc.

Military, U. (1992). *Reliability prediction of electronic equipment*. MIL-HDBK-217F Notice 1.

Misra, K. B. (2008). *Handbook of performability engineering*. Springer Science & Business Media. doi:10.1007/978-1-84800-131-2

Modarres, M. (2006). *Risk analysis in engineering: Techniques, tools, and trends*. CRC Press.

Modarres, M., Kaminskiy, M. P., & Krivtsov, V. (2009). *Reliability engineering and risk analysis: a practical guide*. CRC Press.

O'Connor, A. N. (2011). *Probability distributions used in reliability engineering*. RiAC.

O'Halloran, B. M., Hoyle, C., Stone, R. B., & Tumer, I. Y. (2012). *The early design reliability prediction method*. Paper presented at the ASME 2012 International Mechanical Engineering Congress and Exposition. doi:10.1115/IMECE2012-89592

PAYACLUTCH Corp, . (2013). *QC Laboratory Database*. Rasht, Iran: Author.

Participants, O. (2002). *Offshore reliability data handbook*. Trondheim, Norway: DNV.

Popovic, P., & Ivanovic, G. (2005). Monograph Reliability design of mechanical systems. Belgrade: 'VINCA' Institute of Nuclear Sciences.

Popovic, P., & Ivanovic, G. (2007). A methodology for the design of reliable vehicles in the concept stage. *Strojniski Vestnik*, *53*(3), 173–185.

Popovic, P., & Ivanovic, G. (2010). Methodology design for reliability power transmission systems of vehicles. *International Journal of Vehicle Design*, *54*(3), 262–283. doi:10.1504/IJVD.2010.036130

Popovic, P, Ivanovic, G., Mitrovic, R., & Subic, A. (2011). Design for reliability of a vehicle transmission system. *Proceedings of the Institution of Mechanical Engineers. Part D, Journal of Automobile Engineering*, *226*(2), 194–209. doi:10.1177/0954407011416175

Rao, S. S., & Tjandra, M. (1994). Reliability-based design of automotive transmission systems. *Reliability Engineering & System Safety*, *46*(2), 159–169. doi:10.1016/0951-8320(94)90133-3

ReliaSoft Corporation. (2012). *Blocksim 8 User's Guide*. Retrieved from: http://www.reliasoft.com

ReliaSoft Corporation. (2012). *Reliasoft ALTA 8 User's Guide*. Retrieved from: http://www.reliasoft.com

Salazar, D., Rocco, C. M., & Galvan, B. J. (2006). Optimization of constrained multiple-objective reliability problems using evolutionary algorithms. *Reliability Engineering & System Safety*, *91*(9), 1057–1070. doi:10.1016/j.ress.2005.11.040

Schenkelberg, F. (2012). *Effective reliability program traits and management*. Paper presented at the Reliability and Maintainability Symposium.

Sharirli, M. (1985). *Methodology for system analysis using fault trees, success trees, and importance evaluations*. College Park, MD: Maryland Univ.

Simens, V. (2003). *Worldwide emission standards and related regulations*. Siemens VDO.

Soleimani, M., & Pourgol-Mohammad, M. (2014). Design for reliability of complex system with limited failure data; case study of a horizontal drilling equipment.*Proceedings of the Probabilistic Safety Assessment and Management PSAM*. doi:10.1155/2014/524742

Soleimani, M., Pourgol-Mohammad, M., Rostami, A., & Ghanbari, A. (2014). Design for Reliability of Complex System: Case Study of Horizontal Drilling Equipment with Limited Failure Data. *Journal of Quality and Reliability Engineering*, *1*(1), 1–13. doi:10.1155/2014/524742

Teixeira, C. A. R., & Cavalca, K. L. (2008). Reliability as an added-value factor in an automotive clutch system. *Quality and Reliability Engineering International*, *24*(2), 229–248. doi:10.1002/qre.889

Terninko, J. (1997). *Step-by-Step QFD: Customer-Driven Product Design* (2nd ed.). Taylor & Francis.

Wang, H., & Pham, H. (2006). *Reliability and optimal maintenance*. Springer Science & Business Media.

Contienetal-The Future in Motion. (2013). *Worldwide Emission Standards and Related Regulations*. Author.

Xiao, N., Huang, H.-Z., Li, Y., He, L., & Jin, T. (2011). Multiple failure modes analysis and weighted risk priority number evaluation in FMEA. *Engineering Failure Analysis*, *18*(4), 1162–1170. doi:10.1016/j.engfailanal.2011.02.004

Xie, L., Zhou, J., Wang, Y., & Wang, X. (2005). Load-Strength Order Statistics Interference Models. *International Journal of Performability Engineering*, *1*(1), 23.

Yalaoui, A., Chu, C., & Châtelet, E. (2005). Reliability allocation problem in a series–parallel system. *Reliability Engineering & System Safety*, *90*(1), 55–61. doi:10.1016/j.ress.2004.10.007

Yang, G. (2007). *Life Cycle Reliability Engineering*. Wiley; doi:10.1002/9780470117880

APPENDIX

Nomenclature and Abbreviations

FMEA: Failure mode and effect analysis
MC: Mote Carlo
ARINC: Aeronautical Radio Incorporated
UDC: Urban Driving Cycle
RPM: Revolutions Per Minute
ETC: Emission Test Cycle
RBD: Reliability block diagram
HASS: Highly Accelerated Stress Screening
EDRPM: Early design reliability prediction method
MTTF: Mean Time To Failure
RPM: Rotations Per Minute
CPM: Cycle Per Minute

Chapter 10
Reliability Analysis of Two Dissimilar-Cold Standby Redundant Systems Subject to Inspection with Preventive Maintenance Using Copula

Beena Nailwal
G. B. Pant University of Agriculture and Technology, India

S. B. Singh
G. B. Pant University of Agriculture and Technology, India

ABSTRACT

The objective of this chapter is to study the system reliability through redundant component and additional preventive maintenance of operative and standby units. The considered system consists of two units in cold-standby, one is main unit and other is standby. The configuration of main unit is of 1-out-of-2: G whereas standby unit is of (n-1)-out-of-n: F and (n-2)-out-of-n: D nature. Main unit of the system has two modes viz. operable and failed and standby unit of the system has three modes viz. operable, degraded and failed. Both units after failure go for inspection where the type of the failure is detected and accordingly repairing action is decided. The failure and repair times follow exponential and general time distributions respectively. The system is studied by using the supplementary variable technique, Laplace transformation and Gumbel-Hougaard family of copula to obtain reliability, availability, downtime, busy period, M.T.T.F. and cost effectiveness of the system. At last some special cases of the system have been discussed.

DOI: 10.4018/978-1-5225-1639-2.ch010

INTRODUCTION

Reliability is an important concept at the planning, designing, manufacturing and operation stages of product and systems ranging from simple to complex. Users expect that the product and system they purchase should be reliable. In Practice, we come across with a number of systems consisting of one or more parts where failure of any of the parts results into the complete / partial failure of the system and as a result, the reliability of the system reduces. The better maintenance of such parts originate better reliability which helps in achieving the markets demands of reliability, functionality, price and performance of the system. A unit or system is composed of a number of components or elements and to make the system highly reliable we have to use reliable components. Sometimes either it is not possible to produce highly reliable components or the cost of producing such components is very high, thus introducing redundant parts, providing maintenance and repair at the time of need may achieve better degree of reliability. Also redundancy plays an important role in enhancing system reliability. One of the commonly used forms of the redundancy is the standby redundancy. Standby system often finds applications in various industrial and other setups. In a redundant system, some additional paths are created for the proper functioning of the system. If all the redundant parts start working together at the time of operation, then it is called parallel redundancy. A standby redundant system is the one in which one unit is followed by spare units called standbys. In general there are 3-types of standby i.e. hot, warm and cold. Hot standbys are those which are loaded in exactly the same way as the operating unit. Warm standbys have a diminished load and cold standbys are completely unloaded.

Several researchers like Lasanovsky (1982), Mokaddis and Tawfek (1996), Hatogama (1997), Agarwal et al. (2010), and Kumar and Singh (2016) have studied the reliability behavior of a two/three state system. Many other researchers: Gopalan and Nagarwalia (1985), and Said and EL-Sherbeny (2005) and Haggag (2009) have studied cost analysis of two-unit redundant systems with preventive maintenance assuming two/three states of operation. The purpose of the present chapter is to discuss the reliability characteristics of a two-dissimilar unit cold standby system with preventive maintenance subject to inspection after failure. In this chapter we consider a system which comprises of two unit cold standby system, one is main unit and other is standby unit. The main unit is of 1-out-of-2: G and standby unit is of (n-1)-out-of-n: F and (n-2)-out-of-n: D configurations. The main unit of the system has two modes viz. operable and failed and standby unit of the system has three modes viz. operable, degraded and failed. Two repairmen are involved in repairing of the system. The first repairman repairs the main unit and the second repairman repairs the standby unit. Here we consider two types of failures namely minor and major in main unit. A minor failure is that failure which can be repaired by first repairman and a major failure is that failure which cannot be repaired by first repairman. Thus if the major failure occurs in main unit then second repairman repairs the main unit. After failure, units go for inspection, where the type of the failure is detected and accordingly corrective action is decided. After repair, system becomes as good as new. Preventive maintenance is provided to the system at S_0. At S_6 and S_{10} both the units are completely failed and being repaired with two different repair rates. The repairs from state S_6 to S_0 and S_{10} to S_0 have two types namely exponential and general. Joint probability distribution of repair rate from S_6 to S_0 and S_{10} to S_0 is computed by Gumbel-Hougaard family of copula. The system is studied by using the supplementary variable technique, Laplace transformation and Gumbel-Hougaard family of copula to obtain reliability, availability, downtime, busy period, M.T.T.F. and cost effectiveness of the system as done by Nelson (2006) and Ram and Singh (2010). At last some particular case pertaining to the system have been computed. System state description of the transition diagram is shown in Table.1.

NOTATIONS

- λ_1, α_1 : Failure rates of main unit.
- λ_2, α_2 : Failure rates of standby unit, where $\lambda_2 = \sum_{i=1}^{n-2} \lambda_i$ and $\alpha_2 = \sum_{j=n-1}^{n} \alpha_j$
- W_1 : Inspection time of main unit.
- W_2 : Inspection time of standby unit.
- W_3 : Inspection time of main unit by second repairman.
- W_4 : Inspection time at degraded state.
- $\phi_1(x)$: Repair rate of main unit.
- $\phi_2(x)$: Repair rate of standby unit.
- u : Rate for taking a system into preventive maintenance.
- v : Rate for preventive maintenance.
- $P_i(t)$: Probability that the system is in S_i state at instant t for i=1 to i=10.
- $\overline{P}_i(s)$: Laplace transform of $P_i(t)$.
- $E_p(t)$: Expected profit during the interval (0, t].
- K_1, K_2 : Revenue per unit time and service cost per unit time respectively.
- $\varphi(x)$: Coupled repair rate

Considering $u_1 = e^x$ and $u_2 = \phi_2(x)$ the expression for joint probability (failed state S_6 to good state S_0 and failed state S_{10} to good state S_0) according to Gumbel-Hougaard family of copula is given by

$$\varphi(x) = \exp[(x)^\theta + (\log \phi_2(x))^\theta]^{\frac{1}{\theta}}$$

ASSUMPTIONS

1. The system consists of two dissimilar units, one is main unit and other is standby unit. The main unit is of 1-out-of-2: G and standby unit is (n-2)-out-of-n: D and (n-1)-out-of-n: F configuration.
2. Initially one unit is operative and the other unit in cold standby.
3. Main unit has two modes, normal and complete failure. Standby unit has three modes, normal, degraded and total failure.
4. Preventive Maintenance is provided to this system in state $S_{0.}$
5. Both units on failure go for inspection.
6. The first repairman who repair the main unit can repairs minor failures only. Thus if major failure occurs in main unit then second repairman who repairs the standby unit repairs the main unit.
7. If both units fail, the whole system fails completely.
8. Transition from the completely failed state S_6 to the initial state S_0 and completely failed state S_{10} to the initial state S_0 follows two different distributions.

9. Joint probability distribution of repair rate from S_6 to S_0 and S_{10} to S_0 is computed by Gumbel-Hougaard family of copula.
10. Failure rates of main unit and standby unit are assumed as constant.
11. After repair the system is as good as new. Repairs never damage anything.

BLOCK AND STATE TRANSITION DIAGRAM

Block and state transition diagrams are shown in Figures 1 and 2 respectively.

FORMULATION OF MATHEMATICAL MODEL

By probability consideration and continuity arguments the following difference-differential equations governing the behavior of the system seems to be good.

$$\left(\frac{d}{dt}+\lambda_1+u\right)P_0(t)=vP_9(t)+\int_0^\infty \phi_1(x)P_3(x,t)dx+\int_0^\infty \phi_1(x)P_8(x,t)dx+\int_0^\infty \varphi(x)P_{10}(x,t)dx \\ +\int_0^\infty \varphi(x)P_6(x,t)dx \tag{1}$$

Table 1. System state description

States	Subsystems		System State
	State of Main Unit	State of Standby Unit	
S_0	O	S	O
S_1	F	S	O
S_2	F_i	S	F
S_3	F_r	O	O
S_4	F	D	D
S_5	F	F_i	F
S_6	F_r	F_r	F
S_7	F_i	S	F
S_8	F_r	S	F
S_9	P	P	D_o
S_{10}	F_r	F_r	F

O: Operable; S: Standby; F: Failed; P: Preventive maintenance; F_r: Failed unit under repair; D; degraded state; D_o: Down, both units under preventive maintenance.

Figure 1. Block diagram

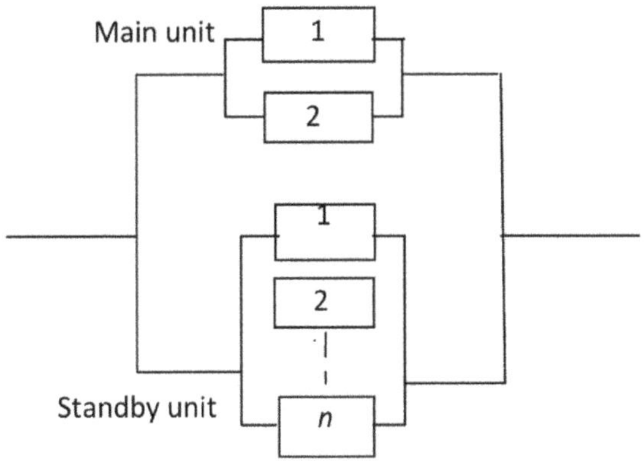

Figure 2. Transition diagram

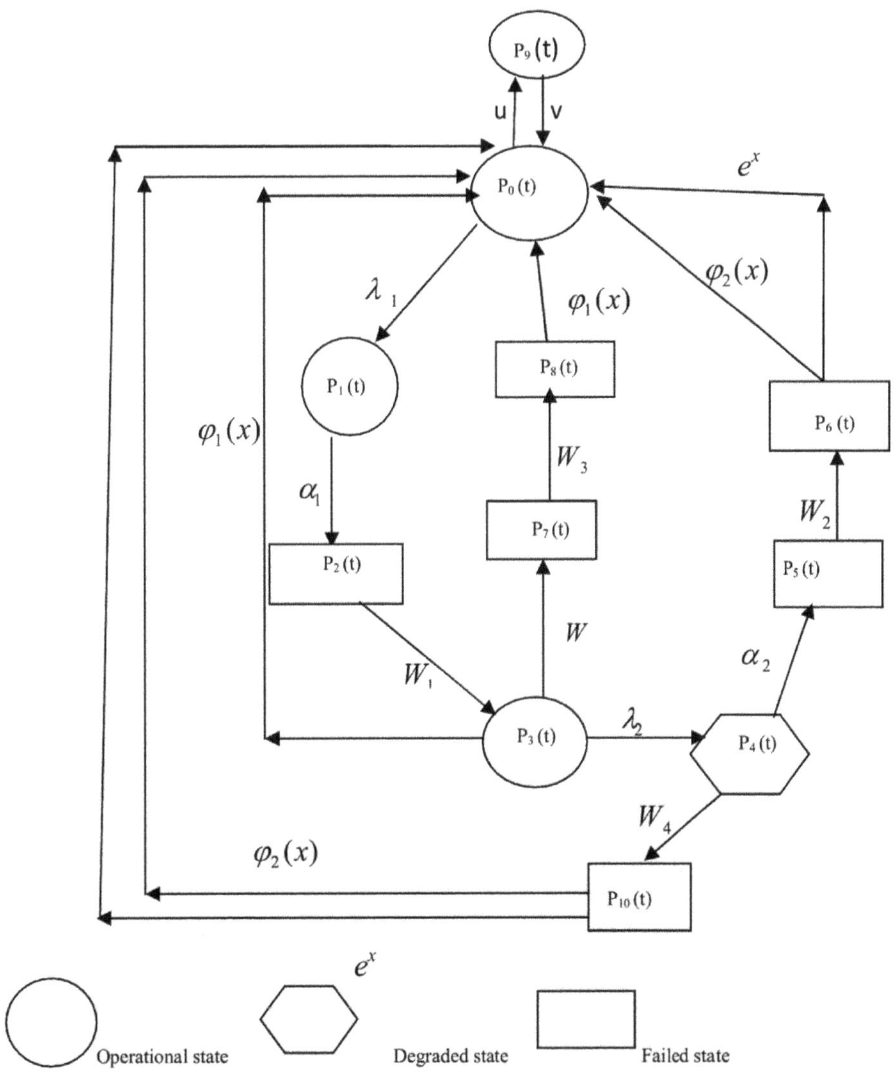

$$\left(\frac{d}{dt}+\alpha_1\right)P_1(t)=\lambda_1 P_0(t) \tag{2}$$

$$\left(\frac{d}{dt}+W_1\right)P_2(t)=\alpha_1 P_1(t) \tag{3}$$

$$\left(\frac{\partial}{\partial t}+\frac{\partial}{\partial x}+\lambda_2+W+\phi_1(x)\right)P_3(x,t)=0 \tag{4}$$

$$\left(\frac{d}{dt}+\alpha_2+W_4\right)P_4(t)=\lambda_2 P_3(t) \tag{5}$$

$$\left(\frac{d}{dt}+W_2\right)P_5(t)=\alpha_2 P_4(t) \tag{6}$$

$$\left(\frac{\partial}{\partial t}+\frac{\partial}{\partial x}+\varphi(x)\right)P_6(x,t)=0 \tag{7}$$

$$\left(\frac{\partial}{\partial t}+\frac{\partial}{\partial x}+\varphi(x)\right)P_{10}(x,t)=0 \tag{8}$$

$$\left(\frac{d}{dt}+W_3\right)P_7(t)=WP_3(t) \tag{9}$$

$$\left(\frac{\partial}{\partial t}+\frac{\partial}{\partial x}+\phi_1(x)\right)P_8(x,t)=0 \tag{10}$$

$$\left(\frac{d}{dt}+v\right)P_9(t)=uP_0(t) \tag{11}$$

Boundary conditions:

$$P_3(0,t) = W_1 P_2(t) \quad (12)$$

$$P_6(0,t) = W_2 P_5(t) \quad (13)$$

$$P_{10}(0,t) = W_4 P_4(t) \quad (14)$$

$$P_8(0,t) = W_3 P_7(t) \quad (15)$$

Initial condition:

$P_0(0) = 1$, and other state probabilities are zero at t=0.

SOLUTION OF THE MODEL

Taking Laplace transformation of (1) to (9) and on further simplification, one may obtain:

$$\overline{P_0}(s) = \frac{1}{A(s)} \quad (16)$$

$$\overline{P_1}(s) = \frac{\lambda_1}{s+\alpha_1} \cdot \frac{1}{A(s)} \quad (17)$$

$$\overline{P_2}(s) = \frac{\alpha_1 . \lambda_1}{(s+W_1).(s+\alpha_1)} \frac{1}{A(s)} \quad (18)$$

$$\overline{P_3}(s) = \frac{W_1 . \alpha_1 . \lambda_1}{(s+W_1).(s+\alpha_1)} \cdot \left[\frac{1-\overline{S\phi_1}(s+\lambda_2+W)}{(s+\lambda_2+W)} \right] \cdot \frac{1}{A(s)} \quad (19)$$

$$\overline{P_4}(s) = \frac{\lambda_{2.}W.\alpha_1.\lambda_1}{(s+\alpha_2+W_4).(s+W).(s+\alpha_1)_1}\left[\frac{1-\overline{S\phi_1}(s+\lambda_2+W)}{(s+\lambda_2+W)}\right].\frac{1}{A(s)} \tag{20}$$

$$\overline{P_5}(s) = \frac{\alpha_2.\lambda_2 W_1.\alpha_1.\lambda_1}{(s+W_2).(s+\alpha_2+W_4).(s+W_1).(s+\alpha_1)}\left[\frac{1-\overline{S\phi_1}(s+\lambda_2+W)}{(s+\lambda_2+W)}\right].\frac{1}{A(s)} \tag{21}$$

$$\overline{P_6}(s) = \frac{W_2.\alpha_{2.}\lambda_2.W_1.\alpha_1.\lambda_1}{(s+W_2).(s+\alpha_2+W_4).(s+W_1).(s+\alpha_1)}\left[\frac{1-\overline{S\phi_1}(s+\lambda_2+W)}{(s+\lambda_2+W)}\right].\left[\frac{1-\overline{S\varphi(s)}}{s}\right].\frac{1}{A(s)} \tag{22}$$

$$\overline{P_7}(s) = \frac{W.W_1.\alpha_1.\lambda_1}{(s+W_3).(s+W_1).(s+\alpha_1)}.\left[\frac{1-\overline{S\phi_1}(s+\lambda_2+W)}{(s+\lambda_2+W)}\right].\frac{1}{A(s)} \tag{23}$$

$$\overline{P_8}(s) = \frac{W_3.W.W_1.\alpha_1.\lambda_1}{(s+W_3).(s+W_1).(s+\alpha_1)}.\left[\frac{1-\overline{S\phi_1}(s+\lambda_2+W)}{(s+\lambda_2+W)}\right].\left[\frac{1-\overline{S\phi_1(s)}}{s}\right].\frac{1}{A(s)} \tag{24}$$

$$\overline{P_{10}}(s) = \frac{W_4.\lambda_2.W_1.\alpha_1.\lambda_1}{(s+\alpha_2+W_4).(s+W_1).(s+\alpha_1)}\left[\frac{1-\overline{S\phi_1}(s+\lambda_2+W)}{(s+\lambda_2+W)}\right].\left[\frac{1-\overline{S\varphi(s)}}{s}\right].\frac{1}{A(s)} \tag{25}$$

where

$$A(s) = (s+\lambda_1+u) - \frac{uv}{s+v} - \frac{W_1.\alpha_1.\lambda_1.\overline{S\phi_1}(s+\lambda_2+W)}{(s+W_1).(s+\alpha_1)} - \frac{W_3.W.W_1.\alpha_1.\lambda_1}{(s+W_3).(s+W_1).(s+\alpha_1)}$$
$$\left[\frac{1-\overline{S\phi_1}(s+\lambda_2+W)}{(s+\lambda_2+W)}\right]\overline{S\phi_1}(s) - \frac{W_4.\lambda_2.W_1.\alpha_1.\lambda_1}{(s+\alpha_2+W_4).(s+W_1).(s+\alpha_1)}\left[\frac{1-\overline{S\phi_1}(s+\lambda_2+W)}{(s+\lambda_2+W)}\right].\overline{S\varphi(s)} \tag{26}$$
$$-\frac{W_2.\alpha_2.\lambda_2.W_1.\alpha_1.\lambda_1}{(s+W_2).(s+\alpha_2+W_4).(s+W_1).(s+\alpha_1)}\left[\frac{1-\overline{S\phi_1}(s+\lambda_2+W)}{(s+\lambda_2+W)}\right]\overline{S\varphi(s)}$$

$$\overline{P}_0(s)+\overline{P}_1(s)+\overline{P}_2(s)+\overline{P}_3(s)+\overline{P}_4(s)+\overline{P}_5(s)+\overline{P}_6(s)+\overline{P}_7(s)+\overline{P}_8(s)+\overline{P}_9(s)+\overline{P}_{10}(s)+\overline{P}_{11}(s)=\frac{1}{s} \tag{27}$$

$$\overline{P}_{up}(s)=\left[1+\frac{\lambda_1}{(s+\alpha_1)}+\frac{W_1.\alpha_1.\lambda_1}{(s+W_1)(s+\alpha_1)}\left\{\frac{1-\overline{S\phi_1}(s+\lambda_2+W)}{(s+\lambda_2+W)}\right\}\right]\frac{1}{A(s)} \tag{28}$$

EVALUATION OF LAPLACE TRANSFORMATION OF UP AND DOWNSTATE PROBABILITIES

The Laplace Transformation of the probabilities that the system is in operable and down state at time‘t’ can be evaluated as follows.

$$\overline{P}_{down}(s)=\overline{P}_2(s)+\overline{P}_4(s)+\overline{P}_5(s)+\overline{P}_6(s)+\overline{P}_7(s)+\overline{P}_8(s)+\overline{P}_9(s)+\overline{P}_{10}(s)$$

$$\begin{aligned}\overline{P}_{down}(s)={}&\frac{\alpha_1.\lambda_1}{(s+W_1).(s+\alpha_1)}.\frac{1}{A(s)}+\frac{\lambda_2.W_1.\alpha_1.\lambda_1}{(s+\alpha_2+W_4).(s+W_1).(s+\alpha_1)}\left\{\frac{1-\overline{S\phi_1}(s+\lambda_2+W)}{(s+\lambda_2+W)}\right\}.\frac{1}{A(s)}\\&+\frac{\alpha_2.\lambda_2.W_1.\alpha_1.\lambda_1}{(s+W_2).(s+\alpha_2+W_4)(s+W_1).(s+\alpha_1)}\left\{\frac{1-\overline{S\phi}(s+\lambda_2+W)}{(s+\lambda_2+W)}\right\}\frac{1}{A(s)}\\&+\frac{W_2.\alpha_2.\lambda_2.W_1.\alpha_1.\lambda_1}{(s+W_2).(s+\alpha_2+W_4).(s+W_1).(s+\alpha_1)}\left\{\frac{1-\overline{S\phi_1}(s+\lambda_2+W)}{(s+\lambda_2+W)}\right\}\left\{\frac{1-\overline{S\varphi}(s)}{s}\right\}\frac{1}{A(s)}\\&+\frac{W.W_1.\alpha_1.\lambda_1}{(s+W_3).(s+W_1).(s+\alpha_1)}\left\{\frac{1-\overline{S\phi_1}(s+\lambda_2+W)}{(s+\lambda_2+W)}\right\}.\frac{1}{A(s)}+\frac{W_3.W.W_1.\alpha_1.\lambda_1}{(s+W_3).(s+W_1).(s+\alpha_1)}\left\{\frac{1-\overline{S\phi_1}(s+\lambda_2+W)}{(s+\lambda_2+W)}\right\}\left\{\frac{1-\overline{S}_{\phi_1}(s)}{s}\right\}\\&\frac{1}{A(s)}+\frac{u}{s+v}.\frac{1}{A(s)}+\frac{W_4.\lambda_2.W_1.\alpha_1.\lambda_1}{(s+\alpha_2+W_4)(s+W_1)(s+\alpha_1)}.\left\{\frac{1-\overline{S\phi_1}(s+\lambda_2+W)}{(s+\lambda_2+W)}\right\}\left\{\frac{1-\overline{S\varphi}(s)}{s}\right\}.\end{aligned} \tag{29}$$

ASYMPTOTIC BEHAVIOR OF THE SYSTEM

Using Abel’s lemma in Laplace transformation viz.

$\lim_{s\to 0}\left\{s\overline{F}(s)\right\}=\lim_{t\to\infty}F(t)=F(say)$ provided that the limit on right hand exists in (28) and (29), the time independent up and down state probabilities are obtained as follows:

$$P_{up} = \left[1 + \frac{\lambda_1}{\alpha_1} + \frac{\lambda_1}{\lambda_2 + W + \phi_1}\right] \cdot \frac{1}{A(0)}$$

$$P_{down} = \frac{\lambda_1}{W_1} + \frac{\lambda_1 . \lambda_2}{(\alpha_2 + W_4)} \cdot \frac{1}{(\lambda_2 + W + \phi_1)} + \frac{\alpha_2 . \lambda_2 . \lambda_1}{W_2 . (\alpha_2 + W_4)} \cdot \frac{1}{(\lambda_2 + W + \phi_1)} + \frac{\alpha_2 . \lambda_2 . \lambda_1}{(\alpha_2 + W_4)} \frac{1}{(\lambda_2 + W + \phi_1)}$$

$$\frac{1}{\varphi} + \frac{W . \lambda_1}{W_3} \cdot \frac{1}{(\lambda_2 + W + \phi_1)} + \frac{W . \lambda_1}{(\lambda_2 + W + \phi_1)} \cdot \frac{1}{\phi_1} + \frac{u}{v} + \frac{W_4 . \lambda_2 . \lambda_1}{(\alpha_2 + W_4)} \cdot \frac{1}{(\lambda_2 + W + \phi_1)} \cdot \frac{1}{\varphi}$$

PARTICULAR CASE

Constant Repair Rates

When repair follows exponential distribution. In this case the result can be derived by putting

$$\bar{S}_{\phi}(s) = \frac{\phi(x)}{s + \phi(x)}, \bar{S}_{\phi_1}(s + W + \lambda_2) = \frac{\phi_1(x)}{s + W + \lambda_2 + \phi_1(x)}$$

$$\bar{S}_{\varphi}(s) = \frac{\exp[x^{\theta} + \{\log \phi_2(x)\}^{\theta}]^{\frac{1}{\theta}}}{s + \exp[x^{\theta} + \{\log \phi_2(x)^{\theta}\}^{\frac{1}{\theta}}}$$

in Equations 16-27, we get $\bar{P}_1(s) = \frac{1}{A(s)}$ (30)

$$\bar{P}_1(s) = \frac{\lambda_1}{s + \alpha_1} . \bar{P}_0(s) \tag{31}$$

$$\bar{P}_2(s) = \frac{\alpha_1 . \lambda_1}{(s + W_1)(s + \alpha_1)} \bar{P}_0(s) \tag{32}$$

$$\bar{P}_3(s) = \frac{W_1 . \alpha_1 . \lambda_1}{(s + W_1) . (s + \alpha_1)(s + \lambda_2 + W + \phi_1)} \bar{P}_0(s) \tag{33}$$

$$\overline{P_4}(s) = \frac{\lambda_{2.}W.\alpha_1.\lambda_1}{(s+\alpha_2+W_4).(s+W).(s+\alpha_1).(s+\lambda_2+W+\phi_1)}\overline{P_0}(s) \tag{34}$$

$$\overline{P_5}(s) = \frac{\alpha_2.\lambda_2 W_1.\alpha_1.\lambda_1}{(s+W_2).(s+\alpha_2+W_4).(s+W_1).(s+\alpha_1).(s+\lambda_2+W+\phi_1)}\overline{P_0}(s) \tag{35}$$

$$\overline{P_6}(s) = \frac{W_2.\alpha_{2.}\lambda_2.W_1.\alpha_1.\lambda_1}{(s+W_2).(s+\alpha_2+W_4).(s+W_1).(s+\alpha_1).(s+\lambda_2+W+\phi_1).(s+\varphi)}\overline{P_0}(s) \tag{36}$$

$$\overline{P_7}(s) = \frac{W.W_1.\alpha_1.\lambda_1}{(s+W_3).(s+W_1).(s+\alpha_1).(s+\lambda_2+W+\phi_1)}\overline{P_0}(s) \tag{37}$$

$$\overline{P_8}(s) = \frac{W_3.W.W_1.\alpha_1.\lambda_1}{(s+W_3).(s+W_1).(s+\alpha_1).(s+\lambda_2+W+\phi_1).(s+\phi_1)}\overline{P_0}(s) \tag{38}$$

$$\overline{P_9}(s) = \frac{u}{(s+v)}\overline{P_0}(s) \tag{39}$$

$$\overline{P_{10}}(s) = \frac{W_4.\lambda_2.W_1.\alpha_1.\lambda_1}{(s+\alpha_2+W_4).(s+W_1).(s+\alpha_1).(s+\lambda_2+W+\phi_1).(s+\varphi)}\overline{P_0}(s) \tag{40}$$

where

$$A(s) = (s+\lambda_1+u) - \frac{uv}{s+v} - \frac{W_1.\alpha_1.\lambda_1.\phi_1}{(s+W_1).(s+\alpha_1)(s+\lambda_2+W+\phi_1)} - \frac{W_3.W.W_1.\alpha_1.\lambda_1.\phi_1}{(s+W_3).(s+W_1).(s+\alpha_1)(s+\lambda_2+W+\phi_1)(s+\phi_1)}$$

$$\frac{W_4.\lambda_2.W_1.\alpha_1.\lambda_1.\varphi}{(s+\alpha_2+W_4).(s+W_1).(s+\alpha_1)(s+\lambda_2+W+\phi_1)(s+\varphi)}. - \frac{W_2.\alpha_2.\lambda_2.W_1.\alpha_1.\lambda_1}{(s+W_2).(s+\alpha_2+W_4).(s+W_1).(s+\alpha_1)}$$

$$\frac{\varphi}{(s+\lambda_2+W+\phi_1)(s+\varphi)}$$

$$\varphi = \exp[(x)^{\theta} + (\log \phi_2(x))^{\theta}]^{\frac{1}{\theta}}$$

Availability of the System for W_1=W_2=0

Let both units do not go for inspection after failure, then $W_1=W_2=0$. Taking $\lambda_1=0.1$, $\alpha_1=0.3$, $\lambda_2=0.2$, $\alpha=0.4$, $W_4=0.6$, $u=0.05$, $W_3=.25$, $W_2=0.4$, repair rates $\Phi=\psi_1=1$, $\theta=1$ and $x=1$. Substituting all these values in equation (28) and taking inverse Laplace transformation, we get

$$P_{up}=0.0498931271e^{(-1.0552343180t)}+1.401719776e^{(0.0947658225t)}-0.4516129032e^{(-0.3000000000t)}\ldots \quad (41)$$

Now in equation (41) setting t=0,1,2,3,4,5,6,7,8,9,10, one can obtain Table 2 and correspondingly Figure 3 which shows the variation of availability with respect to time.

NUMERICAL COMPUTATION

Reliability Analysis

Let the failure rate of the main unit and standby unit is $\lambda_1=0.1$, $\alpha_1=0.3$, $\lambda_2=0.2$, $\alpha_2=0.4$, inspection rate $W_1=0.5$, $W_4=0.6$, waiting rate $W=0.7$, preventive maintenance rate $u=0.05$, repair rates $\Phi=\psi_1=0$, $\theta=1$ and $x=1$. Also assume that the repair follows exponential distribution, i.e.

Figure 3. Time vs. availability

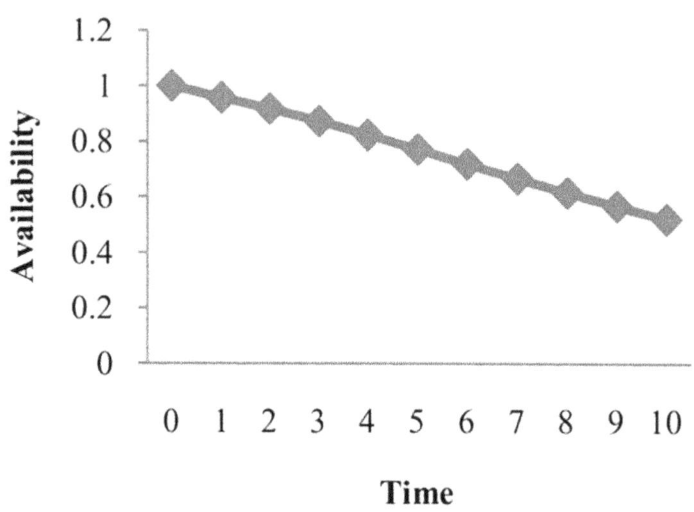

$$\bar{S}_{\varphi}(s)=\frac{\exp[x^{\theta}+\{\log\phi_2(x)\}^{\theta}]^{\frac{1}{\theta}}}{s+\exp[x^{\theta}+\{\log\phi_2(x)^{\theta}\}^{\frac{1}{\theta}}}$$

$$\bar{S}\phi(s)=\frac{\phi(x)}{s+\phi(x)},\ \bar{S}\phi_1(s+W+\lambda_2)=\frac{\phi_1(x)}{s+W+\lambda_2+\phi_1(x)} \qquad \ldots \qquad (42)$$

Putting all these values in equation (28) and taking inverse Laplace transform and varying time, one can obtain Table 3. Also Figure 4 shows how reliability varies with respect to time in this case.

Availability Analysis

Assuming the parameters λ_1=0.1, α_1=0.3, λ_2=0.2, α_2=0.4, inspection rate W_1=0.5, W_4=0.6 waiting rate W=0.7, W_3=.25, W_2=0.4, preventive maintenance rate u=0.05, repair rates $\Phi=\psi_1=1$, $\theta=1$ and $x=1$ and letting that Equation 42 holds. Putting all these values in Equation 28 and taking inverse Laplace transformation, we get

P_{up}=0.7541449550-0.0003725436860exp (-1.904844177t) +0.10442914exp (-0.5584701205t) +0.445694821exp (-0.31738127t) sin (0.1818769175t) +0.04377848645exp (-1.024997109t) sin (0.05345866184 t)+0.04176185432 exp(-1.024997109 t) cos (0.05345866184 t)+0.0642

7442649exp (-0.3173812747t) cos (0.1818769175t) +0.63576215806exp (-0.3519289348t) (43)

Now in equation (43) setting t=0,1,2,3,4,5,6,7,8,9,10, one can obtain Table 4. The same is shown in Figure 5 which shows the variation of availability with respect to time.

Figure 4. Time vs. reliability

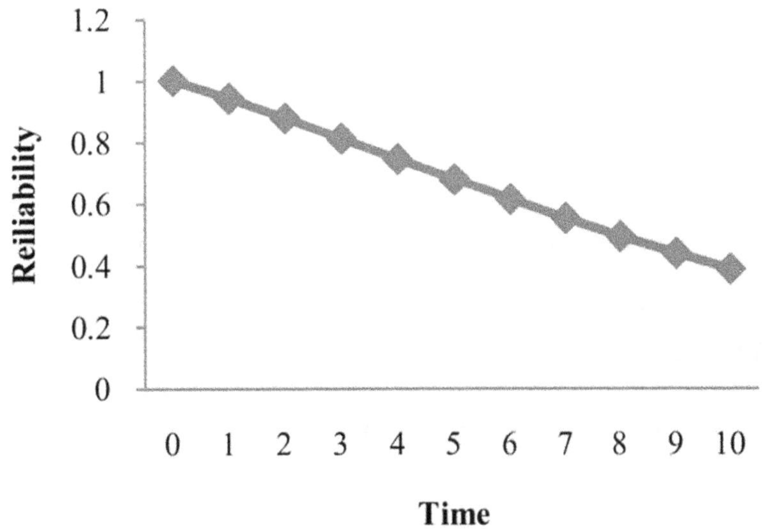

Figure 5. Time vs. availability

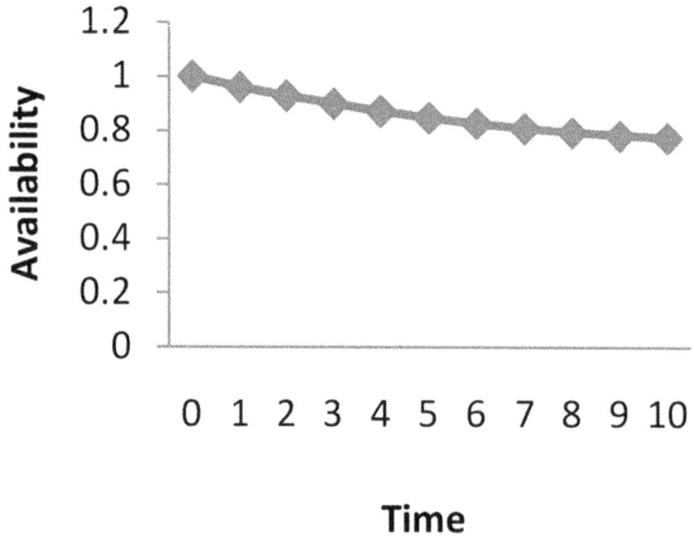

Down Time Analysis

Letting the parameters $\lambda_1=0.1$, $\alpha_1=0.3$, $\lambda_2=0.2$, $\alpha_2=0.4$, inspection rate $W_1=0.5$, $W_4=0.6$, waiting rate $W=0.7$, $W_3=.25$, $W_2=0.4$, preventive maintenance rate $u=0.05$, repair rates $\Phi=\psi_1=1$, $\theta=1$ and $x=1$ and assuming that equation (42) holds. Putting all these values in equation (29) and taking inverse Laplace transformation, we get

$P_{down}=0.2458550450-0.3576215807\ e^{-1}\ \exp\ (-0.3519289348\ t)\ +0.3725439274\ e^{-3}\ \exp\ (-1.904844177$

$t)-0.1044291498\ \exp\ (-0.5584701205\ t)-0.4456948213\ \exp\ (-0.3173812747\ t)$

$\sin\ (0.1818769175\ t)-0.4176185431\ e^{-1}\ \exp(-1.024997109\ t)\ \cos(0.5345866184\ e^{-1}\ t)-$

$0.4377848646\ e^{-1}\exp\ (-1.024997109\ t)\ \sin\ (0.5345866184e^{-1}t)-0.6427442674e^{-1}\exp\ (-$

$0.3173812747t)\cos(0.1818769175t)$ … (44)

Now in equation (44) setting t=0,1,2,3,4,5,6,7,8,9,10, one can obtain Table 5 and correspondingly Figure 6 which shows the change of downtime with respect to time.

Busy Period Analysis

Mean time to repair (M.T.T.R.) of the system is given by

Figure 6. Down time vs. time

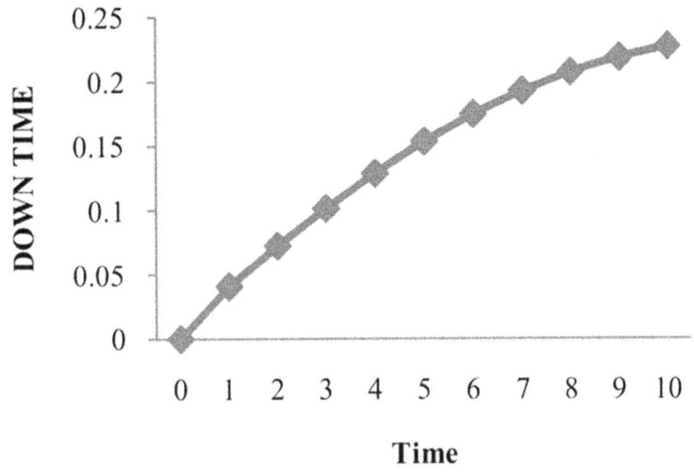

$$M.T.T.R. = \lim_{s \to 0} \overline{P}_{down}(s)$$

When equation (42) holds, we can have following three cases.

1. Assuming ψ_1=Φ=1, x=1, $\lambda_{2=}$0.2, α_1=0.3, α_2=0.4, W_1=0.5, W_4=0.6, W=0.7 W_3=.25. W_2=0.4, u=0.05 and λ_1 as 0.1, 0.2, 0.3, 0.4, 0.5, 0.6, 0.7, 0.8, 0.9, 1.0, one can obtain Table 6. The same is shown in Figure 7 which demonstrates variation of busy period with respect to $\lambda_{1.}$
2. Varying λ_2 as 0.1, 0.2, 0.3, 0.4, 0.5, 0.6, 0.7, 0.8, 0.9, 1.0 and the other parameter are kept fixed as ψ_1=Φ=1, x=1, $\lambda_{1=}$0.1, α_1=0.3, α_2=0.4, W_1=0.5, W_4=0.6, W=0.7, W_3=.25, W_2=0.4, u=0.05, one can obtain Table 7, that is depicted by Figure 8 which demonstrates change of busy period with respect to $\lambda_{2.}$
3. Setting ψ_1=Φ=1, x=1, $\lambda_{1=}$0.1, λ_2=0.2, α_1=0.3, W_1=0.5, W_4=0.6, W=0.7, W_3=.25, W_2=0.4, u=0.05 and varying α_2 as 0.1, 0.2, 0.3, 0.4, 0.5, 0.6, 0.7, 0.8, 0.9, 1.0, one can obtain Table 8. This is shown in Figure 9 which demonstrates how busy period changes with respect to α_2.

M.T.T.F. Analysis

Mean time to failure of the system is given by

$$M.T.T.F = \lim_{s \to 0} \overline{P}_{up}(s)$$

Assuming that repair follows exponential distribution, i.e. equation (42) holds.

Figure 7. Variation of busy period with respect to λ_1

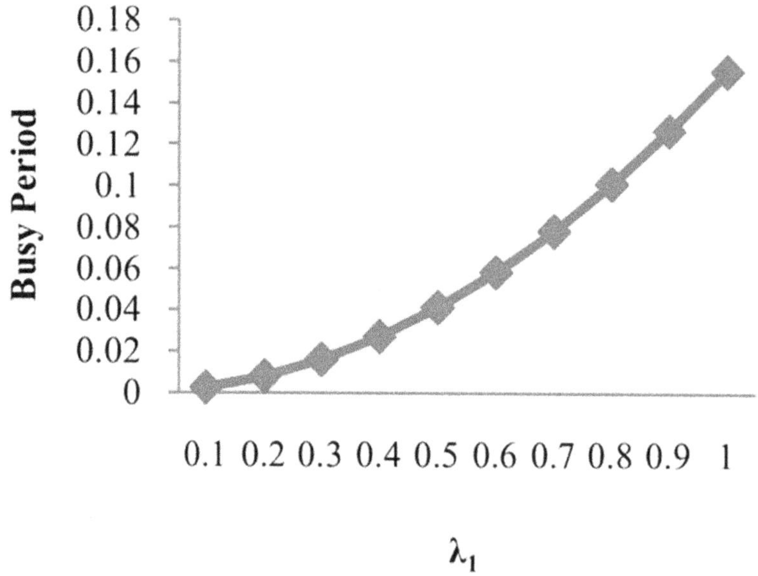

1. Setting ψ_1=Φ=0, x=1, λ_2=0.2, α_1=0.3, α_2=0.4, W_1=0.5, W_4=0.6, W=0.7, u=0.05 and varying λ_1 as 0.1, 0.2, 0.3, 0.4, 0.5, 0.6, 0.7, 0.8, 0.9, 1.0, one can obtain variation of M.T.T.F. with respect to λ_1 which is given in Table 9.
2. Varying λ_2 as 0.1, 0.2, 0.3, 0.4, 0.5, 0.6, 0.7, 0.8, 0.9, 1.0, and keeping other parameter fixed at ψ_1 =Φ=0, x=1, λ_1=0.1, α_1=0.3, α_2=0.4, W_1=0.5, W_4=0.6, W=0.7, u=0.05, one can obtain the change of M.T.T.F. with respect to λ_2 as given in Table 10.
3. Assuming ψ_1=Φ=0, x=1, λ_1=0.1, λ_2=0.2, α_2=0.4, W_1=0.5, W_4=0.6, W=0.7, u=0.05 and α_1 as 0.1, 0.2, 0.3, 0.4, 0.5, 0.6, 0.7, 0.8, 0.9, 1.0. The variation of M.T.T.F. with respect to α_1 is obtained which is given in Table 11.
4. Letting ψ_1 =Φ=0, x=1, λ_1=0.1, λ_2=0.2, α_1=0.3, W_1=0.5, W_4=0.6, W=0.7, u=0.05 and taking different values of α_2 as 0.1, 0.2, 0.3, 0.4, 0.5, 0.6, 0.7, 0.8, 0.9, 1.0, one can obtain Table 12 which shows how M.T.T.F. changes with respect to α_2.

Variations of M.T.T.F. with respect to λ_1, λ_2, α_1 and α_2 of the cases (1), (2), (3) and (4) have been shown by the Figures 10, 11, 12 and 13 respectively.

Setting λ_1=0.1, λ_2=0.2, α_1=0.3, α_2=0.4, W_1=0.5, W_4=0.6, W=0.7, Φ=ψ_1=1, θ=1 and x=1. Furthermore if the repair follows exponential distribution (i.e. equation (42) holds), putting all these values in (28) and taking inverse Laplace transform, one can obtain equation (43). If the service facility is always available, then expected profit during the interval (0, *t*] is given by

$$E_P(t) = K_1 \int_0^t P_{up}(t)dt - K_2 t$$

where K_1 and K_2 are the revenue per unit time and service cost per unit time respectively, then

Figure 8. Change of busy period with respect to λ_2

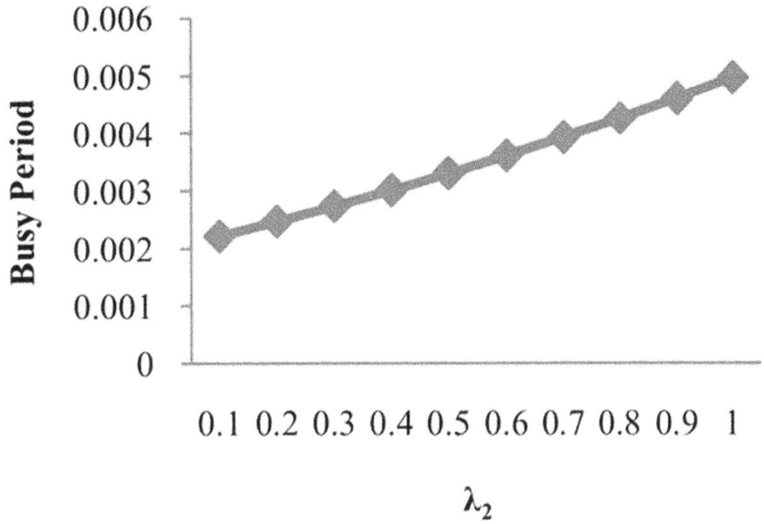

Figure 9. Busy period changes with respect to α_2

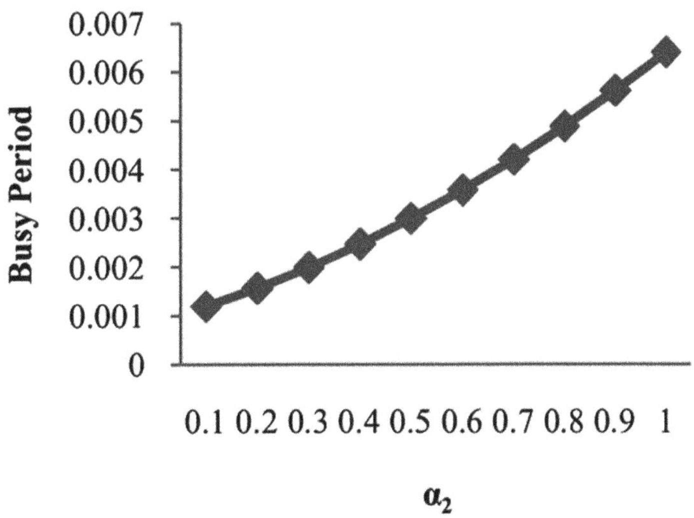

$$E_p(t)=K_1[0.7541449550t+0.0001955769876e(-1.904844177t)-0.1869914718e(-0.5584701205t)-0.7582470217e(-0.3173812747t)\cos(0.1818769175t)-0.9697711079e(-0.3173812747t)\sin(0.1818769175t)-0.04285439877e(-1.024997109t)\cos(0.05345866184t)-0.04047577039e(-1.024997109t)\sin(0.05345866184t)-0.1016175555e(-0.3519289348t)+1.089514871]-K_2t \ldots \quad (45)$$

Keeping $K_1=1$ and changing K_2 at 0.1, 0.2, 0.3, 0.4, 0.5 in Equation (45), one can obtain Table 13. which is depicted by Figure 14.

Figure 10. M.T.T.F. vs. λ_1

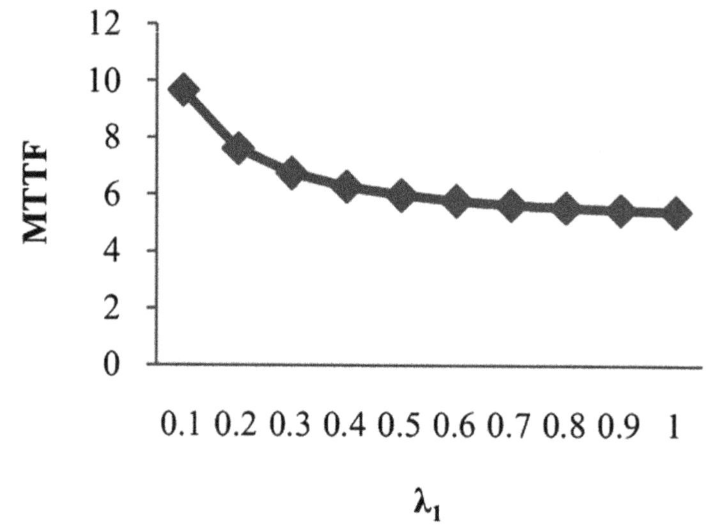

Figure 11. M.T.T.F. vs. λ_2

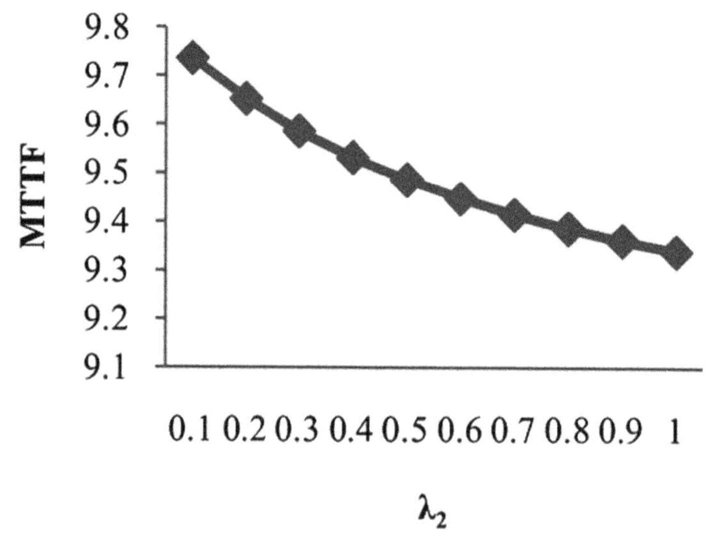

CONCLUSION

The following conclusions may be drawn on the basis of the studies conducted in the present chapter.

1. We can also observe from Figure 4 that reliability of the system decreases with the increment in time. From the examination of Figure 5 one can see that when λ_1=0.1, λ_2=0.2, α_1 =0.3, α_2=0.4, W_1=0.5, W_4=0.6, W=0.7, u=0.05 then availability of the system initially decreases with respect to time and later on it seems to be constant as the time increases. Also from Figure 6 one can conclude that the downtime of the system increases with respect to time.

Figure 12. M.T.T.F. vs. α_1

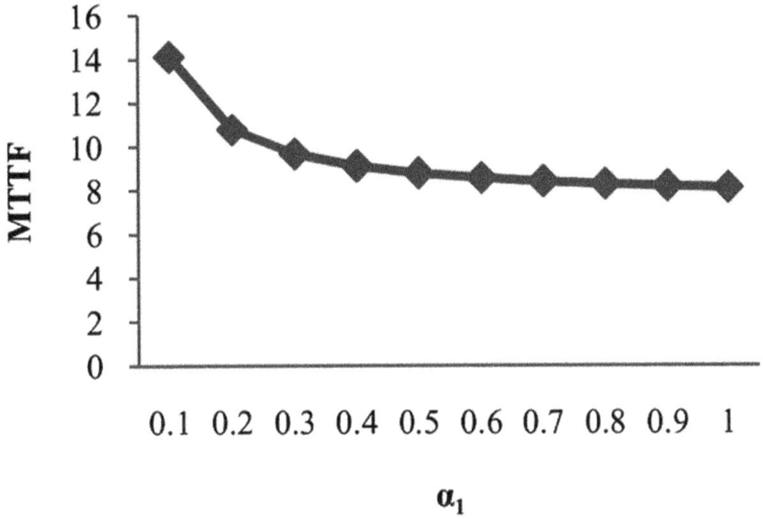

Figure 13. M.T.T.F. vs. α_2 cost analysis

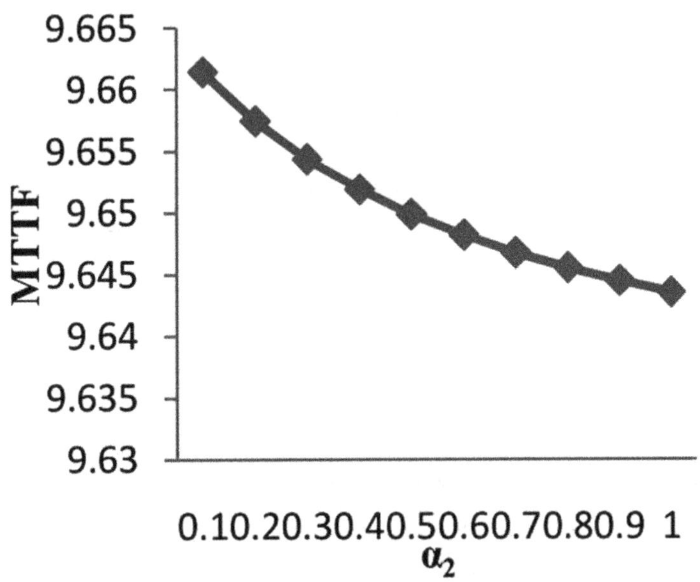

2. Tables 6, 7 and 8 [corresponding Figures 7, 8 and 9] depict the variation of busy period with respect to λ_1, λ_2, and α_2 respectively when other parameters are kept constant. Critical examination of the figures reveals that busy period varies between 0.00246525 and 0.15551025, 0.00221400 and 0.00496125, 0.001190018 and 0.006393120 with respect to λ_1, λ_2 and α_2 respectively. One can easily conclude that as the value of λ_1, λ_2 and α_2 increases, busy period increases.
3. From Figures 10, 11, 12, and 13 one can see the change of M.T.T.F. with respect to λ_1, λ_2, α_1 and α_2 respectively. The figures reveal that M.T.T.F. varies between 9.651 and 5.407, 9.734 and 9.339,

Figure 14. Time vs. expected profit

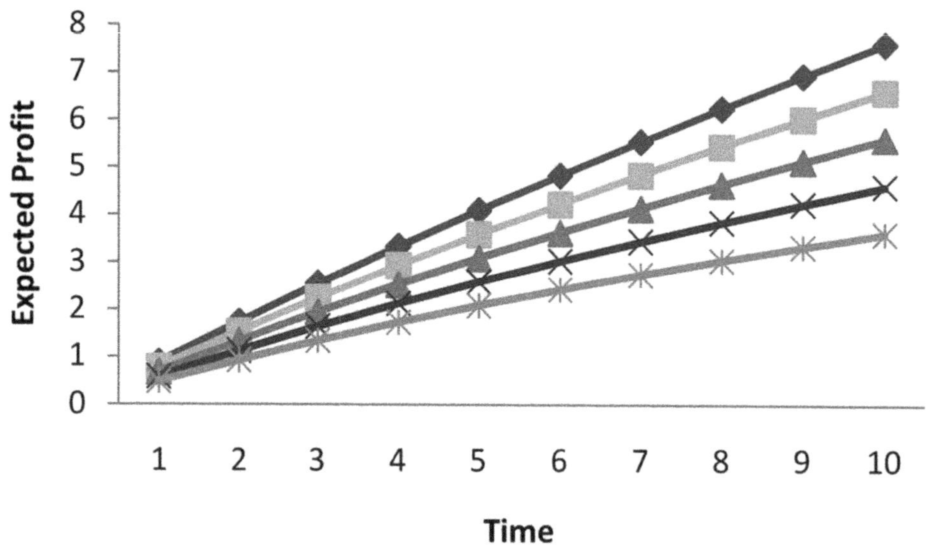

14.096 and 8.096, 9.661 and 9.643 respectively with respect to λ_1, λ_2, α_1 and α_2. It can easily be concluded from the figures that M.T.T.F. of the system decreases as the value of λ_1, λ_2, α_1 and α_2 increases.

4. Keeping revenue cost per unit time K_1 at value 1 and varying service cost K_2 as 0.1, 0.2, 0.3, 0.4, 0.5, Figure 14 is obtained. This graph reveals an important conclusion that increasing service cost leads decrement in expected profit. The highest and lowest values of expected profit are obtained to be 7.595 and 0.478 respectively for the assumed values. Expected profit decreases with the increment in K_2.

REFERENCES

Agarwal, S. C., Sahani, M., & Bansal, S. (2010). Reliability Characteristic of cold-standby redundant system. *International Journal of Research and Reviews in Applied Sciences*, *3*(2), 193–199.

Gopalan, M. N., & Nagarwalia, H. E. (1985). Cost benefit analysis of a one server two-unit cold standby system with repair and preventive maintenance. *Microelectronics and Reliability*, *25*(2), 267–269. doi:10.1016/0026-2714(85)90011-3

Haggag, M. Y. (2009). Cost Analysis of Two-Dissimilar-Unit Cold Standby System with three States and Preventive Maintenance using Linear First Order Differential Equations. *Journal of Mathematics and Statistics*, *5*(4), 395–400. doi:10.3844/jmssp.2009.395.400

Hatogama, Y. (1997). Reliability analysis of three state system. *IEEE Transactions on Reliability*, *28*, 386–393.

Kumar, D., & Singh, S. B. (2016). Stochastic Analysis of Complex Repairable System with Deliberate Failure Emphasizing Reboot Delay. *Communications in Statistics. Simulation and Computation, 45*, 1–20. doi:10.1080/00949655.2016.1232723

Lasanovsky. (1982). Availability of two-unit cold standby system with degraded state. *IEEE Transactions on Reliability, 31*, 123.

Mokaddis, G. S., & Tawfek, M. L. (1996). Some characteristics of a two-dissimilar-unit cold standby redundant system with three modes. *Microelectronics and Reliability, 36*(4), 497–503. doi:10.1016/0026-2714(94)00128-6

Nelson, R. B. (2006). An introduction to Copulas (2nd ed.). Springer.

Ram, M., & Singh, S. B. (2010). Availability, M. T. T. F. and cost analysis of complex system under preemptive repeat repair discipline using Gumbel-Hougaard family of copula. *International Journal of Quality & Reliability Management, 27*(3), 576–595. doi:10.1108/02656711011043544

Said, K. M., & EL-Sherbeny, M. S. (2005). Profit Analysis of a Two Unit Cold Standby System with Preventive Maintenance and Random Change in Units. *Journal of Mathematics and Statistics, 5*(4), 71-77.

Chapter 11
Welding Process under Fault Coverage Approach for Reliability and MTTF

Nupur Goyal
Graphic Era University, India

Mangey Ram
Graphic Era University, India

Pranshu Kumar
Graphic Era University, India

ABSTRACT

This chapter deals with the problems occur in the welding under the consideration of many causes of defects such as cracking, distortion, porosity, inclusion and undercutting. Welding is very useful for every industry. It is used as a fabrication and sculptural that joins the two parts of materials, generally metals or thermoplastic, through a restricted amalgamation from a suitable combination of temperature, pressure and metallurgical conditions. In this chapter, authors study that the welding process is how much reliable when many causes of defects taken account into this process and try to find out the reliability and mean time to failure of welding process with the help of Markov modelling and supplementary variable technique. They derived the generalised expression for the probability of each possible transition state, probability of upstate and downstate of welding, and also studied the sensitivity in reliability and MTTF of welding. Fault coverage technique has been developed to enhance the reliability of the welding process for practical utility in the industry.

INTRODUCTION

Reliability is a measure of performance of any system or process, particularly welding process; it decides that how well an operation will perform or success for a given time period under the specified operating conditions. There are many traditional techniques to analyse the reliability aspects of any process under the consideration of defects occur in it. Todays, many techniques are also available to enhance the

DOI: 10.4018/978-1-5225-1639-2.ch011

reliability of process including the faults occur during process due to design, wrong selection, environmental conditions etc. Welding process is the core and important part for the manufacturing processes of the material engineering. So, the reliability of welding process is a key factor of manufacturing process because it directly influences the reliability of manufacturing of the system or sub-system. The reliability of welding process depends upon the failure rate of some main defects of welding such as cracking, porosity, under cutting, inclusion and distortion. So, maximize the reliability of the welding process through minimize the failure rates of defects is required for manufacturing process. Reliability work on welding process introduces two basic concepts that have central importance namely, modelling of the designed process and their transition state probability. Reliability of the welding process can be enhanced by detecting the number of faults and try to cover the maximum number of faults. The ratio of number of faults that can be covered to the number of total detected faults is called coverage factor (Arnold, 1973; Bouricius et al., 1969), i.e.

Coverage factor= faults that can be covered/total detected faults

The fault coverage that is used to measure the productivity of the system, is imperfect fault coverage because the validation of the system is depend on the efficiency of their fault mechanism (Powell et al., 1965) while a measure that is used to compute the degree of meticulousness of testing in a software is called software test coverage. Software test coverage and other coverage applications are also determined by many tools in various terms such as hunks, divisions, computation-use etc. (Malaiya et al., 2002; Ram et al., 2013; Ram & Manglik, 2014).

As far as we know that the reliability analysis of welding process with coverage factor has never been discussed in the literature while reliability theory has a wide history in the literature. A lot of researchers including Ascher (1986), Smith and Oren (1980), Singh and Jain (2000), Cempel et al. (2000), Singh et al. (2013), Ram et al. (2013) have analysed the reliability of systems. Ke et al. (2007) investigated the reliability and mean time to failure of a system with standby and unreliable server and also analysed the sensitivity and relative sensitivity of these measures under the consideration of exponential distribution for failure and repair rates of the system. They obtained the generalized expression for reliability, MTTF and its sensitivity and relative sensitivity. Gupta and Tewari (2009) proposed a probabilistic model of flue gas and air system. They formulated the model with Markov birth-death process and analyzed the system availability. Authors also discussed the maintenance priority for the sub-system contained in the flue gas and air system of the thermal power plant. Kumar et al. (2015) analyzed the reliability of the casting process and obtained the generalized expression using Markov process. Goyal et al. (2015) proposed a mathematical model of water cooling system and simulated the results using Markov process. They investigated the reliability measures of water cooling system and also seek the cost-benefit analysis in the maintenance of it.

Malaiya et al. (2002) derived the relation between reliability, software test coverage and time duration of testing. Xing (2007) developed an approach to analyse the reliability of a phased mission system with the common cause failure by introducing imperfect fault coverage. Myres (2007) studied the four different models of coverage namely, perfect fault coverage, element level coverage, fault level coverage and one-on-one level coverage and discussed the effects of coverage using two techniques, combinatorial and recursive and the method how it is obtained. Wang et al. (2013) examined the reliability and mean time to failure of a repairable system under the specified service stress condition with imperfect coverage in which the failed unit is immediately replaced by a standby unit with coverage probability c after detection. They also studied the sensitivity and relative sensitivity of reliability and MTTF.

In this problem, a mathematical model for welding process is developed and designed a transition state diagram subjected to the birth and death Markov process (Barlow & Proschan, 1965; Dhillon & Singh, 1981) and supplementary variable techniques (Cox, 1955; Oliveira et al., 2005). After the solution of the formulated differential equations through Laplace transform, the expression for transition state probabilities, upstate and downstate of welding are obtained and discussed the reliability and mean time to failure of the welding process. Authors developed an efficient algorithm for the improvement of reliability and mean time to failure, and they also studied the sensitivity (Gandini, 1990; Saltelli et al., 2008) in reliability and mean time to failure of welding process using imperfect fault coverage technique. The effect of defects existing in welding process which perform an unsuccessful welding or imperfect welding, are analysed on the reliability aspects and demonstrated by sensitivity analysis of reliability aspects.

MATERIALS

Causes of Defects and Repairing Techniques of Welding Defects

Distortion

Uneven heating of work pieces during welding results in the development of welding stresses which often lead to distortion or warpage of the welded structure or component (Khanna, 1993). The distortion in welding process resulted to the process unsuccessful. This defect can be corrected by:

1. Proper sequencing of weld procedure.
2. Proper jigging of joint prior to welding.

Porosity

Porosity is due to gas getting entrapped in solidifying weld metal and due to porosity, welding process would be partially failed (Parmar, 2010). The sources of trapped gas may be:

1. Rust, dirt, grease or primer on the edges of the parent metal or electrode.
2. Damp SAW flux.
3. Wrong grade of electrode wire.

This defect can be minimized by:

1. Using low hydrogen welding process.
2. Using filler metal high in deoxidizers.
3. Increasing the flow of shielding gas.

Inclusion

Slag inclusions are non-metallic solid materials trapped in the weld metal interfaces. Most arc welding leave behind slag inclusions. This fault also leads the welding process into partially failed state.

The common causes are:

1. Failure to remove slag.
2. Improper joint design.
3. Slag flooding ahead of the welding arc.

It can be eliminated by increasing groove angle of the joint, providing proper gas shielding and cleaning the weld surface.

Undercutting

Undercuts are grooves melted into the parent metal adjacent to the toe of a weld and left unfilled by the weld metal. These reduce the cross section of work and may lead to cracking due to notch effect. Undercutting is generally caused due to:

1. Excessive arc voltage and current.
2. Arc blow.

It carried the welding process into partially or complete failed state. If before undercutting, porosity and inclusion occur in the welding process then welding goes to the complete failed state.

Cracking

Presence of crack in HAZ or parent metal or weld metal is known as cracking. These are considered dangerous because they create a serious reduction in strength (Kalpakjian, 2001). Due to cracking, welding is completely unsuccessful. A remedial action is chipping back and welding up again.

Inspection of Welding

Welding is inspected thoroughly to find the presence of internal flaws as well as external defects. The various methods of inspecting welding are:

1. Visual examination.
2. Liquid penetrant testing.
3. Ultrasonic testing.
4. Acoustic emission testing.

Assumptions

The proposed research work has the following assumptions-

1. Initially, the welding is performed successfully without any defect.
2. The analysis of welding process is done under the consideration of various defects.
3. The process covers three types of states, good, degraded and completely failed.
4. In model 2, the faults occur in the welding can be covered imperfectly.
5. The sufficient repair facility is available after the failure of welding process.
6. The failure and repair rates are follow general time distribution.

Nomenclatures

All the notations used in the proposed work and throughout the analysis are shown in Table 1.

PROBLEM STATEMENT

In the present chapter, reliability indices of the welding process incorporating various defects and failures have been investigated. This research is intended to explain how much a defect can affect the welding system and its performance. Welding is done to fabricate or join metal parts or system providing enough strength and rigidity for bearing mechanical loads. To model the system, a number of failure and repair rates are taken to determine system availability and reliability. The main defects generally encountered in welding works are distortion, porosity, inclusion, undercutting and cracking. Although a poor weld job

Table 1. Notations

Notation	Depiction
t / s	Time scale / Laplace transform variable.
$\bar{P}(s)$	Laplace transformation of $P(t)$.
$P_i(t)$	The probability of good or partially failed state S_α of the welding process; i=0, 1, 2,3; α=G, P, PI, U.
$P_j(x,t)$	The probability density function of the complete failed state S_β of the welding process at epoch time t and has an elapsed repair time of x; j=4, 5, 6, 7; β=PIC, PIU, D, CU.
$\lambda_d / \lambda_c / \lambda_o / \lambda_u$	Failure rate of distortion/ cracking/ porosity and inclusion/ undercutting.
$\mu_0 / \mu_1 / \mu_2$	Repair rate from state S_P to S_G / S_{PI} to S_P / S_U to S_G.
μ	Repair rate from complete failed state to good state.
c	Coverage parameter
Model 1	Welding model without fault coverage technique.
Model 2	Welding model with fault coverage technique.
$Rl_1(t) / Rl_2(t)$	The reliability of the welding in model 1/ model 2 at time t.
$MTTF_1 / MTTF_2$	Mean time to failure of the welding in model 1 / model 2.

would suffer from various other defects but authors have assumed only the above mentioned defects to be most frequent and detrimental in nature. Authors designed the two models of welding process subjected to the above defects and draw the transition state diagram for each model as shown in Figure 1 and 2. The states notified in transition state diagram of the proposed models have been described in Table 2.

MATHEMATICAL DETAILS OF THE MODELS

Formulation of the Models

By the probability of consideration and continuity of arguments, the following differential equations possesses the present mathematical models.

Figure 1. Welding model 1

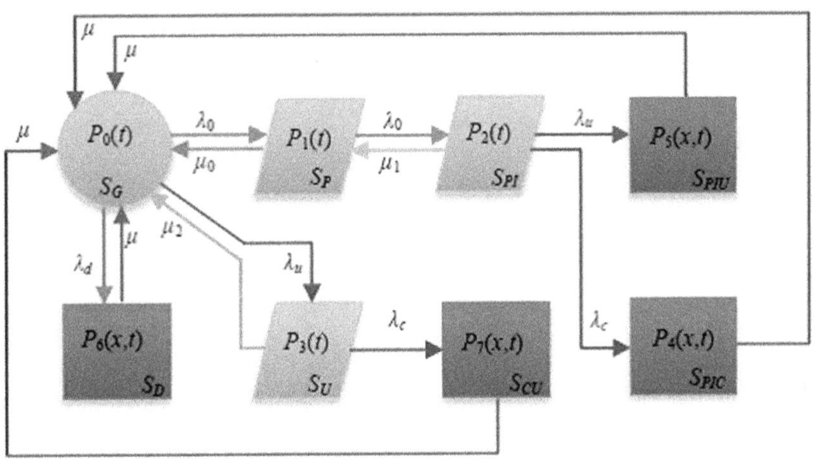

Figure 2. Welding model 2

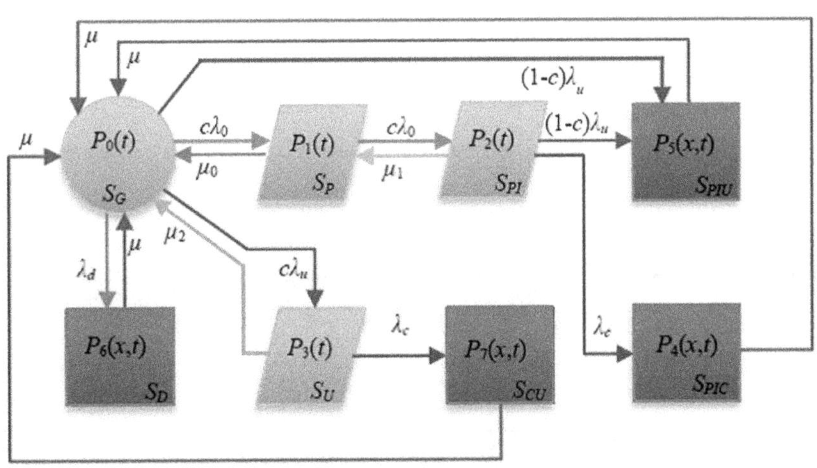

Table 2. State Description

State	Description
S_G	Good working condition of the welding.
S_U	Partially failed state of the welding cause of undercutting defect.
S_P	The welding is partially unsuccessful due to porosity.
S_{PI}	The welding is partially unsuccessful due to porosity and inclusion.
S_D	Unsuccessful state of the welding due to distortion.
S_{CU}	Welding is unsuccessful due to cracking with undercutting defect.
S_{PIC}	Welding is unsuccessful due to porosity, inclusion and cracking.
S_{PIU}	Welding is unsuccessful due to porosity and inclusion with undercutting defect.

Model 1

$$\left[\frac{\partial}{\partial t}+\lambda_d+\lambda_u+\lambda_0\right]P_0(t)=\mu_0P_1(t)+\mu_2P_3(t)+\int_0^\infty \mu(x)\sum_{i=4}^{7}P_i(x,t)dx \tag{1}$$

$$\left[\frac{\partial}{\partial t}+\lambda_0+\mu_0\right]P_1(t)=\lambda_0P_0(t)+\mu_1P_2(t) \tag{2}$$

$$\left[\frac{\partial}{\partial t}+\lambda_c+\lambda_u+\mu_1\right]P_2(t)=\lambda_0P_1(t) \tag{3}$$

$$\left[\frac{\partial}{\partial t}+\lambda_c+\mu_2\right]P_3(t)=\lambda_uP_0(t) \tag{4}$$

$$\left[\frac{\partial}{\partial t}+\frac{\partial}{\partial x}+\mu(x)\right]P_i(x,t)=0; \qquad i=4,5,6,7 \tag{5}$$

Boundary conditions

$$P_i(0,t)=\xi P_2(t); \qquad i=4,5; \qquad \xi=\lambda_c,\ \lambda_u \tag{6}$$

$$P_6(0,t) = \lambda_d P_0(t) \tag{7}$$

$$P_7(0,t) = \lambda_c P_3(t) \tag{8}$$

Initial condition

$$P_i(0) = \begin{cases} 1, & when \quad i = 0 \\ 0, & otherwise \end{cases}$$

Model 2

$$\left[\frac{\partial}{\partial t} + \lambda_d + (1-c)\lambda_u + c\lambda_u + c\lambda_0\right] P_0(t) = \mu_0 P_1(t) + \mu_2 P_3(t) + \int_0^\infty \mu(x) \sum_{i=4}^{7} P_i(x,t)dx \tag{9}$$

$$\left[\frac{\partial}{\partial t} + c\lambda_0 + \mu_0\right] P_1(t) = c\lambda_0 P_0(t) + \mu_1 P_2(t) \tag{10}$$

$$\left[\frac{\partial}{\partial t} + \lambda_c + (1-u)\lambda_u + \mu_1\right] P_2(t) = c\lambda_0 P_1(t) \tag{11}$$

$$\left[\frac{\partial}{\partial t} + \lambda_c + \mu_2\right] P_3(t) = c\lambda_u P_0(t) \tag{12}$$

$$\left[\frac{\partial}{\partial t} + \frac{\partial}{\partial x} + \mu(x)\right] P_i(x,t) = 0; \qquad i = 4,5,6,7 \tag{13}$$

Boundary conditions

$$P_i(0,t) = \xi P_2(t); \qquad i = 4,5; \qquad \xi = \lambda_c,\ (1-c)\lambda_u \tag{14}$$

$$P_6(0,t) = \lambda_d P_0(t) \tag{15}$$

$$P_7(0,t) = \lambda_c P_3(t) \tag{16}$$

Initial condition

$$P_i(0) = \begin{cases} 1, & when \quad i = 0 \\ 0, & otherwise \end{cases}$$

Solution of the Models

Solving both the model by taking Laplace transformation of Equations 1-8 and 9-16 using initial condition, the authors get the solution as shown in Equations 17 to 32

Laplace Transformation of Model 1

$$\left[s + \lambda_d + \lambda_u + \lambda_0\right]\overline{P}_0(s) = 1 + \mu_0 \overline{P}_1(s) + \mu_2 \overline{P}_3(s) + \int_0^\infty \mu(x) \sum_{i=4}^{7} \overline{P}_i(x,s)dx \tag{17}$$

$$\left[s + \lambda_0 + \mu_0\right]\overline{P}_1(s) = \lambda_0 \overline{P}_0(s) + \mu_1 \overline{P}_2(s) \tag{18}$$

$$\left[s + \lambda_c + \lambda_u + \mu_1\right]\overline{P}_2(s) = \lambda_0 \overline{P}_1(s) \tag{19}$$

$$\left[s + \lambda_c + \mu_2\right]\overline{P}_3(s) = \lambda_u \overline{P}_0(s) \tag{20}$$

$$\left[s + \frac{\partial}{\partial x} + \mu(x)\right]\overline{P}_i(x,s) = 0; \qquad i = 4,5,6,7 \tag{21}$$

$$\overline{P}_i(0,s) = \xi \overline{P}_2(s); \qquad i = 4,5; \qquad \xi = \lambda_c, \lambda_u \tag{22}$$

$$\overline{P}_6(0,s) = \lambda_d \overline{P}_0(s) \tag{23}$$

$$\overline{P}_7(0,s) = \lambda_c \overline{P}_3(s) \tag{24}$$

Laplace Transformation of Model 28

$$\left[s + \lambda_d + (1-c)\lambda_u + c\lambda_u + c\lambda_0\right]\overline{P}_0(s) = 1 + \mu_0 \overline{P}_1(s) + \mu_2 \overline{P}_3(s) + \int_0^\infty \mu(x) \sum_{i=4}^{7} \overline{P}_i(x,s)dx \tag{25}$$

$$\left[s + c\lambda_0 + \mu_0\right]\overline{P}_1(s) = c\lambda_0 \overline{P}_0(s) + \mu_1 \overline{P}_2(s) \tag{26}$$

$$\left[s + \lambda_c + (1-u)\lambda_u + \mu_1\right]\overline{P}_2(s) = c\lambda_0 \overline{P}_1(s) \tag{27}$$

$$\left[s + \lambda_c + \mu_2\right]\overline{P}_3(s) = c\lambda_u \overline{P}_0(s) \tag{28}$$

$$\left[s + \frac{\partial}{\partial x} + \mu(x)\right]\overline{P}_i(x,s) = 0; \qquad i = 4,5,6,7 \tag{29}$$

$$\overline{P}_i(0,s) = \xi \overline{P}_2(s); \qquad i = 4,5; \qquad \xi = \lambda_c,\ (1-c)\lambda_u \tag{30}$$

$$\overline{P}_6(0,s) = \lambda_d \overline{P}_0(s) \tag{31}$$

$$\overline{P}_7(0,s) = \lambda_c \overline{P}_3(s) \tag{32}$$

Transition State Probabilities

The transition state probabilities of welding for model 1 and 2 are shown in Equations 33-40 and 41-48.

Model 1

$$\overline{P}_0(s) = \frac{1}{s + c_1 - K} \tag{33}$$

$$\overline{P}_1(s) = \frac{\lambda_0(s+c_4)}{(s+c_3)(s+c_4)-\mu_1\lambda_0}\overline{P}_0(s) \tag{34}$$

$$\overline{P}_2(s) = \frac{(\lambda_0)^2}{(s+c_3)(s+c_4)-\mu_1\lambda_0}\overline{P}_0(s) \tag{35}$$

$$\overline{P}_3(s) = \frac{\lambda_u}{(s+c_2)}\overline{P}_0(s) \tag{36}$$

$$\overline{P}_4(s) = \left(\frac{1-\overline{S}_\mu(s)}{s}\right)\frac{\lambda_c(\lambda_0)^2}{(s+c_3)(s+c_4)-\mu_1\lambda_0}\overline{P}_0(s) \tag{37}$$

$$\overline{P}_5(s) = \left(\frac{1-\overline{S}_\mu(s)}{s}\right)\frac{\lambda_u(\lambda_0)^2}{(s+c_3)(s+c_4)-\mu_1\lambda_0}\overline{P}_0(s) \tag{38}$$

$$\overline{P}_6(s) = \left(\frac{1-\overline{S}_\mu(s)}{s}\right)\lambda_d\overline{P}_0(s) \tag{39}$$

$$\overline{P}_7(s) = \left(\frac{1-\overline{S}_\mu(s)}{s}\right)\frac{\lambda_u\lambda_c}{s+c_2}\overline{P}_0(s) \tag{40}$$

Where,

$c_1 = \lambda_d + \lambda_u + \lambda_0$; $c_2 = \lambda_c + \mu_2$; $c_3 = \lambda_0 + \mu_0$; $c_4 = \lambda_c + \lambda_u + \mu_1$

$$K = \frac{\mu_0\lambda_0}{s+c_3} + \left[\frac{\mu_1\mu_0}{s+c_3} + (\lambda_c+\lambda_u)\overline{S}_\mu(s)\right]\frac{(\lambda_0)^2}{(s+c_3)(s+c_4)-\mu_1\lambda_0} + \lambda_d\overline{S}_\mu(s) + (\lambda_c\overline{S}_\mu(s)+\mu_2)\frac{\lambda_u}{s+c_2}$$

Model 2

$$\overline{P}_0(s) = \frac{1}{s + b_1 - K_1} \tag{41}$$

$$\overline{P}_1(s) = \frac{c\lambda_0(s + b_3)}{(s + b_3)(s + b_2) - \mu_1 c\lambda_0}\overline{P}_0(s) \tag{42}$$

$$\overline{P}_2(s) = \frac{(c\lambda_0)^2}{(s + b_3)(s + b_2) - \mu_1 c\lambda_0}\overline{P}_0(s) \tag{43}$$

$$\overline{P}_3(s) = \frac{c\lambda_u}{(s + b_4)}\overline{P}_0(s) \tag{44}$$

$$\overline{P}_4(s) = \left(\frac{1 - \overline{S}_\mu(s)}{s}\right)\frac{\lambda_c(c\lambda_0)^2}{(s + b_3)(s + b_2) - \mu_1 c\lambda_0}\overline{P}_0(s) \tag{45}$$

$$\overline{P}_5(s) = \left(\frac{1 - \overline{S}_\mu(s)}{s}\right)\frac{(1-c)\lambda_u(c\lambda_0)^2}{(s + b_3)(s + b_2) - \mu_1 c\lambda_0}\overline{P}_0(s) \tag{46}$$

$$\overline{P}_6(s) = \left(\frac{1 - \overline{S}_\mu(s)}{s}\right)\lambda_d \overline{P}_0(s) \tag{47}$$

$$\overline{P}_7(s) = \left(\frac{1 - \overline{S}_\mu(s)}{s}\right)\frac{c\lambda_u\lambda_c}{s + b_4}\overline{P}_0(s) \tag{48}$$

Where,

$b_1 = \lambda_d + (1-c)\lambda_u + c\lambda_u + c\lambda_0$; $b_2 = c\lambda_0 + \mu_0$; $b_3 = \lambda_c + (1-c)\lambda_u + \mu_1$; $b_4 = \lambda_c + \mu_2$

$$K_1 = \left[\mu_0(s+b_3) + c\lambda_0(\lambda_c + (1-c)\lambda_u)\overline{S}_\mu(s)\right]\frac{c\lambda_0}{(s+b_3)(s+b_2)-\mu_1 c\lambda_0} + \lambda_d \overline{S}_\mu(s)$$

$$+(\lambda_c \overline{S}_\mu(s) + \mu_2)\frac{c\lambda_u}{s+b_4}$$

Probability of Upstate and Downstate Situation of the Welding

When the welding is performed successfully, completely or partially, the situation is called upstate of the welding process and if by performing the operation, welding is unsuccessful then the situation is known as downstate of the welding process. Probability of the upstate and downstate of welding is defined as

$$\overline{P}_{up}(s) = \sum_{i=0}^{3} \overline{P}_i(s)$$

$$\overline{P}_{down}(s) = \sum_{i=4}^{7} \overline{P}_i(s)$$

Probability of Upstate and Downstate of Welding in Model 1

$$\overline{P}_{up}(s) = \left(1 + \frac{\lambda_0(s+c_4+\lambda_0)}{(s+c_4)(s+c_3)-\lambda_0\mu_1} + \frac{\lambda_u}{s+c_2}\right)\overline{P}_0(s) \tag{49}$$

$$\overline{P}_{down}(s) = \left(\frac{1-\overline{S}_\mu(s)}{s}\right)\left(\frac{(\lambda_0)^2(\lambda_c+\lambda_u)}{(s+c_4)(s+c_3)-\lambda_0\mu_1} + \frac{\lambda_u\lambda_c}{s+c_2} + \lambda_d\right)\overline{P}_0(s) \tag{50}$$

Probability of Upstate and Downstate of Welding in Model 2

$$\overline{P}_{up}(s) = \left(1 + \frac{c\lambda_0((s+b_3)+c\lambda_0)}{(s+b_2)(s+b_3)-c\lambda_0\mu_1} + \frac{c\lambda_u}{s+b_4}\right)\overline{P}_0(s) \tag{51}$$

$$\overline{P}_{down}(s) = \left(\frac{1-\overline{S}_\mu(s)}{s}\right)\left(\frac{(c\lambda_0)^2(\lambda_c+(1-c)\lambda_u)}{(s+b_2)(s+b_3)-c\lambda_0\mu_1} + \frac{c\lambda_u\lambda_c}{s+b_4} + \lambda_d\right)\overline{P}_0(s) \tag{52}$$

PARTICULAR CASES

Reliability Analysis

Reliability of the welding process is one of the crucial quality appearances which contract with the behaviour of each operation of the process. Reliability represents the probability of non-failure operation to perform their prerequisite functions for a specific time period in definite environmental condition. Repair facility does not taken account into the computation of reliability for any actions (Avizienis, 2004; Ram, 2013). So, to determine the reliability of the designed models of welding, consider the repair rate zero in Equation 49 and 51 and take the inverse Laplace transformation, one can obtain the reliability function of the welding as

$$Rl_1(t) = \frac{\lambda_u e^{(-\lambda_c t)}}{\lambda_d - \lambda_c + \lambda_u + \lambda_o} + \frac{(\lambda_c + \lambda_u)\lambda_o e^{(-\lambda_o t)}}{(\lambda_u + \lambda_d)(\lambda_c + \lambda_u - \lambda_o)} - \frac{\lambda_o^2 e^{(-(\lambda_c + \lambda_u)t)}}{(\lambda_o + \lambda_d - \lambda_c)(\lambda_c + \lambda_u - \lambda_o)}$$

$$+ \frac{((\lambda_d^2 - 2\lambda_c\lambda_d + \lambda_o\lambda_d)(\lambda_d + \lambda_u) - \lambda_c\lambda_o\lambda_u + \lambda_o^2(\lambda_u + \lambda_c) + \lambda_c^2(\lambda_d - \lambda_o + \lambda_u))e^{(-(\lambda_d + \lambda_u + \lambda_o)t)}}{(\lambda_o + \lambda_d - \lambda_c)(\lambda_d - \lambda_c + \lambda_u + \lambda_o)(\lambda_d + \lambda_u)} \tag{53}$$

$$Rl_2(t) = \frac{(-\lambda_c - (1-c)\lambda_u)c\lambda_o e^{(-c\lambda_o t)}}{(\lambda_u + \lambda_d)(-\lambda_c - (1-c)\lambda_u + c\lambda_o)} + \frac{c^2\lambda_o^2 e^{((-\lambda_c - (1-c)\lambda_u)t)}}{(c\lambda_o - (1-c)\lambda_u - \lambda_c)(-\lambda_c + c\lambda_u + c\lambda_o + \lambda_d)}$$

$$+ \frac{e^{(-(\lambda_u + c\lambda_o + \lambda_d)t)}}{(c\lambda_o + c\lambda_u + \lambda_d - \lambda_c)(-\lambda_c + \lambda_u + \lambda_d + c\lambda_o)(\lambda_u + \lambda_d)} \{(c\lambda_o\lambda_c + \lambda_d(\lambda_d - \lambda_c))(-\lambda_c + c\lambda_o + \lambda_d)$$

$$+\lambda_u \left(c\lambda_o(-c^2\lambda_o + c(\lambda_c - \lambda_d + \lambda_o) - \lambda_c + 2\lambda_d) - 3\lambda_c\lambda_d + 2\lambda_d^2 + \lambda_c^2\right)$$

$$+ \lambda_u^2 \left[c(1-c)\left\{\lambda_u + \lambda_d + \lambda_o\right\} - \lambda_c + \lambda_d\right]\} + \frac{c\lambda_u e^{(-\lambda_c t)}}{-\lambda_c + \lambda_u + c\lambda_o + \lambda_d} \tag{54}$$

Mean Time to Failure (MTTF)

The MTTF is the expected elapsed time between essential failures of welding during the operation. MTTF can be calculated as the average time between occurring of faults in welding (Ram & Manglik, 2014). By considering the repair rate zero and taking the limits of s tends to zero in Equation 49 and 51, one can evaluate the MTTF of the welding as

$$MTTF = \lim_{s \to 0} \overline{P}_{up}(s)$$

$$MTTF_1 = \frac{\lambda_c\lambda_o + 2\lambda_c^2 + \lambda_u^2 + 3\lambda_c\lambda_u}{\lambda_c(\lambda_c + \lambda_u)(\lambda_d + \lambda_u + \lambda_o)} \tag{55}$$

$$MTTF_2 = \frac{-c\lambda_c\lambda_o - 2\lambda_c^2 - c(1-c)\lambda_u^2 + (c-2)\lambda_c\lambda_u}{\lambda_c(-\lambda_c - (1-c)\lambda_u)(\lambda_u + c\lambda_o + \lambda_d)} \tag{56}$$

Sensitivity Analysis

Sensitivity analysis is the analysis that finds out how the uncertainty in the output or sensitive an output of a mathematical model can be assigned to different sources of uncertainty in its inputs or to change in an input while keeping other inputs constant. Sensitivity of a measure is defined as the partial derivative of the function with respect to their input factors (Andrews & Moss, 1993; Henley & Kumamoto, 1992). Here, these input factors are failure rates of the faults exist in welding.

Reliability Sensitivity

The sensitivity in the reliability of welding process resulting from the changes in the reliability of the welding subjected to their failure rates of detected faults. It can be examined by the partial differentiation of reliability function as revealed in Equations 53 and 54 with respect to failure rates in model 1 and 2 respectively. One may find the expression for sensitivity in reliability of welding for both the models as revealed in Equations 57-60 and 61-64 in Appendix 1.

MTTF Sensitivity

The sensitivity in MTTF of the welding can be deliberated through the partial differentiation of Equations 55 and 56 with respect to the failure rates of the faults detected in welding in model 1 and 2 respectively. One can compute the expressions for the MTTF sensitivity of welding for both the models as shown in Equations 65-68 and 69-72 in Appendix 2.

CONCLUSION

A well-organized algorithm for the computation of reliability and mean time to failure of welding process, and for its enhancement has been developed in this study. The effect of failure rates of faults on reliability and mean time to failure of welding is analysed by the critical examination of sensitivity. Reliability and mean time to failure are two significant measure of performance analysis. The comparative study of derived expressions has proved the importance of the fault coverage technique in welding. Welding takes an active part in building structure in the current scenario. Now-a-days, different types of welding are used in daily works. In future, welding will be used for strengthens of civil and mechanical engineering. So, it is worthwhile that welding should be highly reliable. This study helps the maintenance personnel and operator to achieve the greatly reliable welding through given more attention to the failure rate precarious faults.

REFERENCES

Andrews, J. D., & Moss, T. R. (2002). *Reliability and risk assessment.* Wiley-Blackwell.

Arnold, T. F. (1973). The concept of coverage and its effect on the reliability model of a repairable system. *IEEE Transactions on Computers*, *100*(3), 251–254. doi:10.1109/T-C.1973.223703

Ascher, H. E. (1968). Evaluation of repairable system reliability using the bad-as-old concept. *IEEE Transactions on Reliability*, *17*(2), 103–110. doi:10.1109/TR.1968.5217523

Avižienis, A., Laprie, J. C., Randell, B., & Landwehr, C. (2004). Basic concepts and taxonomy of dependable and secure computing. *IEEE Transactions on Dependable and Secure Computing*, *1*(1), 11–33. doi:10.1109/TDSC.2004.2

Barlow, R. E., & Proschan, F. (1996). *Mathematical theory of reliability* (Vol. 17). Siam. doi:10.1137/1.9781611971194

Bouricius, W. G., Carter, W. C., & Schneider, P. R. (1969, August). Reliability modeling techniques for self-repairing computer systems. In *Proceedings of the 1969 24th national conference* (pp. 295-309). ACM. doi:10.1145/800195.805940

Cempel, C., Natke, H. G., & Yao, J. T. P. (2000). Symptom reliability and hazard for systems condition monitoring. *Mechanical Systems and Signal Processing*, *14*(3), 495–505. doi:10.1006/mssp.1999.1246

Cox, D. R. (1955, July). The analysis of non-Markovian stochastic processes by the inclusion of supplementary variables. *Mathematical Proceedings of the Cambridge Philosophical Society*, *51*(3), 433–441. doi:10.1017/S0305004100030437

Dhillon, B. S., & Singh, C. (1981). *Engineering Reliability: New Techniques and Applications.* New York: Wiley.

Gandini, A. (1990). Importance and sensitivity analysis in assessing system reliability. *IEEE Transactions on Reliability*, *39*(1), 61–70. doi:10.1109/24.52613

Goyal, N., Kaushik, A., & Ram, M. (2016). Automotive water cooling system analysis subject to time dependence and failure issues. *International Journal of Manufacturing, Materials, and Mechanical Engineering*, *6*(2), 1–22. doi:10.4018/IJMMME.2016040101

Gupta, S. & Tewari, P. C. (2009). Simulation modeling and analysis of a complex system of a thermal power plant. *Journal of Industrial Engineering and Management, 2*(2), 387-406.

Henley, E. J., & Kumamoto, H. (1992). *Probabilistic Risk Assessment.* IEEE Press.

Jain, M., & Singh, C. J. (2000). Reliability of repairable multi-component redundant system. *International Journal of Engineering*, *3*, 107–114.

Kalpakjian, S. (2001). *Manufacturing engineering and technology.* Pearson Education India.

Ke, J. B., Lee, W. C., & Wang, K. H. (2007). Reliability and sensitivity analysis of a system with multiple unreliable service stations and standby switching failures. *Physica A: Statistical Mechanics and its Applications, 380*, 455-469.

Khanna, O. P. (1993). *A Text Book of Welding Technology for Engineering Students*. Dhanpat Rai and Sons.

Kumar, A., Varshney, A., & Ram, M. (2015). Sensitivity analysis for casting process under stochastic modelling. *International Journal of Industrial Engineering Computations*, *6*(3), 419–432. doi:10.5267/j.ijiec.2015.2.001

Malaiya, Y. K., Li, M. N., Bieman, J. M., & Karcich, R. (2002). Software reliability growth with test coverage. *IEEE Transactions on Reliability*, *51*(4), 420–426. doi:10.1109/TR.2002.804489

Myers, A. F. (2007). *k*-out-of-*n*: G system reliability with imperfect fault coverage. *IEEE Transactions on Reliability*, *56*(3), 464–473. doi:10.1109/TR.2007.903229

Oliveira, E. A., Alvim, A. C. M., & Melo, P. F. (2005). Unavailability analysis of safety systems under aging by supplementary variables with imperfect repair. *Annals of Nuclear Energy*, *32*(2), 241–252. doi:10.1016/j.anucene.2004.06.003

Parmar, R. S. (2010). *Welding engineering and technology*. Khanna Publishers.

Powell, D., Martins, E., Arlat, J., & Crouzet, Y. (1995). Estimators for fault tolerance coverage evaluation. *IEEE Transactions on Computers*, *44*(2), 261–274. doi:10.1109/12.364537

Ram, M. (2013). On system reliability approaches: A brief survey. *International Journal of System Assurance Engineering and Management*, *4*(2), 101–117. doi:10.1007/s13198-013-0165-6

Ram, M., & Manglik, M. (2014). Performance evaluation of a multi-state system covering imperfect fault coverage. *Communications in Statistics. Simulation and Computation*, *45*(9), 3259–3280. doi:10.1080/03610918.2014.932802

Ram, M., & Manglik, M. (2014). Stochastic behaviour analysis of a Markov model under multi-state failures. *International Journal of System Assurance Engineering and Management*, *5*(4), 686–699. doi:10.1007/s13198-014-0234-5

Ram, M., Singh, S. B., & Singh, V. V. (2013). Stochastic analysis of a standby system with waiting repair strategy. *IEEE Transactions on Systems, Man, and Cybernetics*, *43*(3), 698–707.

Ram, M., Singh, S. B., & Varshney, R. G. (2013). Performance improvement of a parallel redundant system with coverage factor. *Journal of Engineering Science and Technology*, *8*(3), 344–350.

Saltelli, A., Ratto, M., Andres, T., Campolongo, F., Cariboni, J., Gatelli, D., & Tarantola, S. (2008). *Global sensitivity analysis: the primer*. John Wiley and Sons.

Singh, V. V., Ram, M., & Rawal, D. K. (2013). Cost analysis of an engineering system involving subsystems in series configuration. *IEEE Transactions on Automation Science and Engineering*, *10*(4), 1124–1130. doi:10.1109/TASE.2013.2266134

Smith, S. A., & Oren, S. S. (1980). Reliability growth of repairable systems. *Naval Research Logistics Quarterly*, *27*(4), 539–547. doi:10.1002/nav.3800270403

Wang, K. H., Yen, T. C., & Jian, J. J. (2013). Reliability and sensitivity analysis of a repairable system with imperfect coverage under service pressure condition. *Journal of Manufacturing Systems*, *32*(2), 357–363. doi:10.1016/j.jmsy.2013.01.004

Xing, L. (2007). Reliability evaluation of phased-mission systems with imperfect fault coverage and common-cause failures. *IEEE Transactions on Reliability*, *56*(1), 58–68. doi:10.1109/TR.2006.890900

APPENDIX 1

Reliability Sensitivity in Model 1

$$\frac{\partial Rl_1(t)}{\partial \lambda_u} = \frac{\lambda_o e^{(-\lambda_o t)}}{(\lambda_u+\lambda_d)(\lambda_c+\lambda_u-\lambda_o)}\left[1-\frac{(\lambda_c+\lambda_u)\left(\lambda_d+\lambda_c+2\lambda_u-\lambda_o\right)}{(\lambda_u+\lambda_d)(\lambda_c+\lambda_u-\lambda_o)}\right]+\frac{(\lambda_d-\lambda_c+\lambda_o)e^{(-\lambda_c t)}}{(\lambda_d-\lambda_c+\lambda_u+\lambda_o)^2}$$

$$+\frac{e^{(-(\lambda_d+\lambda_u+\lambda_o)t)}}{(\lambda_o+\lambda_d-\lambda_c)(\lambda_d-\lambda_c+\lambda_u+\lambda_o)(\lambda_d+\lambda_u)}\{(-2\lambda_c\lambda_d+\lambda_o\lambda_d-\lambda_o\lambda_c+\lambda_o^2+\lambda_c^2+\lambda_d^2)-$$

$$\left(\lambda_c^2(\lambda_d-\lambda_o+\lambda_u)+(\lambda_d^2-2\lambda_c\lambda_d+\lambda_o\lambda_d)(\lambda_d+\lambda_u)-\lambda_c\lambda_o\lambda_u+\lambda_o^2(\lambda_u+\lambda_c)\right)\left\{t+\frac{1}{(\lambda_d+\lambda_u)}\right.$$

$$\left.\left.+\frac{1}{(\lambda_d-\lambda_c+\lambda_u+\lambda_o)}\right\}\right\}+\frac{\lambda_o^2 e^{(-(\lambda_c+\lambda_u)t)}(1+t(\lambda_c+\lambda_u-\lambda_o))}{(\lambda_o+\lambda_d-\lambda_c)(\lambda_c+\lambda_u-\lambda_o)^2} \tag{57}$$

$$\frac{\partial Rl_1(t)}{\partial \lambda_c} = -\frac{\lambda_o^2 e^{(-\lambda_o t)}}{(\lambda_u+\lambda_d)(\lambda_c+\lambda_u-\lambda_o)^2}+\frac{\lambda_o^2 e^{(-(\lambda_c+\lambda_u)t)}}{(\lambda_o+\lambda_d-\lambda_c)(\lambda_c+\lambda_u-\lambda_o)}\left\{t+\frac{2(\lambda_o-\lambda_c)+\lambda_d-\lambda_u)}{(\lambda_c+\lambda_u-\lambda_o)(\lambda_o+\lambda_d-\lambda_c)}\right\}$$

$$+\frac{e^{(-(\lambda_d+\lambda_u+\lambda_o)t)}}{(\lambda_o+\lambda_d-\lambda_c)(\lambda_d-\lambda_c+\lambda_u+\lambda_o)(\lambda_d+\lambda_u)}\{(2\lambda_c(\lambda_d+\lambda_u-\lambda_o)-\lambda_o(\lambda_u-\lambda_o)-2\lambda_d(\lambda_d+\lambda_u))$$

$$-\left(\lambda_c^2(\lambda_d-\lambda_o+\lambda_u)+(\lambda_d^2-2\lambda_c\lambda_d+\lambda_o\lambda_d)(\lambda_d+\lambda_u)-\lambda_c\lambda_o\lambda_u+\lambda_o^2(\lambda_u+\lambda_c)\right)$$

$$\left.\left(\frac{(2(\lambda_d+\lambda_o-\lambda_c)+\lambda_u)}{(\lambda_o+\lambda_d-\lambda_c)(\lambda_d-\lambda_c+\lambda_u+\lambda_o)}\right)\right\}+\frac{(1-t(\lambda_d-\lambda_c+\lambda_u+\lambda_o))\lambda_u e^{(-\lambda_c t)}}{(\lambda_d-\lambda_c+\lambda_u+\lambda_o)^2} \tag{58}$$

$$\frac{\partial Rl_1(t)}{\partial \lambda_d} = -\frac{\lambda_o(\lambda_c+\lambda_u)e^{(-\lambda_o t)}}{(\lambda_u+\lambda_d)^2(\lambda_c+\lambda_u-\lambda_o)}-\frac{\lambda_u e^{(-\lambda_c t)}}{(\lambda_d-\lambda_c+\lambda_u+\lambda_o)^2}+\frac{\lambda_o^2 e^{(-(\lambda_c+\lambda_u)t)}}{(\lambda_o+\lambda_d-\lambda_c)^2(\lambda_c+\lambda_u-\lambda_o)}$$

$$+\frac{\left(-4\lambda_c\lambda_d-2\lambda_c\lambda_u+2\lambda_o\lambda_d+2\lambda_d\lambda_u+\lambda_o\lambda_u+\lambda_c^2+3\lambda_d^2\right)e^{(-(\lambda_d+\lambda_u+\lambda_o)t)}}{(\lambda_o+\lambda_d-\lambda_c)(\lambda_d-\lambda_c+\lambda_u+\lambda_o)(\lambda_d+\lambda_u)}$$

$$-\frac{\left(\lambda_c^2(\lambda_d-\lambda_o+\lambda_u)+(\lambda_d^2-2\lambda_c\lambda_d+\lambda_o\lambda_d)(\lambda_d+\lambda_u)-\lambda_c\lambda_o\lambda_u+\lambda_o^2(\lambda_u+\lambda_c)\right)e^{(-(\lambda_d+\lambda_u+\lambda_o)t)}}{(\lambda_o+\lambda_d-\lambda_c)(\lambda_d-\lambda_c+\lambda_u+\lambda_o)(\lambda_d+\lambda_u)}\left\{\frac{1}{(\lambda_d+\lambda_u)}\right.$$

$$\left.+\frac{1}{(\lambda_o+\lambda_d-\lambda_c)}+\frac{1}{(\lambda_d-\lambda_c+\lambda_u+\lambda_o)}+t\right\} \tag{59}$$

$$\frac{\partial Rl_1(t)}{\partial\lambda_o}=\frac{(\lambda_c+\lambda_u)e^{(-\lambda_o t)}}{(\lambda_u+\lambda_d)(\lambda_c+\lambda_u-\lambda_o)}\left[\frac{(\lambda_c+\lambda_u)}{(\lambda_c+\lambda_u-\lambda_o)}-\lambda_o\right]-\frac{\lambda_u e^{(-\lambda_c t)}}{(\lambda_d-\lambda_c+\lambda_u+\lambda_o)^2}$$

$$+\frac{e^{(-(\lambda_d+\lambda_u+\lambda_o)t)}}{(\lambda_o+\lambda_d-\lambda_c)(\lambda_d-\lambda_c+\lambda_u+\lambda_o)(\lambda_d+\lambda_u)}\left\{(2\lambda_c\lambda_o+\lambda_u\lambda_d-\lambda_u\lambda_c+2\lambda_o\lambda_u-\lambda_c^2+\lambda_d^2)-\right.$$

$$\left(\lambda_c^2(\lambda_d-\lambda_o+\lambda_u)+(\lambda_d^2-2\lambda_c\lambda_d+\lambda_o\lambda_d)(\lambda_d+\lambda_u)-\lambda_c\lambda_o\lambda_u+\lambda_o^2(\lambda_u+\lambda_c)\right)\left\{t+\frac{1}{(\lambda_d+\lambda_o-\lambda_c)}\right.$$

$$\left.\left.+\frac{1}{(\lambda_d-\lambda_c+\lambda_u+\lambda_o)}\right\}\right\}+\frac{\lambda_o e^{(-(\lambda_c+\lambda_u)t)}}{(\lambda_o+\lambda_d-\lambda_c)(\lambda_c+\lambda_u-\lambda_o)}\left(\frac{\lambda_o(2(\lambda_c-\lambda_o)+\lambda_u-\lambda_d)}{(\lambda_o+\lambda_d-\lambda_c)(\lambda_c+\lambda_u-\lambda_o)}-2\right) \tag{60}$$

Reliability Sensitivity in Model 2

$$\frac{\partial Rl_2(t)}{\partial\lambda_u}=\frac{c\lambda_o e^{(-c\lambda_o t)}}{(\lambda_u+\lambda_d)(+\lambda_c+(1-c)\lambda_u-c\lambda_o)}\left((\lambda_c+(1-c)\lambda_u)\left[\frac{1-c}{(\lambda_c+(1-c)\lambda_u-c\lambda_o)}+\frac{1}{\lambda_u+\lambda_d}\right]+(1-c)\right)$$

$$-\frac{c^2\lambda_o^2 e^{((-\lambda_c-(1-c)\lambda_u)t)}}{(-c\lambda_o+(1-c)\lambda_u+\lambda_c)(\lambda_c-c\lambda_u-c\lambda_o-\lambda_d)}\left(\frac{c}{(-\lambda_c+c\lambda_u+c\lambda_o+\lambda_d)}+\frac{c-1}{(c\lambda_o-(1-c)\lambda_u-\lambda_c)}-(c-1)t\right)$$

$$-\frac{e^{(-(\lambda_u+c\lambda_o+\lambda_d)t)}}{(c\lambda_o+c\lambda_u+\lambda_d-\lambda_c)(-\lambda_c+\lambda_u+\lambda_d+c\lambda_o)(\lambda_u+\lambda_d)}\left\{(c\lambda_o\lambda_c+\lambda_d(\lambda_d-\lambda_c))(-\lambda_c+c\lambda_o+\lambda_d)\right.$$

$$\left.+\lambda_u\left(c\lambda_o(-c^2\lambda_o+c(\lambda_c-\lambda_d+\lambda_o)-\lambda_c+2\lambda_d)-3\lambda_c\lambda_d+2\lambda_d^2+\lambda_c^2\right)+\lambda_u^2\left[c(1-c)\left\{\lambda_u+\lambda_d+\lambda_o\right\}-\lambda_c+\lambda_d\right]\right\}$$

$$\left\{t+\frac{1}{\lambda_u+\lambda_d}-\frac{1}{(c\lambda_o+c\lambda_u+\lambda_d-\lambda_c)}-\frac{1}{(-\lambda_c+\lambda_u+\lambda_d+c\lambda_o)}\right\}$$

$$+\frac{ce^{(-\lambda_c t)}}{-\lambda_c+\lambda_u+c\lambda_o+\lambda_d}\left(1-\frac{\lambda_u}{-\lambda_c+\lambda_u+c\lambda_o+\lambda_d}\right)+e^{(-(\lambda_u+c\lambda_o+\lambda_d)t)}$$

$$\frac{((1-c)c\left(\lambda_o(2\lambda_u-\lambda_c+c\lambda_o+\lambda_d)+\lambda_u(2\lambda_d+3\lambda_u)\right)+\lambda_d(c\lambda_o+2(\lambda_d+\lambda_u)-3\lambda_c)+\lambda_c(\lambda_c-2\lambda_u))}{(c\lambda_o+c\lambda_u-\lambda_c+\lambda_d)(-\lambda_c+\lambda_u+c\lambda_o+\lambda_d)(\lambda_u+\lambda_d)} \tag{61}$$

$$\frac{\partial Rl_2(t)}{\partial\lambda_c}=-\frac{c\lambda_o e^{(-c\lambda_o t)}}{(\lambda_u+\lambda_d)(-\lambda_c-(1-c)\lambda_u+c\lambda_o)}\left(1-\frac{(-\lambda_c-(1-c)\lambda_u)}{(-\lambda_c-(1-c)\lambda_u+c\lambda_o)}\right)$$

$$+\frac{c^2\lambda_o^2 e^{((-\lambda_c-(1-c)\lambda_u)t)}}{(c\lambda_o-(1-c)\lambda_u-\lambda_c)(-\lambda_c+c\lambda_u+c\lambda_o+\lambda_d)}\left(\frac{1}{(-\lambda_c+c\lambda_u+c\lambda_o+\lambda_d)}+\frac{1}{(c\lambda_o-(1-c)\lambda_u-\lambda_c)}-t\right)$$

$$-\frac{e^{(-(\lambda_u+c\lambda_o+\lambda_d)t)}}{(c\lambda_o+c\lambda_u+\lambda_d-\lambda_c)(-\lambda_c+\lambda_u+\lambda_d+c\lambda_o)(\lambda_u+\lambda_d)}\left\{(c\lambda_o\lambda_c+\lambda_d(\lambda_d-\lambda_c))(-\lambda_c+c\lambda_o+\lambda_d)\right.$$

$$\left.+\lambda_u\left(c\lambda_o(-c^2\lambda_o+c(\lambda_c-\lambda_d+\lambda_o)-\lambda_c+2\lambda_d)-3\lambda_c\lambda_d+2\lambda_d^2+\lambda_c^2\right)+\lambda_u^2\left[c(1-c)\left\{\lambda_u+\lambda_d+\lambda_o\right\}-\lambda_c+\lambda_d\right]\right\}$$

$$\left\{\frac{1}{(c\lambda_o+c\lambda_u+\lambda_d-\lambda_c)}+\frac{1}{(-\lambda_c+\lambda_u+\lambda_d+c\lambda_o)}\right\}+\frac{c\lambda_u e^{(-\lambda_c t)}}{-\lambda_c+\lambda_u+c\lambda_o+\lambda_d}\left(\frac{1}{-\lambda_c+\lambda_u+c\lambda_o+\lambda_d}-t\right)$$

$$-\frac{(c^2\lambda_o^2+c(c-1)\lambda_u\lambda_o-2c\lambda_o\lambda_c+2\lambda_d\lambda_c+2\lambda_u\lambda_c-3\lambda_d\lambda_u-\lambda_u^2-2\lambda_d^2)e^{(-(\lambda_u+c\lambda_o+\lambda_d)t)}}{(c\lambda_o+c\lambda_u+\lambda_d-\lambda_c)(-\lambda_c+\lambda_u+\lambda_d+c\lambda_o)(\lambda_u+\lambda_d)} \tag{62}$$

$$\frac{\partial Rl_2(t)}{\partial\lambda_d}=-\frac{c\lambda_o(-\lambda_c-(1-c)\lambda_u)e^{(-c\lambda_o t)}}{(\lambda_u+\lambda_d)^2(-\lambda_c-(1-c)\lambda_u+c\lambda_o)}-\frac{c\lambda_u e^{(-\lambda_c t)}}{\left(-\lambda_c+\lambda_u+\lambda_d+c\lambda_o\right)^2}$$

$$-\frac{c^2\lambda_o^2 e^{((-\lambda_c-(1-c)\lambda_u)t)}}{(c\lambda_o-(1-c)\lambda_u-\lambda_c)(-\lambda_c+c\lambda_u+\lambda_d+c\lambda_o)^2}$$

$$+\frac{e^{(-(\lambda_u+c\lambda_o+\lambda_d)t)}}{(c\lambda_o+c\lambda_u+\lambda_d-\lambda_c)(-\lambda_c+\lambda_u+\lambda_d+c\lambda_o)(\lambda_u+\lambda_d)}\{(c\lambda_o\lambda_c+\lambda_d(\lambda_d-\lambda_c))(-\lambda_c+c\lambda_o+\lambda_d)$$

$$+\lambda_u\left(c\lambda_o(-c^2\lambda_o+c(\lambda_c-\lambda_d+\lambda_o)-\lambda_c+2\lambda_d)-3\lambda_c\lambda_d+2\lambda_d^2+\lambda_c^2\right)+\lambda_u^2\left[c(1-c)\left\{\lambda_u+\lambda_d+\lambda_o\right\}-\lambda_c+\lambda_d\right]\}$$

$$\left(t+\frac{1}{(-\lambda_c+\lambda_u+\lambda_d+c\lambda_o)}+\frac{1}{(c\lambda_o+\lambda_d+c\lambda_u-\lambda_c)}+\frac{1}{(\lambda_u+\lambda_d)}\right)$$

$$-\frac{((2-c)\lambda_u\lambda_o+2c\lambda_d\lambda_o+\lambda_u^2(1+c-c^2)+\lambda_c^2+3\lambda_d^2-4\lambda_d(\lambda_c-\lambda_u)-3\lambda_c\lambda_u)e^{(-(\lambda_u+c\lambda_o+\lambda_d)t)}}{(c\lambda_o+c\lambda_u+\lambda_d-\lambda_c)(-\lambda_c+\lambda_u+\lambda_d+c\lambda_o)(\lambda_u+\lambda_d)} \tag{63}$$

$$\frac{\partial Rl_2(t)}{\partial\lambda_o}=\frac{c(-\lambda_c-(1-c)\lambda_u)e^{(-c\lambda_o t)}}{(\lambda_u+\lambda_d)(-\lambda_c-(1-c)\lambda_u+c\lambda_o)}\left(1-c\lambda_o\left[\frac{1}{(-\lambda_c-(1-c)\lambda_u+c\lambda_o)}+t\right]\right)-\frac{c^2\lambda_u e^{(-\lambda_c t)}}{\left(-\lambda_c+\lambda_u+\lambda_d+c\lambda_o\right)^2}$$

$$+\frac{c^2\lambda_o e^{((-\lambda_c-(1-c)\lambda_u)t)}}{(c\lambda_o-(1-c)\lambda_u-\lambda_c)(-\lambda_c+c\lambda_u+\lambda_d+c\lambda_o)}\left(2-\frac{c\lambda_o}{(-\lambda_c+c\lambda_u+\lambda_d+c\lambda_o)}-\frac{c\lambda_o}{(c\lambda_o-(1-c)\lambda_u-\lambda_c)}\right)$$

$$+\frac{ce^{(-(\lambda_u+c\lambda_o+\lambda_d)t)}}{(c\lambda_o+c\lambda_u+\lambda_d-\lambda_c)(-\lambda_c+\lambda_u+\lambda_d+c\lambda_o)(\lambda_u+\lambda_d)}\{(c\lambda_o\lambda_c+\lambda_d(\lambda_d-\lambda_c))(-\lambda_c+c\lambda_o+\lambda_d)$$

$$+\lambda_u\left(c\lambda_o(-c^2\lambda_o+c(\lambda_c-\lambda_d+\lambda_o)-\lambda_c+2\lambda_d)-3\lambda_c\lambda_d+2\lambda_d^2+\lambda_c^2\right)+\lambda_u^2\left[c(1-c)\left\{\lambda_u+\lambda_d+\lambda_o\right\}-\lambda_c+\lambda_d\right]\}$$

$$\left(t+\frac{1}{(-\lambda_c+\lambda_u+\lambda_d+c\lambda_o)}+\frac{1}{(c\lambda_o+\lambda_d+c\lambda_u-\lambda_c)}\right)$$

$$-\frac{((1-c)(2c^2\lambda_c\lambda_o-c\lambda_u\lambda_c+c\lambda_u^2)+2c^2\lambda_u\lambda_o-c\lambda_d\lambda_u(c-2)-\lambda_c^2+c\lambda_d^2)e^{(-(\lambda_u+c\lambda_o+\lambda_d)t)}}{(c\lambda_o+c\lambda_u+\lambda_d-\lambda_c)(-\lambda_c+\lambda_u+\lambda_d+c\lambda_o)(\lambda_u+\lambda_d)} \tag{64}$$

APPENDIX 2

MTTF Sensitivity for Model 1

$$\frac{\partial MTTF_1}{\partial \lambda_u} = \frac{1}{\lambda_c(\lambda_c+\lambda_u)(\lambda_d+\lambda_u+\lambda_o)}\left[(3\lambda_c+2\lambda_u) - \frac{(3\lambda_u\lambda_c+2\lambda_c^2+\lambda_u^2+\lambda_u\lambda_c)(\lambda_d+2\lambda_u+\lambda_o+\lambda_c)}{(\lambda_c+\lambda_u)(\lambda_d+\lambda_u+\lambda_o)}\right] \tag{65}$$

$$\frac{\partial MTTF_1}{\partial \lambda_c} = \frac{1}{\lambda_c(\lambda_c+\lambda_u)(\lambda_d+\lambda_u+\lambda_o)}\left[(3\lambda_u+4\lambda_c+\lambda_o) - \frac{(3\lambda_u\lambda_c+2\lambda_c^2+\lambda_u^2+\lambda_o\lambda_c)(\lambda_u+2\lambda_c)}{\lambda_c(\lambda_c+\lambda_u)}\right] \tag{66}$$

$$\frac{\partial MTTF_1}{\partial \lambda_d} = -\frac{\lambda_c\lambda_o+2\lambda_c^2+\lambda_u^2+3\lambda_c\lambda_u}{\lambda_c(\lambda_c+\lambda_u)(\lambda_d+\lambda_u+\lambda_o)^2} \tag{67}$$

$$\frac{\partial MTTF_1}{\partial \lambda_o} = \frac{1}{\lambda_c(\lambda_c+\lambda_u)(\lambda_d+\lambda_u+\lambda_o)} - \frac{\lambda_c\lambda_o+2\lambda_c^2+\lambda_u^2+3\lambda_c\lambda_u}{\lambda_c(\lambda_c+\lambda_u)(\lambda_d+\lambda_u+\lambda_o)^2} \tag{68}$$

MTTF Sensitivity for Model 2

$$\frac{\partial MTTF_2}{\partial \lambda_u} = \frac{(c\lambda_c\lambda_o+2\lambda_c^2+c(1-c)\lambda_u^2+(2-c)\lambda_c\lambda_u)\left[(1-c)(\lambda_d+c\lambda_o)-\lambda_c\right]}{\lambda_c(-\lambda_c-(1-c)\lambda_u)^2(\lambda_u+c\lambda_o+\lambda_d)^2}$$

$$+\frac{-2c(1-c)\lambda_u+(c-2)\lambda_c}{\lambda_c(-\lambda_c-(1-c)\lambda_u)(\lambda_u+c\lambda_o+\lambda_d)} \tag{69}$$

$$\frac{\partial MTTF_2}{\partial \lambda_c} = -\frac{(-c\lambda_c\lambda_o-2\lambda_c^2-c(1-c)\lambda_u^2+(c-2)\lambda_c\lambda_u)\left((1-c)\lambda_u+2\lambda_c\right)}{\lambda_c^2(-\lambda_c-(1-c)\lambda_u)^2(\lambda_u+c\lambda_o+\lambda_d)}$$

$$+\frac{(c-2)\lambda_u-4\lambda_c-c\lambda_o}{\lambda_c(-\lambda_c-(1-c)\lambda_u)(\lambda_u+c\lambda_o+\lambda_d)} \tag{70}$$

$$\frac{\partial MTTF_2}{\partial \lambda_d} = \frac{(-c\lambda_c\lambda_o - 2\lambda_c^2 - c(1-c)\lambda_u^2 + (c-2)\lambda_c\lambda_u)}{\lambda_c(-\lambda_c - (1-c)\lambda_u)(\lambda_u + c\lambda_o + \lambda_d)} \tag{71}$$

$$\frac{\partial MTTF_2}{\partial \lambda_o} = -\frac{c}{(-\lambda_c - (1-c)\lambda_u)(\lambda_u + c\lambda_o + \lambda_d)} + \frac{(-c\lambda_c\lambda_o - 2\lambda_c^2 - c(1-c)\lambda_u^2 + (c-2)\lambda_c\lambda_u)c}{\lambda_c(-\lambda_c - (1-c)\lambda_u)(\lambda_u + c\lambda_o + \lambda_d)^2} \tag{72}$$

Chapter 12
Electrical Impedance Modeling of Electrochemical Spark Micromachining Process

Anjali Vishwas Kulkarni
Indian Institute of Technology Kanpur, India

ABSTRACT

Electrochemical spark micromachining (ECSMM) process, an advanced machining process, has been used for micromachining. Various kinds of machining power sources, different workpiece materials and different tool materials have been used to study the performance of the process. The nature of the sparking has been investigated by studying the online, transient process current in synchronization with the machining supply voltage. Based on the study of the transient current, the spark formation process is identified as discrete and repetitive. The spark formation comprises of many physical and chemical phases in a single spark cycle. The online process current has been mathematically analyzed to formulate the electrical model of the process to understand the overall process mechanism. The electrical impedance in each phase of a single spark cycle is modeled. This semiempirical modeled impedance is compared against the impedance computed by taking the ratio of the measured online voltage and current data during single spark formation.

INTRODUCTION

The electrochemical spark micromachining (ECSMM) process is an advanced microfabrication process, suitable for machining of wide range of materials. The modeling of the process in the light of electrochemistry (Fascio, Wuthrich, & Bleuer, 2004), heat source (Basak & Ghosh, 1992; Jain, Dixit, & Pandey, 1999; Yadav, Jain, & Dixit, 2002) have been performed by various researchers. However the holistic approach is lacking in understanding the process mechanism as the material removal mechanism in ECSMM is a complex phenomenon. Moreover, sparking is not a continuous process as it is considered; rather it is a discrete phenomenon. Complexity arises primarily due to the presence of various physio-chemical phases involved in forming the discrete sparking and secondly due to the non-thermal nature

DOI: 10.4018/978-1-5225-1639-2.ch012

of these sparks. In the existent literature, the spark energy is considered to be of thermal nature. Also the thermal analysis and material removal are considered to be due to this thermal source (Basak & Ghosh, 1992; Jain, Dixit, & Pandey, 1999). But this is not the case as it has been found experimentally that the spark is a non thermal discharge. This has been established (Kulkarni, 2009) while making an attempt to measure the spark temperature by a pyrometer. Pyrometer failed to measure the temperature as the radiation is a non thermal type. Instead it is a repeated discharge process similar to that of the breakdown of the hydrogen gas bubble isolating the tool tip from the surrounding electrolyte. Moreover electro chemical systems are known to exhibit complex non-linear behavior. These nonlinearities arise due to electro hydro dynamism, ionic reactions, bubble generation, their growth and their breakdown phenomena. The spark formation cycle is a series of such activities. The present mechanism has been studied by presenting the electrical impedance during each phase of one cycle of the spark formation. The electrical impedance in each stage of a single spark cycle is modeled using MATLAB software. This semiempirical modeled impedance is compared against the impedance computed by taking the ratio of the measured on line, transient and synchronized voltage and current data during single spark formation.

EXPERIMENTAL

The ECSMM micro fabrication equipment has been designed and developed in house at Indian Institute of Technology Kanpur (Kulkarni & Jain, 2015). It is an automated 4 axes experimental set-up with following subsystems: Machining Chamber, Exhaust System, LabVIEW based Control Station, and Power Supply System. Machining chamber is the main sub system of the set-up. Exhaust system is provided to take away the fumes generated during the electrochemical spark process. The control station is established on the personal computer (PC) using motion control hardware (NI 7358) and supporting accessories (MID 7604, and UMI 7774). Power supply system incorporates the power supplies used for machining, control hardware and for driving the stepper motors. Various kinds of machining power sources have been experimented with. The equipment has been successfully used for micromachining of wide range of materials such as copper, platinum, molybdenum, tungsten, tantalum, soda lime glass, quartz, and silicon (Kulkarni, Jain, & Misra, 2011a, 2011b, 2013). Microholes and microchannels (as the microfeatures) have been fabricated. Microholes on the metals have applications as apertures, filters, etc. Microchannels have applications in biological industries for mixing of the liquids, etc. Soda lime glass, glass pellets and quartz are the suitable materials for these kinds of applications as they react less with the fluids. Silicon is a ‘hot’ material in electronics industry. Silicon technology is the most sophisticated and researched fabrication technology for semiconductors. To have an alternative process for μm sized features fabrication on silicon, the present ECSMM process has been studied and established. Microholes are fabricated on metals. To begin with, copper has been used as workpiece material which is easily available. Platinum and molybdenum have been used to form the microholes as they are the suitable materials for vacuum related operations. Though copper was machinable with pulsed power source, tantalum and molybdenum could not be machined using pulsed power source. Hence experiments have been carried out with DC power source for machining of metals. Many researchers have attempted the performance studies of glass, and quartz using DC power source. An attempt has been made to study the response of ECSMM for silicon micromachining. Microchannels on silicon find use in power electronics for heat dissipation. No research study has been performed, in the literature, using HWR (Half wave rectified) DC, AC and triangular pulsating power sources. A limited study has been performed by researchers using square pulsating power source. Glass and silicon have been chosen to study the performance of ECSMM using HWR DC and AC power sources. With AC power source, deposition of tool material (copper in these cases) has been envisaged. Copper deposition on glass finds applications in making sensors and copper on silicon has applications in having the Ohmic contacts and Schottky diodes. During the pilot experiments using pulsating power source, it was found that shallow microchannels result mainly due to the low rms value of voltage and less machining time. Hence, the study is limited to glass while micromachining with pulsating power source. Thus, materials and power source type have been decided on the basis of the application of the microfeatures in various industry applications. In view of the discrete nature of the process, the transient current is measured on line while micromachining is performed using different machining power source, as described in the following section.

SYNCHRONOUS, TRANSIENT MEASUREMENT SCHEME

The time varying on line process current is measured in synchronization with the machining supply voltage using digital storage oscilloscope (DSO, Hameg 1008) by 'resistive shunt method'. For this, a 1 Ω resistance is connected in series with cathode and ground of the power supply to the ECS cell. The time varying voltage across this resistance is the direct measure of the time varying process current. These synchronized current and voltage wave forms are saved on the control PC via RS 232 connectivity module of the DSO. Each waveform contains 2000 samples of the voltage and current readings. The current waveform contains many spark cycles. Figure 1 shows the block diagram of the measurement scheme. Channel 1 (Ch 1) measures the voltage and channel 2 (Ch 2) measures the transient current in synchronization (Kulkarni, 2013).

Figure 1. Block diagram of synchronous, on-line current and voltage measurement scheme
(DSO- Digital storage oscilloscope, MPS- Machining power supply)

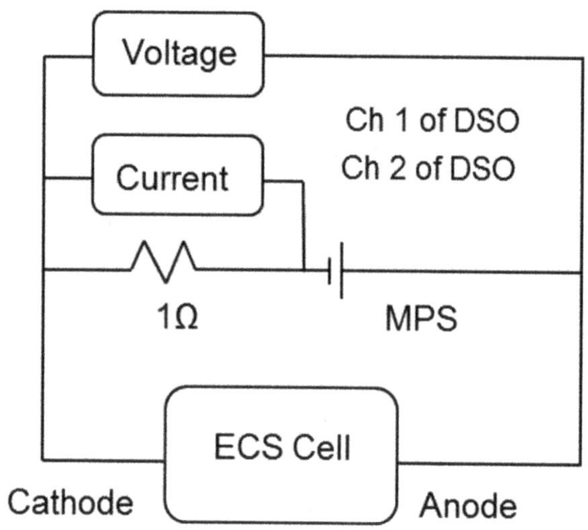

INSPECTION OF INSTANTANEOUS ON LINE CURRENT

The inspection of the on line, instantaneous current shows a typical pattern, which is repetitive and at the same time, stochastic. Some of the representative measured current snapshots are presented in Figure 1 below to visualize and understand the pattern.

Irrespective of the workpiece material and the machining power source used (DC/HWR DC/AC) for machining, it is to be noted that the nature of the current spikes is discrete and repetitive, though different in magnitude and time scale. This leads to an understanding of formation of series of events or phases during the occurrence of a single, complete spark cycle. From the sampled waveforms one can estimate the sparking frequency, On time, Off time, etc. which are otherwise theoretically estimated in the literature.

As stated earlier, the interesting process phenomenon is obvious from visualizing the time varying current. The processes' discrete nature can be only revealed by studying the transient current. The transient current waveforms reveal the process complexity and help in understanding the holistic and time changing phenomena during the single entire current cycle of ECSMM process. The actual machining is occurring during the very short time of the instantaneous current pulses (called as spark) carrying high energy density. Out of many current spikes in Figure 2, a single current spike is highlighted to explain the sparking phenomenon. Various phases of operation can be best viewed in the zoomed current waveform.

Figure 3 shows a snapshot of the on line, transient current waveform sampled in synchronization with the DC machining voltage while micromachining on glass workpiece. Many current pulses are seen in the current waveform representing the occurrences of the sparks during the micromachining of glass.

Figure 2. Digitally stored transient, on line current highlighting single current spike to understand the nature of the spark

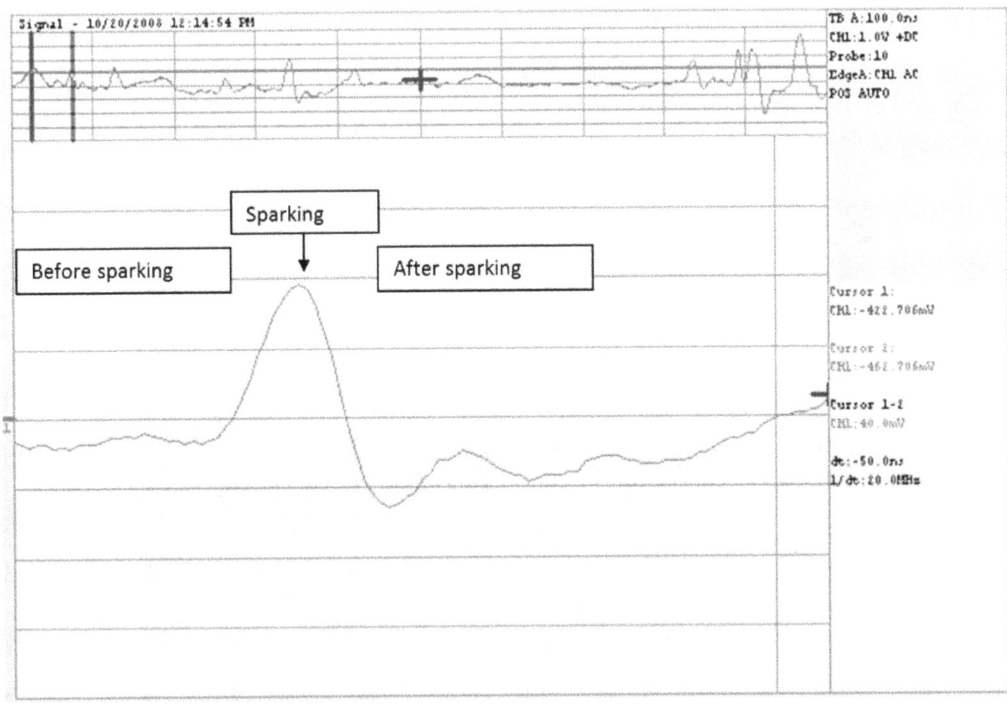

Figure 3. Snap shot of the on line, transient current waveform together with DC machining supply during micromachining of glass sample

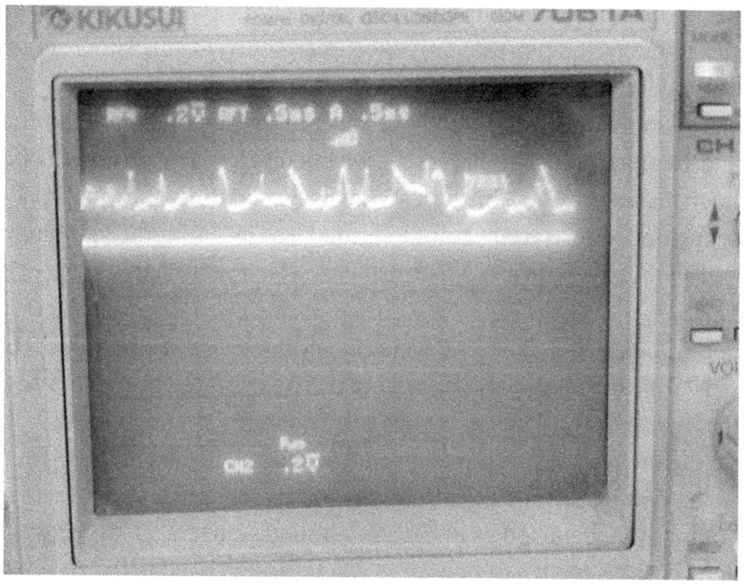

It emphasizes the discrete nature of the process. The on time of the pulses is ranging from 0.1 ms – 0.5 ms, the time between the two consecutive sparks is ranging from less than 0.1 ms- 0.5 ms. The sparking frequency is in the range of 1 kHz – 5 kHz.

Figure 4 gives the snapshots of the transient, instantaneous on line current while micromachining of silicon. In Figure 4, Channel 1 of the oscilloscope is showing 60 V DC excitation (machining power

Figure 4. Snapshot of on line, transient instantaneous current waveform during micromachining of silicon using DC excitation, sampled at 100 ns

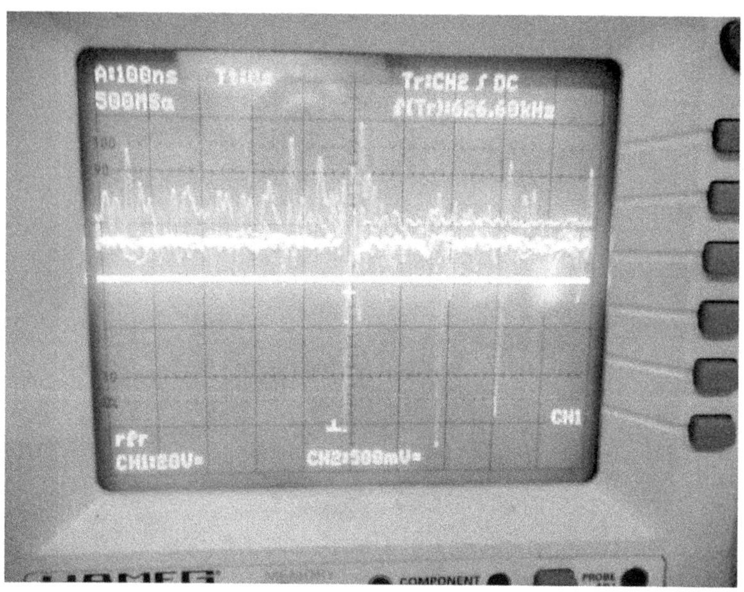

source) and channel 2 is showing the respective process current sampled at 100 ns time base. The average value of the current is 400 mA – 600 mA, as seen by the bright line buried in the transient current waveform. Many short duration pulses are seen pertaining to the instantaneous process current. Each current pulse is representing the occurrence of the spark. Theses pulses are of varying height. The largest seen in the figure is around 1.5 A.

Figure 5 shows the snapshot of the transient current for HWR DC type of excitation during silicon micromachining during one full half cycle of 10 ms duration. It is seen that during the rise time of

the input pulse, sparking is taking place after the input is achieved the critical voltage (around 35 Vrms) for sparking to take place. During the next 3-9 ms of time, repetitive sparking is taking place with different time duration and current amplitude. Negative pulses are also visible. It is to be noted that during the fall of the input pulse, sparking is taking place at a lower voltage level than that in rising pulse. This is due to the fact that once the sparking starts then it sustains at a lower voltage level. The sparking is ceasing at around 30 V rms. Around 25 (positive and negative) pulses are observed during about 7 ms time. The average current is about 100 – 200 mA.

Figure 6 shows the snapshot of the transient current during micromachining of silicon using AC excitation. It shows the random positive and negative pulses during the positive and negative half cycles of excitation sampled at 10 ms. The pattern for the instantaneous current is repetitive with different instantaneous values.

Figure 7 shows the another snapshot of the waveform. The upper waveform is the machining power supply (square waveform) and the corresponding on line, transient current waveform in the lower waveform. The peak value of input square wave is 25 V with T_{ON} of 750 μs and T_{OFF} of 300 μs. Many transient current spikes corressponding to discrete sparks can be seen during the on time of the machining voltage.

Figure 5. Snapshot of on line current sampled at 2 ms for silicon workpiece while micromachining with HWR DC excitation (approx. 25 pulses in 5 ms time duration)

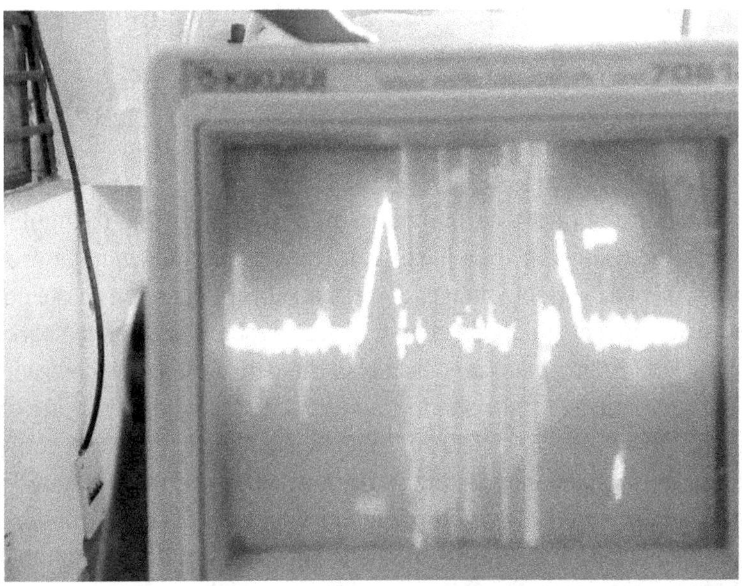

Figure 6. Snap shot of the on line, transient current during micromachining of silicon using AC excitation

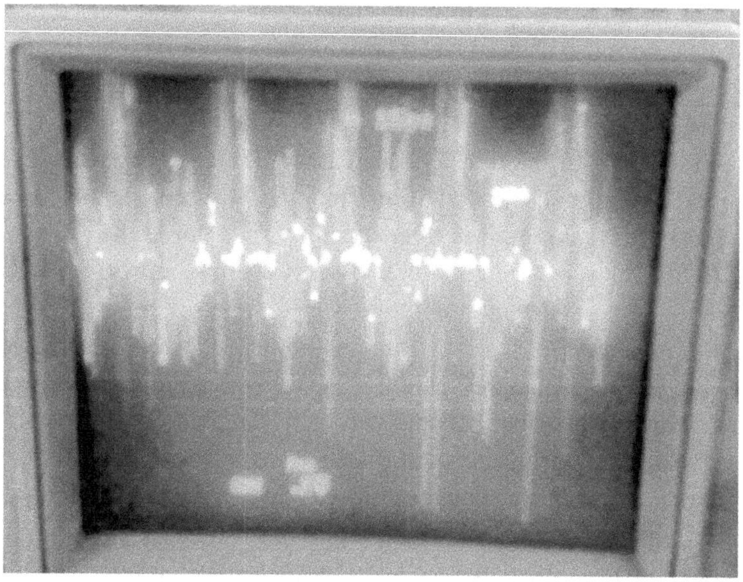

Figure 7. The snapshot of the synchronized voltage and current waveform during the micromachining of glass using pulsating power source

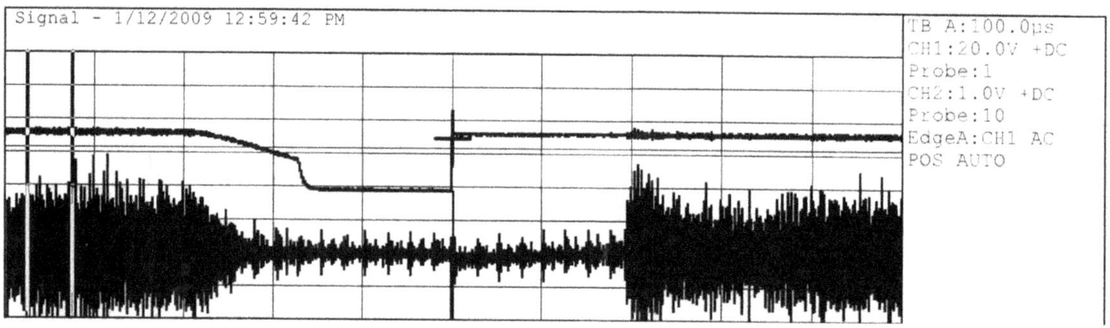

Figure 8 shows the snapshot of the on line transient current w.r.t. the triangular supply waveform. It is to be noted that as the triangular wave peak voltage is rising, the average process current is also increasing. At about 20 V peak supply voltage, there is an abrupt decrease in the current. It again soon rises and there after many current spikes are seen which represent the discrete sparking till the decreasing voltage is about 18 V. Current spikes are absent till the peak voltage reaches about 20 V in the next triangular cycle. The onset voltage for the sparking is about 20 V and the sparking ceases at about 18 V.

The inspection of the transient current, from Figures 2-8 shows that irrespective of the machining power source used, the transient, on line current is discrete and repetitive in nature. The time period and amplitude and their frequency may vary with different workpiece and tool material combination but their nature is similar. This leads to a repeated sequence of events in the formation of spark. Following section describes this time sequence which is occurring repeatedly.

Figure 8. Snashot of the on line, transient current waveform w.r.t. the triangular supply waveform while micromachining of glass

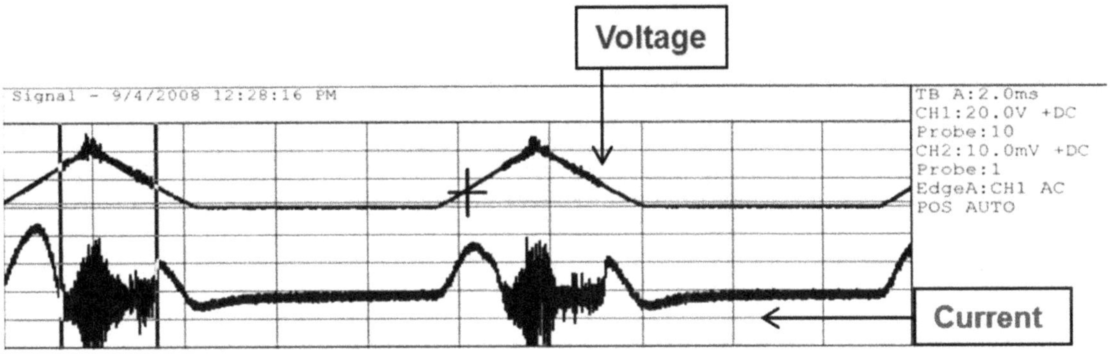

OPERATIONAL PHASES OF ECSMM PROCESS IN THE LIGHT OF ON LINE, TRANSIENT CURRENT

The formation of spark happens through series of chemical and physical processes. When a voltage (enough to form spark) is applied across the cell, mainly following series of events take place (Kulkarni, 2012, 2013).

1. Electrochemical phase giving rise to oxidation and hydrogen gas generation.
2. Hydrogen and vapor bubble formation, coalescence and bubble growth during pool boiling, slowly covering the tool tip.
3. Tool tip isolation, momentary 'virtual switch off' phase.
4. Instantaneous generation of high electric field which causes sparking due to hydrogen breakdown.
5. Reestablishment of tool-electrolyte contact leading to 'virtual switch on' phase.
6. Drifting of energetic electrons towards the workpiece due to potential gradient and subsequent material removal from the workpiece kept near the tool tip due to this 'electron gun'.
7. Steps 1-6 begin all over again.

First phenomenon contributes to the process current comprising the ionic current and the electronic current respectively. Pool boiling of hydrogen gas bubbles occurs due to the intense ohmic heating of the copper tool. This stage poses a variable reactance to the process current. A momentary zero current due to 'virtual switch off' occurs. Soon after in the fourth stage of sparking, a sudden increase in the current is observed. This is due to the breakdown of the hydrogen gas bubble by the momentary high electric field getting generated due to the isolation of cathode tip from the surrounding electrolyte. The electrons generated during the ionization bombard on the workpiece surface kept near the cathode tip. This bombardment of energetic, accelerated electrons causes machining of the workpiece. During the entire process, splashing and heating of electrolyte take place. Machining is the fruitful result while splashing and heating of the electrolyte are the unwanted phenomena reducing the efficiency of the ECSMM process. Soon after the sparking takes place, the cathode – electrolyte contact establishes and electrochemical action takes place followed by bubble formation, pool boiling, sparking and machining, as explained above. Based on this qualitative understanding of the process, the situation can be pictorially described as depicted in Figure 9. To better understand the situation during the spark formation, Figure 10 shows the time sequence of the hydrogen bubble growth, application of high electric field and the ensuing process current.

The overall material mechanism of the ECSM process can be understood in the light of the electrochemistry, heat transfer, ionization theory and electrical response of the system. Subsequently, the ECSMM process is analyzed by representing the instantaneous electrical impedance in each phase of operation.

Following section establishes the theoretical background behind each phase and identification of the major electrical parameters responsible for the instantaneous current contribution in each phase of operation during the single spark formation.

ELECTROCHEMICAL PHASE

When the machining supply to the electrolyte cell is applied in the proper polarity, (i.e. positive terminal connected to anode and negative terminal to cathode) electrochemical action starts. The electrons move from the cathode–electrolyte interface, and go to the solution. At the anode–electrolyte interface, equal numbers of electrons are discharged from the solution to the anode. Electrochemical reactions that occur at the electrode–electrolyte interface continuously supply electrons from cathode to solution and solution to anode. The type of reaction depends on the characteristics of electrodes, electrolyte and applied

Figure 9. Pictorial representation of the current contributions due to sequential events
(1. current due to electrochemical action during time interval 1,
2. current due to varying impedance of growing hydrogen bubble during time interval 2,
3. current during 'Virtual switch off' due to isolation during time interval 3 and
4. ionization current due to bubble breakdown during time interval 4)

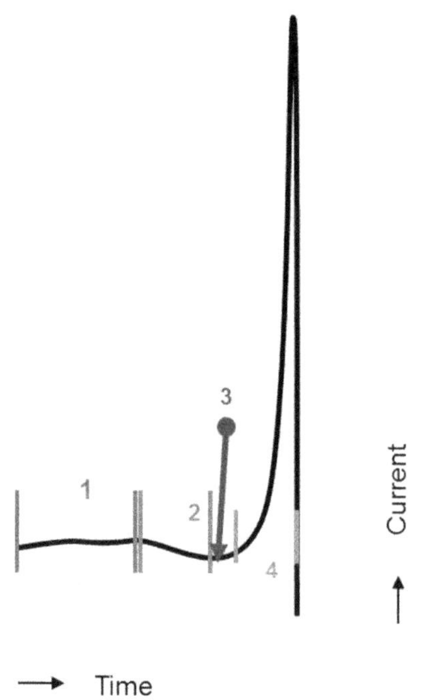

Figure 10. Time sequence of the hydrogen bubble growth, generation of high electric field and the resulting process current

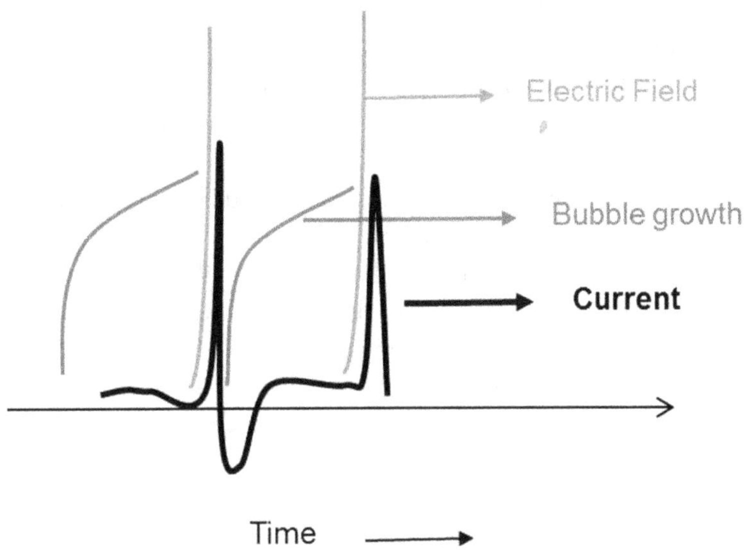

voltage J.A. Mcgeough, Principles of Electrochemical Machining, Chapman and Hall, London (19 J.A. Mcgeough, Principles of Electrochemical Machining, Chapman and Hall, London (19 J.A. Mcgeough, Principles of Electrochemical Machining, Chapman and Hall, London (19. This is called as the 'migration' state of the ECSMM process. The anodic and cathodic reactions occur together with reduction in electrolyte. These liberated positive ions move towards cathode and negative ions move towards anode. In the external circuit, electrons move towards the cathode–electrolyte interface, and go to the solution. At the anode–electrolyte interface equal numbers of electrons are discharged from the solution to the anode. Electrochemical reactions that occur at the electrode–electrolyte interface continuously supply electrons from cathode to solution and solution to anode. This ionic and electronic current is the average current of the order of 100 – 200 mA.

Current Contributing Elements during Electrochemical Phase

Electrolyte resistance, double layer capacitance, polarization resistance, and charge transfer resistance are the elements those contribute to the overall electrochemical reactions during electrochemical phase and hence the instantaneous current, as described below.

Electrolyte Resistance

In an electrochemical cell, bulk solution resistance is a significant factor. Any solution resistance between the electrodes must be considered while modeling the cell. The resistance of an ionic solution depends on the ionic concentration, types of ions, and the geometry of the area in which the current is carried. Unfortunately, most electrochemical cells do not have uniform current distribution through a definite electrolyte area. The major problem in calculating the solution resistance therefore concerns determination of the current flow path and the geometry of the electrolyte that carries the current.

Double Layer Capacitance

An electrical double layer exists on the interface between an electrode and its surrounding electrolyte. This double layer is formed as ions from the solution 'stick on' the electrode surface. The charged electrode is separated from the charged ions. The separation is very small, of the order of angstroms. Charges separated by an insulator form a capacitor. The capacitance such formed is of the order of 20 to 60 μF per cm2 area of the electrode.

The value of the double layer capacitance depends on electrode potential, ionic concentrations, oxide layers, electrode roughness, and many other factors.

Figure 11 shows the process of the double layer formation. All the electrolyte processes take place at the electrode-electrolyte interface. The structure of the double layer is similar to an electrical capacitor formed by a dielectric of thickness of about an ionic radius, i.e. 50 nm. Figure 11 a shows the formation of inner Helmholtz plane (IHP), outer Helmholtz plane (OHP) and the electrode with an excess of negative charge. Figure 11 b shows the localization of the excess charges and Figure 11 c shows the potential gradient formed.

Polarization Resistance

Whenever the potential of an electrode is forced away from its value at 'open-circuit', it is referred to as polarizing the electrode. When an electrode is polarized, it causes current to flow through electrochemical reactions that occur at the electrode surface. The amount of current is controlled by the kinetics of the reactions and the diffusion of reactants both towards and away from the electrode. In the case of ECS cell, the diffusion of reactants is absent. Hence only the kinetics of the reactions plays the major role.

Charge Transfer Resistance

Another resistance is formed by a single kinetically controlled electrochemical reaction. In the case of ECS, the reactions occurring at the cathode-electrolyte interface are of concern and form the basis of the analysis. The kinetics of the charge transfer depends on the kind of reaction, the concentration of reactants, the potential, and many other factors.

Figure 12 shows the various potential drops formed during a general electrochemical cell with an applied supply voltage. The supply voltage overcomes the following potentials:

- The electrode potential,
- The activation over potential,
- Ohmic potential drop,
- Concentration overpotential, (not of much relevance in ECS) and
- The potential due to resistance of electrolyte.

Equivalent Circuit during Electrochemical Phase

Figure 13 shows the circuit parameters for a general electrochemical cell. In the case of ECS, the kinetics at the anode electrode is neglected due to the non consuming electrode material and neglecting the

Figure 11. Process of the double layer formation

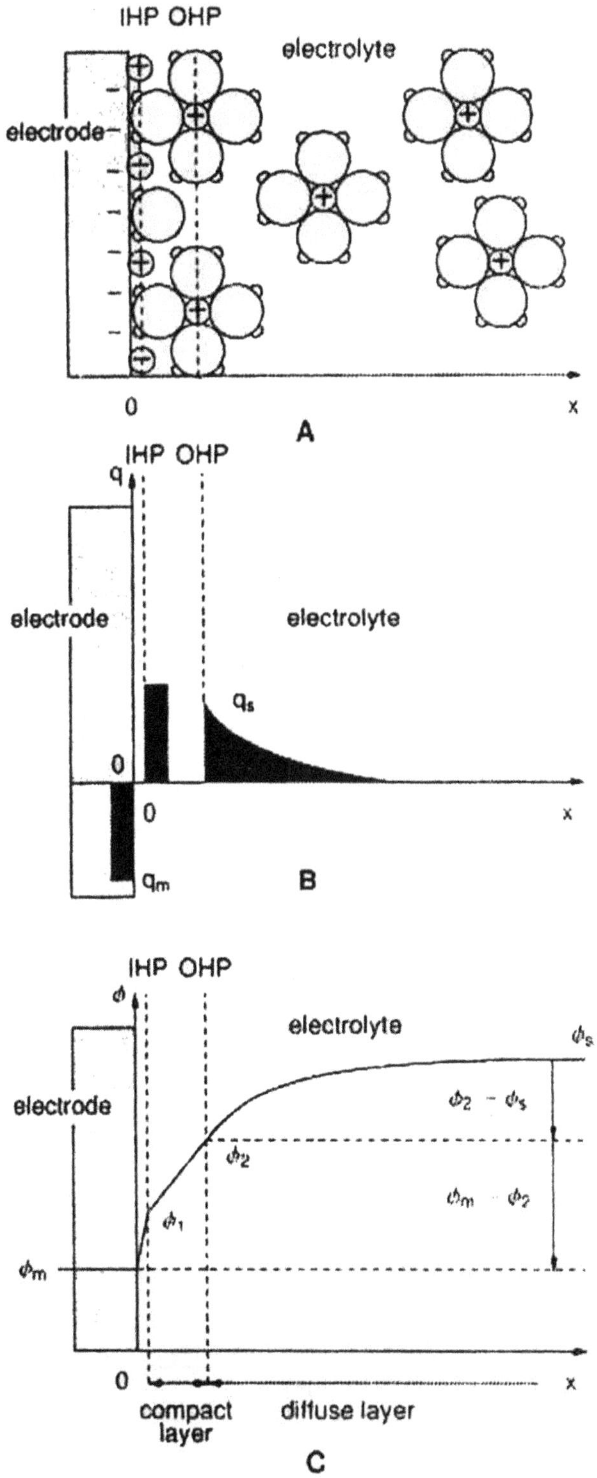

Figure 12. Various potential drops during a general electrochemical cell

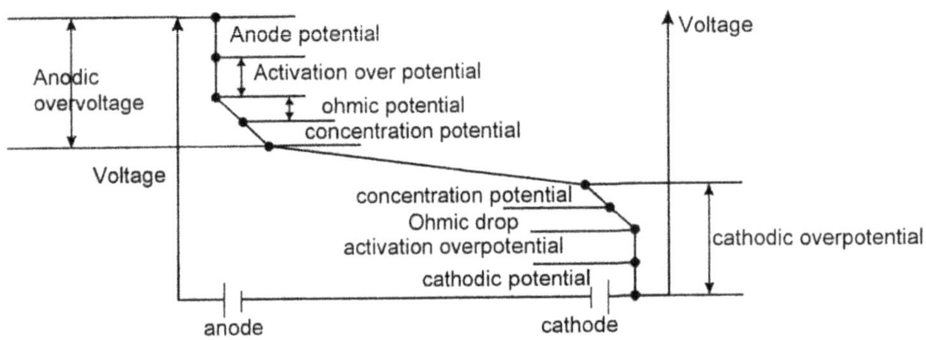

wiring inductance, bulk contact resistance, Z_w, the effective equivalent circuit during electrochemical phase is as shown in Figure 14.

Time Varying Impedance during Electrochemical Phase

The effective circuit elements those contribute to the current and form the equivalent circuit are derived as described above. The time varying impedance $Z_1(t)$ during electrochemical phase (neglecting R_{ext}) from Figure 14 is shown by Equation 1 below.

$$Z1(t) = Rsol + \frac{Rct}{1 + j\omega RctCdl} \tag{1}$$

Butler-Volmer and Tafel expressions (Das, 2005) are used to derive the electrode kinetics for the charge transfer reactions.

Figure 13. Circuit parameters for a general electrochemical cell

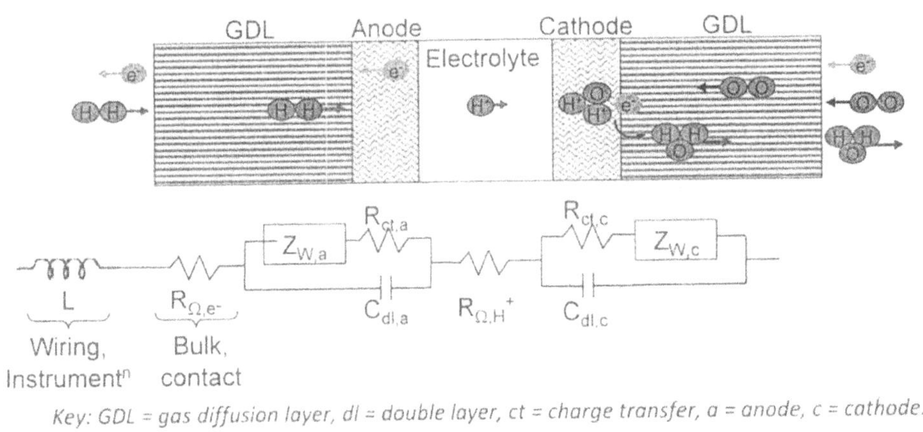

Figure 14. Electrical equivalent circuit in the electrochemical stage of ECS process

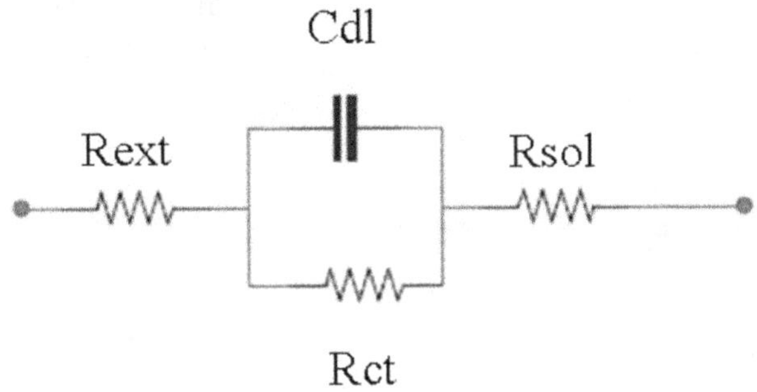

Impedance Spectroscopy Using COMSOL Multiphysics

The 'electrochemical model' in COMSOL is used to solve the 1-d simulation of the electrode kinematics in the ECS cell using the 'secondary current distribution' physics built-in in COMSOL. Figure 15 shows the simulated electrolyte potential distribution in the cell. The electrolyte voltage in the cell is around 75 x 10^{-6} V at a distance of 22 μm distance away from cathode.

Figure 15. Electrolyte potential distribution in the cell

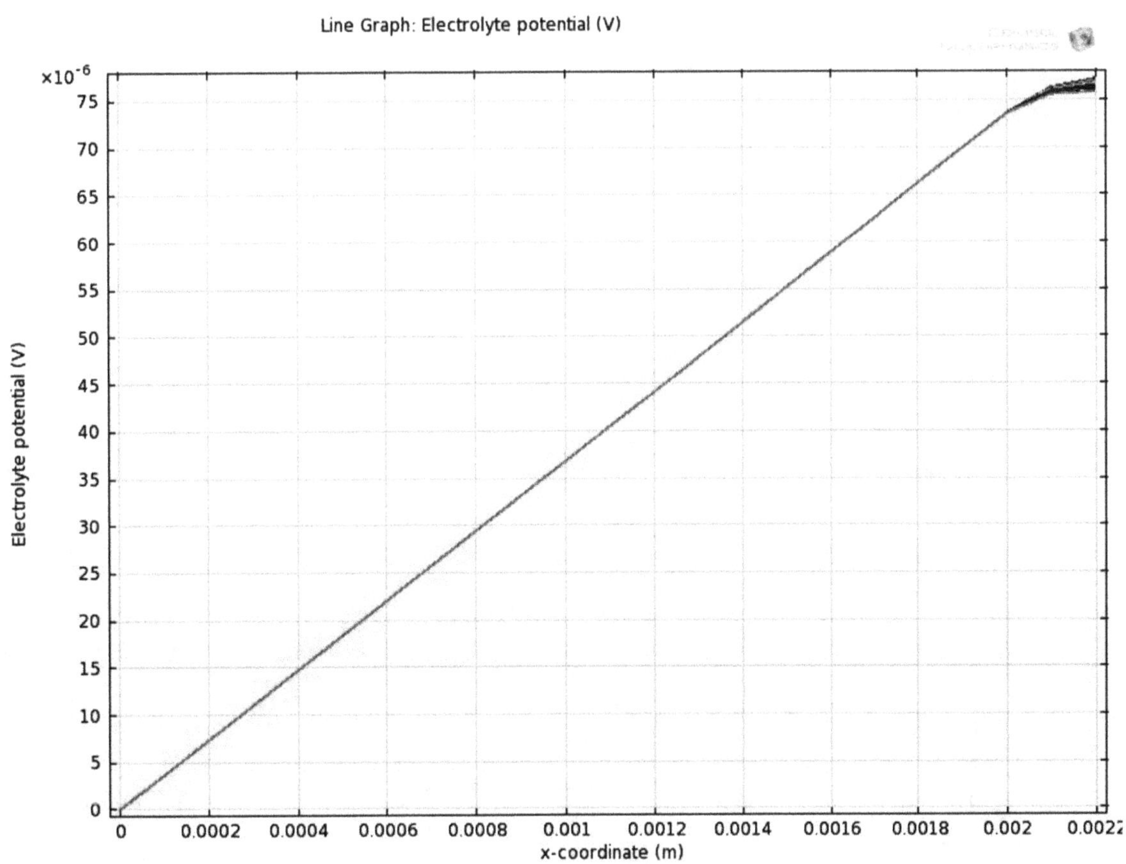

Figure 16 gives the Nyquist plot generated by the COMSOL software by using the perturbation current of 5 mA. It gives the Rct about 15 x 10^{-6} Ω. The Rct x Cdl time constant is very small of around few hundreds of ns. That means the electrochemical reactions are very fast. The hydrogen bubble growth due to pool boiling phenomena soon takes over.

HYDROGEN BUBBLE GROWTH PHASE

The tip of the cathode gets heated up due to ohmic heating. This causes nucleate pool boiling that leads to development and formation of an isolation chamber of hydrogen (H_2) gas bubble. The bubble starts growing till it reaches its departure diameter, reaching which the bubble gets detached from the lower surface of the tool. An isolating film of hydrogen gas bubble covers the cathode tip portion in the electrolyte, abruptly a large dynamic resistance appears and the current through the circuit becomes almost zero. At the same moment, a high electric field of the order of 10^7 V/m gets generated due to bubble's proximity with negatively charged cathode tip and positively charged electrolyte, with a separation distance of a few micro meters. This high electric field causes the bubble breakdown which results in sparking. This leads to generation of energetic electros. These electrons in turn generate secondary electrons. These electrons get drifted towards the workpiece surface due to potential gradient in the form of an energetic electron gun, which causes material removal from the workpiece surface after its bombardment.

Figure 16. Nyquist plot generated by the COMSOL software

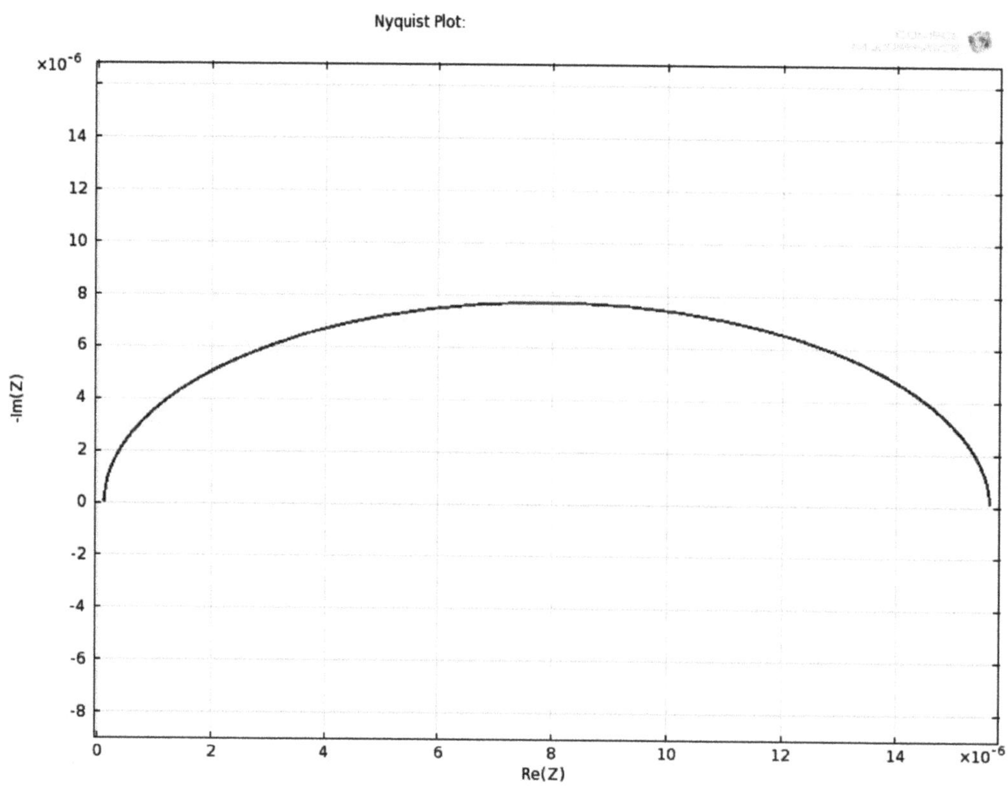

Current Contributing Elements during Hydrogen Bubble Growth Phase

As described above the radius of the bubble varies as a function of time. The growing spherical bubble poses time variable impedance to the current flow. As the radius of the bubble is changing, there is a formation of a varying capacitance similar to a parallel plate capacitor with a growing hydrogen bubble. Thus R(t) and C(t) are the current contributing elements during the hydrogen bubble growth phase.

Equivalent Circuit during Hydrogen Bubble Growth Phase

This impedance is formed as the parallel combination of the variable resistance R(t) and capacitance C(t) those are getting formed due to the variable dielectric due to hydrogen gas. Hence the resultant equivalent circuit in this stage is as shown in Figure 17.

Time Varying Impedance during Hydrogen Bubble Growth Phase

The effective circuit elements those contribute to the current and form the equivalent circuit are derived as described above. The time varying impedance $Z_2(t)$ during the hydrogen bubble growth phase derived from Figure 18 is shown by Equation 13 below.

$$Z2(t) = Rsol + \frac{Rct}{1 + j\omega Rct Cdl} + \frac{R(t)}{1 + j\omega R(t)C(t)} \tag{2}$$

The ECS process is stochastic due to many factors contributing to the bubble growth. The important parameters are the bubble generation rate, their size and site near the cathode tip, bubble growth time, time of its survival, etc. Following assumptions are made while computing the R(t) and C(t) due to hydrogen bubble growth (inverse functions of time dependent radius of bubble, r(t)), pictorially presented in Figure 18:

- It is assumed that a single bubble appears and grows in size following the growth rate proportional to $(\text{time})^{1/2}$.
- Bubbles are hemispherical in size.

Figure 17. Variable impedance posed by the growing hydrogen bubble

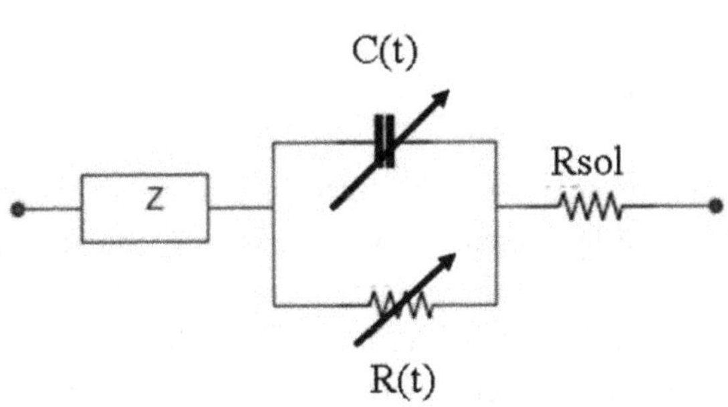

Figure 18. Simplified picture of the ECSMM cell for determining R(t) and C(t)

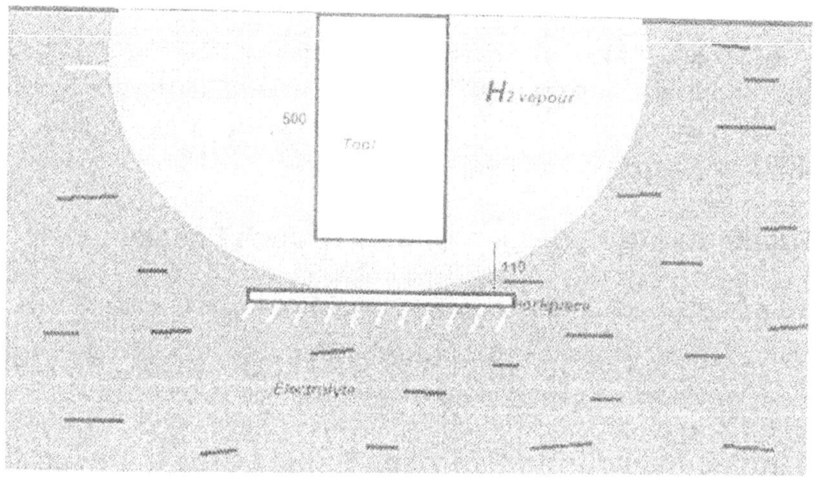

- The nucleation site of the bubble is uniformly distributed over the entire surface of the electrode.
- Current distribution is uniform over the entire surface of the electrode in spite of the constriction of the current due to the bubble stiction.
- Electrolyte temperature and concentration do not change over the time.

The bubble growth can be modeled as per the 'heat transfer controlled growth model' (Karl Stephen, 1992). The time dependent radius of the bubble is written as:

$$r(t) = 1.739[t^{\frac{1}{2}} - 1.091\{(t + 0.076)^{\frac{1}{2}} - 0.2757\}] \quad (3)$$

'VIRTUAL SWITCH OFF' STAGE

As soon as the isolating film of hydrogen gas bubble covers the cathode tip portion in the electrolyte, abruptly a large dynamic resistance is present and the current through the circuit becomes almost zero. This situation is represented by an open switch and the impedance is denoted by $Z_3(t)$. The effective resistance is considered high, around 60-100 MΩ. This assumed range of resistance is close to the real situation.

SPARKING OR IONIZATION

At the same instant of 'virtual off' stage, a high electric field of the order of 10^7 V/m gets generated. This high electric field causes the hydrogen bubble breakdown and sparking takes place. This leads to a formation of conducting channel and the impedance is denoted by $Z_4(t)$. This is manifested by an instantaneous current spike as recorded by the digital oscilloscope.

The resistance during ionization of the conducting channel is assumed to be low, around 10-100 Ω.

The guess values of $Z_3(t)$ and $Z_4(t)$ are very simplistic but are realistic. The physics involved in these phenomena are very complex. The assumption of these values does not, by any means, hamper the analysis of the ECSMM process.

'VIRTUAL SWITCH ON' STAGE

The bubble bursts during the sparking and the contact between the cathode-electrolyte interface reestablishes. Electrochemical action takes over again and the cycle repeats till the machining supply voltage is given to the ECS cell.

MATERIAL REMOVAL

The electrons generated due to the breakdown get accelerated due to voltage gradient. These electrons in the form of an energetic 'electron gun' bombard on the workpiece. This causes intense heating of the workpiece material and material removal takes place.

SEMIEMPERICAL MODEL OF 'SPARK' CYCLE

The formation of spark happens through a series of chemical and physical processes as discussed above. Impedance assigned with each phase w.r.t. Figure 19, has been denoted as below:

1. Z1(t): Impedance during electrochemical phase
2. Z2(t): Impedance during hydrogen bubble growth phase
3. Z3(t): Impedance during 'Virtual Switch Off' phase / 'Open' phase
4. Z4(t): Impedance during Ionization or Sparking phase

Equation 4 describes the impedance of the ECSMM process.

$$Z(t)\text{ECSMM} = \{Z1(t) \quad for\, 0 < t < t1$$

$$\{Z2(t) \quad for\, t1 < t < t2$$

$$\{Z3(t) \quad for\, t2 < t < t3$$

$$\{Z4(t) \quad for\, t3 < t < t4 \tag{4}$$

Where, t1 + t2 + t3 + t4 = T, form a complete cycle period through which a single spark takes place.

Figure 19. Flowchart for impedance modeling in ECSMM process

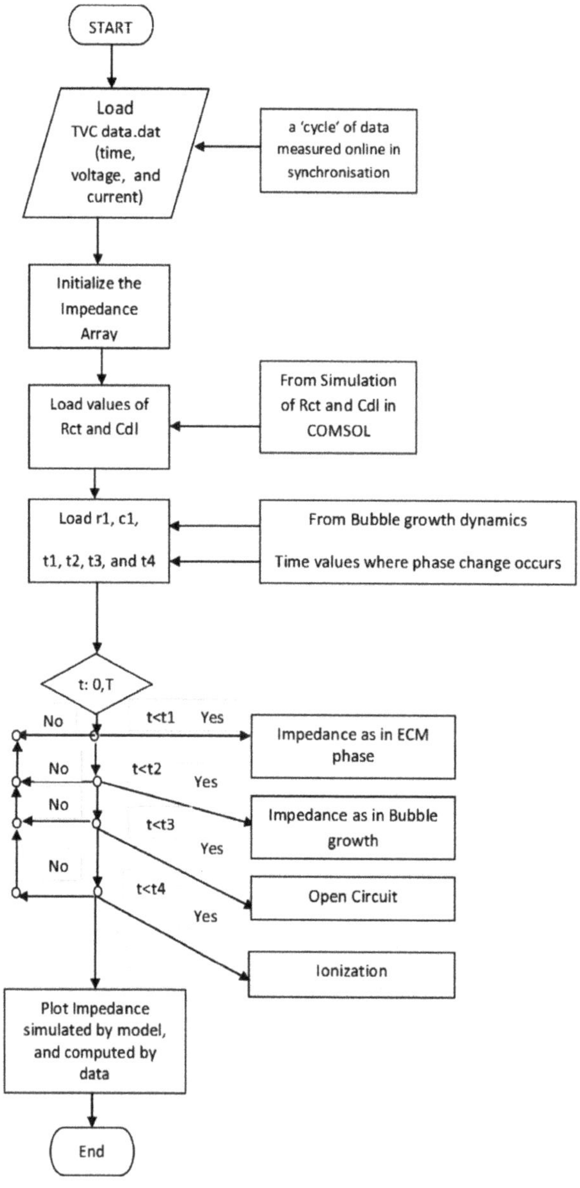

PROCEDURE FOR MODELING

Equation 4 presents the mathematical model of the ECSMM process. This is solved using MATLAB software (Kulkarni, 2013). Figure 19 presents the flow chart of modeling procedure.

The important steps involved in it are:

1. Load time, voltage, and current data (in array form) from the on line, synchronized measurements, digitally stored waveform. For this the digitally stored waveform in the *.mes format is saved in *.csv format using the HAMEG oscilloscope supporting software. The data comprising of one 'cycle' is

cut from the entire data (2000 samples) and is saved as 'TVCdata.dat'. It is suitably manipulated to have count start from this newly defined array (Kulkarni et al, 2013a). Appendix I present the data used for calculation.

2. Initialize Impedance array.
3. Load values of Rsol derived from the conductivity measurement of the NaOH electrolyte. Load values of Ropen and Rionization.
4. Load Rct estimated from the simulated impedance spectroscopy using COMSOL software. Load Cdl value.
5. Load the time intervals t1-t4 from the 'TVCdata.dat' approximately defining the start of each phase of operation, such that $t1 + t2 + t3 + t4 = T$. Where T is the complete period of cycle in which sparking takes place (as seen in Figure 9).
6. Estimate the impedance in each phase of operation during t1, t2, t3, and t4.
7. Plot the modeled impedance.
8. Compute the impedance by measured voltage and current (voltage / current) data at each instant for a single spark ycle.
9. Compare the results.

SIMULATED IMPEDANCE RESULTS

The execution of the MATLAB file 'TVC.m' gives the impedance plots for modeled values and those derived from the measured data (as in TVCdata.dat file, dividing column 2 by 3). Figure 20 presents impedance plots generated by the simulated model. The plot in blue color represents the modeled impedance and that in red color represents that derived from the data. The nature of these plots is same. The impedance in both the cases is very low during the electrochemical and bubble growth phases. It is peaking at the time of isolation. The difference is vast hence the minor details during the bubble growth phase are not remarkable. The impedance is attaining a smaller value after the ionization has occurred. The difference in the peak values at isolation or 'virtual switch off' phase is seen due to various approximations during modeling. The major approximations made during the modeling are:

- Estimation of NaOH solution resistance: It is simply taken as the inverse of the conductivity. But in real situations, it depends on the geometry of the solution (container), its temperature, etc. It is found that the effect of it is dominant during electrochemical and bubble growth phases.
- Resistance of the 'virtual switch off phase': The 'open' resistance during this phase is assumed to be between the 60 – 100 MΩ range. This is a very simplistic assumption.
- Resistance during ionization: An Ionization phenomenon is highly complicated. Again a simplistic assumption is made and the resistance during this phase is assumed to be around 10 - 100 Ω.

CONCLUSION

- The important intermittent phases of the process have been identified and analyzed. The formation of spark happens through series of chemical and physical processes in phases. These are: electrochemical action, bubble formation, coalescence and growth of bubble during pool boiling

Figure 20. Impedance plots generated by MATLAB

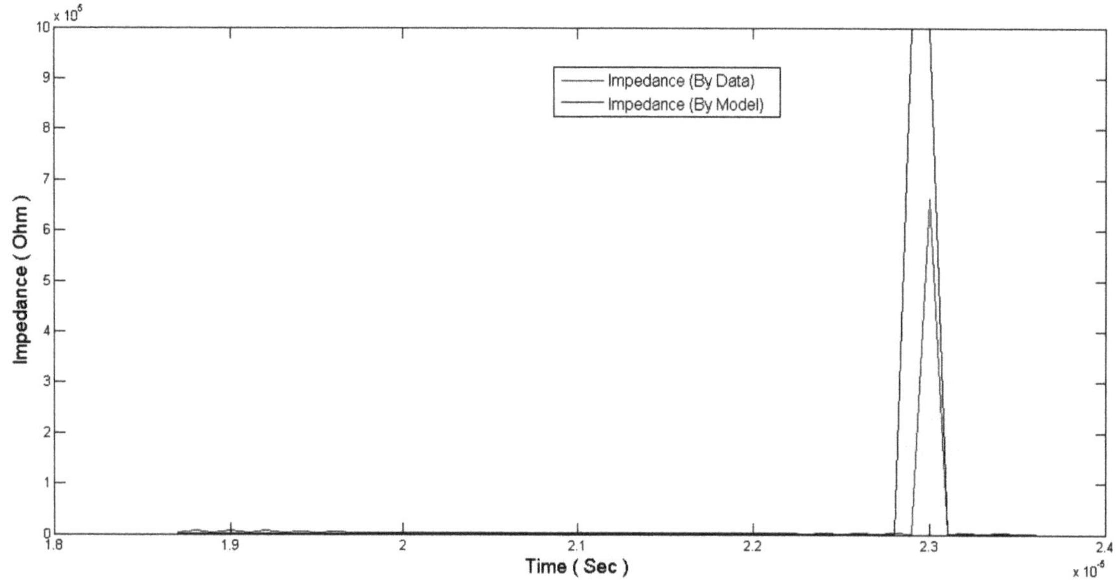

slowly covering the tool tip, tool tip isolation leading to a momentary 'virtual switch off' situation, instantaneous application of high electric field causes sparking of hydrogen breakdown, reestablishment of tool-electrolyte contact leading to 'virtual switch on' situation, drifting of energetic electrons towards the workpiece due to potential gradient.

- The instantaneous current comprises of current due to these sequential events. The electrical impedance during each phase of operation varies. The ECSMM process has been sequentially modeled with electrical impedance in each phase. The electrical impedance during electrochemical phase ($Z_1(t)$), bubble growth phase ($Z_2(t)$) and ionization phase ($Z_4(t)$, are small than that compared with the impedance during the 'Virtual Switch Off' or isolation phase ($Z_3(t)$). $Z_3(t)$ should be as large as possible. The condition of isolation due to a single large bubble is favorable to have the high energy spikes for machining to take place. The time taken for the bubble to grow its fullest dimensions is the dominating parameter of the frequency of the sparking. The semiemperically calculated impedance and the simulated impedance match well.
- It is observed that at low to moderate sparking frequency, of the order of few hundred Hz to few hundred kHz, machining takes place. It is also observed that at very high frequencies of sparking, of the order of MHz range, machining does not take place.
- The instantaneous spark energy is of the order of few hundreds of kJ/m^2.
- Of these phases, the electrochemical action is the fast process, hydrogen growth phase is a slow process, again the Virtual switching On and Off processes are instantaneous. Break down and drifting are fast processes.
- Sparking is a discrete phenomenon due to the repeated breakdown of the hydrogen bubble due to cascade of the actions as detailed.
- It is for the first time, that the process is studied from a holistic view point by developing an electrical equivalent circuit in electrochemical phase and hydrogen growth during pool boiling phase,

of operation. Simplified electrical model in different stages of the ECSMM process are presented. These are consistent with the observed on line transient current.

- Charge transfer kinetics at anode electrode is neglected. This is minimized by the design of the anode electrode.
- The charge transfer resistance in the equivalent circuit during the electrochemical phase is found by performing the impedance spectroscopy. It is done using COMSOL Multiphysics modeling software. For this the 1-d model of the electrochemical process is developed using the secondary current distribution. The time constant due to the parallel combination of the double layer capacitor and the charge transfer resistance is of the order of thousands of ns. That means the formation of the double layer capacitance and charge transfer resistance happens within thousands of ns time.
- The MATLAB software is used to compute simplified variable impedance due to time varying R(t) and C(t) during bubble growth and simulate the overall impedance.

ACKNOWLEDGMENT

I am indebted to Prof. V. K. Jain for his immense guidance and support throughout my academic life at IIT Kanpur. I am thankful to Prof. K. A. Misra for his guidance in carrying out the work. My thanks are also due to Prof. B. Dasgupta for the foundation for mathematical modeling. Financial support for this work from Department of Science and Technology, Government of India, New Delhi, is gratefully acknowledged (Grant no. SR/S3/MERC-079/2004). Thanks are due to the staff at Manufacturing Science Lab and Centre for Mechatronics, at IIT, Kanpur. Thanks are also due to the staff at Glass Blowing section of IIT, Kanpur.

REFERENCES

Basak, I., & Ghosh, A. (1992). Mechanism of material removal in electrochemical discharge machining: A theoretic model and experimental verification. *Journal of Materials Processing Technology*, *71*(3), 350–359. doi:10.1016/S0924-0136(97)00097-6

Basak, I., & Ghosh, A. (1996). Mechanism of spark generation during electrochemical discharge machining: A theoretical model and experimental investigation. *Journal of Materials Processing Technology*, *62*(1-3), 46–53. doi:10.1016/0924-0136(95)02202-3

Comsol. (n.d.). Batteries Fuel Cells. *Comsol.* Retrieved from: http://www.comsol.co.in/products/batteries-fuel-cells/

Das, S. K. (2005). *Process heat transfer*. Narosa Publishing House.

Fascio, V., Wüthrich, R., & Bleuler, H. (2004). Spark assisted chemical engraving in the light of electrochemistry. *Electrochimica Acta*, *49*(22-23), 3997–4003. doi:10.1016/j.electacta.2003.12.062

Jain, V. K., Dixit, P. M., & Pandey, P. M. (1999). On the analysis of electro chemical spark machining process. *International Journal of Machine Tools & Manufacture*, *39*(1), 165–186. doi:10.1016/S0890-6955(98)00010-8

Karl, S. (1992). *Heat transfer in boiling and condensation*. Berlin: Springer Verlag.

Kulkarni, A. V. (2009). Systematic analysis of electrochemical discharge process. *Int. J. Machining and Machinability of Materials*, *6*(3/4), 194–211. doi:10.1504/IJMMM.2009.027324

Kulkarni, A. V. (2012). Electrochemical Spark Micromachining Process. In M. Kahrizi (Ed.), Micromachining techniques for fabrication of micro and nano structures, (pp. 235-252). Academic Press.

Kulkarni A.V. & Jain, V.K. (2015). Design and development of an electrochemical spark micro manufacturing equipment. *International Journal of Mechanical Engineering and Robotics Research, 4*(4), 368-372. DOI: 10.18178/ijmerr.4.4

Kulkarni, A. V., Jain, V. K., & Misra, K. A. (2011a). Electrochemical spark micromachining microchannels and microholes of metals and non-metals. *Int. Journal of Manufacturing Technology Management*, *22*(2), 107–123. doi:10.1504/IJMTM.2011.038713

Kulkarni, A. V., Jain, V. K., & Misra, K. A. (2011b). Electrochemical spark micromachining: Present scenario. *IJAT*, *5*(1), 52–59. doi:10.20965/ijat.2011.p0052

Kulkarni, A.V., Jain, V.K., & Misra, K.A. (2013). Study of microstructure and transient, instantaneous current in different excitations during silicon micromachining. *J. Inst. Eng. India Ser. B.*

Kulkarni Anjali, V. (2013). *Performance investigations into microfabrication using electrochemical spark* (Doctoral dissertation). GBTU, Lucknow, India.

Peng, W. Y., & Liao, Y. S. (2004). Study of electrochemical discharge machining technology for slicing non-conductive brittle materials. *Journal of Materials Processing Technology*, *149*(1-3), 363–369. doi:10.1016/j.jmatprotec.2003.11.054

Scribner. (n.d.). *Electrochemical Impedance Spectroscopy (EIS): A Powerful and Cost-Effective Tool for Fuel Cell Diagnostics*. Scribner. Retrieved from: http://www.scribner.com/files/tech-papers/Scribner%20Associates%20-%20Electrochemical%20Impedance%20Spectroscopy%20for%20Fuel%20Cell%20Research.pdf

Sorkhel, S. K., Bhattacharyya, B., Mitra, S., & Doloi, B. 1996. Development of electrochemical discharge machining technology for machining of advanced ceramics.*International Conference on Agile Manufacturing*, (pp. 98-103).

Yadava, V., Jain, V. K., & Dixit, P. M. (2002). Temperature Distribution during Electro-Discharge Abrasive Grinding. *Machinin. Science and Technology*, *6*(1), 97–127.

APPENDIX 1

Table 1. TVCdata.dat and X array

Time	Voltage	Current
1.87E-05	6.48E+01	2.61E-02
1.88E-05	6.32E+01	1.01E-02
1.89E-05	6.48E+01	2.01E-02
1.90E-05	6.32E+01	8.09E-03
1.91E-05	6.48E+01	1.81E-02
1.92E-05	6.32E+01	8.09E-03
1.93E-05	6.48E+01	1.81E-02
1.94E-05	6.32E+01	1.21E-02
1.95E-05	6.48E+01	2.01E-02
1.96E-05	6.32E+01	1.41E-02
1.97E-05	6.48E+01	2.41E-02
1.98E-05	6.32E+01	1.61E-02
1.99E-05	6.48E+01	2.41E-02
2.00E-05	6.32E+01	1.81E-02
2.01E-05	6.48E+01	2.41E-02
2.02E-05	6.32E+01	1.81E-02
2.03E-05	6.48E+01	2.41E-02
2.04E-05	6.32E+01	1.81E-02
2.05E-05	6.48E+01	2.41E-02
2.06E-05	6.32E+01	1.81E-02
2.07E-05	6.48E+01	2.81E-02
2.08E-05	6.32E+01	1.81E-02
2.09E-05	6.48E+01	3.01E-02
2.10E-05	6.32E+01	2.01E-02
2.11E-05	6.48E+01	2.81E-02
2.12E-05	6.32E+01	2.21E-02
2.13E-05	6.48E+01	3.01E-02
2.14E-05	6.40E+01	2.41E-02
2.15E-05	6.56E+01	3.01E-02
2.16E-05	6.40E+01	2.61E-02
2.17E-05	6.56E+01	3.01E-02
2.18E-05	6.24E+01	2.21E-02
2.19E-05	6.56E+01	3.01E-02
2.20E-05	6.32E+01	2.41E-02
2.21E-05	6.56E+01	3.21E-02
2.22E-05	6.40E+01	2.61E-02
2.23E-05	6.56E+01	3.41E-02
2.24E-05	6.24E+01	2.61E-02
2.25E-05	6.56E+01	3.41E-02
2.26E-05	6.32E+01	2.41E-02
2.27E-05	6.64E+01	3.61E-02
2.28E-05	6.32E+01	2.81E-02
2.29E-05	6.56E+01	3.41E-02
2.30E-05	6.24E+01	9.40E-05
2.31E-05	6.80E+01	5.21E-02
2.32E-05	6.32E+01	2.61E-02
2.33E-05	6.56E+01	3.81E-02
2.34E-05	6.24E+01	2.61E-02
2.35E-05	6.46E+01	3.26E-02
2.36E-05	6.46E+01	3.30E-02

continued in next column

Table 1. Continued

X = [25.6e-7,5e-6,28.69e-8,6.6e-5,1.9e-5,2.04e-5,2.26e-5,2.3e-5,2.33e-5]

APPENDIX 2

List of Symbols, Abbreviations and Nomenclature

Ω: Ohm
T: Time period of complete cycle, s

Acronyms

AC: Alternating current
Cdl: Double layer capacitance, F
C(t): Time varying capacitance, F
Cu: Copper
DC: Direct current
HWR: Half wave rectified
ECS: Electrochemical Spark
ECSM: Electrochemical Spark Machining
ECSMM: Electrochemical Spark Micromachining
IEG: Inter electrode gap, μm
Rct: Charge transfer resistance, Ω
R(T): Time varying resistance, Ω
R_{sol}: Sol Resistance of electrolyte between two electrodes, Ω
$Z_1(t)$: Time varying impedance during electrochemical stage, Ω
$Z_2(t)$: Time varying impedance during bubble growth stage, Ω
$Z_3(t)$: Time varying impedance during ionization stage, Ω
$Z_4(t)$: Time varying impedance during 'virtual switch off' stage, Ω

Chapter 13
Development of a Material Constitutive Model and Simulation Technique to Predict Nonlinearities in Piezoelectric Materials under Weak Electric Fields

M. K. Samal
Bhabha Atomic Research Centre, India & Homi Bhabha National Institute, India

ABSTRACT

Piezoceramic materials exhibit different types of nonlinearities depending upon the magnitude of the mechanical and electric field strength in the continuum. Some of the nonlinearities observed under weak electric fields are: presence of superharmonics in the response spectra and jump phenomena etc. especially if the system is excited near resonance. It has also been observed by many researchers that, at weak alternating stress fields, the relationship between the piezoelectrically induced charge and applied stress in ferroelectric ceramics, has the same form as the Rayleigh law (for magnetization versus magnetic field) in ferromagnetic materials. Applicability of the Rayleigh law to the piezoelectric effect has been demonstrated for Lead Zirconate Titanate ceramics by many researchers and their experimental results indicate that the dominant mechanism responsible for piezoelectric hysteresis and the dependence of the piezoelectric coefficient on the applied alternating stress is the pinning of non-180° domain walls. In this chapter, the Rayleigh law for ferromagnetic hysteresis has been modified and incorporated in a nonlinear electric enthalpy function and then applied in the analysis of hysteresis behavior of piezoelectric continua. Analytical solutions have been derived for a cantilever beam actuated by two piezo-patches attached to the top and bottom of the beam and excited by opposite electric fields. Analysis has been carried out at different electric field excitations of varying amplitude and frequencies and the results have been compared with the available experimental results from literature.

DOI: 10.4018/978-1-5225-1639-2.ch013

INTRODUCTION

Piezoceramic materials find wide applications in many different areas of modern technology (e.g. aerospace structures) and their use is increasing day by day (Crawley, 1994). Prediction of performance of the structures with piezoelectric sensors and actuators (for various control applications such as active shape, vibration and noise control etc.) requires thorough understanding of the piezoelectric material behaviour at different applied electrical and mechanical fields. Though nonlinearities under strong electric fields such as dielectric and butterfly hysteresis have drawn attention of many authors in recent times (Viswamurthy et al., 2007; Viswamurthy & Ganguli, 2007; Hwang & McMeeking, 1998; Thakkar & Ganguli, 2004, 2006, 2007), modeling nonlinear behaviour of piezoceramics under weak electric fields is limited in literature.

With a number of piezoelectric finite element (FE) models existing in the literature to cope with complicated boundary problems of smart structures, it is very important to develop a simple procedure to obtain analytical solutions for a few simple geometries for verification of finite element formulations and also to gain confidence on the procedure. Moreover it is also necessary to develop analytical solutions to determine the various material parameters including those of the nonlinear model to be used in the finite element analysis. The coupling of mechanical and electrical properties of piezoelectric materials makes the analytical solution complicated. Piezoelectric plates composed of orthotropic or anisotropic layers are usually analyzed using laminated composite plate theory. For the analytical solution of these plates, Pagano (1970) first presented the exact solution of a laminated plate under cylindrical bending. Ray et al. (1992, 1993a, b, 1998), Heyliger and Brooks (1996) extended this methodology to develop the exact solutions for the piezoelectric laminated plates with the help of Eshelby-Stroh formalism. Later, taking the thermal effect into account, Xu et al. (1995) presented a 3-D analysis for the coupled thermo-electro-elastic response of multilayered hybrid composite plates with four simply supported edges by using the state space method. Dube et al. (1996) developed a series solution method for a single layer thermo-electro-elastic plate subjected to cylindrical bending using the approach of Pagano (1970).

To solve the dynamic problem of piezoelectric array elements, several simplified solutions for modeling the linear piezoelectric behaviour in 1-D, 2-D and 3-D domains were tried. Ray et al. (1998) presented an exact solution for analyzing dynamics of composite plates with piezoelectric layers bonded at the top and bottom surfaces. Later, Ray (1998) also presented the closed-form solution for the optimal control of flexural vibration of a simply supported symmetric laminated plate. Wolf and Gottlieb (2001) derived the nonlinear equations of motion for a cantilever beam covered by PZT layers in symmetric and asymmetric configuration. By defining a nonlinear enthalpy function with nonlinear strain and applying Hamilton's principle, they derived the equations of motions and solved by perturbation analysis and multiple scales method.

Beige and Schmidt (1982) modeled the nonlinearities under weak electric fields using higher order quadratic and cubic elastic terms in the electric enthalpy function. von Wagner and Hagedorn (2002) studied the nonlinearity in a piezo-beam system. Later von Wagner (2003, 2004) and Neumann (2002) predicted the amplitude of displacement and current responses of piezo-rods of different diameters using a one-dimensional nonlinear formulation. Neumann (2002) used the nonlinear damping formulation to derive the virtual work done by the dissipative damping forces. A generalized 3-D formulation for modeling the nonlinear behaviour of piezoelectric materials under weak electric fields was presented by Samal et al. (2006a) and it was used to derive a 3-D finite element model.

The application of the formulation was demonstrated in Samal et al. (2006b) using different types of examples. The parameter identification procedure used by von Wagner (2003, 2004) and Neumann (2002) is restricted to those of 1-D continuum. In order to predict the nonlinear response of actual 3-D piezoelectric systems, the material properties of the 3-D piezoelectric continuum are required. This requires a closed-form solution for a simple geometry. In this paper, such closed-form solutions for prediction of displacement and current responses of simple rectangular and cylindrical geometries have been developed using a simple proportional damping formulation for the material damping.

In an earlier work, Samal et al. (2007) have derived a closed-form solution for a rectangular piezoeceramic plate of a soft piezoceramic PIC 255. However, a nonlinear damping formulation was used in the above model which requires identification of a large number of nonlinear model parameters which is many times time-consuming, complicated and cumbersome. On the other hand, Usher and Sim (2005), von Wagner (2003, 2004), Wang. et al. (1999) etc. demonstrated that the simple proportional damping theory is able to predict the response of wide varieties of piezoecramics satisfactorily. There is a need to develop an efficient nonlinear model using a simple damping theory, so that the electric field dependent damping coefficient can be evaluated easily from experimental measurements and at the same time both the mechanical and electrical responses of the piezoceramic are satisfactorily predicted. Another important aspect of this model is the use of an approach where the numerical shape functions (obtained from a prior FE analysis) can be used in place of analytical shape functions in the Rayleigh-Ritz discretisation. This approach is helpful especially for geometries where the evaluation of analytical shape functions is difficult. In this work, experiments have been conducted on plate and cylindrical geometries of a PIC 181 material (which is a hard piezoceramic) and the predicted results have been compared with those of experiment.

The piezoelectric response of ferroelectric ceramics consists of two parts. The first part is the lattice contribution, which originates from the piezoelectric effect within every individual ferroelectric domain (Figure 1(a) and 1(b)). The second part, often called the extrinsic contribution, is largely due to the displacement of non-180° domain walls (Cady, 1964; Ikeda, 1990). In many ferroelectric ceramics, such as lead zirconate titanate, $Pb(Zr,Ti)O_3$, and barium titanate, $BaTiO_3$, the piezoelectric response is dominated by domain-wall contributions. Other processes that are not necessarily related to the domain-wall motion may also contribute to the apparent piezoelectric response of a ferroelectric material. The domain-wall contribution to dielectric, elastic, and piezoelectric properties of ferroelectric ceramics has been the subject of many studies (Cross, 1993). Most of the early information on domain-wall processes was obtained from experiments and models that were focused on nucleation, growth and switching of domain walls. The domain-wall response to weak dynamic fields (subcoercive electric fields or stresses) has been less thoroughly investigated.

Under weak field conditions, the external field causes only vibration or small displacements of the walls, but the overall structure of the domain walls does not change, i.e., there is no switching of domain walls and no nucleation of new walls. Typically, ferroelectric materials exhibit a response (mechanical strain, dielectric polarization or piezoelectrically induced strain and charge) that is a nonlinear and hysteretic function of subcoercive external ac stress or electric field. One approach to separating the domain-wall and intrinsic lattice contributions to the properties of ferroelectric ceramics is to calculate the lattice contribution by using, for example, the Landau–Ginzburg–Devonshire phenomenological thermodynamic theory (Damjanovic, 1997).

The domain-wall contribution is assumed to be equal to the difference between the measured and the calculated response. Whereas this approach is useful in separating the two contributions, a general

Figure 1. (a) Lattice contribution of piezoelectric response; (b) domain contribution of piezoelectric response

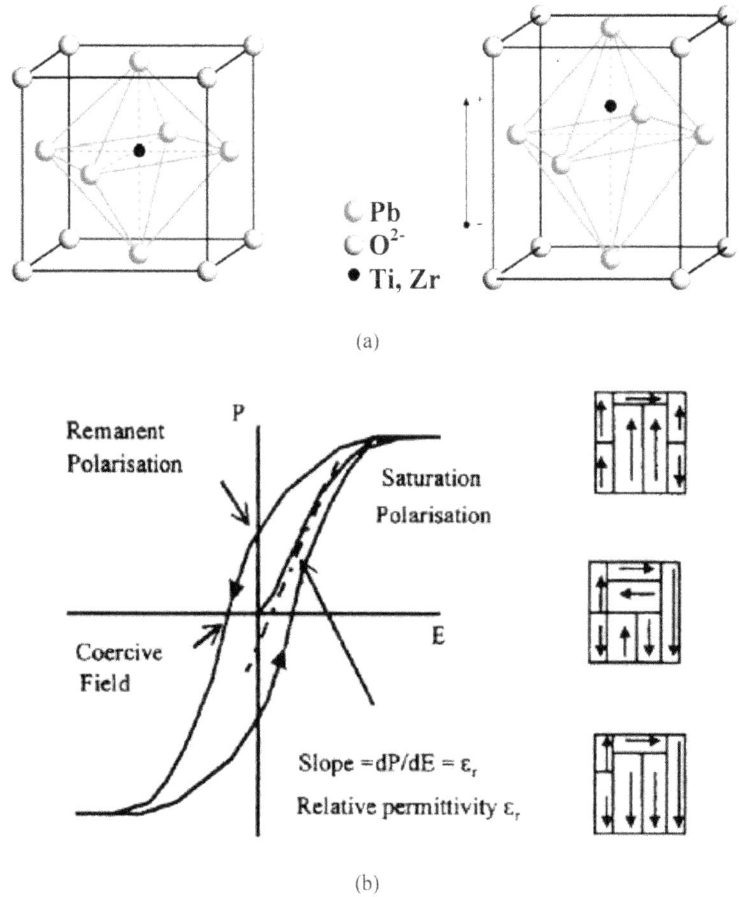

theory of domain-wall displacement under weak field conditions still does not exist for ferroelectric ceramics. This lack of a successful model is partly due to the complexity of domain-wall motion, which depends on lattice defects, dopants, composition, crystal structure and microstructure of the ceramics.

The understanding of domain wall processes in magnetic materials is much more advanced than in ferroelectrics. From both the theoretical and practical point of view, emphasis has been placed on understanding the interaction of mobile domain walls with defects that are randomly distributed in an (anti) ferromagnetic material and which act as pinning centers. One consequence of weak-field domain-wall pinning in magnetic materials is the field dependence of the dynamic susceptibility, which may be described by the classical Rayleigh law. Recent studies have shown that the pinning of magnetic domain walls may, in many cases, be treated quite generally as interface pinning.

The theoretical analysis appears to be similar whether the interface is a ferromagnetic domain wall, a liquid–liquid interface, or a ferroelectric domain wall. The applicability of this general approach to weak field displacement of ferroelectric domain walls has recently been confirmed experimentally (Damjanovic & Demartin, 1997; Damjanovic et al., 1996) by showing that the direct piezoelectric effect in several ferroelectric ceramics with different domain wall structure may be described by relations analogous to those of the Rayleigh law. The piezoelectrically active non-180° domain walls are simultaneously ferroelastic and ferroelectric, which may imply universal validity of the Rayleigh law for the domain wall displacements in all ferroic (i.e., ferromagnetic, ferroelectric and ferroelastic) materials.

In this chapter, the Rayleigh law has been used to modify the electric enthalpy of the piezoelectric continuum (to take care of dielectric hysteresis) and it has been applied to predict the nonlinear response of a catilever beam (actuated by two piezoceramic actuators attached to the top and bottom surface at the center of the beam). Hamilton principle has been used to derive the governing nonlinear differential equation of the beam. Rayleigh-Ritz discretization along with harmonic balance method has been used to solve for the nonlinear response of the piezo-beam system. Experimental results of responses of the cantilever beam with hysteresis and structural mode type nonlinearity has been taken from literature and the results have been compared with the predicted results of the suggested model.

Modeling of Ferroelectric Hysteresis in Piezoelectric Materials Using Rayleigh Law

Rayleigh (1887) observed that in certain ferromagnetic materials, under weak field conditions, the magnetic susceptibility χ during the first magnetization is linearly related to the magnetic field H by:

$$\chi\left(H\right)=\chi_{init}+\alpha H \tag{1}$$

where χ_{init} is the initial susceptibility and α is called the Rayleigh coefficient. The susceptibility is defined as:

$$\chi=M/H \tag{2}$$

where M is the magnetization. The initial magnetization curve is given by:

$$M\left(H\right)=\chi_{init}H+\alpha H^{2} \tag{3}$$

If the magnetic field is cycled, M lags in phase behind H. Rayleigh showed that the weak field hysteresis loop can be described by the following non-analytical relationship:

$$M\left(H\right)=\left(\chi_{init}+\alpha H_{\max}\right)H\pm\frac{\alpha}{2}\left(H_{\max}^{2}-H^{2}\right) \tag{4}$$

where the sign "-" stands for increasing field, "+" for decreasing field, and $H_{\max}$ designates the maximum field reached during the cycle. Equation 4 for the hysteresis may be derived from Equations 1 and 3 with the assumption that the magnetization versus field curves have differential susceptibility ($\chi=\partial M/\partial H$) at the loop tips (at $H=H_{\max}$ for the decreasing field and at $H=-H_{\max}$ for the increasing field) that is equal to χ_{init}. According to Rayleigh, the linear term in Equation 3 represents the reversible or instantaneous component of the magnetization and the quadratic term, which includes the Rayleigh coefficient, represents the irreversible component. Figure 2(a) illustrates the so-called Rayleigh loop and its parameters. The Rayleigh coefficient α determines the irreversible part of the loop. The residual magnetization at zero field, $M\left(H=0\right)$, is equal to ($\pm\frac{\alpha}{2}H_{\max}^{2}$).

From Equation 4, it is seen that the tip of the Rayleigh loop, i.e., $M_{\max}=M\left(H=H_{\max}\right)$, traces the curve of initial magnetization when the amplitude $H_{\max}$ of the ac field is changed. Hence, the plot of

Figure 2. (a) Schematic of Rayleigh curve for magnetic moment vs. applied magnetization; (b) variation of potential energy with position of domain wall

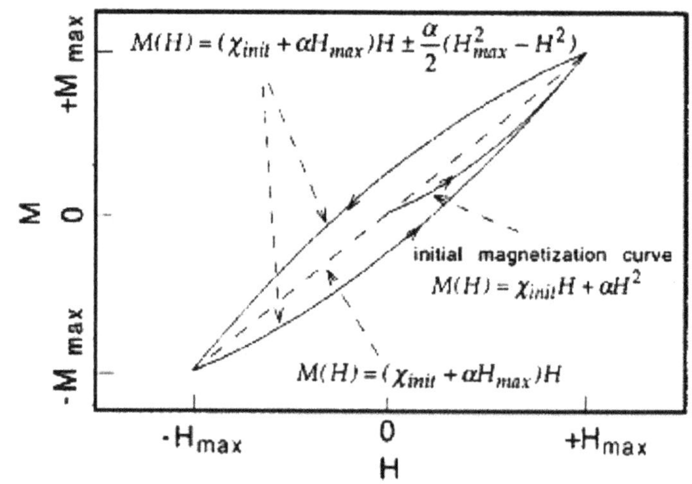

(a)

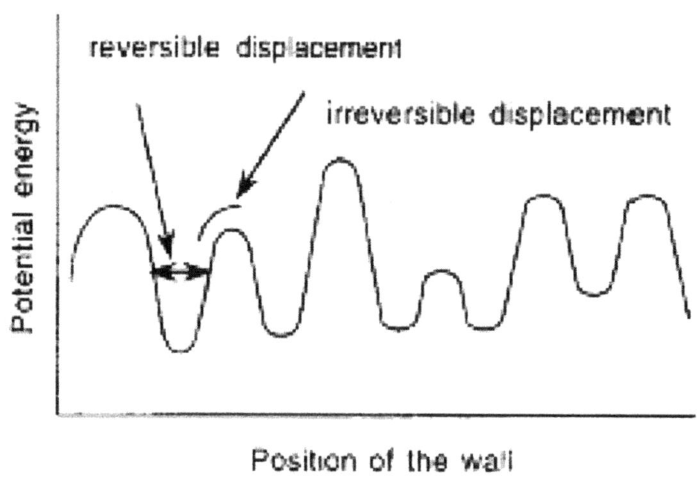

(b)

M_{max} vs H_{max} is described by Eq., where H and M are replaced by H_{max} and M_{max} respectively. Similarly, the plot of $\chi(H_{max})$ vs H_{max} is given by Equation 1, where H is replaced by H_{max}. The first theory, which attempted to explain the physical origins of the Rayleigh law, was proposed by Ne´el in 1942. Ne´el (1942) analyzed displacements of a single domain wall in a medium containing defects. These defects introduced irregular perturbations of the potential energy for domain wall motion (Fig. 2(b)). It was shown that displacement of a wall in such a medium leads to magnetization that has the main characteristics of the Rayleigh law, namely a reversible linear component and irreversible quadratic component.

The irreversible component of the magnetization and the dependence of the magnetic susceptibility on the external field are due to the movement of the domain wall between two local potential energy minima. The irreversible domain wall motion occurs when the applied force is sufficiently large to overcome the potential barrier separating two local energy minima. The Rayleigh coefficient α is thus a measure of the irreversible displacement of the domain wall. The origin of the hysteresis loss is in pinning and unpinning of the wall as the field is cycled and the wall moves from one potential well into another. A statistical theory of the Rayleigh law, was later given by Kronmueller (1970). The theory is based on Ne´el's (1942) assumption that the potential energy of the domain walls is a statistical function of their position. Kronmueller (1970) further assumed that the domain walls are decoupled, and that no other forces except those associated with lattice defects are acting on the walls. An important result of the theory was a relationship between the Rayleigh coefficient and microstructural features that act as pinning centers. Kronmueller (1970) showed that the Rayleigh coefficient is inversely proportional to the density of dislocations. It was later shown experimentally that some predictions of the theory are not valid; however, the Rayleigh coefficient α is generally found to be dependent on the microstructure of the material. The magnetic properties of materials have been extensively analyzed in terms of the Rayleigh law.

Proposed Model for Application of Rayleigh Law To Predict Ferroelectric Hysteresis

Analogous to Equation 4, the hysterectic relation between electric polarization D_z and electric field E_z (both in polarization direction), is proposed to be modeled as:

$$D_z(E_z) = c_1 E_z + c_2 S_{xx} \pm \frac{\alpha_1}{2}\left(E_{z\mathrm{m}}^2 - E_{z0}^2\right) \tag{5}$$

where c_1 and c_2 are constants. α_1 is Rayleigh's constant, E_{zm} is the maximum value of the electric field amplitude during cyclic application (+ve sign for increasing and –ve sign for decreasing) of sinusoidal electric field and E_{z0} is the electric field amplitude of sinusoidal excitation. Using the above relation [i.e., Eq.] for D_z, the nonlinear electric enthalpy of the piezoelectric continuum can be modified as given in next section.

Constitutive Equations of The Piezoelectric Continuum

Linear Electric Enthalpy Density Function

The electric enthalpy density function H_{lin} is generally used to derive the governing equations of the coupled piezoelectric continuum and for the linear piezoelectric behavior; it is given as (IEEE standards on piezoelectricity, 1988)

$$H_{lin} = \frac{1}{2} C^{E}_{ijkl} S_{ij} S_{kl} - e_{mij} E_m S_{ij} - \frac{1}{2} \varepsilon^{S}_{mn} E_m E_n \tag{6}$$

where $\left(i = 1,6;\ j = 1,6;\ m = 1,3;\ n = 1,3\right)$, C^{E}_{ijkl} is the fourth order elastic tensor under constant electric field, e_{mij} is the 3rd order piezoelectric tensor, ε^{S}_{mn} is the second order dielectric tensor, S_{ij} is the strain tensor and E_m is the electric field vector. The above expression for H_{lin} can be expressed in simple matrix form as given in (Maugin, 1985)

$$H_{nonl} = \frac{1}{2}\{S\}^T [C]\{S\} - \{E\}^T [d][C]\{S\} - \frac{1}{2}\{E\}^T \left[\left[\varepsilon^T\right] - [d][C][d]^T\right]\{E\} \tag{7}$$

Nonlinear Electric Enthalpy Density Function (Without Hysteresis)

The above linear electric enthalpy function can be expanded to the nonlinear electric enthalpy function using higher order terms (quadratic and cubic terms in strain and electric field tensor) and hence the generalized expression for H_{lin} is written as (Samal, 2003)

$$\begin{aligned}
H_{nonl} &= \frac{1}{2}\{S\}^T [C]\{S\} - \{E\}^T [d][C]\{S\} - \frac{1}{2}\{E\}^T \left[\left[\varepsilon^T\right] - [d][C][d]^T\right]\{E\} \\
&+ \frac{1}{3}\{S\}^T [C_1]\{S^2\} + \frac{1}{4}\{S^2\}^T [C_{21}]\{S^2\} + \frac{1}{4}\{S\}^T [C_{22}]\{S^3\} \\
&- \frac{1}{2}\{E\}^T [\gamma_{11}]\{S^2\} - \frac{1}{2}\{E^2\}^T [\gamma_{12}]\{S\} - \frac{1}{3}\{E\}^T [\upsilon_1]\{E^2\} \\
&- \frac{1}{3}\{E\}^T [\gamma_{21}]\{S^3\} - \frac{1}{2}\{E^2\}^T [\gamma_{22}]\{S^2\} \\
&- \frac{1}{3}\{E^3\}^T [\gamma_{23}]\{S\} - \frac{1}{4}\{E\}^T [\upsilon_{21}]\{E^3\} \\
&- \frac{1}{4}\{E^2\}^T [\upsilon_{22}]\{E^2\}
\end{aligned} \tag{8}$$

where $\{S\}$ is the strain vector, $[C]$ is linear elasticity matrix, $[d]$ is the linear piezoelectric coefficient matrix, $\{E\}$ is the electric field vector, $[\varepsilon^{T}]$ is the dielectric coefficient matrix, $[C_1]$ is quadratic elasticity matrix, $[C_{21}]$ and $[C_{21}]$ are cubic elasticity matrices at constant electric field respectively. The superscript E and S (for the matrices at constant electric field and constant strain) are omitted for clarity. $[\gamma_{11}]$ and $[\gamma_{12}]$ are quadratic piezoelectric matrices, $[\gamma_{21}]$, $[\gamma_{22}]$ and $[\gamma_{23}]$ are cubic piezoelectric matrices, $\left[\upsilon_0\right] = \left[\left[\varepsilon^T\right] - \left[d\right]\left[C\right]\left[d\right]^T\right]$, $\left[\upsilon_1\right], \left[\upsilon_{21}\right] and \left[\upsilon_{22}\right]$ are linear, quadratic and cubic dielectric matrices at constant strain field S respectively. The vectors with superscripts 2 and 3 are defined as the vectors with square and cube of individual terms of the vectors. For example, the vector $\{S^2\}$ is written as $\left[S_1^2, S_2^2, S_3^2, S_4^2, S_5^2, S_6^2,\right]^T$.

Estimation of Current

Once we have determined the amplitude of mechanical displacement, the actual displacement components can be determined. The electric displacement vector in thickness direction (i.e., D_z) is a function of mechanical strain vector $\left\{S\right\}$ and electric field in thickness direction, E_z which is available in standard references (IEEE standards on piezoelectricity, 1988) and hence the electric current can be calculated from the electric displacement, D_z, in thickness direction as follows.

The current flowing through the piezoelectric plate can be calculated using the expression for the electric displacement as follows.

$$I\left(t\right) = \int_{A_s} \frac{\partial D_z\left(t\right)}{\partial t} dA_s \tag{9}$$

where A_s is the area through which current is flowing and z is the direction perpendicular to the area A_s. The electric displacement vector D_z can be calculated from the generalized nonlinear electric enthalpy density function H_{nonl} Equation 8 as follows.

$$\left\{D\right\} = -\frac{\partial H_{nonl}}{\partial\left\{E\right\}} \tag{10}$$

Modification of Nonlinear Electric Enthalpy Function to Take Care of Ferroelectric Hysteresis

Based on proposed expression for D_z from Equation 5, the generalized nonlinear electric enthalpy function of Equation 8 is further modified to take into account the effect of ferroelectric hysteresis as follows. The new nonlinear electric enthalpy function with hysteresis is denoted as H_{nl_hy}. It is written as follows, i.e., using the relation $\left[D_z = -\partial H / \partial E_z\right]$,

$$
\begin{aligned}
H_{nl\_hy} &= \frac{1}{2}\{S\}^T[C]\{S\} - \{E\}^T[d][C]\{S\} - \underline{\left(\pm\frac{\alpha_1}{2}\left[E_{zm}^2 - E_{z0}^2\right]\right)\{E\}^T[d][C]\{S\}} \\
&- \frac{1}{2}\{E\}^T\left[\left[\varepsilon^T\right] - [d][C][d]^T\right]\{E\} + \frac{1}{3}\{S\}^T[C_1]\{S^2\} \\
&+ \frac{1}{4}\{S^2\}^T[C_{21}]\{S^2\} + \frac{1}{4}\{S\}^T[C_{22}]\{S^3\} - \frac{1}{2}\{E\}^T[\gamma_{11}]\{S^2\} \\
&- \frac{1}{2}\{E^2\}^T[\gamma_{12}]\{S\} - \frac{1}{3}\{E\}^T[\upsilon_1]\{E^2\} - \frac{1}{3}\{E\}^T[\gamma_{21}]\{S^3\} \\
&- \frac{1}{2}\{E^2\}^T[\gamma_{22}]\{S^2\} - \frac{1}{3}\{E^3\}^T[\gamma_{23}]\{S\} - \frac{1}{4}\{E\}^T[\upsilon_{21}]\{E^3\} \\
&- \frac{1}{4}\{E^2\}^T[\upsilon_{22}]\{E^2\}
\end{aligned}
\tag{11}
$$

Based on the above generalized expression of H_{nl_hy} [Eq.] and by use of Hamilton's principle, the nonlinear hysterectic response of a cantilever beam (actuated by piezoelectric patches) is derived analytically in the following section.

Analytical Solutions of Nonlinear Response of The Piezo-Beam System

As the piezo-beam system (Figure 3) falls into the category of beam bending theory, we essentially consider here the 1-D formulation for the nonlinear behaviour of the piezo-beam system. Initially, we derive the expressions of the nonlinear electric enthalpy density function without consideration of hysteresis. Later, the nonlinear electric enthalpy density function will be suitably modified to take care of hysteresis as discussed in previous section. The nonlinear constitutive equations for the d_{31} effect in beam theory (without hysteresis) are given by

$$T_{xx} = E_c S_{xx} - d_{31} E_c E_z \tag{12}$$

$$D_z = d_{31} E_c S_{xx} + \left(\varepsilon_{33}^T - d_{31}^2 E_c\right) E_z \tag{13}$$

where T_{xx} and S_{xx} are the stress and strain respectively in x-direction, E_c is the elastic modulus of the piezoceramic orthogonal to the polarization direction and E_z is the electric field in z-direction.

The parameter d_{31} corresponds to the piezoelectric 31-effect and ε_{33}^T is the dielectric constant measured at constant tension. Based on the experimental results (von Wagner and Hagedorn, 2002), the dependence of E_c and d_{31} on S_{xx} (i.e., including higher order terms of strain dependence) can be approximated by (using only mechanical nonlinear terms):

$$E_c = E_c^{(0)} + E_c^{(1)} S_{xx} + E_c^{(2)} S_{xx}^2 \tag{14}$$

Figure 3. Sketch and parameters of the piezo-beam system studied in this work

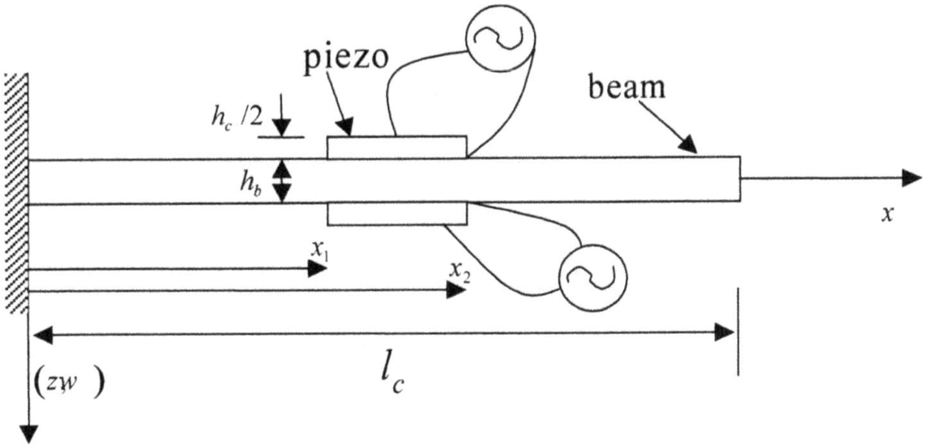

$$d_{31} = d_{31}^{(0)} + d_{31}^{(1)} S_{xx} + d_{31}^{(2)} S_{xx}^2 \tag{15}$$

Assuming a linear relationship between D_z and E_z and retaining terms only upto the third order, the nonlinear constitutive equation for T_{xx} and D_z can be written as

$$T_{xx} = E_c^{(0)} S_{xx} + E_c^{(1)} S_{xx}^2 + E_c^{(2)} S_{xx}^3 - \gamma_0 E_z - \gamma_1 S_{xx} E_z - \gamma_2 S_{xx}^2 E_z \tag{16}$$

$$D_z = \gamma_0 S_{xx} + \frac{1}{2}\gamma_1 S_{xx}^2 + \frac{1}{3}\gamma_2 S_{xx}^3 + \nu_0 E_z \tag{17}$$

where

$$\begin{aligned} \nu_0 &= \varepsilon_{33}^T - \left(d_{33}^{(0)}\right)^2 E_c^{(0)} \\ \gamma_0 &= E_c^{(0)} d_{31}^{(0)},\ \gamma_1 = E_c^{(0)} d_{31}^{(1)} + E_c^{(1)} d_{31}^{(0)},\ \gamma_2 = E_c^{(0)} d_{31}^{(2)} + E_c^{(2)} d_{31}^{(0)} + E_c^{(1)} d_{31}^{(1)} \end{aligned} \tag{18}$$

One must satisfy the compatibility equations, i.e.,

$$T_{xx} = \frac{\partial H}{\partial S_{xx}}$$

and (19)

$$D_z = -\frac{\partial H}{\partial E_z}$$

Considering the above expressions, the expression for the nonlinear electric enthalpy function can be derived as

$$H = H_{nonl} = \frac{1}{2}E_c^{(0)}S_{xx}^2 + \frac{1}{3}E_c^{(1)}S_{xx}^3 + \frac{1}{4}E_c^{(2)}S_{xx}^4 - \gamma_0 S_{xx}E_z \\ -\frac{1}{2}\gamma_1 S_{xx}^2 E_z - \frac{1}{3}\gamma_2 S_{xx}^3 E_z - \frac{1}{2}\nu_0 E_z^2 \tag{20}$$

This expression for the nonlinear electric enthalpy function also satisfies the necessary and sufficient condition for the existence of the electric enthalpy function (IEEE standards on piezoelectricity, 1988) i.e.,

$$\frac{\partial^2 H}{\partial S_{xx}\partial E_z} = \frac{\partial T_{xx}}{\partial E_z} = -\frac{\partial D_z}{\partial S_{xx}} = \frac{\partial^2 H}{\partial E_z \partial S_{xx}} \tag{21}$$

The above expression for electric enthalpy density function can now be modified to take care of ferroelectric hysteresis and the corresponding expression for H_{nl_hy} can be written as:

$$H = H_{nl\_hy} = \frac{1}{2}E_c^{(0)}S_{xx}^2 + \frac{1}{3}E_c^{(1)}S_{xx}^3 + \frac{1}{4}E_c^{(2)}S_{xx}^4 - \gamma_0 S_{xx}E_z \\ -\underline{\left[\pm\frac{\alpha_1}{2}\left(E_{zm}^2 - E_{z0}^2\right)\right]\gamma_0 S_{xx}E_z} - \frac{1}{2}\gamma_1 S_{xx}^2 E_z - \frac{1}{3}\gamma_2 S_{xx}^3 E_z \\ -\frac{1}{2}\nu_0 E_z^2 \tag{22}$$

The above expression for H_{nl_hy} is used in the Hamilton's principle to derive the governing differential equation for the piezo-beam system.

Hamilton's Principle

The dynamic equation of motion of the cantilever beam with the attached piezo-patch can be derived using Hamilton's principle as:

$$\delta\int_{t_0}^{t_1} L dt + \int_{t_0}^{t_1} \delta W dt = 0 \tag{23}$$

where the Lagrangian 'L' is written as

$$L = \int_V \left(T - H\right) dV \tag{24}$$

where T and H denote kinetic energy and electric enthalpy density respectively. The kinetic energy T of the piezo-beam system is written as:

$$\begin{aligned}\int_V T dV &= \frac{1}{2}\int_{V_b} \rho \dot{w}^2 dV + 2.\frac{1}{2}\int_{V_c} \rho \dot{w}^2 dV \\ &= \frac{1}{2}\rho_b A_b \int_0^l \dot{w}^2\left(x,t\right) dx + \rho_c A_c \int_{x_1}^{x_2} \dot{w}^2\left(x,t\right) dx\end{aligned} \tag{25}$$

where V_b and V_c are volume of the beam and the piezoceramic respectively. The virtual work done by damping forces δW, assuming linear (proportional) damping, is given as:

$$\delta W = \int_0^l -d\dot{w}\left(x,t\right)\delta w dx \tag{26}$$

where d is the linear or proportional damping constant. The electric enthalpy density H of the piezo-beam system consists of the enthalpy density of the beam and the two piezo-patches attached to the beam respectively. Integrating H over the whole volume (i.e., V_b and V_c), we get:

$$\begin{aligned}\int_V H dV &= \int_{V_b} H dV + \int_{V_c} H dV \\ &= \frac{1}{2}\int_{V_b} E_b S_{xx}^2 dV \\ &+ \int_{V_c}\begin{pmatrix}\frac{1}{2}E_c^{(0)}S_{xx}^2 + \frac{1}{3}E_c^{(1)}S_{xx}^3 + \frac{1}{4}E_c^{(2)}S_{xx}^4 - \gamma_0 S_{xx}E_z \\ -\left[\pm\frac{\alpha_1}{2}\left(E_{zm}^2 - E_{z0}^2\right)\right]\gamma_0 S_{xx}E_z \\ -\frac{1}{2}\gamma_1 S_{xx}^2 E_z - \frac{1}{3}\gamma_2 S_{xx}^3 E_z - \frac{1}{2}\upsilon_0 E_z^2\end{pmatrix} dV\end{aligned} \tag{27}$$

where E_b is the elastic modulus of the beam. The terms containing $E_c^{(1)}$ and γ_1 vanish due to the symmetry of the system (upon integration over $\pm\frac{h}{2}$). Now, the strain S_{xx} in the beam in the x-direction can be expressed in terms of the deflection w of the beam as:

$$S_{xx}\left(x,z,t\right) = -z\frac{\partial^2 w\left(x,t\right)}{\partial x^2} \tag{28}$$

Assuming that the electric field E_z does not change with respect to x, the electric enthalpy of the piezo-beam system is expresses as:

$$\int_V HdV = \frac{1}{2}E_bI_b\int_0^l \left(w''\right)^2 dx + 2a_1\int_{x_1}^{x_2}\left(w''\right)^2 dx + 2a_3\int_{x_1}^{x_2}\left(w''\right)^4 dx + 2b_1\left(\int_{x_1}^{x_2} w''dx\right)E_z$$
$$+ 2b_3\left(\int_{x_1}^{x_2}\left(w''\right)^3 dx\right)E_z - 2b_0E_z^2 \tag{29}$$

where the constants a_1, a_3, b_0, b_1 and b_3 are given as:

$$a_1 = \frac{1}{6}E_c^{(0)}b\left[\left(\frac{h}{2}+h_c\right)^3 - \left(\frac{h}{2}\right)^3\right]$$
$$a_3 = \frac{1}{20}E_c^{(2)}b\left[\left(\frac{h}{2}+h_c\right)^5 - \left(\frac{h}{2}\right)^5\right]$$

$$b_0 = \frac{1}{2}\upsilon_0 bh_cl_c$$
$$b_1 = \frac{1}{2}\gamma_0\left[1 \pm \frac{\alpha_1}{2}\left(E_{zm}^2 - E_{z0}^2\right)\right]b\left[\left(\frac{h}{2}+h_c\right)^2 - \left(\frac{h}{2}\right)^2\right] \tag{30}$$
$$b_3 = \frac{1}{12}\gamma_2 b\left[\left(\frac{h}{2}+h_c\right)^4 - \left(\frac{h}{2}\right)^4\right]$$

Substituting Equations 25, 26 and 29 in Equation 23, we get:

$$\delta\int_{t_0}^{t_1} Ldt + \int_{t_0}^{t_1}\delta Wdt = \delta\int_{t_0}^{t_1}\left\{\begin{aligned}&\left[\frac{1}{2}\rho_bA_b\int_0^l\left(\dot{w}\right)^2 dx + \rho_cA_c\int_{x_1}^{x_2}\left(\dot{w}\right)^2 dx\right]\\ &-\left\{\begin{aligned}&\left[\frac{1}{2}E_bI_b\int_0^l\left(w''\right)^2 dx + 2a_1\int_{x_1}^{x_2}\left(w''\right)^2 dx + 2a_3\int_{x_1}^{x_2}\left(w''\right)^4 dx\right]\\ &+\left[2b_1\int_{x_1}^{x_2} w''dx\right]E_z + \left[2b_3\int_{x_1}^{x_2}\left(w''\right)^3 dx\right]E_z - 2b_0E_z^2\end{aligned}\right\}\end{aligned}\right\}dt$$
$$+\int_{t_0}^{t_1}\int_0^l -d\dot{w}\delta wdxdt \tag{31}$$
$$= 0$$

Now, using Rayleigh-Ritz discretization for the deflection $w\left(x,t\right)$, we can write the deflection as

$$w\left(x,t\right)=\sum_{k=1}^{n}\phi_k\left(x\right)p_k\left(t\right) \tag{32}$$

where $\phi_k\left(x\right)$ are the eigen-functions of the beam without the piezo-ceramics. The electric field is assumed to be given as

$$E_z=\frac{U\left(t\right)}{h_c} \tag{33}$$

where h_c is the thickness of the piezoceramic and $U\left(t\right)$ is the applied electric potential across the thickness of the piezoceramic. It may be seen from Equation 33 that electric field is assumed to be constant over the thickness of the piezoceramic. Now, substituting expression for $w\left(x,t\right)$ from Equation 32 in Equation 31 and performing variation w.r.t. $p_k\left(t\right)$, we obtain [using only one eigenfunction corresponding to the mode of excitation of the system] the differential equation of the piezo-beam system as:

$$m_k\ddot{p}_k\left(t\right)+d_k\dot{p}_k\left(t\right)+\left(c_k^{*}+c_k^{**}\right)p_k\left(t\right)+\varepsilon_k^{(1)}p_k^3\left(t\right)=f_kU\left(t\right)-\varepsilon_k^{(2)}p_k^2\left(t\right)U\left(t\right) \tag{34}$$
where $k=1,2,3,\ldots$ and

$$m_k=\rho_bA_b\int_0^l\phi_k^2\left(x\right)dx+2\rho_cA_c\int_{x_1}^{x_2}\phi_k^2\left(x\right)dx$$

$$d_k=d\int_0^l\phi_k^2\left(x\right)dx$$

$$c_k^{*}=E_bI_b\int_0^l\left[\phi_k''\left(x\right)\right]^2dx$$

$$c_k^{**}=\frac{2}{3}E_cb\left[\left(\frac{h_b}{2}+h_c\right)^3-\left(\frac{h_b}{2}\right)^3\right]\int_{x_1}^{x_2}\left[\phi_k''\left(x\right)\right]^2dx$$

$$\varepsilon_k^{(1)}=\frac{4}{5}E_c^{(2)}b\left[\left(\frac{h_b}{2}+h_c\right)^5-\left(\frac{h_b}{2}\right)^5\right]\int_{x_1}^{x_2}\left[\phi_k''\left(x\right)\right]^4dx$$

$$f_k=E_cb\left(h_b+h_c\right)\left[1\pm\frac{\alpha_1}{2}\left(E_{zm}^2-E_{z0}^2\right)\right]\left[\phi_k'\left(x_2\right)-\phi_k'\left(x_1\right)\right]d_{31} \tag{35}$$

$$\varepsilon_k^{(2)}=\frac{1}{2}\left[d_{31}^{(1)}E_c^{(1)}+d_{31}^{(2)}E_c^{(0)}+d_{31}^{(0)}E_c^{(2)}\right]b\left[\left(\frac{h_b}{2}+h_c\right)^4-\left(\frac{h_b}{2}\right)^4\right]\frac{1}{h_c}\int_{x_1}^{x_2}\left[\phi_k''\left(x\right)\right]^3dx$$

The excitation potential is expressed in sinusoidal form as:

$$U(t) = U_0 \cos(\Omega t) \tag{36}$$

where Ω is the frequency of excitation for the piezoceramic actuator attached to the cantilever beam.

Solution of Nonlinear Equation of Motion by Harmonic Balance Method

In general, the forcing function (here, the electrical excitation) will have a phase difference φ with the deflection (response) of the beam. Hence, Equation 34 can be written as:

$$\begin{aligned} m_k \ddot{p}_k(t) + d_k \dot{p}_k(t) + \left(c_k^* + c_k^{**}\right) p_k(t) + \varepsilon_k^{(1)} p_k^3(t) &= f_k(t) U_0 \cos(\Omega t + \varphi) \\ &- \varepsilon_k^{(2)} p_k^2(t) U_0 \cos(\Omega t + \varphi) \\ &\text{where } k = 1, 2, 3, \ldots \end{aligned} \tag{37}$$

Now, we can express the function $p_k(t)$ as

$$p_k(t) = A \cos(\Omega t) \tag{38}$$

where A is the amplitude of response of the piezo-beam system. Substituting Equation 38 in Equation 37 and using the following relationships for $\cos^3(\Omega t)$ and $\sin^3(\Omega t)$, i.e.,

$$\begin{aligned} \cos^3(\Omega t) &= \frac{1}{4}\left[3\cos(\Omega t) + \cos(3\Omega t)\right] \\ \sin^3(\Omega t) &= \frac{1}{4}\left[3\sin(\Omega t) - \sin(3\Omega t)\right] \end{aligned} \tag{39}$$

we get the following equation.

$$\begin{aligned} &-A m_k \Omega^2 \cos(\Omega t) - A d_k \Omega \sin(\Omega t) + A\left(c_k^* + c_k^{**}\right)\cos(\Omega t) + \varepsilon_k^{(1)} A^3 \frac{3}{4}\cos(\Omega t) \\ &+\varepsilon_k^{(1)} A^3 \frac{1}{4}\cos(3\Omega t) \\ &= f_k U_0 \cos(\Omega t)\cos\phi - f_k U_0 \sin(\Omega t)\sin\phi - \varepsilon_k^{(2)} U_0 A^2 \frac{3}{4}\cos(\Omega t)\cos\phi \\ &-\varepsilon_k^{(2)} U_0 A^2 \frac{1}{4}\cos(3\Omega t)\cos\phi + \varepsilon_k^{(2)} U_0 A^2 \frac{1}{4}\sin(\Omega t)\sin\phi + \varepsilon_k^{(2)} U_0 A^2 \frac{1}{4}\sin(3\Omega t)\sin\phi \end{aligned} \tag{40}$$

Extracting terms of $\cos\left(\Omega t\right)$ and $\sin\left(\Omega t\right)$ from both sides of the equation and using the identity, $\cos^2\left(\Omega t\right)+\sin^2\left(\Omega t\right)=1$, we get the following polynomial expression in terms of the amplitude A, i.e.,

$$a_5\left(A^2\right)^5+a_4\left(A^2\right)^4+a_3\left(A^2\right)^3+a_2\left(A^2\right)^2+a_1A^2+a_0=0 \tag{41}$$

where the coefficients are given as

$$\begin{aligned}
a_0 &= -\left(f_kU_0\right)^4\\
a_1 &= \left[\left(c_k-m_k\Omega^2\right)^2+\left(d\Omega\right)^2\right]\left(f_kU_0\right)^2+2\varepsilon_k^{(2)}\left(f_k\right)^3\left(U_0\right)^4\\
a_2 &= \left[\frac{3}{2}\left(c_k-m_k\Omega^2\right)\varepsilon_k^{(1)}-\frac{11\left(U_0\right)^2}{8}\left(\varepsilon_k^{(2)}\right)^2\right]\left(f_kU_0\right)^2\\
&\quad -\frac{1}{2}\left[\left(c_k-m_k\Omega^2\right)^2+3\left(d\Omega\right)^2\right]\varepsilon_k^{(2)}f_k\left(U_0\right)^2\\
a_3 &= \frac{1}{16}\left[\left(c_k-m_k\Omega^2\right)^2+9\left(d\Omega\right)^2\right]\left(\varepsilon_k^{(2)}\right)^2\left(U_0\right)^2+\frac{9}{16}\left(\varepsilon_k^{(1)}\right)^2\left(f_k\right)^2\left(U_0\right)^2\\
&\quad -\frac{3}{4}\left(c_k-m_k\Omega^2\right)f_k\varepsilon_k^{(1)}\varepsilon_k^{(2)}\left(U_0\right)^2\\
a_4 &= -\frac{9}{32}f_k\left(\varepsilon_k^{(1)}\right)^2\varepsilon_k^{(2)}\left(U_0\right)^2-\frac{9}{256}\left(\varepsilon_k^{(2)}\right)^4\left(U_0\right)^4+\frac{3}{32}\left(c_k-m_k\Omega^2\right)\varepsilon_k^{(1)}\left(\varepsilon_k^{(2)}\right)^2\left(U_0\right)^2\\
a_5 &= \frac{9}{256}\left(\varepsilon_k^{(1)}\varepsilon_k^{(2)}U_0\right)^2
\end{aligned} \tag{42}$$

Evaluation of Eigen-Function of the Cantilever Beam

To evaluate the nonlinear response of the cantilever piezo-beam system (Figure 3), the eigen-function of the mode nearest to the frequency of excitation is required if we use only one function in the Rayleigh-Ritz expression. If one neglects the effect of small piezo-patches attached to the cantilever beam and evaluates the mode shapes, the mode shape functions can be expressed as:

$$\phi(x)=\left(\cosh\beta x-\cos\beta x\right)-c_3\left(\sinh\beta x-\sin\beta x\right) \tag{43}$$

where the constant c_3 is given as:

$$c_3=\frac{\cosh\left(\beta l\right)+\cos\left(\beta l\right)}{\sinh\left(\beta l\right)+\sin\left(\beta l\right)} \tag{44}$$

The length of the beam is l and the value for (βl) can be found as $\sqrt{9.87}$, $\sqrt{39.5}$ and $\sqrt{88.9}$ for the first, second and third modes respectively. Using the above expression for the eigen-function $\phi(x)$, the nonlinear response of two cantilever piezo-beam systems has been predicted and the results of the analysis compared with experimental results from literature (von Wagner & Hagedorn, 2002; Raghavan et al., 2003)

Results and Discussion

The experimentals results have been taken from von Wagner & Hagedorn (2002) and Raghavan et al. (2003) for comparison with predicted results of the proposed model. Some of the relevant nonlinear hysteresis models for piezoelectric actuators and their application in nonlinear PID control are discussed in Aljanaideh et al. (2016), Tang & Li (2015), and Hassani & Tjahjowidodo (2013). The details of the materials property used, geometrical dimensions and the results are given in the following sections. The sketch of the piezo-beam system is shown in Figure 3. The actual picture of the piezo-beam system on which experiments were carried out by von Wagner and Hagedorn (2002) at TU Darmstadt is shown in Figure 4.

Solution of Piezo-Beam System-I

Table 1 gives the details of the geometry and material properties used by von Wagner and Hagedorn (2002) for the piezo-beam system. Figure 5 shows the stiffness proportional non-dimensional damping parameter used for the piezo-beam system. It can be seen from this figure that the damping parameter is linearly dependent upon the electric field amplitude. The piezo-patches are excited by application sinusoidal electric field across the thickness of the piezo-patches. The electric field directions are op-

Figure 4. Picture of the piezo-beam system studied in this chapter for their nonlinear response behavior

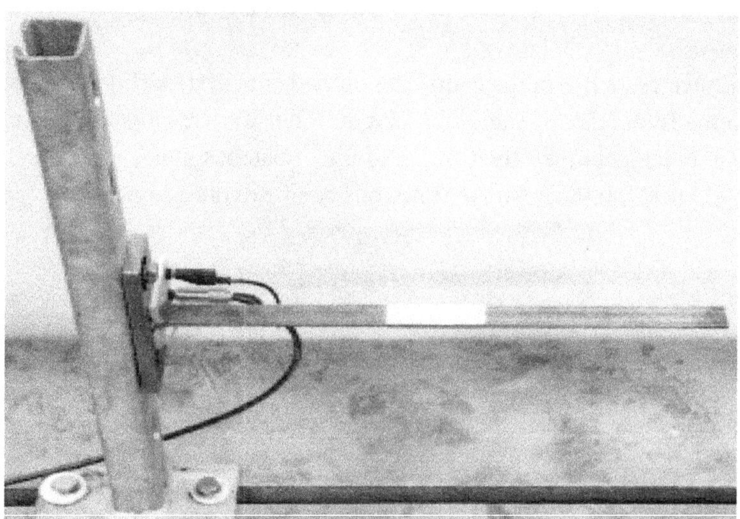

Table 1. Details of geometry and material properties of piezo-beam system

Parameters	Beam (Material: Steel SAE 1040)	Piezo (PIC 181)
Length (l)	0.4 m	0.07 m
Width (b)	0.025 m	0.025 m
Thickness (h)	0.003 m	0.001 m
Density (ρ)	8850.5 kg/m^3	8337.4 kg/m^3
Young's modulus (E)	2.718 e11 N/m^2	0.696 e11 N/m^2
d_{31}	--	-2.1 e-10 m/volt
$E_c^{(1)}$	--	6.76 e-14 N/m^2
$E_c^{(2)}$	--	-2.4 e16 N/m^2
$d_{31}^{(1)}$	--	-0.0311 m/volt
$d_{31}^{(2)}$	--	-0.000012 m/volt

Figure 5. Variation of non-dimensional damping parameter with electric field amplitude

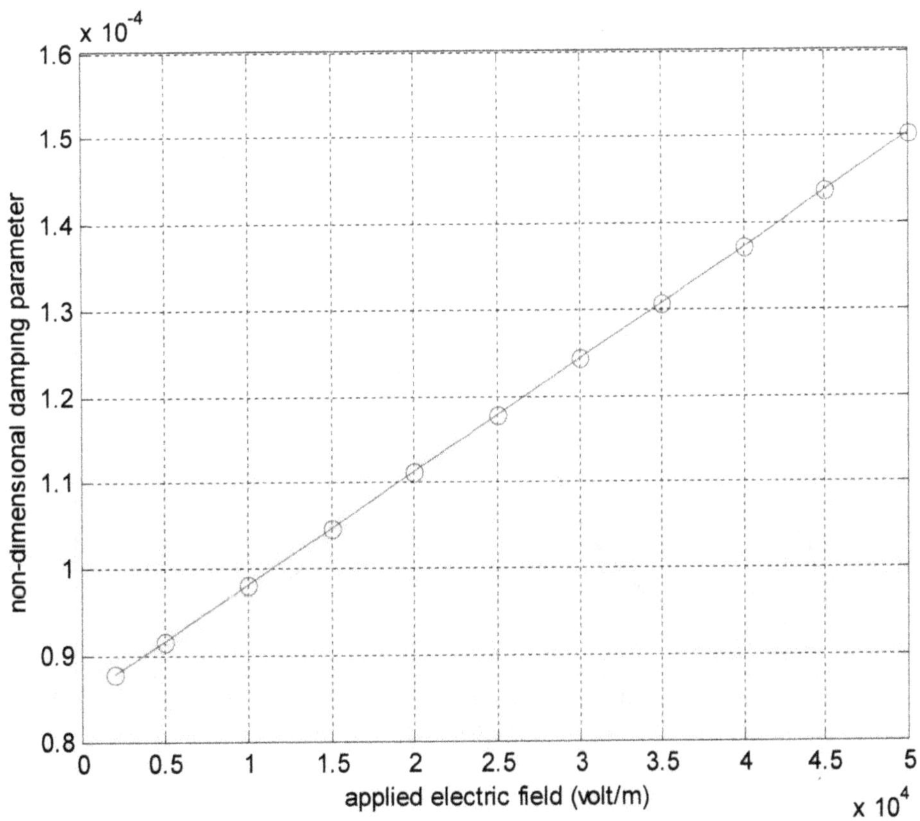

posite to each other for the top and bottom piezo-patches so that a bending moment can be applied to the beam. Experiments were conducted near the first natural frequency of the piezo-beam system, which is 17.073 Hz. The natural frequency of the cantilever beam only is 16.8 Hz. So, there is change of 1.62% of the first natural frequency due to presence of piezo-patches.

The piezo-patches are attached to the beam exactly at the center and hence the value of x_1 and x_2 to be used for integration in Equation 31 are given as 0.165 and 0.235 m respectively. Experiment were conducted in both sweep up (increasing frequency) and sweep down mode (decreasing frequency) and at different electric field amplitudes to observe the nonlinear jump phenomena and dependence of resonance frequency on applied electric field amplitude. It was seen from Figure 6 that when the system is excited at very low electric field of 2 kV/m, the response of the system is linear. Figure 6 shows the predicted responses using both linear and nonlinear theory (sweep up and sweep down modes) where the linear and nonlinear responses are nearly identical. However, with increasing electric fields, the difference in predictions of linear and nonlinear theory widens.

The results of frequency response at an electric field amplitude of 50 kV/m is shown in Figure 7 for both linear and nonlinear theory. As can be seen from the above figure, the difference between the responses between linear theory and nonlinear theory is significant and hence one needs to consider the nonlinearity in the piezo-beam system while designing it for any application. Figure 7 also shows the nonlinear responses of the piezo-beam system where one can see the prominent jump phenomena. The other feature of this nonlinear behaviour of piezo-systems when excited near the structural mode is the dependence of resonance electric field and the displacement amplitude. These features are represented by Figures 8 and 9 respectively. Figure 8 shows that the maximum value of displacement amplitude

Figure 6. Frequency response at an electric field of 2 kV/m for piezo-beam system-I

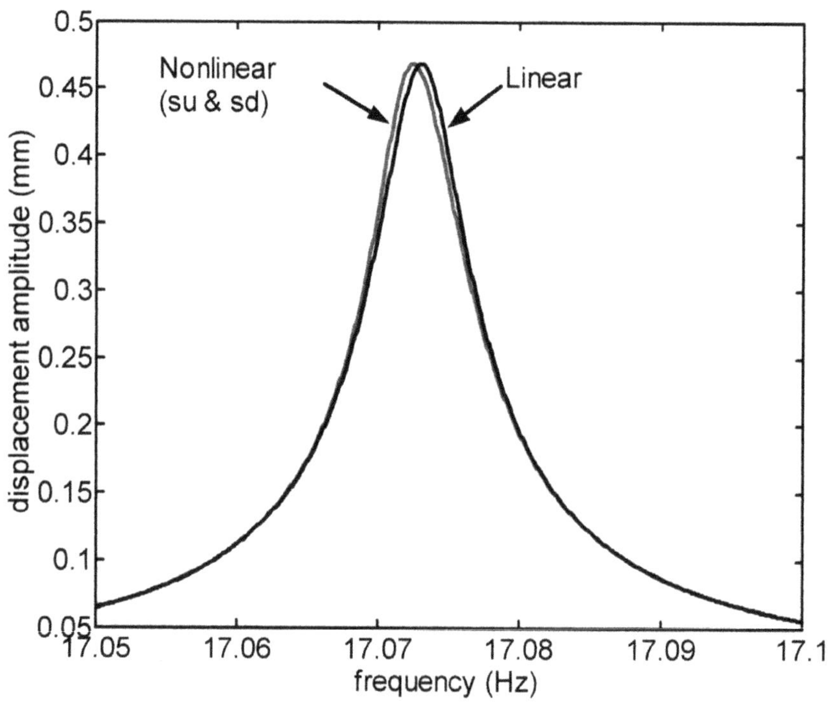

Figure 7. Frequency response at an electric field of 50 kV/m for piezo-beam system-I

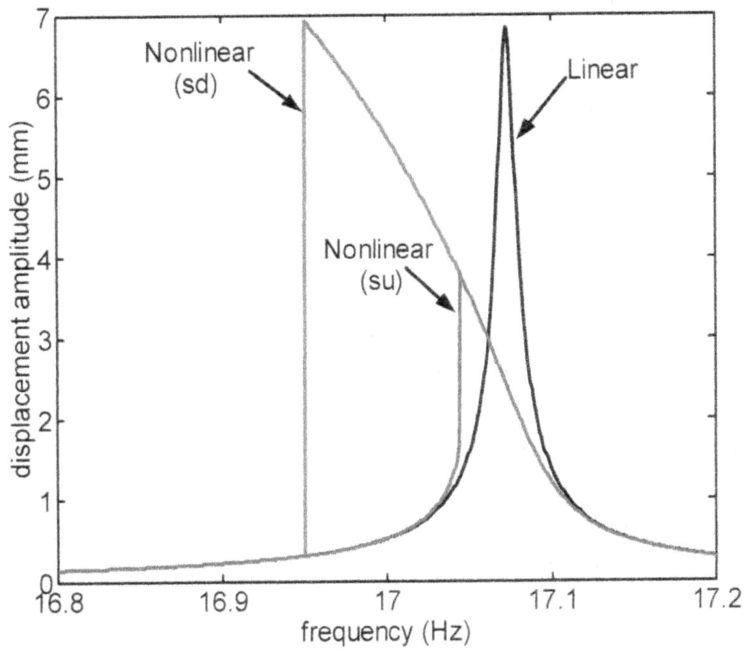

Figure 8. Displacement amplitude per unit electric field vs. frequency at varying electric fields for the piezo-beam system

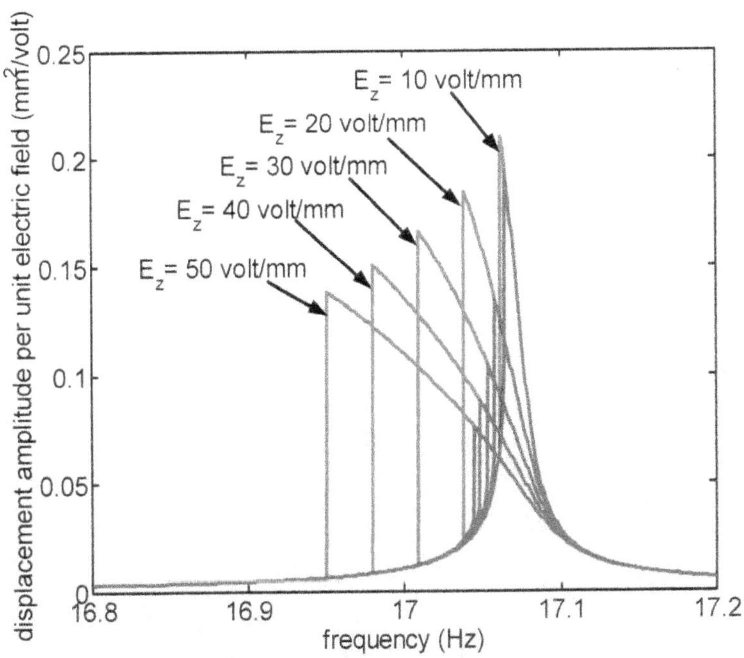

per unit electric field decreases with increasing field (this should have been constant in case of linear response). Figure 9 shows that the displacement amplitude varies in a nonlinear manner with increasing electric field. The experiments with varying electric field amplitudes were conducted by von Wagner and Hagedorn (2002). Figure 9 compares the simulated results with the experimental results and a very good agreement between the two results can be observed.

Experimental results for hysteresis behaviour of the piezo-beam systems are not available. However, it is well known that when the electric field excitation is high and it is applied in a cyclic manner (increasing and decreasing electric field amplitudes), different types of hysteresis, such as dielectric and butterfly hysteresis, are observed in piezoelectric systems. The hysteresis behaviour is different at high (near saturation polarization region) and sub-coercive regions. When the electric field is lower than the saturation region, the hysteresis loops are called minor or sub-coercive hysteresis loops.

In thin films (Damjanovic, 1997), hysteresis loops are observed even at very small electric field excitations. In the absence of experimental results, simulations have been carried out for the piezo-beam system. The Rayleigh paremeter has been taken as $\alpha_1 = 1.0e - 6 / E_{z0}$ where E_{z0} amplitude of electric field excitation. Figure 10 shows the variation of displacement amplitude with electric potential for the linear piezo-beam system without consideration of hysteresis and structural mode nonlinearity when the system is excited at 17.1 Hz, which is very close to the first natural frequency of the system (17.073 Hz). It can be seen from Figure 10 that displacement amplitude is proportional to the applied electric potential as expected for a linear system. The same system was simulated; by taking into account of the hysteresis and the results of displacement amplitude with electric potential is shown in Figure 11.

Figure 9. Nonlinear variation of displacement amplitude w.r.t. electric field amplitude for piezo-beam system

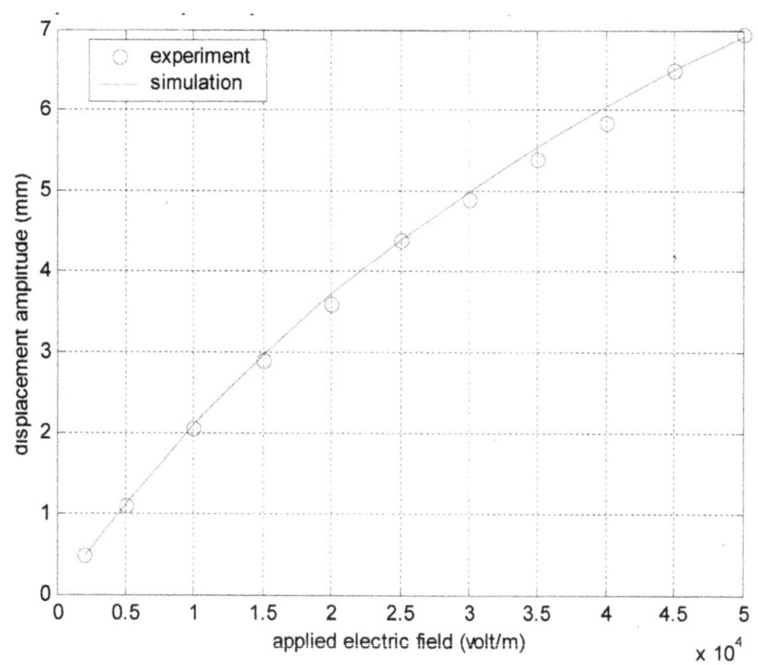

Figure 10. Linear variation of displacement amplitude w.r.t. applied electric potential when excited at 17.1 Hz (without hysteresis and structural mode nonlinearity)

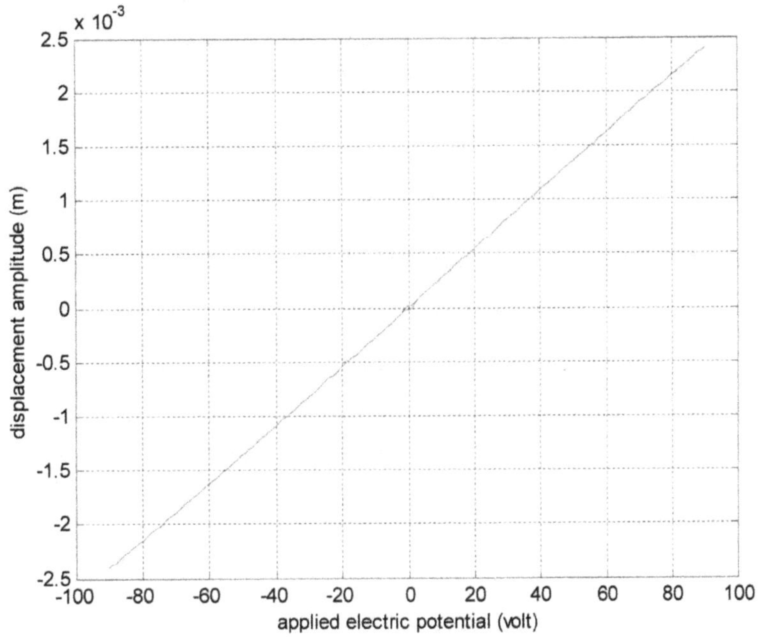

Figure 11. Variation of displacement amplitude w.r.t. applied electric potential when excited at 17.1 Hz (with hysteresis and without structural mode nonlinearity)

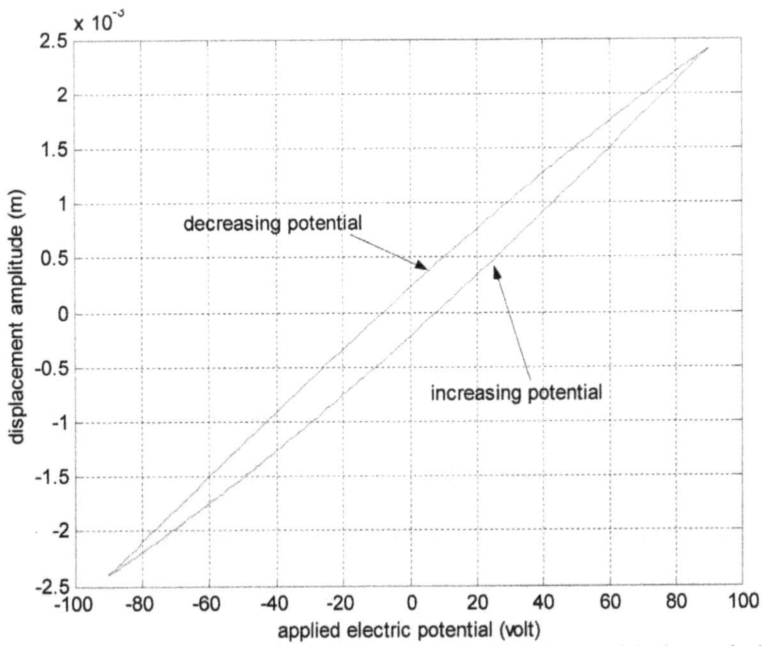

It is known that when the piezoelectric systems are excited close to a structural mode, additional nonlinearity (structural mode nonlinearity) also comes into picture alongwith hysterectic behaviour (von Wagner, 2003, 2004). The same system was simulated, by taking into account the combined nonlinear behaviour (hysteresis and structural mode nonlinearity) and results are shown in Figure 12. The comparison of the hysterectic behaviours with and without consideration of structural mode nonlinearity are shown in Figure 13. It can be observed from the above figure that structural mode nonlinearities affect the system response significantly and hence these should be modeled accurately when designing a system to operate near a structural mode (e.g. in ultrasonic motors).

The difference between hysterectic behaviour with and without consideration of nonlinearity becomes significant as we approach the natural frequency of the system and the results are shown in Figure 14 (system is excited at linear natural frequency of 17.073 Hz). Again as one excites the system away from resonance region, the structural mode nonlinearity becomes insignificant and it can be seen from Figure 15 (electric field amplitude is 15kV/m) that the hysterectic behaviours are same with and without consideration of structural mode nonlinearity. The simulated results are very encouraging and there is a need for a detailed analytical and experimental study of the piezo-beam systems, which will be the scope of further research

Solution of Piezo-Beam System-II

Table 2 gives the details of the geometry and material properties used by Raghavan et. al. (2003) for the piezo-beam system. Figure 16 shows the stiffness proportional non-dimensional damping parameter for

Figure 12. Variation of displacement amplitude w.r.t. applied electric potential when excited at 17.1 Hz (with both hysteresis and structural mode nonlinearity)

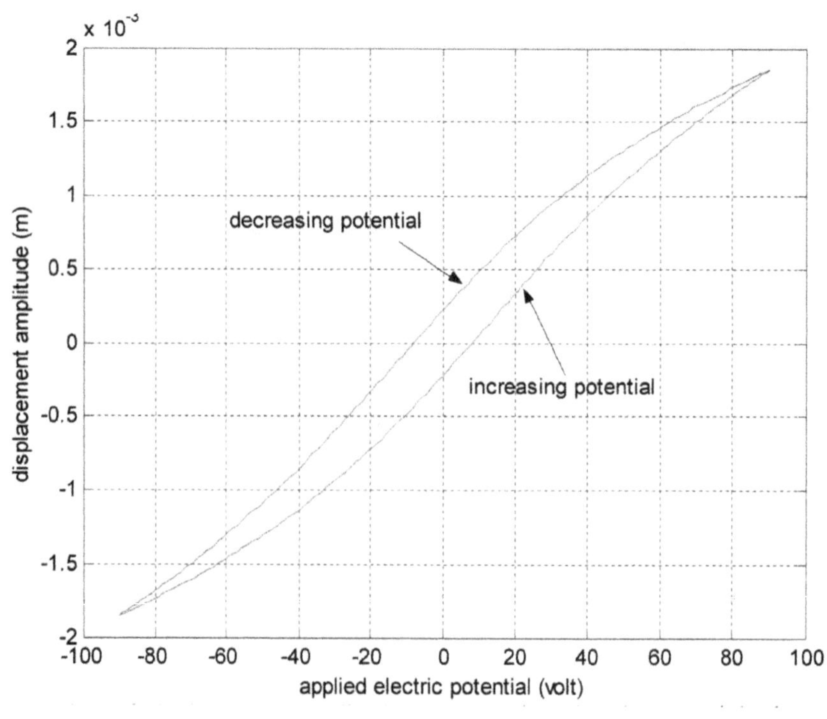

Figure 13. Comparison of hysterectic response with and without consideration of structural mode non-linearity when excited at 17.1 Hz

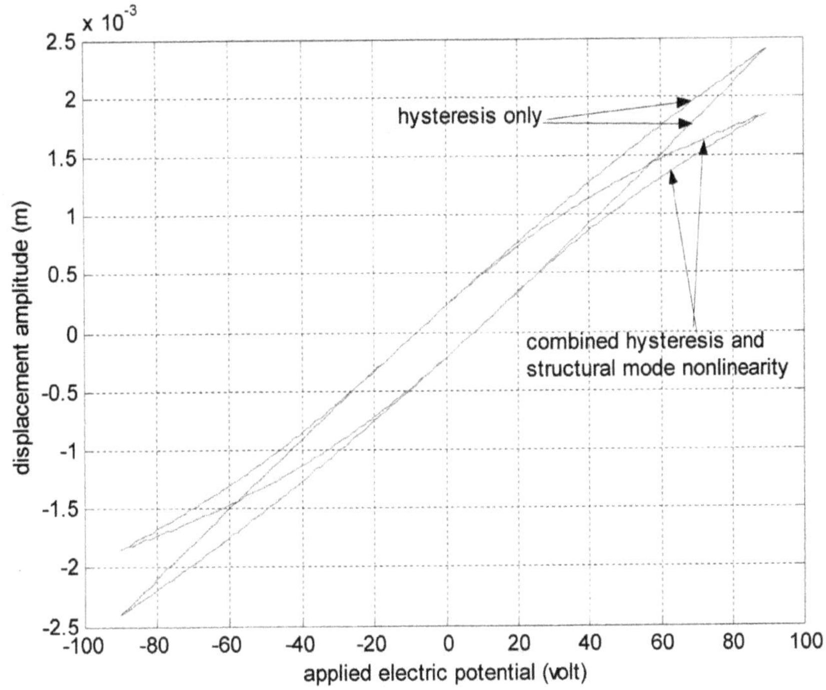

Figure 14. Comparison of hysterectic response with and without consideration of structural mode non-linearity when excited at first natural frequency of 17.073 Hz

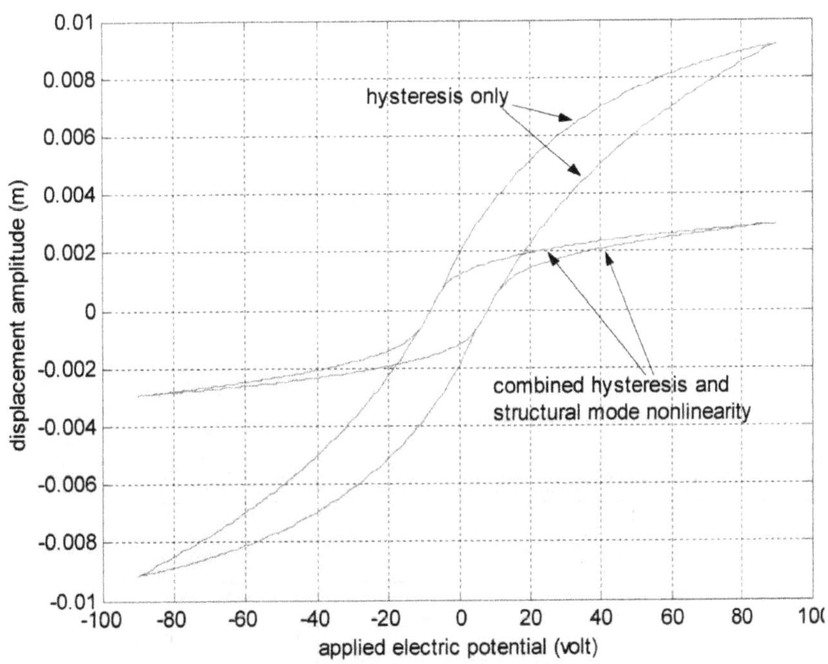

Figure 15. Comparison of hysterectic response with and without consideration of structural mode non-linearity when excited at away from first natural frequency (i.e. 15 Hz)

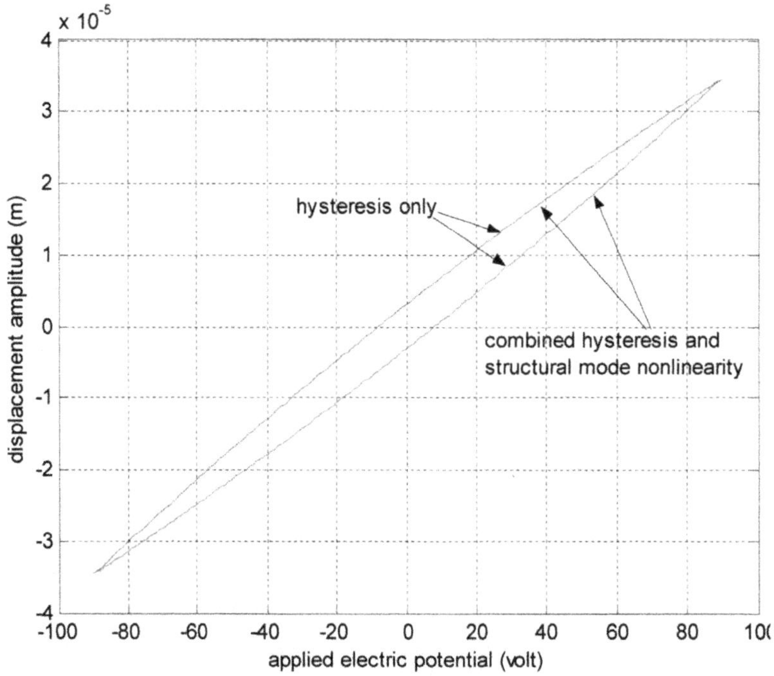

Table 2. Details of geometry and material properties of piezo-beam system-II

Parameters	Beam (material:-steel SAE 1040)	Piezo (PSI-5A 54ENH of piezo-systems inc.)
Length (l)	0.25 m	0.0724 m
Width (b)	0.0245 m	0.0245 mm
Thickness (h)	0.00082 m	0.000267 m
Density (ρ)	7777.0 kg/m³	7800.0 kg/m³
Young's modulus (E)	1.676 e11 N/m²	0.660 e11 N/m²
d_{31}	--	-2.1 e-10 m/volt (guess value)
$E_c^{(1)}$	--	6.76 e-14 N/m²(guess value)
$E_c^{(2)}$	--	-2.4 e16 N/m²(guess value)
$d_{31}^{(1)}$	--	-0.0311 m/volt (guess value)
$d_{31}^{(2)}$	--	-0.000012 m/volt (guess value)

the piezo-beam system used in the analysis. The same data of piezo-beam system-I has been taken in the absence of actual experimental data.

The piezo-patches are excited by application sinusoidal electric field across the thickness of the piezo-patches. The electric field directions are opposite to each other for the top and bottom piezo-patches so that a bending moment can be applied to the beam. Experiments were conducted at frequency of 0.1591 Hz, which is small, compared to the first natural frequency of the piezo-beam system, which is 10.454 Hz. The natural frequency of the cantilever beam only is 9.848 Hz. So, there is change of 6.15% of the first natural frequency due to presence of piezo-patches. In the absence of experimental data near first natural frequency (structural mode nonlinearities are dominant here), simulations have been carried out for both sweep up and sweep down modes of excitation at different electric field amplitudes.

Figure 16 shows the frequency response curve of the piezo-beam system when excited at an electric field of 18.72 kV/m. As can be seen from the above figure, the behaviour is almost linear in the presence of small electric field. However, the nonlinearity becomes significant when the electric field amplitude is high as can be seen from Figure 17 (the magnitude of electric field amplitude is 187.2 kV/m).

Experimental results for hysteresis behaviour of the piezo-beam systems are available for this piezo-beam system from Raghavan et al. (2003). Simulations have been carried out using the proposed model discussed in the previous chapter. The Rayleigh paremeter has been taken as $\alpha_1 = 1.0e - 6 / E_{z0}$ where E_{z0} amplitude of electric field excitation which is same as the piezo-beam system-I. Fig. 18 shows the variation of displacement amplitude with electric potential for the linear piezo-beam system without

Figure 16. Comparison of linear and nonlinear frequency response at an electric field of 18.72 kV/m for piezo-beam system-II

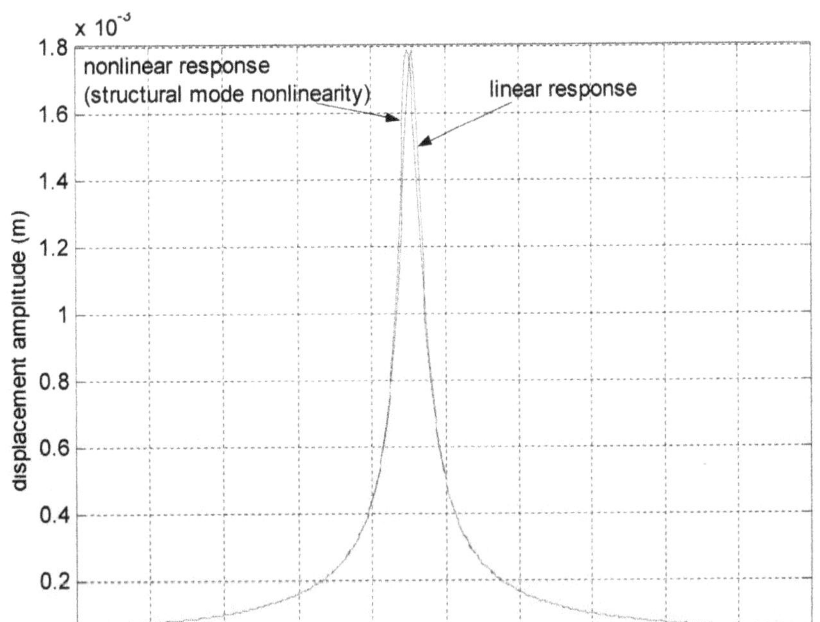

Figure 17. Comparison of linear and nonlinear frequency response at an electric field of 187.2 kV/m for piezo-beam system-II

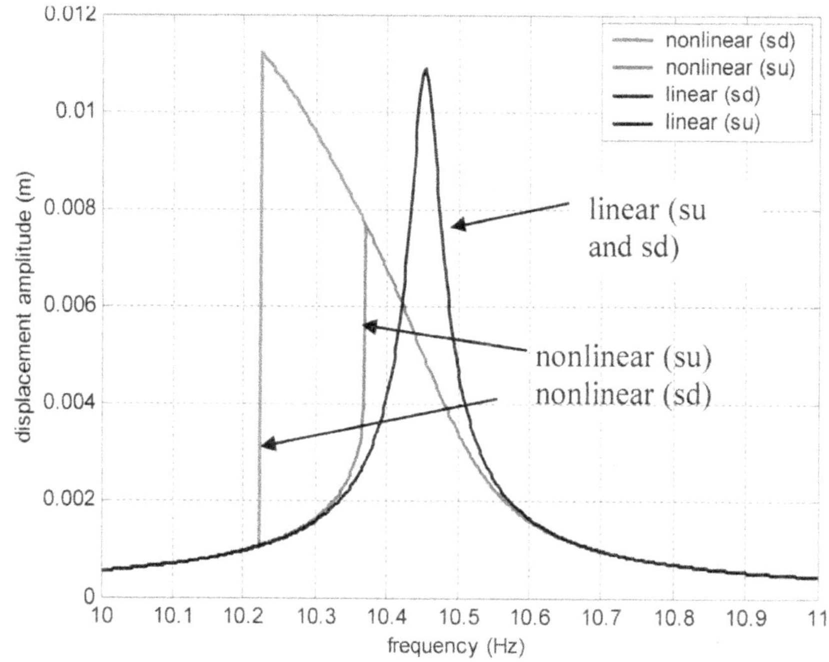

Figure 18. Linear variation of displacement amplitude w.r.t. applied electric potential (without hysteresis) for piezo-beam system-II

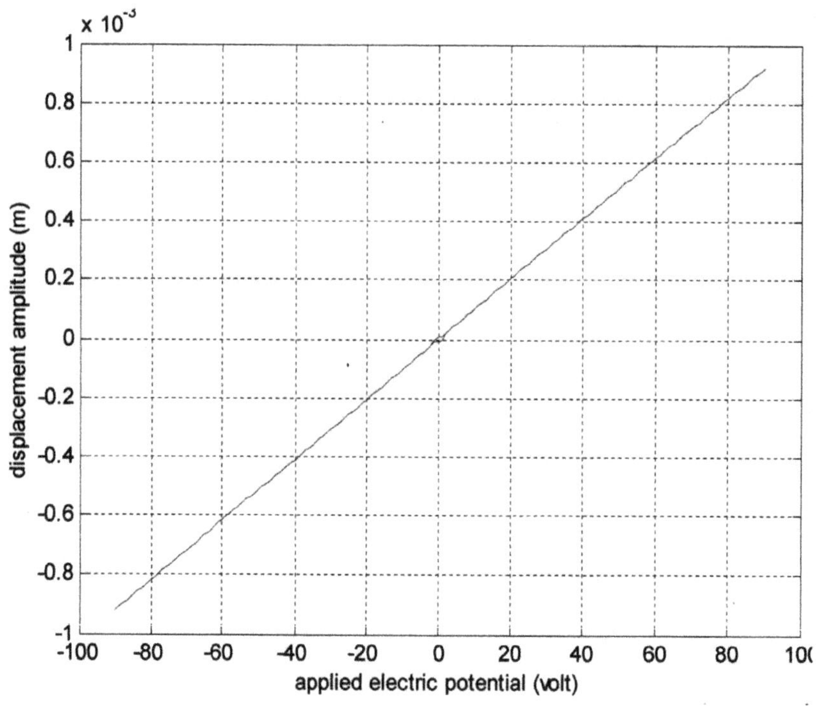

consideration of hysteresis when the system is excited at 0.1591 Hz, which is away from the first natural frequency of the system (10.454 Hz) (hence structural mode nonlinearity is absent)

It can be seen from the above Figure 18 that displacement amplitude is proportional to the applied electric potential as expected for a linear system. The same system was simulated, by taking into account of the hysteresis and the results of displacement amplitude with electric potential is shown in Figure 19. Figure 19 also shows the comparisons of simulated results and experimental results and the agreement between the two are very satisfactory. The piezo-patches are attached to the beam exactly at the center and hence the value of x_1 and x_2 to be used for integration in Equation 31 are given as 0.0888 and 0.1612 m respectively.

CONCLUSION

Modeling of hysteresis in piezoelectric continua is very important not only for control applications, but also for designing piezoelectric components against excessive heating. While nonlinearties such as jump phenomena can cause system instability, the combined effect of nonlinearity and hysteresis can be detrimental to system performance and hence should be modeled accurately. In this work, the nonlinear electric enthalpy density function (without hysteresis) has been modified to take into account of hysteresis effect at subcoercive fields (minor hysteresis loops). Simulations have been carried out for two

Figure 19. Variation of displacement amplitude w.r.t. applied electric potential (with hysteresis) for piezo-beam system-II

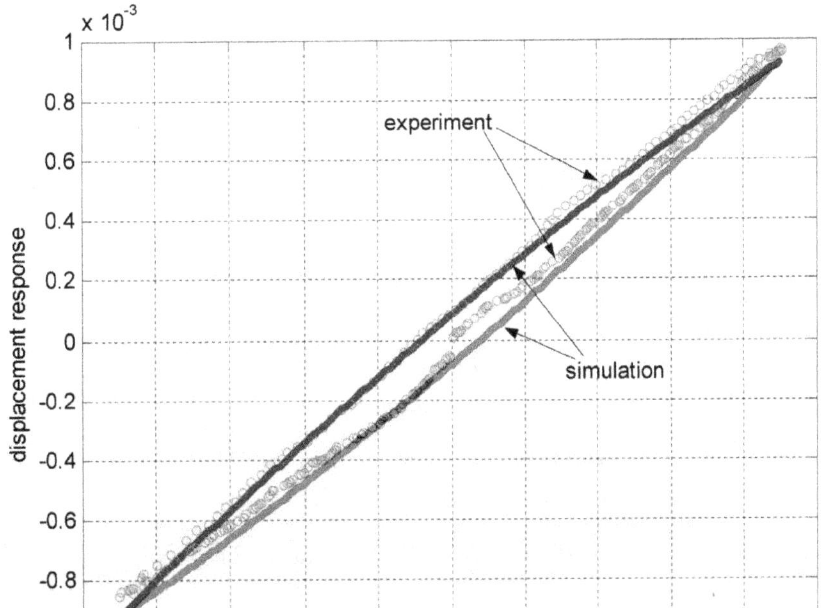

different piezo-beam systems and results have been compared with experimental results from literature. The predicted results show a good agreement with those of experiment wherever available. Hence, this model is expected to be of great use in designing of piezoelectric components under varying conditions of applied electrical and mechanical fields.

REFERENCES

Aljanaideh, O., Habineza, D., Rakotondrabe, M., & Janaideh, M. A. (2016). Experimental comparison of rate-dependent hysteresis models in characterizing and compensating hysteresis of piezoelectric tube actuators. *Physica B, Condensed Matter*, *486*, 64–68. doi:10.1016/j.physb.2015.10.021

Beige, H., & Schmidt, G. (1982). Electromechanical Resonances for investigating linear and nonlinear properties of Dielectrics. *Ferroelectrics*, *41*(1), 39–49. doi:10.1080/00150198208210608

Cady, W. G. (1964). *Piezoelectricity*. Dover.

Crawley, E. F. (1994). Intelligent structures for aerospace: A technology overview and assessment. *AIAA Journal*, *32*(8), 1689–1699. doi:10.2514/3.12161

Cross, L. E. (1993). *Ferroelectric Ceramics* (N. Setter & E. Colla, Eds.). Basel, Switzerland: Birkhaeuser.

Damjanovic, D. (1997). Stress and frequency dependence of the direct piezoelectric effect in ferroelectric ceramics. *Journal of Applied Physics*, *82*(4), 1788–1797. doi:10.1063/1.365981

Damjanovic, D., Demartin, M., Chu, F., & Setter, N. (1996). *Proceedings of the Tenth IEEE International Symposium on Applications of Ferroelectrics*. Rutgers University Press.

Damjanovic, D., & Demartin, M. J. (1996). Article. *Physica D. Nonlinear Phenomena*, *29*, 2057–2069.

Dube, G. P., Kapuria, S., & Dumir, P. C. (1996). Exact piezothermoelastic solution of simply supported orthotropic flat panel in cylindrical bending. *International Journal of Mechanical Sciences*, *38*(11), 1161–1177. doi:10.1016/0020-7403(96)00020-3

Hassani, V., & Tjahjowidodo, T. (2013). Structural response investigation of a triangular-based piezoelectric drive mechanism to hysteresis effect of the piezoelectric actuator. *Mechanical Systems and Signal Processing*, *36*(1), 210–223. doi:10.1016/j.ymssp.2012.10.011

Heyliger, P., & Brooks, S. (1996). Exact solutions for laminated piezoelectric plates in cylindrical bending.*Transactions of ASME Journal of Applied Mechanics, 63*(4), 903-910. doi:10.1115/1.2787245

Hwang, S. C., & McMeeking, R. M. (1998). A finite element model of ferroelectric polycrystals. *Ferroelectrics*, *211*(1), 177–194. doi:10.1080/00150199808232342

IEEE standards on piezoelectricity. (1988), *ANSI/IEEE standard*. The Institute of Electrical and Electronic Engineers.

Ikeda, T. (1990). *Fundamentals of piezoelectricity*. New York: Oxford University Press.

Kronmueller, H. (1970). Article. *Zeitschrift fur Angewandte Physik*, *30*, 9–15.

Maugin, G. A. (1985). Nonlinear electromechanical effects and applications. World Scientific Publishing Company.

Ne'el, N. (1942). Article. *Cahiers Phys.*, *12*, 1–6.

Neumann, N. (2002). *Nichtlineare Effekte bei Längsschwingungen axial polarisierter piezokeramischer Staabe: Experimentelle Untersuchungen und Parameteridentifikation. Diplomarbeit, Institut für mechanik*. Technische Universität Darmstadt.

Pagano, N. J. (1970). Exact solutions for rectangular bidirectional composites and sandwich plates. *Journal of Composite Materials*, *4*(1), 20–34.

Raghavan, A., Seshu, P., & Gandhi, P. S. (2003). Hysteresis modeling in piezoceramic actuator systems. *Proceedings of the Society for Photo-Instrumentation Engineers*, *5062*, 560–568. doi:10.1117/12.514647

Ray, M. C. (1998). Closed-form solution for optimal control of laminated plate. *Computers & Structures*, *69*(2), 283–290. doi:10.1016/S0045-7949(97)00093-X

Ray, M. C., Bhattacharya, R., & Samanta, B. (1993a). Exact solutions for static analysis of intelligent structures. *AIAA Journal*, *31*(9), 1684–1691. doi:10.2514/3.11831

Ray, M. C., Bhattacharya, R., & Samanta, B. (1998). Exact solutions for dynamic analysis of composite plates with distributed piezoelectric layers. *Computers & Structures*, *66*(6), 737–743. doi:10.1016/S0045-7949(97)00126-0

Ray, M. C., Rao, K. M., & Samanta, B. (1992). Exact analysis of coupled electro-elastic behaviour of a piezoelectric plate under cylindrical bending. *Computers & Structures*, *45*(4), 667–677. doi:10.1016/0045-7949(92)90485-I

Ray, M. C., Rao, K. M., & Samanta, B. (1993b). Exact solution for static analysis of an intelligent structure under cylindrical bending. *Computers & Structures*, *47*(6), 1031–1042. doi:10.1016/0045-7949(93)90307-Y

Rayleigh, R. (1887). XXV. Notes on electricity and magnetism. —III. On the behaviour of iron and steel under the operation of feeble magnetic forces. *Philosophical Magazine*, *23*(142), 225–232. doi:10.1080/14786448708628000

Samal, M. K. (2003). *Modelling of nonlinear behaviour of piezoelectric materials under weak electric fields* (Masters Thesis). Department of Mechanical Engineering, Indian Institute of Technology, Bombay, India.

Samal, M. K., Seshu, P., & Dutta, B. K. (2007). An analytical formulation in 3D domain for the nonlinear response of piezoelectric slabs under weak electric fields. *International Journal of Solids and Structures*, *44*(14-15), 4656–4672. doi:10.1016/j.ijsolstr.2006.11.045

Samal, M. K., Seshu, P., Parashar, S. K., von Wagner, U., Hagedorn, P., Dutta, B. K., & Kushwaha, H. S. (2006a). Nonlinear Behaviour of Piezoceramics under Weak Electric Fields, Part-I: 3-D Finite Element formulation. *International Journal of Solids and Structures*, *43*(6), 1422–1436. doi:10.1016/j.ijsolstr.2005.06.084

Samal, M. K., Seshu, P., Parashar, S. K., von Wagner, U., Hagedorn, P., Dutta, B. K., & Kushwaha, H. S. (2006b). Nonlinear Behaviour of Piezoceramics under Weak Electric Fields, Part-II: Numerical results and validation with experiments. *International Journal of Solids and Structures*, *43*(6), 1437–1458. doi:10.1016/j.ijsolstr.2005.06.067

Tang, H., & Li, Y. (2015). Feed forward nonlinear PID control of a novel micromanipulator using Preisach hysteresis compensator. *Robotics and Computer-integrated Manufacturing*, *34*, 124–132. doi:10.1016/j.rcim.2014.11.006

Thakkar, D., & Ganguli, R. (2004). Helicopter vibration reduction in forward flight with induced shear based piezoceramic actuation. *Smart Materials and Structures*, *13*(3), 599–608. doi:10.1088/0964-1726/13/3/019

Thakkar, D., & Ganguli, R. (2006). Use of single crystal and soft piezoceramics for alleviation of flow separation induced vibration in smart helicopter rotor. *Smart Materials and Structures*, *15*(2), 331–341. doi:10.1088/0964-1726/15/2/013

Thakkar, D., & Ganguli, R. (2007). Induced shear actuation of helicopter rotor blade for active twist control. *Thin-walled Structures*, *45*(1), 111–121. doi:10.1016/j.tws.2006.11.001

Usher, T. & Sim, A. (2005). Nonlinear dynamics of piezoelectric high displacement actuators in cantilever mode. *Journal of Applied Physics, 98*(6).

Viswamurthy, S. R., & Ganguli, R. (2007). Modeling and analysis of piezoceramic actuator hysteresis for helicopter vibration control. *Sensors and Actuators. A, Physical*, *35*(2), 801–810. doi:10.1016/j.sna.2006.09.020

Viswamurthy, S. R., Rao, A. K., & Ganguli, R. (2007). Dynamic hysteresis of piezoceramic stack actuators used in helicopter vibration control: Experiments and simulations. *Smart Materials and Structures*, *16*(4), 1109–1119. doi:10.1088/0964-1726/16/4/020

von Wagner, U. (2003). Non-linear longitudinal vibrations of piezoceramics excited by weak electric fields. *International Journal of Non-linear Mechanics*, *38*(4), 565–574. doi:10.1016/S0020-7462(01)00113-5

von Wagner, U. (2004). Non-linear longitudinal vibrations of non-slender piezoceramic rods. *International Journal of Non-linear Mechanics*, *39*(4), 673–688. doi:10.1016/S0020-7462(03)00108-2

von Wagner, U., & Hagedorn, P. (2002). Piezo-beam-systems subjected to weak electric field: Experiment and modelling of nonlinearities. *Journal of Sound and Vibration*, *256*(5), 861–872. doi:10.1006/jsvi.2002.5024

Wang, Q.-M., Zhang, Q., Xu, B., Liu, R., & Cross, L. E. (1999). Nonlinear piezoelectric behavior of ceramic bending mode under strong electric fields. *Journal of Applied Physics*, *86*(6), 3352–3360. doi:10.1063/1.371213

Wolf, K., & Gottlieb, O. (2001). *Nonlinear dynamics of a cantilever beam actuated by piezoelectric layers in symmetric and asymmetric configuration, Technical Report, No. ETR-2001-02*. Faculty of Mechanical Engineering, Israel Institute of Technology.

Xu, K. M., Noor, A. K., & Tang, Y. (1995). Three-dimensional solutions for coupled thermoelectroelastic response of multilayered plates. *Computer Methods in Applied Mechanics and Engineering*, *126*(3-4), 355–371. doi:10.1016/0045-7825(95)00825-L

Chapter 14
Walking Control of Humanoid Robots on Uneven Ground Using Fuzzy Algorithm

Saeed Abdolshah
University of Padova, Italy

Majid Abdolshah
University of Tehran, Iran

Mohammad Abdolshah
Islamic Azad University – Semnan, Iran

S. Vahid Hashemi
Semnan University, Iran

ABSTRACT

Walking control of humanoid robots is a challenging issue. In this chapter, a method for modeling humanoid robots is presented considering the human being indices such as DOFs, mass and the moment of inertia of the segments. In the next step, a walking pattern on the flat ground is generated and the robot motion is simulated in the MSC. Visual Nastran 4D™ software. ZMP trajectory of the simulated humanoid robot in walking cycle has been obtained. An uneven ground is generated in the software, where the robot falls down during the motion. A fuzzy algorithm is employed to balance the robot; input is defined as the differences between the projections of ZMP in flat and uneven ground and output is a compensative signal to make the robot follow the flat ground ZMP pattern to refuse the robot falling. Output signal is distributed in different joints to make faster and more effective compensation. Although the type of uneven ground can be important, but the robot could successfully pass the designed uneven ground in MSC.Visual Nastran 4D.

INTRODUCTION

Recently there are many research projects continuously in progress for control, planning and motion generation of humanoid robots. In order to accomplish complicated jobs, humanoid robots must possess intelligent abilities. A humanoid robot should be capable of both interacting with human beings and moving within human living environments. A humanoid robot capable of cooperating with human beings needs the ability to move between objects, walk on different types of ground and communicate.

DOI: 10.4018/978-1-5225-1639-2.ch014

Controlling the balance of a humanoid robot in complicated environments is an important issue. Vukobratovic proposed ZMP (Zero Moment Point) as an index for the stability of humanoid robots (Vukobratovic & Juricic, 1969). The ZMP stability criteria states that the biped will not fall down as long as the ZMP maintains inside the convex hall of the support polygon of the foot. Since that time, intelligent control techniques such as genetic algorithms, neural networks, and fuzzy logic and hybrid forms such as neuro-fuzzy networks, neuro-genetic and fuzzy-genetic algorithms were applied in humanoid robot control techniques (Katić & Vukobratović, 2003).

Some researchers have used fuzzy logic in tuning of gains of PID controller of humanoid robot (Vukobratovic & Timcenko, 1995). Zhou and Meng (2000) used a combination of reinforcement learning with the fuzzy gait synthesizer to track the desired trajectory. Chi- Cheng, Chen, Hu and Wong (2009) proposed a fuzzy algorithm for balancing a small-size humanoid robot to adjust the robot's posture based on an accelerometer sensor. Fuzzy logic has been used for controlling humanoid robots in facing with obstacles and uneven grounds too. Wong, Cheng, Huang and Yang (2008) implemented fuzzy system on a humanoid robot to avoid obstacles based on the obtained information of infrared (IR) sensors and electronic com-pass. Wong, Hwang, Huang, Hu and Cheng (2011) also designed a vision-based obstacle avoidance method to avoid humanoid robot from obstacles based on fuzzy control. Park, Han and Hahn (2009) proposed a fuzzy balance control algorithm using 3D (Dimensional) geometric information. A few research projects have focused on fuzzy control of humanoid robots on uneven surfaces; however, the ability of robot to move on human-base environments is an important point. In addition, the necessity of building capable humanoid robots for exploration on unknown environments is a significant issue.

Kim, Park, and Oh (2007) proposed an online control algorithm that considers local and global inclinations of the surface by which a biped humanoid robot can adapt to the floor conditions. Kong, Lee, and Kim (2008) proposed a real-time walking stabilization method utilizing a fuzzy algorithm under uneven terrain. They used ankle motors as an active joint to balance the robot using fuzzy algorithm. In this method, the ankle joints were producing great amount of torque to compensate the mass deviation of robot to refuse falling. The algorithm was verified by simulation and walking experiments on a 24-DOFs humanoid robot. When a robot is falling in sagittal plane direction, different DOFs such as ankle, upper body, hip and arms can refuse it; meanwhile that research has just focused on ankle, which increases the response time and risk of falling greatly. Human beings use different joints to balance when they are falling (e.g. Upper body as well as lower body segments).

In this chapter, we propose a method to design a humanoid robot based on human characteristics, and then a walking pattern is generated for flat ground and the ZMP trajectory of robot is obtained. In the next step, a fuzzy control algorithm has been designed, input is the difference between the projections of ZMP in the flat and uneven ground; however, output is the proper motion of the robot to control the ZMP considering the flat ground ZMP pattern as a reference to refuse the robot falling. The weighted average method is used to determine the final output of the fuzzy system. Compensation value is shared between different DOFs. The model is designed in MSC.visualNastran 4D™ software and the fuzzy algorithm has been implemented on the robot. Although the uneven ground type is so much important, but the robot could successfully pass the designed uneven ground in MSC.visualNastran 4D™.

HUMANOID ROBOT MODEL

To have the best design of humanoid robot capable of imitating human motion, a bio-inspired model of human body is used. The mechanical construction of humanoid robots consists of several joints and segments. Human dimensions and the mass of body segments were considered as the main reference to design a humanoid robot model. Such design can result in stable walking and optimal distribution of the forces in walking cycle. It should be emphasized that human joints are significantly more flexible than humanoid robot joints.

Characteristic values of body segments such as height, mass and moments of inertia of each segment are roughly depended on the race, sex and age of human beings. Abdolshah, S., Abdolshah M. and Tang (2014) considered four main group segments in human body including:

1. Head and neck,
2. Torso,
3. Total arm, and
4. Total leg.

Each one has some subparts as it can be seen in Table 1. In this table, the body segment weights were calculated based on an 80 kilograms subject which has been applied in simulation model of humanoid robot.

Human body has many degrees of freedom. In this study, the essential degrees of freedom were considered. The robot has been simulated in MSC.visualNastran (Figure 1). Two degrees of freedom in each ankle, one degree in each knee, two degrees in each pelvis, one degree in waist, two degrees in each arm and one degree in the neck were considered in this robot. Each leg of the humanoid robot has 6 degrees of freedom.

Table 1. Body segments weight

part no.	Grouped Segments	Group Weight Percentage	Individual Segments	Individual Weight Percentage	Weight(Total 80 Kg)
1	Head and Neck	8.4	Head	73.8	4.96
			Neck	26.2	1.76
2	Torso	50	Thorax	43.8	17.52
			Lumber	29.4	11.76
			Pelvis	26.8	10.72
3	Total Arm	5.1	Upper Arm	54.9	2.24
			Forearm	33.3	1.36
			Arm	11.8	0.48
4	Total Leg	15.7	Thigh	63.7	8.00
			Shank	27.4	3.44
			Foot	8.9	1.12

Figure 1. Humanoid robot model

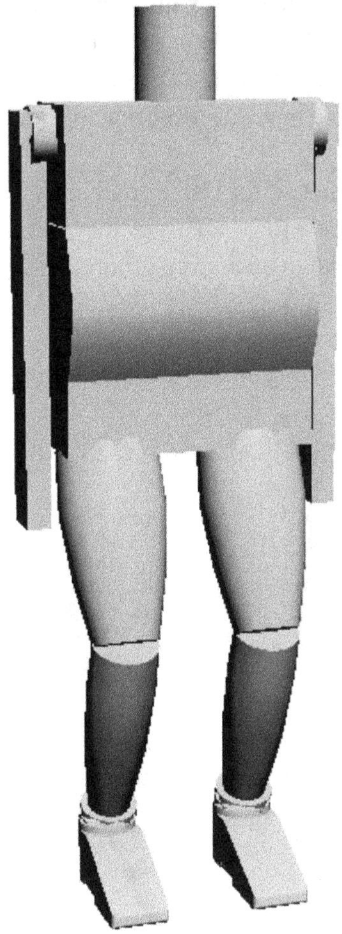

WALKING PATTERN GENERATION

The human walking motion was studied to obtain the appropriate trajectory motion of body segments during the walking cycle. Generating a walking trajectory was implemented in this study (Abdolshah, Abdolshah, & Tang, 2014).

To make a walking motion trajectory, different walking phases were combined together. We had proposed some different walking patterns like moving left leg, moving right leg and so on. A mixture of these motions within nine steps makes a full walking cycle.

The extracted motion phases from human pattern were applied in this section. The Denavit-Hartenberg method was used for deriving the relations of the 16 degrees of freedom robot. At the global form for solving the motion equations we have 16 unknown quantities in each step of walking. To gain these 16 unknowns in different movement phases from a set of constraints such as symmetry, constant length of two joints of each side of hip, movement direction, and place of extracted ZMP point of human subject was used. Inverse kinematics approach was used to solve the problems.

As an example, at the first phase, the robot should lift one of the legs, to prevent falling robot, fist the projection of ZMP should be placed in convex hull, so the projection of ZMP should be transferred to one of the legs, therefore the robot have to lean on one of the legs, to implement this phase just 4 motors- two in wrists and two in hip- were used. Regarding to constraints and ZMP position, equations have been solved. In other phases some motors have been considered 'off' and other motions have been estimated by equations.

The dynamic simulation of walking gait for flat ground in Visual Nastran for a half cycle is shown in Figure 2. The robot will be able to follow the ZMP trajectory if the ground remains smooth, and the projection of ZMP remains in convex hall, but on uneven ground, fuzzy balancing control should be applied.

Figure 2. Dynamic simulation of walking gait in a half cycle

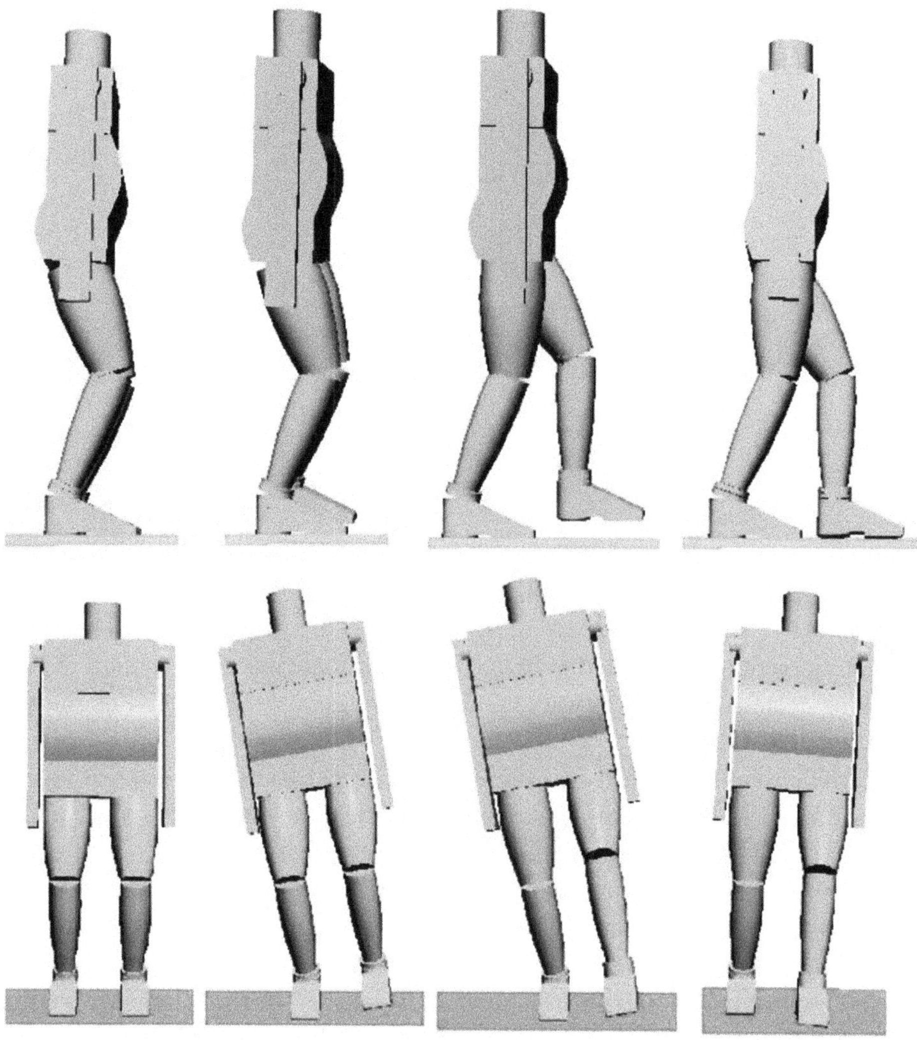

FUZZY BALANCING CONTROL OF ROBOT ON UNEVEN GROUND

In fuzzy balancing control of the humanoid robot, five issues were considered to follow the desired ZMP trajectory by returning the projection of ZMP to the convex hall:

1. Forward Compensation,
2. Backward Compensation,
3. Left Compensation,
4. Right Compensation, and
5. None Compensation.

Four set of motors are responsible for a, b compensations, ankles, knees, pelvises and arms which are totally 8 motors. These motors can swing robot to forward or backward, other motors are useful in c, d compensations, which are set of motors in ankles, pelvises and arms. These motors can move the robot to right or left side. Waist and neck motors were not used in this study. Figure 3 illustrates mentioned DOFs.

The rule-base model with the result of fuzzification based on the robot states is defined as:

If x is A1 and Δx is A2 then u is B,

If y is A1 and Δy is A2 then u is B.

Where A1 is the distance between the ZMP position of on uneven ground and flat surface, A2 is the variation of the A1, which is sign of growing the distance, and B is the compensation. HN, LN, Z, LP, and HP denote High Negative, Low Negative, Zero, Low Positive and High Positive. The inference engine of robot condition is obtained using this language rule based on the rule-base.

Table 2 represents the total rule-base. Where Negative High Motion (NHM), Negative Low Motion (NLM), Without Motion (WM), Positive Low Motion (PLM), and Positive High Motion (PHM) are for the angle of compensation. As an example when the ZMP error from the flat surface pattern are zero, it means the robot is following the predesigned trajectory of flat ground, so WM(Without motion) is the best option and etc.

The weighted average method is used to determine the final output of the fuzzy system. As it mentioned before, 4 set of motors can generate variations in sagittal plane and 3 others in coronal plane, but contribution of these motors in motion is the most important point. The ZMP projection of robot

Table 2. Rule-base of fuzzy algorithm

		Δx, Δy				
		HN	**LN**	**Z**	**LP**	**HP**
x,y	HN	PHM	PHM	PHM	PLM	WM
	LN	PHM	PLM	PLM	WM	WM
	Z	WM	WM	WM	WM	WM
	LP	WM	WM	NLM	NLM	NHM
	HP	WM	NLM	NHM	NHM	NHM

Figure 3. a) DOFs of ankles, pelvises and arms for c, d compensations, b) DOFs of ankles, knees, pelvises and arms for a, b compensations

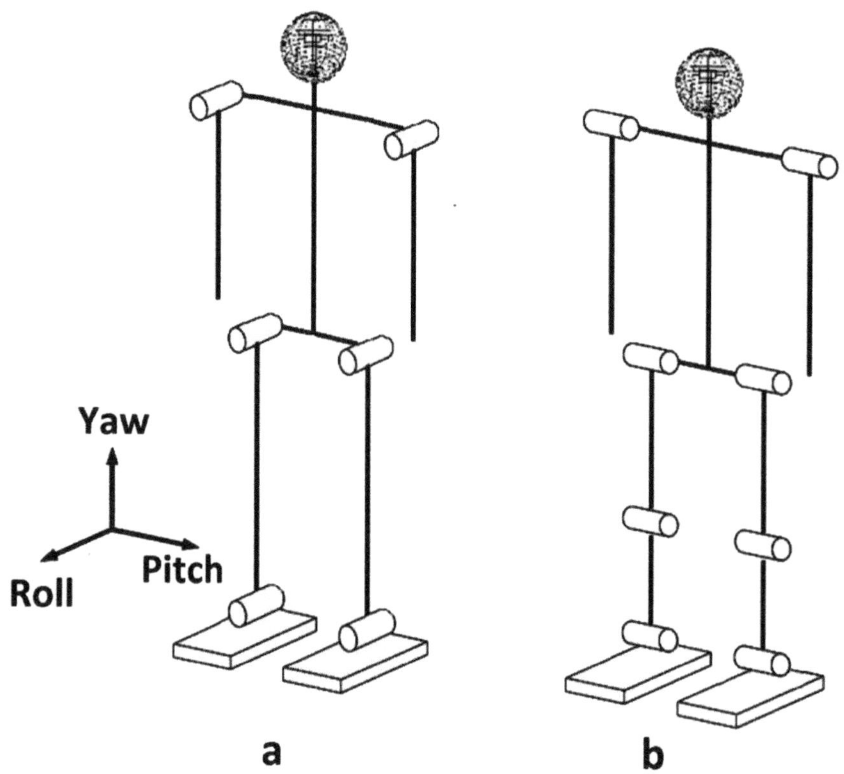

should be used to calculate the contribution of motors. As the robot on uneven ground has slow motion, we used COG as an important index. In this method, contribution of motors in COG variation has been estimated via effective weight, which is carrying with the motors motion. Quantity of motors rotations are a few, so it is not an important factor; however, quantity of the rotated weight plays a significant role. Consequently, it is reasonable to take the share of motors in COG variation as an effective parameter.

As an example the ankle motor rotates whole body instead of foot at the bottom of the ankle, knee motors can rotate whole body instead of shank and foot at the bottom of the knee. Obviously; the influence of ankle motors are more than knees and knees are more than pelvises. Estimation showed that effective weight for ankle motors to rotate in an 80 Kg model is about 78.89 Kg, which is the weight of total body instead of one foot. For knee, it is about 75.45 Kg, which is total body weight instead of one shank and foot. For pelvis, it is 67.45, which is total body weight instead of one foot, shank and thigh, and for arms, it is just 4.06 Kg, which has the least influence in variation of COG.

RESULTS

In order to illustrate the efficiency of the proposed method for the humanoid robots motion control, an uneven ground has been designed randomly to test the robot stability in Visual Nastran. The ground has made of some spheres, cubic and some random shapes. The robot stand in the front of the ground for testing the balance ability as shown in Figure 4.a. Then it started walking without fuzzy control, where the robot fails during walking on the uneven ground and results in falling down (Figure 4b). Figure 4c shows how the robot can successfully pass the uneven surface.

A comparison between the projection of ZMP on flat surface and uneven ground is shown in Figure 5. As it is obvious, ZMP of the robot on the flat surface is followed during the motion on the uneven surface.

CONCLUSION

In this chapter, a bio-inspired method for modeling humanoid has been proposed. The method uses human body segments characteristics such as mass and dimensions. The robot is simulated in MSC. visualNastran 4D™ and a walking pattern for the robot on the flat ground is generated. The robot mo-

Figure 4. a) Initial condition, b) motion without fuzzy control, c) motion with fuzzy control

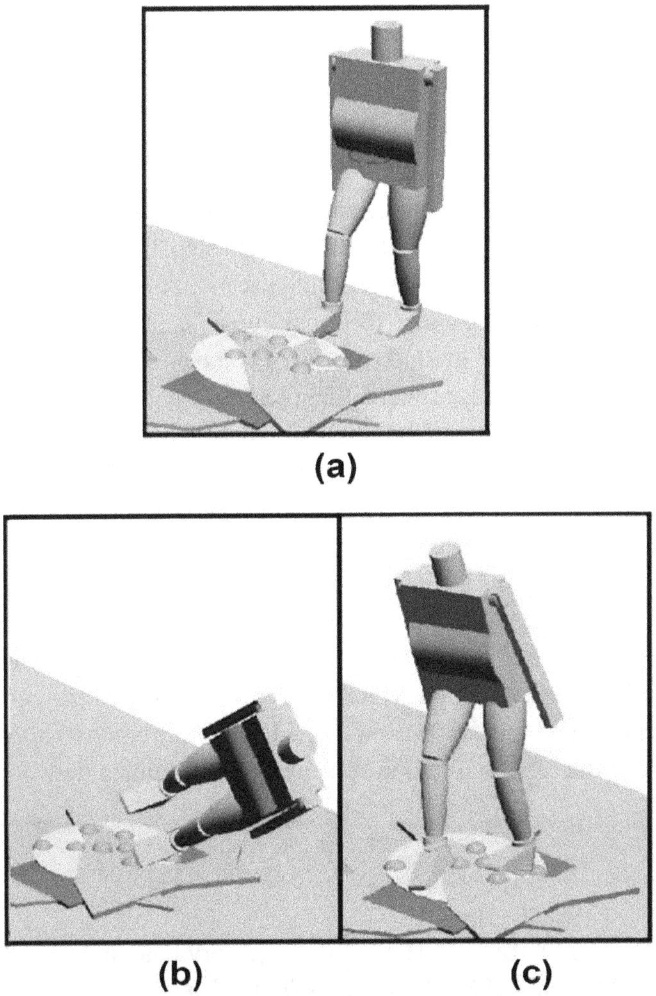

Figure 5. ZMP projection of robot

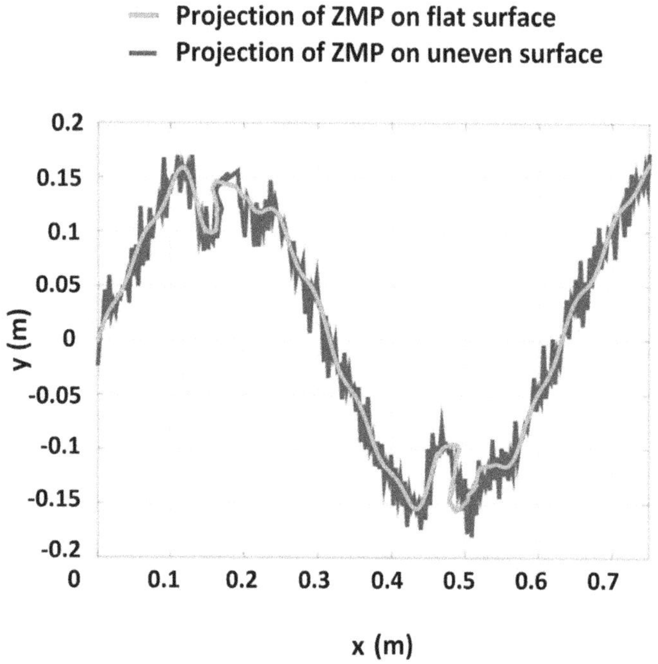

tion on a flat surface is simulated in the MSC.visualNastran 4D™ software. ZMP trajectory motion of the simulated humanoid robot has been extracted. In the next step, the motion of the robot on an uneven ground is simulated, where the robot falls down during the motion.

A fuzzy algorithm is employed to balance the robot; input is defined as the differences between the projections of ZMP in flat and uneven ground and output is the proper motion of the robot to follow the ZMP pattern of the flat ground to refuse robot falling. Output signal is distributed in different joints to make faster and more effective compensation. Although the type of uneven ground can be important, but the robot could successfully pass the designed uneven ground in MSC.visualNastran 4D™.

A fuzzy control system has been proposed and implemented on a sample humanoid robot and uneven ground in visual Nastran. The proposed fuzzy control method makes the robot capable to pass the uneven ground and uses different robot joints as the human does.

From the simulation results, we can see that the robot can effectively pass the ground; however it should be mentioned that the robot could pass this uneven ground just with fuzzy control system, but different uneven grounds can be designed which the robot falls with fuzzy control too. So in future research, difficulty of uneven ground can be determined and fuzzy control should be tested in different grounds. It is obvious that the robot will fail in some tests as the human may not pass them too.

REFERENCES

Abdolshah, S., Abdolshah, M., & Tang, S. H. (2014). Trajectory Planning and Walking Pattern Generation of Humanoid Robot Motion. *IAES International Journal of Robotics and Automation, 4*(2).

Cheng, C. T., Chen, H. C., Hu, Y. Y., & Wong, C. C. (2009). Fuzzy balancing control of a small-size humanoid robot based on accelerometer. *International Journal of Fuzzy Systems*, *11*(3), 146–153.

Katić, D., & Vukobratović, M. (2003). Survey of intelligent control techniques for humanoid robots. *Journal of Intelligent & Robotic Systems*, *37*(2), 117–141. doi:10.1023/A:1024172417914

Kim, J. Y., Park, I. W., & Oh, J. H. (2007). Walking control algorithm of biped humanoid robot on uneven and inclined floor. *Journal of Intelligent & Robotic Systems*, *48*(4), 457–484. doi:10.1007/s10846-006-9107-8

Kong, J. S., Lee, E. H., Lee, B. H., & Kim, J. G. (2008). Study on the real-time walking control of a humanoid robot using fuzzy algorithm. *International Journal of Control, Automation, and Systems*, *6*(4), 551–558.

Park, S., Han, Y., & Hahn, H. (2009). Fuzzy controller based biped robot balance control using 3D image. *International Journal of Fuzzy Systems*, *11*(3), 202–212.

Vukobratovic, M., & Timcenko, O. (1995). Stability analysis of certain class of bipedal walking robots with hybridization of classical and fuzzy control. In *Proc. International Conference on Advanced Robotics and Intelligent Automation* (pp. 290-295).

Vukobratovic, M., & Juricic, D. (1969). Contribution to the synthesis of biped gait. *Biomedical Engineering, IEEE Transactions on*, (1), 1-6.

Wong, C. C., Cheng, C. T., Huang, K. H., & Yang, Y. T. (2008). Fuzzy control of humanoid robot for obstacle avoidance. *International Journal of Fuzzy Systems*, *10*(1), 1.

Wong, C. C., Hwang, C. L., Huang, K. H., Hu, Y. Y., & Cheng, C. T. (2011). Design and implementation of vision-based fuzzy obstacle avoidance method on humanoid robot. *International Journal of Fuzzy Systems*, *13*(1), 45–54.

Zhou, C., & Meng, Q. (2000). Reinforcement learning with fuzzy evaluative feedback for a biped robot. In *Robotics and Automation, 2000. Proceedings. ICRA'00. IEEE International Conference on* (Vol. 4, pp. 3829-3834). IEEE.

Chapter 15
Development of Path Tracking Control Algorithm for a 4 DOF Spatial Manipulator Using PID Controller

Pradeep Reddy Bonikila
Indian Institute of Technology Bhubaneswar, India

Ravi Kumar Mandava
Indian Institute of Technology Bhubaneswar, India

Pandu Ranga Vundavilli
Indian Institute of Technology Bhubaneswar, India

ABSTRACT

The path tracking phenomenon of a robotic manipulator arm plays an important role, when the manipulators are used in continuous path industrial applications, such as welding, machining and painting etc. Nowadays, robotic manipulators are extensively used in performing the said tasks in industry. Therefore, it is essential for the manipulator end effector to track the path designed to perform the task in an effective way. In this chapter, an attempt is made to develop a feedback control method for a 4-DOF spatial manipulator to track a path with the help of a PID controller. In order to design the said controller, the kinematic and dynamic models of the robotic manipulator are derived. Further, the concept of inverse kinematics has been used to track different paths, namely a straight line and parabolic paths continuously. The effectiveness of the developed algorithm is tested on a four degree of freedom manipulator arm in simulations.

DOI: 10.4018/978-1-5225-1639-2.ch015

INTRODUCTION

From the past few decades, the usage of robotic manipulators in various applications such as in medicine, aerospace, automotive and high precision manufacturing of components and parts become more significant and therefore their study is important. These manipulators consist of links which are joined by joints mainly revolute and prismatic. Revolute joint allow rotation about that axis while prismatic joint allows relative translation about that axis. The linkage between the links make them to move relative to one another in order to achieve the required end effector position. In the present chapter, the path decided to perform a specific path has been traced by dividing it into small segments, and the corresponding joint angles made by these links is determined by using the concept of inverse kinematics.

The inverse kinematics problem can be solved by various methods such as, pseudo inverse, iterative and geometric approaches. Whitney (1969), introduced the pseudo inverse approach and showed that joint velocities obtained from the pseudo-inverse approach represent a particular path. In the literature, it was clearly demarcated that the trajectory planning is different from path planning and the later precedes the former. Path planning is defined by generating geometric path irrespective of the time law, while trajectory planning attaches a time law with path planning. In order to perform the job as required, the end effector of the robotic manipulator must follow predefined path between the initial and final (goal) points, which prevents any damage caused due to deviation from the path. Thus, it makes path planning an important task to be executed by the robotic manipulator. Several efforts were made by various researchers to develop an efficient path and design suitable control schemes for effective path planning.

Determination of inverse kinematics for a mechanism is a common technique in mechanical engineering, particularly in robotics community (Craig, 1986). However, analytical solutions for inverse kinematics for robotic manipulators are highly intricate and are difficult to apply for high-dimensional redundant manipulators (Chevallereau & Khalil, 1988). Instead, iterative and numerical techniques based on the calculation of the pseudo-inverse or the weighted pseudo-inverse of the Jacobian are introduced (Klein and C. Huang, 1983). Even though, the researchers propose many approaches, most of them lack in generality. Therefore in (Zhao & Badler, 1994), a new and feasible approach to solve the inverse kinematics had been proposed based on nonlinear programming methods. This nonlinear function was successfully applied to position the articulated manipulator and this helped in achieving satisfactory results. Deepak et al. (2000), implemented inverse kinematics algorithms using a combination of analytical and numerical methods to solve generalized inverse kinematics problems after including position, orientation and aiming constraints. It resulted in faster and reliable performance than the conventional methods of solving this problem. This methodology is applied to an anthropomorphic arm or leg. Dinesh et al. (1992), implemented an algorithm for the inverse kinematics of a 6R manipulator, which solve the inverse kinematics problem using multi-variable calculus that used symbol techniques, matrix computations and numerical methods.

Latombe et al. (1991), defined three important approaches such as: road-map, cell decomposition and potential field methods for sorting path planning problems. The authors explained the basics of path planning, like the Configuration space, free space and obstacles in C-space. In that study, they considered collision avoidance as a high level motion planning problem, which allowed the operation of manipulators in a hazardous and complex ambiance. Khatib (1985) and Volpe et al. (1990) developed artificial potential field method to represent the robots in C-space. With this method, the points in space was assumed to move under artificial potential created by the object configuration and the obstacles present in the C-space. Later on, (Faverjon et al., 1987) developed an alternative for artificial potential field method in which they clearly differentiated the difference between task and constraints of anti-collision. This algorithm was applied for control of high number of degrees of freedom manipulators. Once the path is obtained, a controller commands the manipulator to traverse the trajectory by tracking values from the above method.

Another important problem to be taken care of is designing a suitable controller which helps the manipulator to follow desired path. Qu (1994), for the first time developed PD feedback control with time-varying gains to achieve stability for the path following problem of a manipulator. Even though dynamics of the robot are highly nonlinear, higher order nonlinear terms are not necessary in tracking controls, revealing the fact that the PD control performs as good as complicated nonlinear controls. Then, (Dung et al., 2010) developed dynamic modeling of a simple 2-DOF robotic manipulator, using PD controller. The authors observed a better performance while using online gravity compensation rather than constant gravity compensation. Later on, (Prasenjit et al., 2013) implemented a classical PID controller to control a six-link, six-joint flexible robotic manipulator. The controller was tuned after utilizing three different methods, and best result was observed in RRT method and emphasis was paid for developing proper mechanism for tuning PID gains. After choosing a particular controller to perform the task, tuning of the controller is necessary for the better performance of the controller. The tuning of the controller can be done in many ways, such as manual tuning, Ziegler and Nichols (1942) method and online tuning. David (2012), developed a dynamic model and manually tuned PID controller for 2-DOF planar manipulator. While conducting the experiment the author observed only a slight change in proportional, integral and derivative parameters resulted in more overshoots and oscillations. Ravi et al., (2015) discussed the design of a PID controllers for 4-DOF spatial and planar manipulators. Here, the authors used manual tuning method to tune kp, kd and ki of the controller. These parameters were tested in simulations to verify the performance and found that the torque required for base joint was more when compared with others as it was carrying the remaining links.

The Chapter is organized as follows:

- Section 2 includes mathematical formulation of the problem i.e., kinematics and dynamics of the robotic manipulator.
- Section 3 describes the design of PID controller for 4-DOF spatial manipulator.
- Section 4 explains the results and discussions.
- Finally, section 5 concludes the chapter.

MATHEMATICAL FORMULATION

In the present study, an attempt is made to develop a PID controller based path tracking algorithm for a 4-DOF spatial manipulator. The schematic diagram showing the structure of the manipulator is given in Figure 1. This section deals with assignment of D-H parameters, deducing forward kinematics, dynamic modeling of four link manipulator, moving towards inverse kinematics and finally an approach to solve the problem of path tracking. The links of the manipulators are considered to have masses and lengths equal to (m_1, m_2, m_3, m_4) kg and (L_1, L_2, L_3, L_4) m, respectively. Let the included angles made by the links (L_1 through L_4) of the manipulators are given as θ_1, θ_2, θ_3 and θ_4 in order to define a position for the end effector. The D-H parameters obtained for the spatial manipulator are given in Table 1.

Figure 1. Schematic diagram showing the structure of 4-DOF spatial manipulator.

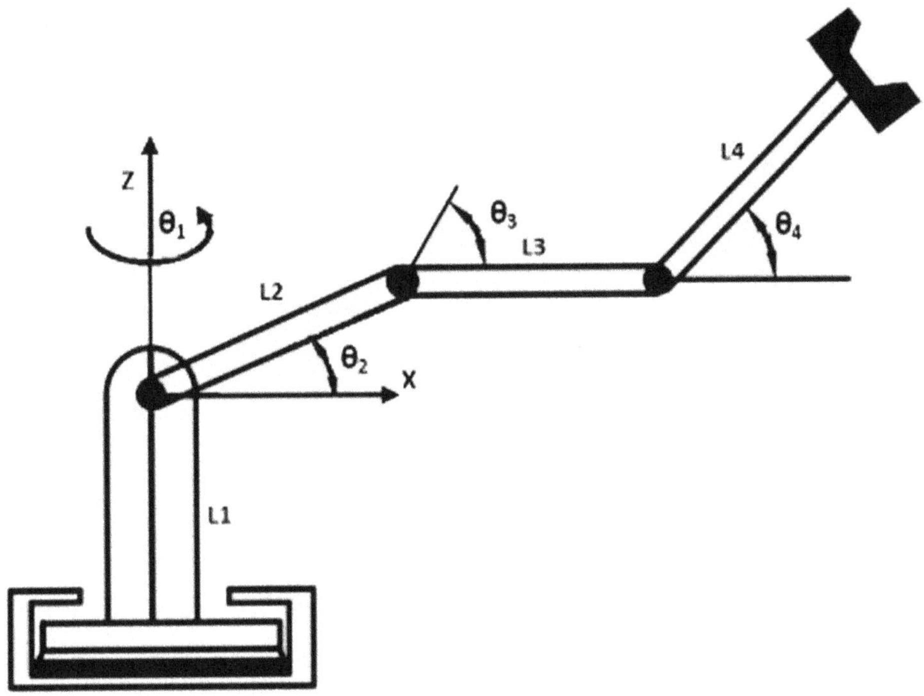

Table 1. The D-H parameters of spatial manipulator

Link(i)	A_i	α_i	d_i	θ
1	0	90°	0	θ_1
2	L2	0	0	θ_2
3	L3	0	0	θ_3
4	L4	0	0	θ_4

Forward Kinematics

Forward kinematics helps in deducing the position of the end effector given by the joint variables. D-H parameters (refer to Figure 2) are assigned to the manipulator and individual transformation matrices are formed.

Then these individual transformation matrices are multiplied to obtain the overall transformation 0T4 matrix for the given manipulator configuration. The overall transformation matrix for the 4-DOF spatial manipulator is given by Equation 1.

$$^{0}T_{4} = {}^{0}T_{1} * {}^{1}T_{2} * {}^{2}T_{3} * {}^{3}T_{4} \tag{1}$$

Figure 2. Coordinate frame assignment for 4-DOF spatial manipulator.

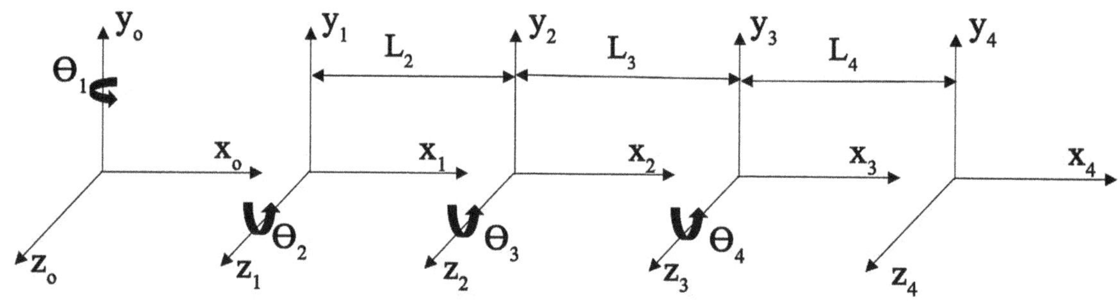

$$0_{T1}=\begin{bmatrix} C_1 & 0 & S_1 & 0\\ S_1 & 0 & -C_1 & 0\\ 0 & 1 & 0 & 0\\ 0 & 0 & 0 & 1\end{bmatrix} \quad 1_{T2}=\begin{bmatrix} C_2 & -S_2 & 0 & L_2C_2\\ S_2 & C_2 & 0 & L_2S_2\\ 0 & 0 & 1 & 0\\ 0 & 0 & 0 & 1\end{bmatrix}$$

$$2_{T3}=\begin{bmatrix} C_3 & -S_3 & 0 & L_3C_3\\ S_3 & C_3 & 0 & L_3S_3\\ 0 & 0 & 1 & 0\\ 0 & 0 & 0 & 1\end{bmatrix} \quad 0_{T4}=\begin{bmatrix} C_4 & -S_4 & 0 & L_4C_4\\ S_4 & C_4 & 0 & L_4S_4\\ 0 & 0 & 1 & 0\\ 0 & 0 & 0 & 1\end{bmatrix}$$

$$0_{T4}=\begin{bmatrix} C_1C_{234} & -C_1S_{234} & S_1 & d_1\\ S_1C_{234} & -S_1S_{234} & -C_1 & d_2\\ S_{234} & C_{234} & 0 & d_3\\ 0 & 0 & 0 & 1\end{bmatrix} \tag{2}$$

where the meaning of the terms d_1, d_2 and d_3 are given in Equations 3, 4, and 5, respectively.

$$d_1 = C_1\,[L_2C_2 + L_3C_{23\,+}\,L_4C_{234]} \tag{3}$$

$$d_2 = S_1\,[L_2C_2 + L_3C_{23\,+}\,L_4C_{234]} \tag{4}$$

$$d_3 = [L_2\,S_2 + L_3S_{23\,+}\,L_4\,S_{234}] \tag{5}$$

the significance of the term (S and C) are given below

$S_1 = \sin(\theta_1)$; $C_1 = \cos(\theta_1)$;

$C_{12} = \cos(\theta_1 + \theta_2)$; $S_{12} = \sin(\theta_1 + \theta_2)$;

$S_{123} = \sin(\theta_1 + \theta_2 + \theta_3)$; $C_{123} = \cos(\theta_1 + \theta_2 + \theta_3)$;

$C_{1234} = \cos(\theta_1 + \theta_2 + \theta_3 + \theta_4)$; $S_{1234} = \sin(\theta_1 + \theta_2 + \theta_3 + \theta_4)$

Inverse Kinematics

The solution for the real time path planning is solved using inverse kinematics approach. The step by step procedure used for tracking the path is given below.

Step 1: Find the Overall Transformation Matrix:

The individual transformation matrices of various links that are obtained from the D-H parameters has been used to find the overall transformation matrix.

Step 2: Evaluation and Comparison:

The overall transformation matrix obtained in step 1 is equated to the position and orientation matrix of the end effector. By careful scrutinizing and comparing the corresponding elements of these two matrices, the joint angle information is extracted directly. For other angles certain mathematical simplifications need to be performed.

Step 3: Inverse Transform Method:

Sometimes, it is very difficult and impossible to solve the joint variables even after employing the mathematical manipulations. In such case, inverse transform approach is employed by Multiplying with transformation matrix on both sides of the matrix equation obtained in step 2.

Step 4: Satisfying Boundary Conditions:

In the present problem, four variables that correspond to the 4-DOF of the manipulator are needed to be solved. There are only 2 boundary conditions that correspond to initial and final positions of the end effector. The other boundary conditions like, sum of last 3 joint angles is pi (π) and the first link is fixed, making it spatial manipulator which relaxed one condition. After following the above four steps, the inverse kinematics solution has been obtained for the 4-DOF spatial manipulator and the solution is given below.

$$\theta_1 = \pi/2 \tag{6}$$

Xn = u − (L_{4*} cos (tha)), yn = v − L_1 − (L_{4*} sin (tha))

where, tha =π, and the expression for θ_3 is given in Eqn. (7)

$$\theta_3 = \text{atan2}\,(a_{s,}\ a_c) \tag{7}$$

where $a_c = \dfrac{\left(\left(x_n\right)^2 + \left(y_n\right)^2 - (L_2^2 + L_3^2)\right)}{2 * L_2 * L_3}$, $a_s = \sqrt{\left(1 - a_c^2\right)}$, further θ_2 can be calculated by using Eqn. (8).

$$\theta_2 = \text{atan2}\,((k_1 * y_n - k_2 * x_{n)}, (k_1 * x_n + k_2 * y_{n)}) \tag{8}$$

where, $k_1 = L_2 + L_3 * \cos(\theta_3)$, $k_2 = L_3 * \sin(\theta_3)$. Finally, θ_4 can be determined by using Eqn. (9).

$$\theta_4 = \text{tha} - \theta_3 - \theta_2 \tag{9}$$

where the terms *tha* indicates the sum of joint angles of joints 2, 3 and 4. Moreover, u and v represent the temporary variables that store the values of position of end effector.

Dynamics of Manipulator

The dynamics of manipulator is the time rate of change of links configuration in relation to the torques exerted by the actuators. The actuators need to balance the torque requirement of four different sources namely motion, friction, gravity and external tasks. The mathematical equations that describes the torques required at the various joints of the manipulators has been derived after considering Lagrange-Euler formulation. The Equation 10 that represent the torque required at joint 'i' can be given as

$$\tau_i = \sum_{j=1}^{n} M_{ij}\left(q\right)\ddot{q}_j + \sum_{j=1}^{n}\sum_{k=1}^{n} h_{ijk}\dot{q}_j\dot{q}_k + G_i \qquad i, j = 1, 2, \ldots\ldots\ldots n \tag{10}$$

here τ_i represent force applied at joint i due to motion of links. q, $\dot{q}_j$,$\ddot{q}_j$ indicates the displacement, velocity and acceleration of the joint j respectively. Where the terms M_{ij}, h_{ijk} and G_i indicate inertia, Coriolis and gravity terms and can be given by Equations 11, 12, and 13, respectively.

$$M_{ij} = \sum_{p=\max\left(i,j\right)}^{n} Tr[d_{pj} I_p d_{pi}^T] \qquad i, j = 1, 2, \ldots\ldots\ldots n \tag{11}$$

$$h_{ijk} = \sum_{p=\max\left(i,j,k\right)}^{n} Tr\left[\frac{\partial\left(d_{pk}\right)}{\partial q_p} I_p d_{pi}^T\right] \qquad i, j1, 2, \ldots\ldots\ldots n \tag{12}$$

$$G_i = -\sum_{p=i}^{n} m_p g d_{pi}\, {}^{p}_{e}\overline{r}_p \qquad i,j = 1,2,\ldots\ldots\ldots n \tag{13}$$

DESIGN OF PID CONTROLLER

There is a necessity to control the end effector to reach the final position without any deviations from the desired path by reducing the error. The most commonly used control technique to achieve the said task is PID controller due to its robust and reliable performance, if the parameters are tuned properly. While solving the problem, the PID algorithm assumes the robotic manipulator as controller plant. The error signal, which is the difference between the set point and the output e (t) is used to drive the control action. The large proportional term gives fast rising time with large overshoot and oscillatory response. The addition of derivative term gives a slower response and less overshoot. The main function of integral term is to reduce the effect of steady state error. These three terms are amplified by multiplying with the individual gains and then sum of all three terms will be applied to the plant to adjust the process. The control equation used to control the process is given in Equation 14.

$$F = K_P e + K_D \dot{e} + K_I \int e\,dt \tag{14}$$

where Kp, Kd and Ki represent proportional, derivative and integral gains of the controller, respectively. The modified form of Equation 14, after introducing the error term is given below.

$$f_i = K_{Pi}\left(\theta_{if} - \theta_i\right) - K_{Di}\dot{\theta}_i + K_{Ii}\int e\left(\theta_i\right)dt \tag{15}$$

$$x_i = \int e\left(\theta_i\right)dt \Rightarrow \dot{x}_i = \theta_{if} - \theta_i \tag{16}$$

RESULTS AND DISCUSSIONS

In the present manuscript, the developed PID path tracking algorithm has been tested in simulations by considering a straight line (that is, x + y = 2) and parabolic paths (that is, x2 + y = 2) between assumed start and goal points. The parameters of the manipulator that are used in the simulations are given in Table 2.

The developed PID controller has been tuned with the help of Ziegler and Nicholas method to identify the values of proportional, integral and derivative gains. The values of gains obtained after the said tuning process is given in Table 3.

Once the developed PID controller is tuned, the performance of the controller in tracing straight line and parabolic paths are tested.

Figure 3 shows the variation of error for different joints of the spatial manipulator while tracking a linear path. It is important to note that the path to be traced is divided into ten equal intervals to have a

Table 2. Parameters used for spatial manipulator

S. No.	Link No.	Length (m)	Mass (kg)
1	1	1	1
2	2	1	1
3	3	1	1
4	4	1	1

Table 3. The gain values of the designed PID controller

JOINTS	K_p	K_i	K_d
1	100	35	90
2	90	35	75
3	80	30	65
4	70	30	55

Figure 3. Variation of error at different joints for tracking linear path, (a) joint 1, (b) joint 2, (c) joint 3, (d) joint 4

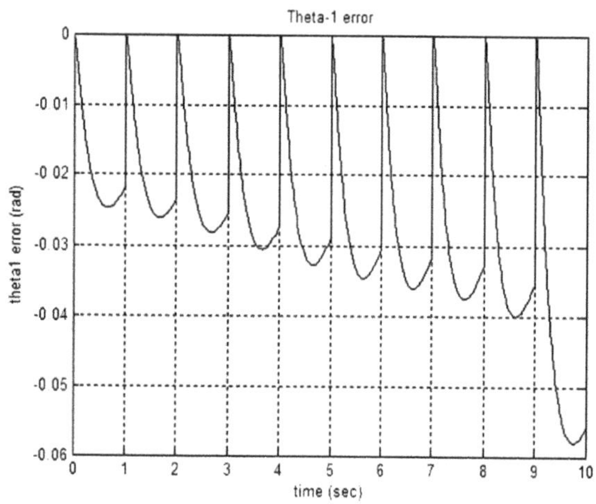

3 (a)

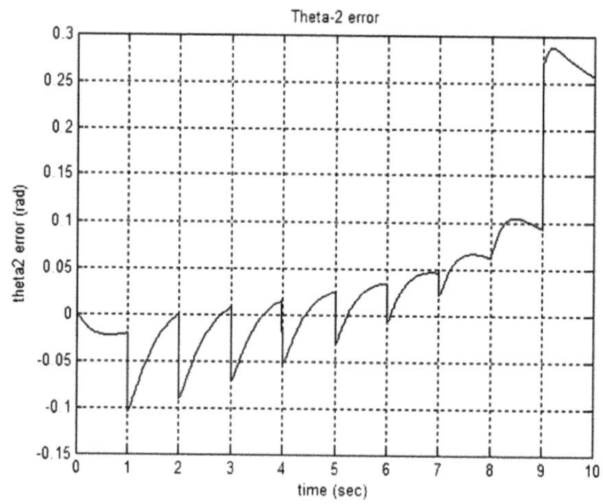

3 (b)

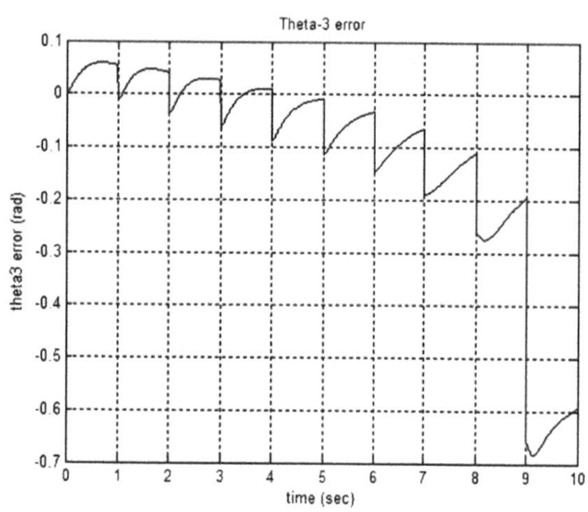

3 (c)

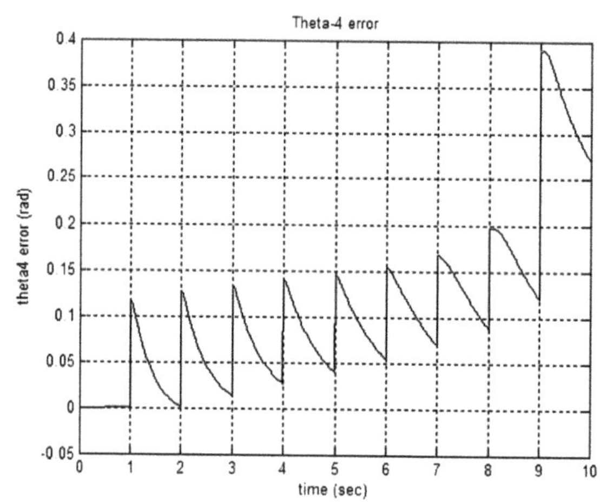

3 (d)

better control while tracing the decided path. From Figure 3 (a), (b), (c) and (d), it can be observed that the value of the error is oscillating on either sides of zero and converging towards zero in all the intervals of joints 1, 2, 3 and 4, respectively. This reveals the fact that the PID controller helped the manipulator reducing the error while tracking the path.

Figure 4 (a), (b), (c) and (d) show the variation of torques at joints 1, 2, 3 and 4 respectively. From the graphs, it can be observed that the peak torques for joint 1, 2, 3 and 4 are seen to be equal to 38, 14, -18, 11 N-m, respectively. From the data, it is interesting to note that the joint 1 has consumed more torque when compared with other joints. This may be due to the fact that joint 1 supports the rest of the links while moving from start to goal points.

Figure 4. Variation of torque at different joints for tracking linear path, (a) joint 1, (b) joint 2, (c) joint 3, (d) joint 4

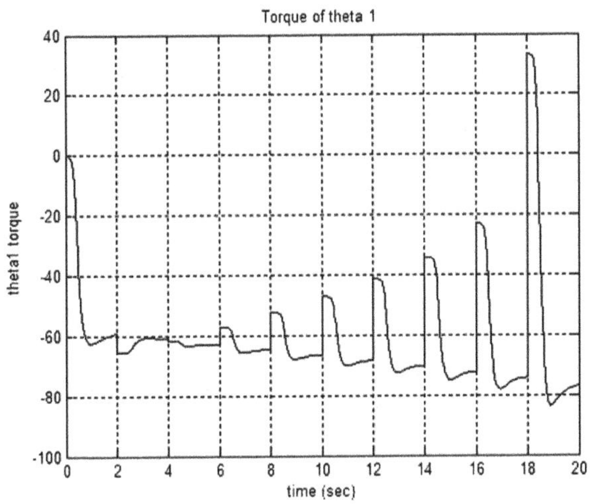

4 (a)

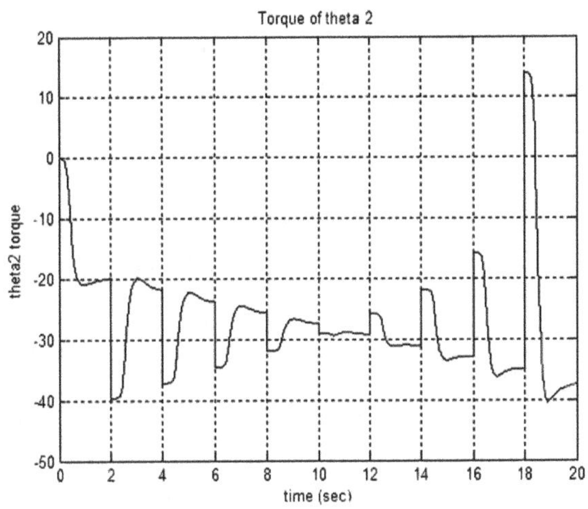

4 (b)

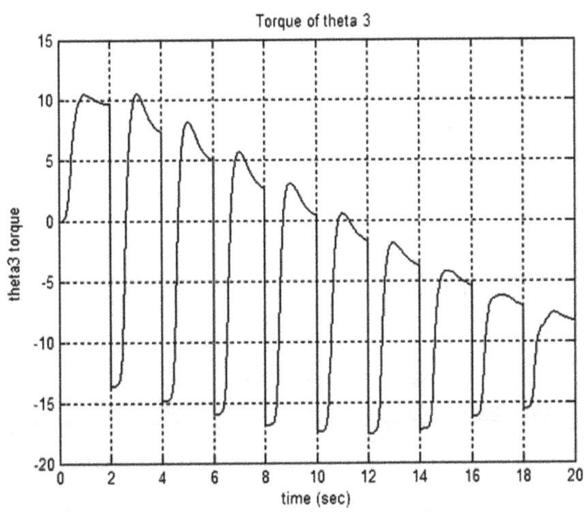

4 (c)

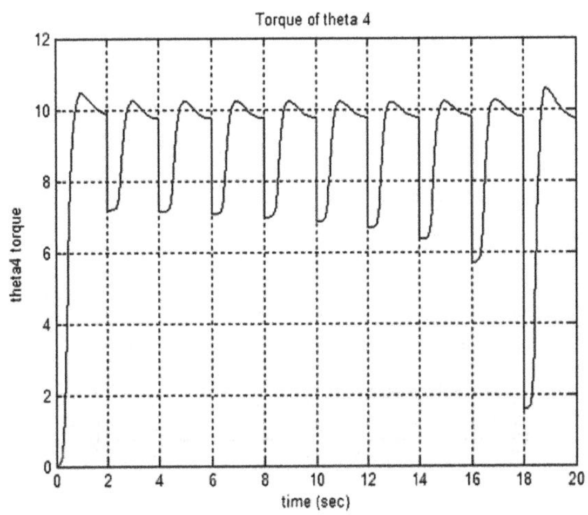

4 (d)

Figure 5 describes the error of various joints of the spatial manipulator while following a parabola path between start and goal points. Aiming to have a better control over the manipulator the path was segmented into ten equal parts. From the Figure 5 (a), (b), (c) and (d), it can be deduced that error is moving near to zero, which is reasonable in all the joints 1, 2, 3 and 4, respectively. Finally, it is observed that designed PID controller helped in reducing the error while following the parabolic path.

Figure 6 (a), (b), (c) and (d) describes the behavior of joint variable torque at joints 1, 2, 3 and 4, respectively. From the graphs it can be observed that the peak torques for joints 1, 2, 3 and 4 are seen to be equal to -82, -42, -24 and 11 N-m, respectively. From the data, the key point that should be noted

Figure 5. Variation of error at different joints for tracking parabolic path, (a) joint 1, (b) joint 2, (c) joint 3, (d) joint 4

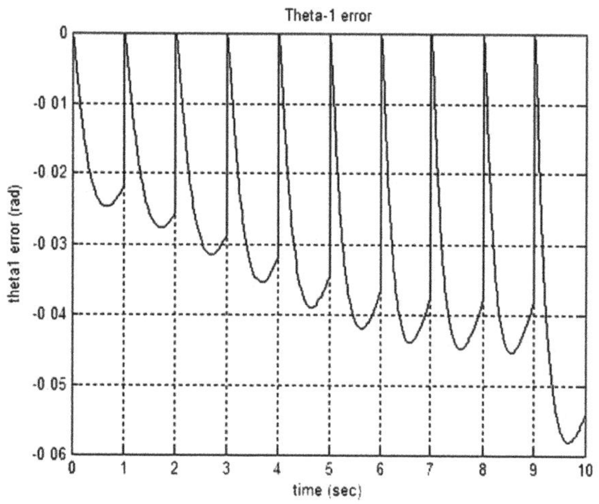

5 (a)

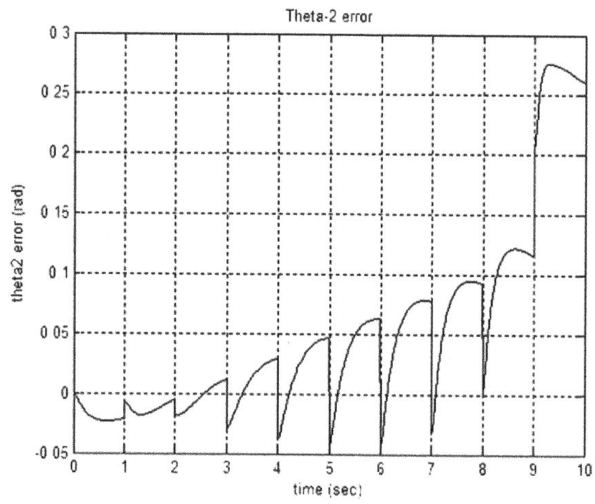

5 (b)

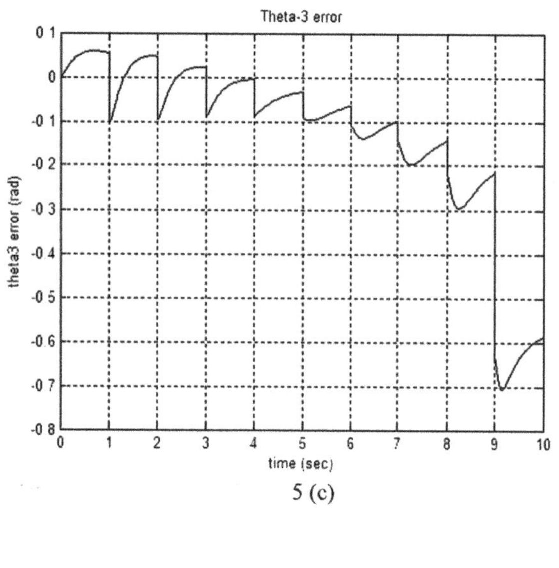

5 (c)

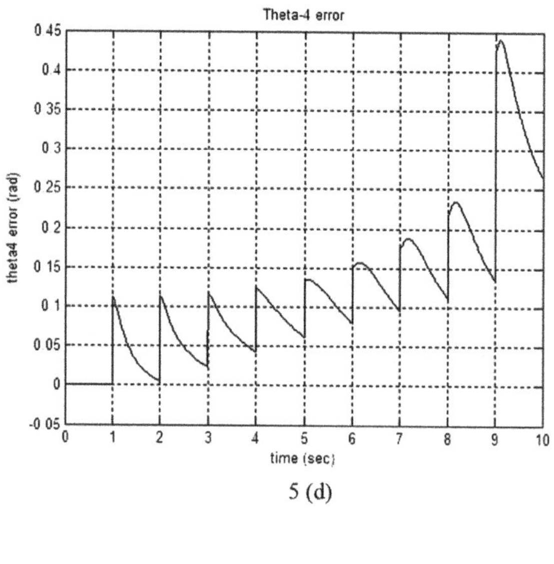

5 (d)

Figure 6. Variation of torque at joints for tracking parabolic path, (a) joint 1, (b) joint 2, (c) joint 3, (d) joint 4

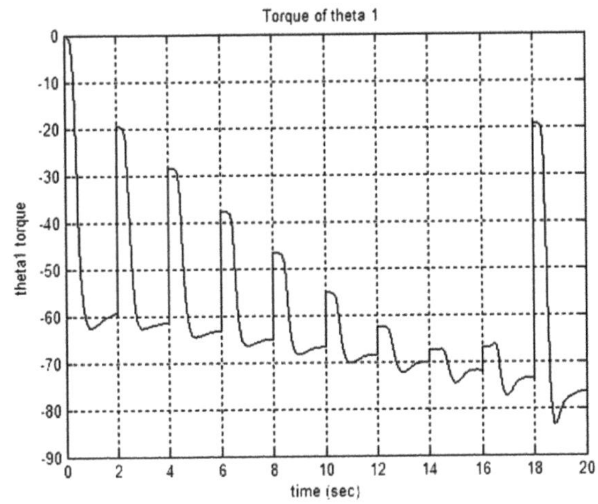

6 (a)

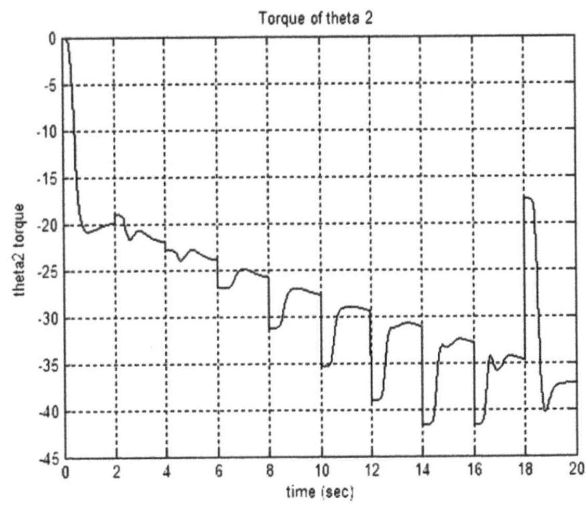

6 (b)

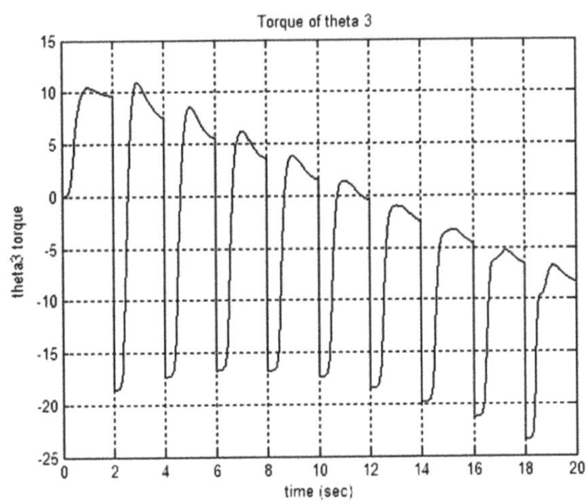

6 (c)

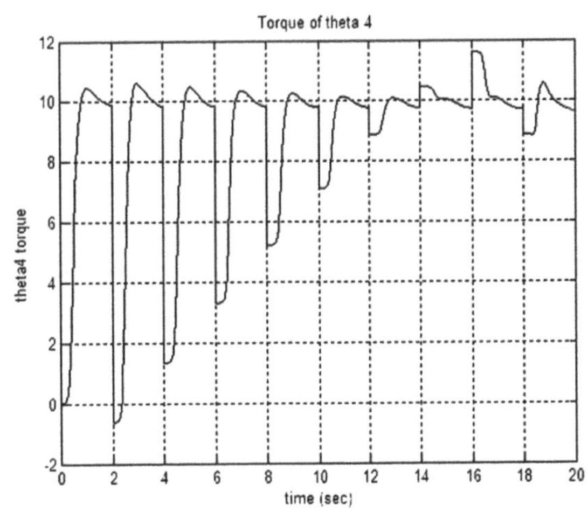

6 (d)

is, joint 1 has consumed more torque when compared with other joints, which can be explained as first link is used to move the remaining links from rest. The positive and negative sign shows the clockwise and anti-clockwise rotation of the motors, respectively.

Figure 7 (a) and (b) show the simulation of 4-DOF manipulator tracing linear and parabolic paths. From both the figures it can be observed that the developed PID controller had traced both linear and parabolic paths with reasonably good accuracy. However, the inaccuracies that are present while tracking the path using PID controller can be eliminated by turning the gains Kp, Ki and Kd of the controller after using the global optimization algorithms, namely genetic algorithm, particle swarm optimization, differential evolution, artificial bee colony algorithm and etc.

Figure 7. Simulations showing the path traced by 4-DOF spatial manipulator (a) linear path (b) parabolic path

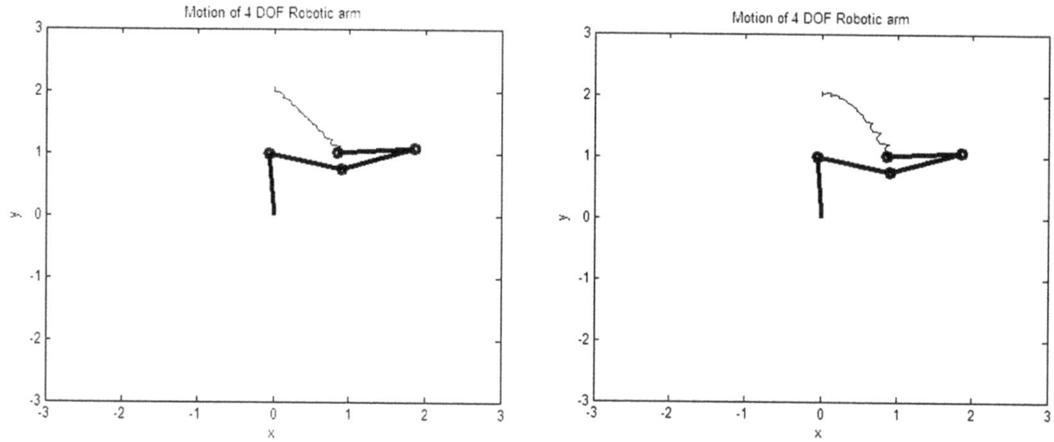

7 (a) 7 (b)

CONCLUSION

In this article, an attempt has been made to design a PID controller, that helps the 4-DOF spatial manipulator to track the desired path successfully. The included angles made by various links to track the path has been derived from pseudo-inverse approach. The developed algorithm has been tested in computer simulations to track a linear and parabolic paths. The results of the simulation shows that the 4-DOF manipulator has traced the paths with reasonably good accuracy. However, the path tracking capability of the developed PID controller can be enhanced by tuning the gains of the controller with the help of global optimization algorithms.

REFERENCES

Chevallereau, C., & Khalil, W. (1988). A new method for the solution of the inverse kinematics of redundant robots.*Proceedings. IEEE International Conference on Robotics and Automation*. IEEE Computer Society Press. doi:10.1109/ROBOT.1988.12020

Craig John, J. (1986). Introduction to Robotics: Mechanics and Control (3rd ed.). Addision-Wesley.

David, I., & Robles, G. (2012). *PID Control Dynamics of a Robotic Arm Manipulator with Two Degrees of Freedom*. Retrieved from http://www.slideshare.net/popochis/pid-control-dynamics-of-a-robotic-arm-manipulator-with-two-degrees-of-freedom

Deepak, A. G., & Badler, N. I. (2000). Real-Time Inverse Kinematics Techniques for Anthropomorphic Limbs. *Graphical Models*, *62*(5), 353–388. doi:10.1006/gmod.2000.0528 PMID:12143897

Dinesh, M., John, F., & Canny. (1992). Real Time Inverse Kinematics for General 6R Manipulators. In *Proceedings. IEEE International. Conference on Robotics and Automation* (pp. 383-389).

Dung, Kang, & Ro. (2010). Robot Manipulator Modeling in Matlab Simmechanics with PD Control and Online Gravity Compensation. *Proceedings of the IEEE.*

Faverjon, B., & Tournassoud, P. (1987). A local based approach for path planning of manipulators with a high number of degrees of freedom. *Robotics and Automation. Proceedings.IEEE International Conference* (*Vol. 4*). doi:10.1109/ROBOT.1987.1087982

Khatib, O. (1985). Real-time obstacle avoidance for manipulators and mobile robots. In *Proceedings. IEEE International. Conference on Robotics and Automation* (pp. 500-505). doi:10.1109/ROBOT.1985.1087247

Klein, C., & Huang, C. (1983). Review on Pseudo inverse Control for Use with kinematically re- dundant manipulators. *IEEE Transactions on Systems, Man, and Cybernetics*, *13*(3), 245–250. doi:10.1109/TSMC.1983.6313123

Latombe, J. C. (1991). *Robot Motion Planning*. Norwell, MA: Kluwer Academic Publishers. doi:10.1007/978-1-4615-4022-9

Mandava, R. K., & Vundavilli, P. R. (2015). Design of PID Controllers for 4-DOF Planar and Spatial Manipulators. *International Conference on Robotics, Automation, Control and Embedded Systems.* Hindustan University.

Qu, Z. (1994). Global Stability of Trajectory Tracking of Robot under PD Control. *Dynamics and Control*, *4*(1), 59–71. doi:10.1007/BF02115739

Sarkhel, Banerjee, & Hui. (2013). *Analysis and Control of a Six Link Serial Manipulator with Flexible Joints.* 142-Inacomm.

Volpe, R. A., & Khosla, P. K. (1990). Manipulator control with super quadric artificial potential functions: Theory and experiments. *IEEE Transactions on Systems, Man, and Cybernetics*, *20*(6), 1423–1436. doi:10.1109/21.61211

Whitney, D. (1969). Resolved motion rate control of manipulators and human prostheses. *IEEE Transactions on Man-Machine Systems*, *10*(2), 47–53. doi:10.1109/TMMS.1969.299896

Zhao, J., & Badler, N. (1994). Inverse kinematics positioning using nonlinear programming for highly articulated figures. *ACM Transactions on Computer Graphics*, *13*(4), 313–336. doi:10.1145/195826.195827

Ziegler, J. G., & Nichols, N. B. (1942). Optimum Setting for Automatic Controllers. *Trans. ASME*, *64*, 759–768.

Chapter 16
Efficient Planning:
In-Service Inspections of Fatigued Structures under Parametric Uncertainty

N. A. Nechval
Baltic International Academy, Latvia

K. N. Nechval
Transport and Telecommunication Institute, Latvia

ABSTRACT

The main aim of this chapter is to present more accurate stochastic fatigue models for solving the fatigue reliability problems, which are attractively simple and easy to apply in practice for situations where it is difficult to quantify the costs associated with inspections and undetected cracks. From an engineering standpoint the fatigue life of a structure consists of two periods: (i) crack initiation period, which starts with the first load cycle and ends when a technically detectable crack is presented, and (ii) crack propagation period, which starts with a technically detectable crack and ends when the remaining cross section can no longer withstand the loads applied and fails statically. Periodic inspections of aircraft, which are common practice in order to maintain their reliability above a desired minimum level, are based on the conditional reliability of the fatigued structure. During the period of crack initiation, when the parameters of the underlying lifetime distributions are not assumed to be known, for efficient in-service inspection planning (with decreasing intervals as alternative to constant intervals often used in practice for convenience in operation), the pivotal quantity averaging (PQA) approach is offered. During the period of crack propagation (when the damage tolerance situation is used), the approach, based on an innovative crack growth equation, to efficient in-service inspection planning (with decreasing intervals between sequential inspections) is proposed to construct more accurate reliability-based inspection strategy in this case. To illustrate the suggested approaches, numerical examples are given.

DOI: 10.4018/978-1-5225-1639-2.ch016

INTRODUCTION

Fatigue in engineering structures involves crack initiation and growth processes. Current life-prediction approaches for fatigue-sensitive aircraft structure components address both crack initiation and propagation lives. Typically, the components are designed to have a minimum low-cycle fatigue (LCF) crack-initiation life exceeding the total specified service life. The current life-prediction systems may be improved by development of better crack-initiation life models and probabilistic treatment of the variability of crack-initiation life. One of the limitations of current crack-initiation life models is that they are incapable of predicting the crack size at fatigue-crack initiation. Consequently, an initial crack size must be assumed in the prediction of crack-growth life.

In spite of decades of investigation, fatigue response of materials is yet to be fully understood. While most industrial failures involve fatigue, the assessment of the fatigue reliability of industrial components being subjected to various dynamic loading situations is one of the most difficult engineering problems. The traditional analytical method of engineering fracture mechanics (EFM) usually assumes that crack size, stress level, material property and crack growth rate, etc. are all deterministic values which will lead to conservative or very conservative outcomes. According to many experimental results and field data, even in well-controlled laboratory conditions, crack growth results usually show a considerable statistical variability.

In this chapter, sequential inspections are obtained to ensure that the conditional fatigue reliability is at the required level. Sequential inspections are important, especially during the set-up and installation stage. Frequent inspection leads to a high cost and infrequent inspection will lead to low fatigue reliability of the system upon demand. Although the cost might be an issue in this type of analysis, the focus here is to meet the fatigue reliability requirement with an appropriate time to next inspection. The sequential inspection procedure and decision making procedure studied in this chapter allows an appropriate level of fatigue reliability to be reached with minimum cost as well. Furthermore, we do not assume the distribution of system lifetime to be completely known which is usually the case. The information from the inspection can be used to determine the parameters of the underlying system lifetime distribution. Hence, such a combined estimation and decision-making analysis is important and useful in practice. The models are to allow both the understanding of physical phenomena and the practical application in engineering design with maintaining a clear physical interpretation of the stochastic crack growth.

Fatigue is one of the most important problems of aircraft arising from their nature as multiple-component structures, subjected to random dynamic loads. The analysis of fatigue crack growth is one of the most important tasks in the design and life prediction of aircraft fatigue-sensitive structures (for instance, wing, fuselage) and their components (for instance, aileron or balancing flap as part of the wing panel, stringer, etc.). From an engineering standpoint the fatigue life of a structure (or component) consists of two periods (this concept is shown schematically in Figure 1):

1. Crack initiation period, which starts with the first load cycle and ends when a technically detectable crack is present, and
2. Crack propagation period, which starts with a technically detectable crack and ends when the remaining cross section can no longer withstand the loads applied and fails statically.

Periodic inspections of aircraft are common practice in order to maintain their reliability above a desired minimum level.

Figure 1. Schematic fatigue crack growth curve (Crack initiation period (A-B); Crack propagation period (B-C))

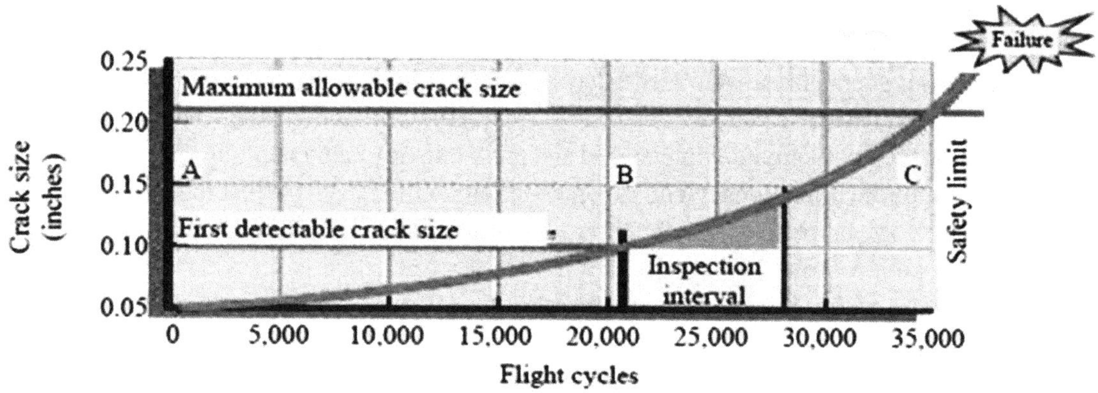

For guaranteeing safety, the structural life ceiling limits of the fleet aircraft are defined from three distinct situations: Safe-Life, Damage Tolerance, and Fail-Safe situations. The common objectives to define fleet aircraft lives by the three situations are to ensure safety while at the same time reducing total ownership costs. The Safe-Life situation is based on the concept that significant damage, i.e. fatigue cracking, will not develop during the service life of a structure (or component). The life is initially determined from fatigue test data (*S–N* curves) and calculations using a cumulative damage "law". Then the design Safe-Life is obtained by applying a safety factor. When the service life equals the design Safe-Life the component must be replaced. However, there are two major drawbacks to this situation:

1. Components are taken out of service even though they may have substantial remaining lives;
2. Despite all precautions, cracks sometimes occur prematurely. This latter fact led airlines to introduce the Damage Tolerance situation (Iyyer et al., 2007).

The Damage Tolerance situation recognizes that damage can occur and develop during the service life of a structure (or component). Also, it assumes that cracks or flaws can be present in new structures. Safety is obtained from this situation by the requirements that either:

1. Any damage will be detected by routine inspection before it results in a dangerous reduction of the static strength (inspectable components), or
2. Initial damage shall not grow to a dangerous size during the service life (non-inspectable components).

For Damage Tolerance analysis to be successful it must be possible to:

1. Define either a minimum crack length that will not go undetected during routine inspections, or else an initial crack length, nominally based on pre-service inspection capability;
2. Predict crack growth during the time until the next inspection or until the design service life is reached.

The Fail-Safe situation assumes initial damage as manufactured and its subsequent growth during service to detectable crack sizes or greater. Service life in fail-safe structures can thus be defined as the time to a service detectable damage. Inspection intervals are determined by using appropriate safety factors on calculated crack growth time interval from service detectable cracks to critical crack sizes. The prediction of crack growth is similar to that for Damage Tolerance situation, except that a much smaller initial crack length is used.

Many important fatigued structures (for instance, Transportation Systems and Vehicles: aircraft, space vehicles, trains, ships; Civil Structures: bridges, dams, tunnels; and so on) for which extremely high reliability is required are maintained by in-service inspections to prevent the reliability degradation due to fatigue damage. However, temporal transition of the reliability is significantly affected by the inspection strategy selected. Thus, to keep structures reliable against fatigue damage by inspections, it is clearly important in engineering to examine the optimal inspection strategy. In particular, it should be noticed that periodical inspections with predetermined constant intervals are not always effective, since a fatigue crack growth rate is gradually accelerated as fatigue damage grows, i.e. the intervals between inspections should be gradually smaller in order to restrain the reliability degradation by repeated inspections. Therefore, we need to construct the inspection strategy by paying attention to this case. Barlow et al. (1963) tackled this problem by assuming a known, fixed cost of making an inspection and a known fixed cost per unit time due to undetected failure. They then found a sequence of inspection times for which the expected cost is a minimum. Their results have been extended by various authors (Luss & Kander, 1974; Sengupta, 1977). Unfortunately, it is difficult to compute optimal checking procedures numerically, because the computations are repeated until the procedures are determined to the required degree by changing the first check time. To avoid this, Munford and Shahani (1972) suggested a sub-optimal (or nearly optimal) but computationally easier inspection policy. This policy was used for Weibull and gamma failure distribution cases (Munford & Shahani, 1973; Tadikamalla, 1979). Numerical comparisons among certain inspection policies are given by Munford (1981) for the case of Weibull failure times. This case under parametric uncertainty was considered by Nechval et al. (2008, 2009, 2011).

The pivotal quantity averaging (PQA) technique proposed in this chapter represents a simple and computationally attractive statistical method based on the constructive use of the invariance principle in mathematical statistics. This technique allows one to eliminate unknown parameters from the problem and to use the past lifetime data for planning future inspections as completely as possible. The unbiasedness and minimum variance criteria, which are proposed in the chapter for constructing inspection strategies under fatigue crack initiation such as the unbiased inspection strategy (UIS) and minimum variance inspection strategy (MVIS), respectively, represent the novelty of the work. It is clear that these inspection strategies, which have such properties as unbiasedness and minimum variance, are preferable as compared to the maximum likelihood inspection strategy (MLIS). We have illustrated the methods of in-service inspection planning for log-location-scale distributions (such as the Exponential, Gumbel or Weibull distribution). Applications to other distributions could follow directly. Under fatigue crack growth, the damage tolerance situation is considered. As a result of our investigations of the experimental data of fatigue crack growth, we have found that for planning in-service inspections of fatigued structures under crack propagation it can be used the approach based on a novel crack growth equation to construct more accurate reliability-based inspection strategy in this case. The new technique proposed for planning in-service inspections of fatigued structures under crack propagation requires a quantile of the t-distribution and is conceptually simple and easy to use.

PLANNING INSPECTIONS UNDER FATIGUE CRACK INITIATION

Complete Information

In this chapter we look at inspection strategies for items or structures that can be described as being in one of two states, one of which is preferable to the other. This preferred state might be described as 'working' whilst the other may represent some sort of 'fatigue damage'. The structures are originally known to be in a working state but may subsequently have fatigue damage. In other words, at $t_0 = 0$ the structure is in state $S(t_0)$ (working) but at a later time, t_1, the structure will move into state $S(t_1)$ (fatigue damage). We suppose that we do not know when the transition from $S(t_0)$ into $S(t_1)$ will occur, and that fatigue damage (crack) can only be detected through inspection. We deal with situations, where it is difficult to quantify the costs associated with inspections and undetected fatigue damage, or when these costs vary in time.

Inspection Strategy

The inspection strategy defined is based on the conditional reliability of the structure. It is given as follows. Fix $0 < \gamma < 1$ and let

$$\tau_1 = \arg\ (\Pr\{X > \tau_1\} = \gamma), \tag{1}$$

$$\tau_j = \arg\ (\Pr\{X > \tau_j \mid X > \tau_{j-1}\} = \gamma), \quad j \geq 2, \tag{2}$$

where $\{\tau_j\}_{j=1,2,\ldots}$ are inspection times, X is a random variable representing the lifetime of the component (structure). This is named as 'reliability-based inspection'. The above inspection strategy makes use of the information about the remaining life that is inherent in the sequence of previous inspection times. The value of γ can be seen as 'minimum fatigue reliability required' (or 'fatigue reliability index') during the next period when the structure was still operational at last inspection time, that is, in other words, the conditional probability that the failure (fatigue crack) occurs in the time interval (τ_{j-1},τ_j) without failure at time τ_{j-1} is always assumed $1-\gamma$. It is clear that if F_θ, the structure lifetime distribution with the parameter θ (in general, vector), is continuous and strictly increasing, the definition of the inspection strategy is equivalent to

$$\tau_j = \arg(\bar{F}_\theta(\tau_j) = \gamma^j), \quad j \geq 1, \tag{3}$$

or equivalent to

$$\tau_j = \arg\min_{\tau_j}\ [\bar{F}_\theta(\tau_j) - \gamma^j]^2, \quad j \geq 1, \tag{4}$$

where

$$\bar{F}_\theta(\tau_j) = 1 - F_\theta(\tau_j). \tag{5}$$

Optimization of Fatigue Reliability Index Under Complete Information

If it is known that each inspection costs c_1 and the cost of leaving an undetected failure (fatigue crack) is c_2 per unit time, then the total expected cost per inspection cycle is given by

$$C(\tau(\theta,\gamma)) = \sum_{j=1}^{\infty} \int_{\tau_{j-1}}^{\tau_j} [jc_1 + c_2(\tau_j - x)] f_\theta(x)dx$$

$$= c_1 \sum_{j=1}^{\infty} j\,[F_\theta(\tau_j) - F_\theta(\tau_{j-1})] + c_2 \sum_{j=1}^{\infty} \tau_j [F_\theta(\tau_j) - F_\theta(\tau_{j-1})] - c_2 \int_0^{\infty} x f_\theta(x)dx$$

$$= c_1 \sum_{j=0}^{\infty} \bar{F}_\theta(\tau_j) + c_2 \sum_{j=1}^{\infty} \tau_j [\bar{F}_\theta(\tau_{j-1}) - \bar{F}_\theta(\tau_j)] - c_2 E_\theta\{X\}$$

$$= c_1 \sum_{j=0}^{\infty} \gamma^j + c_2 \sum_{j=1}^{\infty} \varphi(\theta) j \ln\gamma [\gamma^{j-1} - \gamma^j] - c_2 E_\theta\{X\}$$

$$= \frac{c_1}{1-\gamma} + c_2 \varphi(\theta) \ln\gamma \left(\frac{1-\gamma}{\gamma}\right) \sum_{j=1}^{\infty} j\gamma^j - c_2 E_\theta\{X\}$$

$$= \frac{c_1}{1-\gamma} + c_2 \varphi(\theta) \ln\gamma \left(\frac{1-\gamma}{\gamma}\right) \frac{\gamma}{(1-\gamma)^2} - c_2 E_\theta\{X\}$$

$$= \frac{c_1}{1-\gamma} + \frac{c_2 \varphi(\theta) \ln\gamma}{1-\gamma} - c_2 E_\theta\{X\}, \tag{6}$$

where $f_\theta(x)$ is the probability density function of the structure lifetime X,

$$E_\theta\{X\} = \int_0^{\infty} x f_\theta(x)dx, \tag{7}$$

$$\tau_j = \arg\left(\ln \bar{F}_\theta(\tau_j) = \ln\gamma^j\right) = \varphi(\theta) j \ln\gamma. \tag{8}$$

Thus, we can choose the fatigue reliability index γ such that $C(\tau(\theta,\gamma))$ as defined in (6) is minimized. The appropriate value of γ is determined from $\partial C(\tau(\theta,\gamma))\} / \partial\gamma = 0,$ and we get the optimal value of the fatigue reliability index as

$$\gamma^* = \arg\left(\frac{1-\gamma}{\gamma} + \ln\gamma = -\frac{c_1}{c_2\varphi(\theta)}\right). \tag{9}$$

If $-\infty < \varphi(\theta) < 0$, it follows from (9) that the optimal value of the fatigue reliability index γ exists and is unique. To make sure that the optimal solution (9) represents a minimum of (6), not a maximum, the second derivative of the performance index (6) is calculated and evaluated at $\gamma = \gamma^*$:

$$\frac{\partial^2 C(\tau(\theta,\gamma))}{\partial\gamma^2} = c_2\varphi(\theta)\frac{1}{\gamma}\left(1-\frac{1}{\gamma}\right) > 0. \tag{10}$$

which is positive (i.e., (9) represents a minimum point of (6)), as desired. Thus, the following theorem has been proven.

Theorem 1. The fatigue reliability index γ (such that $C(\tau(\theta,\gamma))$ as defined in (6) is minimized) exists and is unique if and only if $\varphi(\theta)$ as defined in (8) satisfies the inequality: $-\infty < \varphi(\theta) < 0$.

Index of Improvement Percentage in Effectiveness of Inspection Strategy

The index of improvement percentage in effectiveness of the optimal inspection strategy (with $\gamma = \gamma^*$) as compared with the standard inspection strategy (with $\gamma = \gamma_{st}$) is given by

$$I_{\text{imp.per}}(\gamma^*, \gamma_{st}) = \frac{C(\tau(\theta,\gamma_{st}) - C(\tau(\theta,\gamma^*)}{C(\tau(\theta,\gamma_{st})}100\%. \tag{11}$$

Numerical Example 1

Let

$$X \sim f_\theta(x) = (1/\theta)\exp\left(-x/\theta\right), \quad x \ge 0, \quad \theta > 0; \quad c_1 = 1, \quad c_2 = 15, \quad \theta = 2000. \tag{12}$$

It follows from (9) and (12) that

$$\gamma^* = \arg\left(\frac{1-\gamma}{\gamma} + \ln\gamma = \frac{c_1}{c_2\theta}\right) = 0.991879. \tag{13}$$

It follows from (6) and (13) that

$$C(\tau(\theta,\gamma^*) = \frac{c_1}{1-\gamma^*} - \frac{c_2\theta \ln \gamma^*}{1-\gamma^*} - c_2\theta = \frac{c_1}{1-\gamma^*} - c_2\theta\left(\frac{\ln \gamma^*}{1-\gamma^*} + 1\right) = 245.6161. \quad (14)$$

If the standard fatigue reliability index $\gamma_{st} = 0.95$ is used, then

$$C(\tau(\theta,\gamma_{st}) = \frac{c_1}{1-\gamma^*} - c_2\theta\left(\frac{\ln \gamma^*}{1-\gamma^*} + 1\right) = 795.9766. \quad (15)$$

Index of Improvement Percentage in Effectiveness of Inspection Strategy

It follows from (11) that the index of improvement percentage in effectiveness of the optimal inspection strategy (with $\gamma = \gamma^* = 0.991879$) as compared with the standard inspection strategy (with $\gamma = \gamma_{st} = 0.95$) is given by

$$I_{\text{imp.per}}(\gamma^*,\gamma_{st}) = \frac{C(\tau(\theta,\gamma_{st}) - C(\tau(\theta,\gamma^*)}{C(\tau(\theta,\gamma_{st})}100\% = 69.1428\%. \quad (16)$$

Parametric Uncertainty

To construct the reliability-based inspection strategy under parametric uncertainty, the two criteria are proposed.

The first criterion, which takes into account (3) and the random sample of the past lifetime data (X_1, …, X_n) of the components of the same type, allows one to construct the inspection strategy given by

$$\tau_j^{\text{unb}} = \arg(E_\theta\{\bar{F}_\theta(\tau(S))\} = \gamma^j), \quad j \ge 1, \quad (17)$$

where S represents some sample statistic (say, either the maximum likelihood estimator of θ or sufficient statistic for θ). This criterion is named as ‘unbiasedness criterion’.

The second criterion, which takes into account (4) and the random sample of the past lifetime data (X_1, …, X_n) of the components of the same type, allows one to construct the inspection strategy given by

$$\tau_j^{\text{mv}} = \arg\min_{\tau(S)} E_\theta\{[\bar{F}_\theta(\tau(S)) - \gamma^j]^2\}, \quad j \ge 1, \quad (18)$$

This criterion is named as ‘minimum variance criterion’.

It will be noted that in practice, under parametric uncertainty, the criterion,

$$\tau_j^{\text{ml}} = \arg\ (\bar{F}_{\hat{\theta}}(\tau) = \gamma^j), \quad j \ge 1, \quad (19)$$

is usually used, where $\hat{\theta}$ is the maximum likelihood estimate of θ. This criterion is named as 'maximum likelihood criterion'.

When there is a fatigue reliability requirement, the problem is usually to develop an inspection strategy that meets the reliability requirements. It is assumed that only the functional form of the underlying invariant distribution of time to crack detection is specified, but some or all of its parameters are unspecified. The pivotal quantity averaging (PQA) approach proposed in this chapter allows one to construct an optimal inspection strategy under parametric uncertainty.

Inspection Strategies

Theorem 2. Let $X_1, \ldots, X_n$ be the random sample of the past independent observations of time to crack detection from the fatigued structures (components) of the same type, which follow the exponential distribution with the probability density function

$$f_\theta(x) = (1/\theta)\exp\left(-x/\theta\right), \quad x \geq 0, \quad \theta > 0, \tag{20}$$

where the parameter θ is unknown. Then the reliability-based inspection strategies for a new fatigued component of the same type are given as follows.

The unbiased inspection strategy:

$$\tau_j^{\text{unb}} = [\gamma^{-j/n} - 1]\, S, \quad j \geq 1. \tag{21}$$

The minimum variance inspection strategy:

$$\tau_j^{\text{mv}} = \frac{1 - \gamma^{\frac{j}{n+1}}}{2\gamma^{\frac{j}{n+1}} - 1} S, \quad j \geq 1, \tag{22}$$

where $S = \sum_{i=1}^{n} X_i$ is the sufficient statistic for θ.

The maximum likelihood inspection strategy:

$$\tau_j^{\text{ml}} = j \ln \gamma^{-1} \frac{S}{n}, \quad j \geq 1. \tag{23}$$

Proof. Using the pivotal quantity averaging approach (Nechval et al. 2008a, 2011a), the unbiased inspection strategy can be obtained from (17) as

$$\tau_j^{\text{unb}} = \arg[E_\theta\{\bar{F}_\theta(\tau(S))\} = \gamma^j] = \arg\left[E_\theta\left\{\exp\left(-\frac{\tau_j}{S}\frac{S}{\theta}\right)\right\} = \gamma^j\right]$$

$$= \arg\left[E\left\{\exp\left(-\rho_j V\right)\right\} = \gamma^j\right] = \arg\left[\frac{1}{(1+\rho_j)^n} = \gamma^j\right] = [\gamma^{-j/n} - 1]\; S, \quad j \geq 1, \tag{24}$$

where the ancillary factor ρ_j is given by

$$\rho_j = \tau_j \;/\; S, \tag{25}$$

the pivotal quantity V is given by

$$V = S \;/\; \theta \sim f(v) = \frac{1}{\Gamma(n)} v^{n-1} \exp(-v), \quad v \geq 0. \tag{26}$$

The minimum variance inspection strategy is obtained from (18) as

$$\tau_j^{\text{mv}} = \arg\min_{\tau(S)} \; E_\theta\{[\bar{F}_\theta(\tau(S)) \;-\gamma^j]^2\} = \arg\min_{\tau(S)} \; E_\theta\left\{[\exp\left(-\tau(S)\;/\;\theta\right) \;-\gamma^j]^2\right\}$$

$$= \arg\min_{\rho_j} \; E\{\exp(-2\rho_j V) - 2\exp(-\rho_j V)\;\gamma^j + \gamma^{2j}\} = \arg\min_{\rho_j} \left(\frac{1}{(1+2\rho_j)^n} - \frac{2\gamma^j}{(1+\rho_j)^n} + \gamma^{2j}\right)$$

$$= [(1-\gamma^{j/(n+1)})/(2\gamma^{j/(n+1)} - 1)]S, \quad j \geq 1. \tag{27}$$

The maximum likelihood inspection strategy follows immediately from (19):

$$\tau_j^{\text{ml}} = \arg(\bar{F}_{\hat{\theta}}(\tau) = \gamma^j) = \arg\left[\exp\left(-\frac{\tau}{\hat{\theta}}\right) = \gamma^j\right] = j \ln \gamma^{-1} \frac{S}{n}, \quad j \geq 1, \tag{28}$$

where the maximum likelihood estimator of θ is $\hat{\theta} = S \;/\; n.$ This ends the proof.

Theorem 3. If (under conditions of Theorem 2) the unbiased inspection strategy is used, where each inspection costs c_1 and the cost of leaving an undetected failure (fatigue crack) is c_2 per unit time, then the fatigue reliability index γ minimizing the total expected cost per inspection cycle is given by

$$\gamma_{\text{unb}}^* == \arg\min_\gamma \left(\frac{c_1 n}{S} \frac{1}{1-\gamma^{(n+1)/n}} + c_2 n \frac{\gamma^{-1/n} - 1}{1-\gamma} - c_2\right). \tag{29}$$

Proof. It follows from (20) and (26) that $C(\tau(\theta, \gamma))$ can be transformed as

$$C(\tau(\theta,\gamma)) = c_1\sum_{j=0}^{\infty}\bar{F}_\theta(\tau_j) + c_2\sum_{j=1}^{\infty}\tau_j[\bar{F}_\theta(\tau_{j-1}) - \bar{F}_\theta(\tau_j)] - c_2E_\theta\{X\}$$

$$= c_1\sum_{j=0}^{\infty}\exp\left(-\frac{\tau_j}{\theta}\right) + c_2\sum_{j=1}^{\infty}\tau_j\left[\exp\left(-\frac{\tau_{j-1}}{\theta}\right) - \exp\left(-\frac{\tau_j}{\theta}\right)\right] - c_2\theta$$

$$= \theta\left(c_1\sum_{j=0}^{\infty}\frac{1}{S}V\exp\left(-\frac{\tau_j}{S}V\right) + c_2\sum_{j=1}^{\infty}\tau_j\left[\frac{1}{S}V\exp\left(-\frac{\tau_{j-1}}{S}V\right) - \frac{1}{S}V\exp\left(-\frac{\tau_j}{S}V\right)\right] - c_2\right)$$

$$= \theta C(\tau(S,\gamma),V). \tag{30}$$

Using the pivotal quantity averaging approach, it follows from (30) that

$$E\{C(\tau(S,\gamma),V\}$$

$$= E\left\{\left(c_1\sum_{j=0}^{\infty}\frac{1}{S}V\exp\left(-\frac{\tau_j}{S}V\right) + c_2\sum_{j=1}^{\infty}\tau_j\left[\frac{1}{S}V\exp\left(-\frac{\tau_{j-1}}{S}V\right) - \frac{1}{S}V\exp\left(-\frac{\tau_j}{S}V\right)\right] - c_2\right)\right\}$$

$$= \frac{c_1}{S}\sum_{j=0}^{\infty}n\left(1+\frac{\tau_j}{S}\right)^{-1} + \frac{c_2}{S}\sum_{j=1}^{\infty}\tau_j\left[n\left(1+\frac{\tau_{j-1}}{S}\right)^{-1} - n\left(1+\frac{\tau_j}{S}\right)^{-1}\right] - c_2 = C(\tau(S,\gamma)). \tag{31}$$

It follows from (24) and (31) that

$$C(\tau^{\text{unb}}(S,\gamma)) = \frac{c_1}{S}\sum_{j=0}^{\infty}n\left(1+\frac{\tau_j^{\text{unb}}}{S}\right)^{-1} + \frac{c_2}{S}\sum_{j=1}^{\infty}\tau_j^{\text{unb}}\left[n\left(1+\frac{\tau_{j-1}^{\text{unb}}}{S}\right)^{-1} - n\left(1+\frac{\tau_j^{\text{unb}}}{S}\right)^{-1}\right] - c_2$$

$$= \frac{c_1 n}{S}\sum_{j=0}^{\infty}\gamma^{j(n+1)/n} + c_2 n\sum_{j=1}^{\infty}\left(\gamma^{-j/n}-1\right)\left[\gamma^{(j-1)(n+1)/n} - \gamma^{j(n+1)/n}\right] - c_2$$

$$= \frac{c_1 n}{S}\frac{1}{1-\gamma^{(n+1)/n}} + c_2 n\sum_{j=1}^{\infty}\left(\gamma^{-j/n}-1\right)\left[\gamma^{(j-1)/n} - \gamma^{j/n}\right] - c_2$$

$$= \frac{c_1 n}{S} \frac{1}{1-\gamma^{(n+1)/n}} + c_2 n \sum_{j=1}^{\infty} \left(\gamma^{j-1}\gamma^{-1/n} - \gamma^{(j-1)(n+1)/n}\right)\left[1-\gamma^{(n+1)/n}\right] - c_2$$

$$= \frac{c_1 n}{S} \frac{1}{1-\gamma^{(n+1)/n}} + c_2 n \frac{\gamma^{-1/n}-1}{1-\gamma} - c_2. \tag{32}$$

Thus, we can choose the fatigue reliability index γ such that $C(\tau^{\mathrm{unb}}(S,\gamma))$ as defined in (32) is minimized. The optimal value of the fatigue reliability index γ is determined as

$$\gamma_{\mathrm{unb}}^{*} = \arg\min_{\gamma} C(\tau^{\mathrm{unb}}(S,\gamma))$$

$$= \arg\min_{\gamma} \left(\frac{c_1 n}{S} \frac{1}{1-\gamma^{(n+1)/n}} + c_2 n \frac{\gamma^{-1/n}-1}{1-\gamma} - c_2 \right). \tag{33}$$

This ends the proof.

Index of Improvement Percentage in Effectiveness of Inspection Strategy

The index of improvement percentage in effectiveness of the optimal unbiased inspection strategy (with $\gamma = \gamma_{\mathrm{unb}}^{*}$) as compared with the standard unbiased inspection strategy (with $\gamma = \gamma_{\mathrm{st}}$) is given by

$$I_{\mathrm{imp.per}}(\gamma_{\mathrm{unb}}^{*}, \gamma_{\mathrm{st}}) = \frac{C(\tau(S,\gamma_{\mathrm{st}}) - C(\tau(S,\gamma_{\mathrm{unb}}^{*})}{C(\tau(S,\gamma_{\mathrm{st}})} 100\%. \tag{34}$$

Numerical Example 2

Let $X_1, \ldots, X_n$ be the random sample of the past independent observations of time to crack detection (from the same fatigued structures) which follow the exponential distribution (20), where $n = 2$ and the parameter θ is unknown. The sufficient statistic for θ is $S = 4000$ hours. In order to construct the reliability-based inspection strategy for a new fatigued structure of the same type, the unbiased inspection strategy (21) will be used. Let us assume that each inspection costs c_1=1 (in terms of money) and the cost of leaving an undetected failure (fatigue crack) is c_2=15 (in terms of money) per unit time. Then it follows from (33) that

$$\gamma_{\mathrm{unb}}^{*} == \arg\min_{\gamma} \left(\frac{c_1 n}{S} \frac{1}{1-\gamma^{(n+1)/n}} + c_2 n \frac{\gamma^{-1/n}-1}{1-\gamma} - c_2 \right) = 0.994581. \tag{35}$$

It follows from (32) and (33) that

$$C(\tau^{\text{unb}}(S,\gamma_{\text{unb}}^*)=\frac{c_1 n}{S}\frac{1}{1-(\gamma_{\text{unb}}^*)^{(n+1)/n}}+c_2 n\frac{(\gamma_{\text{unb}}^*)^{-1/n}-1}{1-\gamma_{\text{unb}}^*}-c_2=0.122836. \quad (36)$$

If the standard fatigue reliability index $\gamma_{\text{st}}=0.95$ is used, then

$$C(\tau^{\text{unb}}(S,\gamma_{\text{st}})=\frac{c_1 n}{S}\frac{1}{1-(\gamma_{\text{st}})^{(n+1)/n}}+c_2 n\frac{(\gamma_{\text{st}})^{-1/n}-1}{1-\gamma_{\text{st}}}-c_2=0.593763. \quad (37)$$

Index of Improvement Percentage in Effectiveness of Inspection Strategy

It follows from (34) that the index of improvement percentage in effectiveness of the optimal unbiased inspection strategy (with $\gamma=\gamma^*=0.994581$) as compared with the standard unbiased inspection strategy (with $\gamma=\gamma_{\text{st}}=0.95$) is given by

$$I_{\text{imp.per}}(\gamma_{\text{unb}}^*,\gamma_{\text{st}})=\frac{C(\tau(S,\gamma_{\text{st}})-C(\tau(S,\gamma_{\text{unb}}^*)}{C(\tau(S,\gamma_{\text{st}})}100\%=79.3123\%. \quad (38)$$

Numerical Example 3

Consider, for example, the data of fatigue tests on a particular type of structural components (stringer) of aircraft IL-86. The data are for a complete sample of size n = 5, with observations of time to crack initiation (in number of 10^4 flight hours): X_1=5, X_2=6.25, X_3=7.5, X_4=7.9, X_5=8.1.

Goodness-of-fit Testing. It is assumed that X_i, i=1(1)5, follow the two-parameter Weibull distribution with the probability distribution function

$$\Pr\{X\le x\}=1-\exp\left[-\left(\frac{x}{\beta}\right)^{\delta}\right], x\ge 0, \quad (39)$$

where the parameters β and δ are unknown. We assess the statistical significance of departures from the Weibull model by performing empirical distribution function goodness-of-fit test. The K statistic (Kapur & Lamberson, 1977) is used. For censoring (or complete) datasets, the K statistic is given by

$$K=\frac{\sum_{i=[r/2]+1}^{r-1}\left(\frac{\ln(x_{i+1}/x_i)}{M_i}\right)}{\sum_{i=1}^{r-1}\left(\frac{\ln(x_{i+1}/x_i)}{M_i}\right)}=\frac{\sum_{i=3}^{4}\left(\frac{\ln(x_{i+1}/x_i)}{M_i}\right)}{\sum_{i=1}^{4}\left(\frac{\ln(x_{i+1}/x_i)}{M_i}\right)}=0.184, \quad (40)$$

where $[r/2]$ is a largest integer $\le r/2$, the values of M_i are given in Table 13 (Kapur & Lamberson, 1977). The rejection region for the α level of significance is $\{K > K_{n;\alpha}\}$. The percentage points for $K_{n;\alpha}$ were given by Kapur and Lamberson (1977). For this example,

$$K = 0.184 < K_{n=5;\alpha=0.05} = 0.86. \tag{41}$$

Thus, there is not evidence to rule out the Weibull model.

Theorem 4. Let $X_1 \le ... \le X_r$ be the first r ordered past observations of time to crack detection from the n fatigued structures of the same type, which follow the Weibull distribution with the probability density function

$$f_\theta(x) = \frac{\delta}{\beta}\left(\frac{x}{\beta}\right)^{\delta-1} \exp\left[-\left(\frac{x}{\beta}\right)^{\delta}\right], x \ge 0, \tag{42}$$

where $\theta = (\beta, \delta)$, $\beta>0$ is the scale parameter and $\delta>0$ is the shape parameter of the distribution. Then the unbiased reliability-based strategy of inspections of a new fatigued structure of the same type is given as follows:

$$\tau_j = \arg[\Pr\{Y_1 > \tau_j \mid \mathbf{w}\} = \gamma^j]$$

$$= \arg\left[\frac{\int_0^\infty v_2^{r-2}\prod_{i=1}^{r} w_i^{v_2}\left(w_{\tau_j}^{v_2} + \sum_{i=1}^{r} w_i^{v_2} + (n-r)w_r^{v_2}\right)^{-r} dv_2}{\int_0^\infty v_2^{r-2}\prod_{i=1}^{r} w_i^{v_2}\left(\sum_{i=1}^{r} w_i^{v_2} + (n-r)w_r^{v_2}\right)^{-r} dv_2} = \gamma^j\right], j \ge 1, \tag{43}$$

where

$$V_2 = \frac{\delta}{\widehat{\delta}}, \mathbf{w} = (w_1, w_2, ..., w_r), W_i = \left(\frac{X_i}{\widehat{\beta}}\right)^{\widehat{\delta}}, \quad i = 1, ..., r, w_{\tau_j} = \left(\frac{\tau_j}{\widehat{\beta}}\right)^{\widehat{\delta}}, \tag{44}$$

$\widehat{\beta}$ and $\widehat{\delta}$ are the maximum likelihood estimators of β and δ based on the first r ordered past observations ($X_1 \le ... \le X_r$) from a sample of size n from the two-parameter Weibull distribution (39), which can be found from solution of

$$\widehat{\beta} = \left(\left[\sum_{i=1}^{r} x_i^{\widehat{\delta}} + (n-r)x_r^{\widehat{\delta}}\right]\Big/ r\right)^{1/\widehat{\delta}}, \tag{45}$$

and

$$\hat{\delta} = \left[\left(\sum_{i=1}^{r} x_i^{\hat{\delta}} \ln x_i + (n-r) x_r^{\hat{\delta}} \ln x_r \right) \left(\sum_{i=1}^{r} x_i^{\hat{\delta}} + (n-r) x_r^{\hat{\delta}} \right)^{-1} - \frac{1}{r} \sum_{i=1}^{r} \ln x_i \right]^{-1}. \tag{46}$$

Proof. The proof is similar to that of Theorem 2 and so it is omitted here.

Inspection planning under fatigue crack initiation. Using (43), for $\gamma = 0.95$, we have obtained the following inspection time sequence (see Table 1). Graphical representation of inspection intervals is shown in Figure 2.

PLANNING INSPECTIONS UNDER FATIGUE CRACK PROPAGATION (GROWTH)

To capture the statistical nature of fatigue crack growth, different stochastic models have been proposed in the literature. Some of the models are purely based on direct curve fitting of the random crack growth data, including their mean value and standard deviation (Bogdanoff & Kozin, 1985). These models, however, have been criticized by other researchers, because less crack growth mechanisms have been included in them. To overcome this difficulty, many probabilistic models adopted the crack growth equations proposed by fatigue experimentalists, and randomized the equations by including random factors into

Table 1. Inspection time sequence under fatigue crack initiation

Inspection Time (In Terms Of Flight Hours) τ_j, j=0(1)9										
τ_0	τ_1	τ_2	τ_3	τ_4	τ_5	τ_6	τ_7	τ_8	τ_9	...
0	25549	32569	36975	40212	42775	44898	46708	48287	49685	...
Interval (In Terms Of Flight Hours) $\tau_{j+1}-\tau_j$, j=0(1)8										
–	25549	7020	4406	3237	2563	2123	1810	1579	1398	...

Figure 2. Graphical representation of inspection intervals under fatigue crack initiation

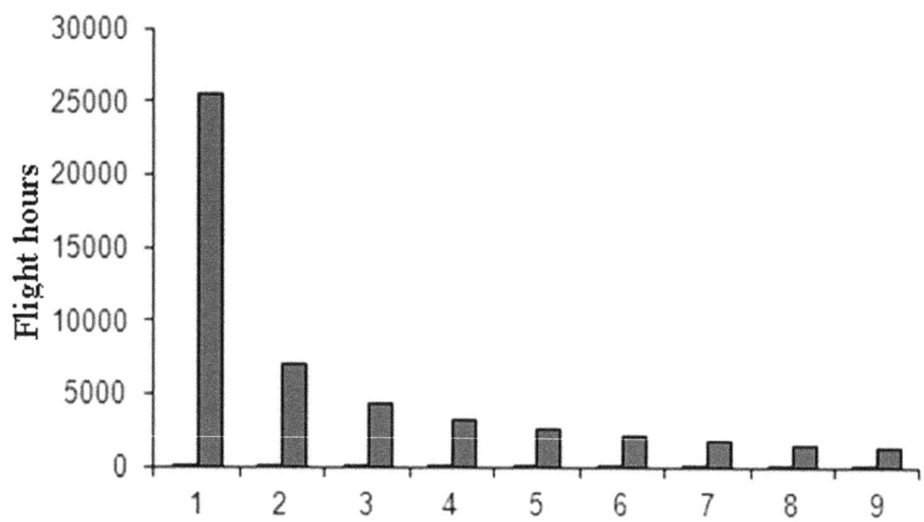

them (Lin & Yang, 1985; Yang et al., 1985; Yang & Manning, 1990; Nechval et al., 2003, 2004; Straub & Faber, 2005). The random factor may be a random variable, a random process of time, or a random process of space. It then creates a random differential equation. The solution of the differential equation reveals the probabilistic nature as well as the scatter phenomenon of the fatigue crack growth. To justify the applicability of the probabilistic models mentioned above, fatigue crack growth data are needed. However, it is rather time-consuming to carry out experiments to obtain a set of statistical meaningful fatigue crack growth data. To the writers' knowledge, there are only a few data sets available so far for researchers to verify their probabilistic models. Among them, the most famous data set perhaps is the one produced by Virkler et al. (1979). More frequently used data sets include one reported by Ghonem and Dore (1987). Itagaki and his associates have also produced some statistically meaningful fatigue crack growth data, but have not been mentioned very often (Itagaki et al., 1993). In fact, many probabilistic fatigue crack growth models are either lack of experimental verification or just verified by only one of the above data sets. It is suspected that a model may explain a data set well but fail to explain another data set. The universal applicability of many probabilistic models still needs to be checked carefully by other available data sets.

Many probabilistic models of fatigue crack growth are based on the deterministic crack growth equations. The most well known equation is

$$\frac{da(t)}{dt} = q(a(t))^b \tag{47}$$

in which q and b are constants to be evaluated from the crack growth observations. The independent variable t can be interpreted as stress cycles, flight hours, or flights depending on the applications. It is noted that the power-law form of $q(a(t))^b$ at the right hand side of (47) can be used to fit some fatigue crack growth data appropriately and is also compatible with the concept of Paris–Erdogan law (Paris & Erdogan, 1963).

Innovative Stochastic Model of Fatigue Crack Propagation

This model of fatigue crack propagation (growth) is based on the following deterministic crack growth equation,

$$\frac{dt(a)}{da} = \frac{q_1 q_0}{a^2} \exp\left(-\frac{q_0}{a}\right), \tag{48}$$

in which q_1 and q_0 are constants to be evaluated from the crack growth observations. The variable t can be interpreted as stress cycles, flight hours, or flights depending on the applications. The service time for a crack to grow from size a_0 to a (where $a > a_0$) can be found by performing the necessary integration

$$\int_{t(a_0)}^{t(a)} dt = q_1 \int_{a_0}^{a} \frac{q_0}{v^2} \exp\left(-\frac{q_0}{v}\right) dv \tag{49}$$

to obtain

$$t(a) - t(a_0) = q_1 \left[\exp\left(-\frac{q_0}{a} \right) - \exp\left(-\frac{q_0}{a_0} \right) \right]. \tag{50}$$

If a_0=0, then it follows from (50) that

$$t(a) = q_1 \exp\left(-\frac{q_0}{a} \right) \tag{51}$$

or

$$a \ln t(a) = \beta_0 + \beta_1 a, \tag{52}$$

where $\beta_1 = \ln q_1$, $\beta_0 = -q_0$. Including a stochastic factor ε (say, $\varepsilon \sim N(0,\sigma^2)$) into (52), we obtain the innovative stochastic model of fatigue crack propagation

$$a \ln t(a) = \beta_0 + \beta_1 a + \varepsilon. \tag{53}$$

New Equation for Planning In-Service Inspections Under Fatigue Crack Propagation

Let us assume that we have a sample of k data points consisting of pairs of observed values of a and τ, say (τ_1, a_1), (τ_2, a_2), …, (τ_k, a_k), where $k > 2$, τ_j is the time of the jth inspection, a_j is the crack size detected by means of the jth inspection, j=1(1)k. Then, taking into account (53), it can be shown that the time τ_{k+1} of the next inspection is determined as

$$\tau_{k+1} = \exp\left(\frac{h_{k+1}^{(1-\alpha)}}{\tilde{a}_{k+1}} \right), \tag{54}$$

where

$$\tilde{a}_{k+1} = a_k + \Delta_k \tag{55}$$

represents an assumed value of future crack size a_{k+1} (not yet observed), $\Delta_k = \tilde{a}_{k+1} - a_k$ represents an increment of a_k,

$$h_{k+1}^{(1-\alpha)} = \widehat{\beta}_0 + \widehat{\beta}_1\tilde{a}_{k+1} - t_{k-2;1-\alpha}\widehat{\sigma}\left[1 + \frac{1}{k} + \frac{\left(\tilde{a}_{k+1} - \frac{1}{k}\sum_{j=1}^{k} a_j\right)^2}{(k-1)\left[\sum_{j=1}^{k} a_j^2 - \frac{1}{k}\left(\sum_{j=1}^{k} a_j\right)^2\right]}\right]^{1/2} \tag{56}$$

represents the 100(1−α)% lower prediction limit for a single future value of $a_{k+1}\ln\tau_{k+1}$, $t_{n-2;1-\alpha}$ denotes the (1−α) quantile of the *t*-distribution with (n – 2) degrees of freedom,

$$\widehat{\beta}_0 = \frac{1}{k}\sum_{j=1}^{k} a_j \ln\tau_j - \widehat{\beta}_1\frac{1}{k}\sum_{j=1}^{k} a_j,\ \widehat{\beta}_1 = \frac{\sum_{j=1}^{k} a_j^2 \ln\tau_j - \frac{1}{k}\left(\sum_{j=1}^{k} a_j\right)\left(\sum_{j=1}^{k} a_j \ln\tau_j\right)}{\sum_{j=1}^{k} a_j^2 - \frac{1}{k}\left(\sum_{j=1}^{k} a_j\right)^2}, \tag{57}$$

$$\widehat{\sigma} = \left[\frac{1}{k-2}\sum_{j=1}^{k}(a_j \ln\tau_j - \widehat{\beta}_0 - \widehat{\beta}_1 a_j)^2\right]^{1/2}. \tag{58}$$

Numerical Example 4

For illustration, the procedure of inspection planning based on the innovative model (53) was used for the upper longeron of RNLAF F-16 aircraft (Figure 3).

Figure 4 shows the deterministic damage tolerance inspection requirements for the RNLAF longerons Source: Military Specification, 1974

These requirements led to the following inspection scheme which includes $N_{\text{const.int.}}^{\text{inspections.}} = 42$ inspections:

Figure 3. Inspection points of the upper longeron of RNLAF F-16 aircraft
Source: Military Specification, 1974

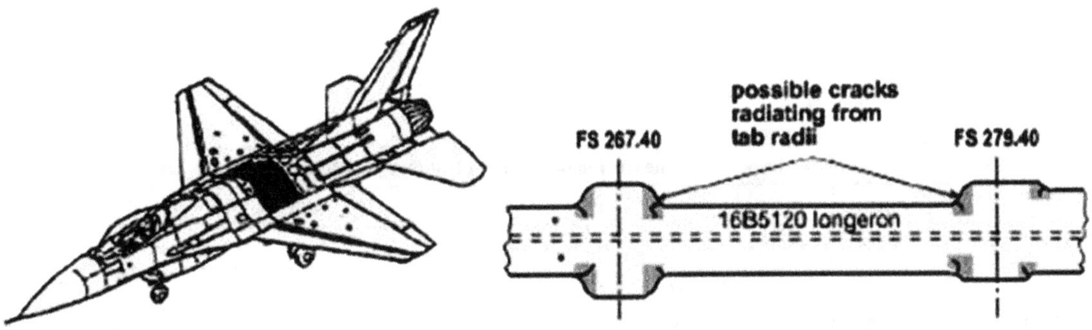

Figure 4. Deterministic damage tolerance inspection requirements for the RNLAF longerons
Source: Military Specification, 1974

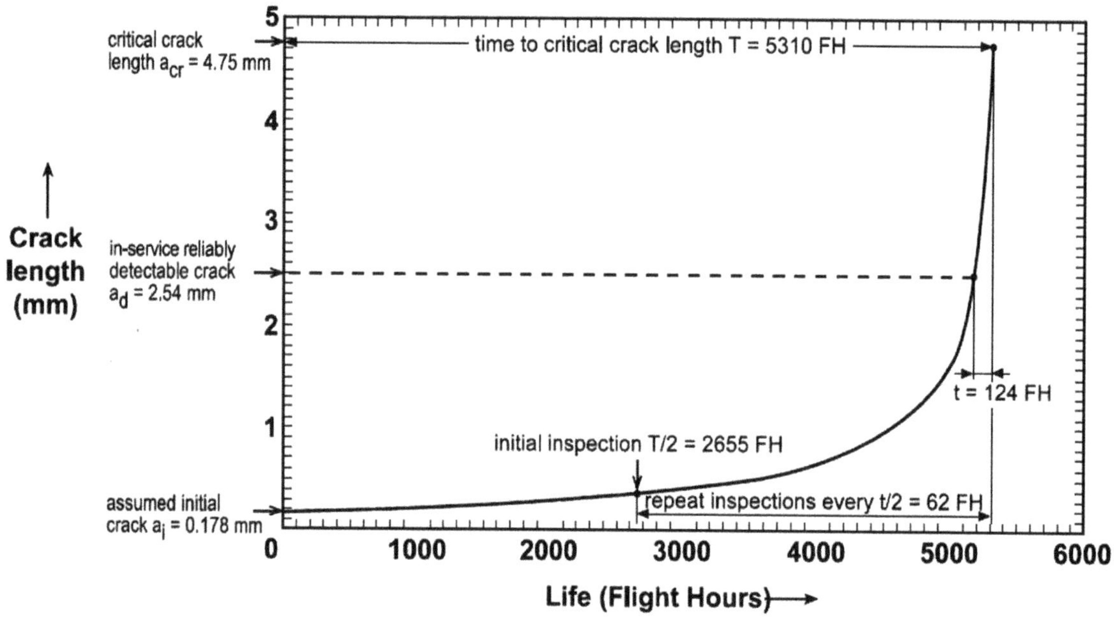

1. Initial inspection after 2655 flights hours;
2. Repeat inspections every 62 flight hours. It will be noted that this inspection scheme has an unknown safety level.

Using the RNLAF longeron mean crack growth curve (Figure 4) and the innovative stochastic model of fatigue crack propagation (53), we have obtained (from (54), where the safety level was specified to be $1-\alpha = 0.99$, Δ_k=4 mm for all k) the following in-service inspection time sequence with decreasing intervals (see Table 2).

Graphical representation of decreasing in-service inspection intervals under crack propagation for the proposed inspection scheme (Table 2) is shown in Figure 5. Thus, the proposed inspection scheme contains only $N^{\text{inspections}}_{\text{decr.int.}}$ =12 inspections.

The index of relative efficiency of the inspection scheme with constant intervals as compared with the inspection scheme with decreasing intervals is given by

Table 2. Inspection time sequence under fatigue crack propagation

Inspection Time (In Terms Of Flight Hours) τ_j, j=1(1)12											
τ_1	τ_2	τ_3	τ_4	τ_5	τ_6	τ_7	τ_8	τ_9	τ_{10}	τ_{11}	τ_{12}
2655	4200	4600	4800	4930	5030	5110	5170	5220	5260	5290	5310
Interval (In Terms Of Flight Hours) $\tau_{j+1}-\tau_j$, j=0(1)11, $\tau_0 = 0$											
2655	1545	400	200	130	100	80	60	50	40	30	20

Figure 5. Graphical representation of inspection intervals under fatigue crack propagation

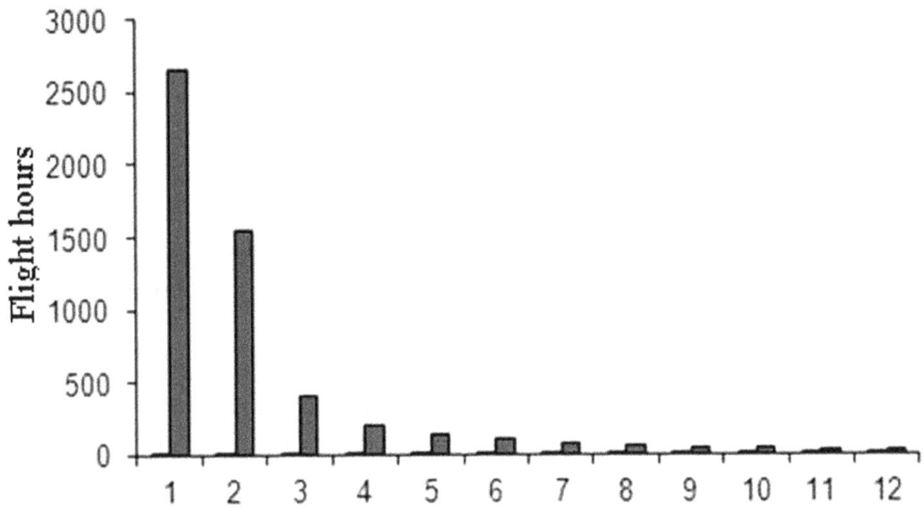

$$I_{\text{rel.eff.}}(N_{\text{const.int.}}^{\text{inspections}}, N_{\text{decr.int.}}^{\text{inspections}}) = N_{\text{decr.int.}}^{\text{inspections}} \;/\; N_{\text{const.int.}}^{\text{inspections}} = 12 \;/\; 42 = 0.29. \tag{59}$$

The index of reduction percentage in the inspection costs for the inspection scheme with decreasing intervals as compared with the inspection scheme with constant intervals is given by

$$I_{\text{red.per.}}(N_{\text{decr.int.}}^{\text{inspections}}, N_{\text{const.int.}}^{\text{inspections}}) = [1 - I_{\text{rel.eff.}}(N_{\text{const.int.}}^{\text{inspections}}, N_{\text{decr.int.}}^{\text{inspections}})]100\% = 71\%. \tag{60}$$

CONCLUSION

The results obtained in this work can be used to solve the service problems of the following important engineering structures:

1. Transportation Systems and Vehicles – aircraft, space vehicles, trains, ships;
2. Civil Structures – bridges, dams, tunnels;
3. Power Generation – nuclear, fossil fuel and hydroelectric plants;
4. High-Value Manufactured Products – launch systems, satellites, semiconductor and electronic equipment;
5. Industrial Equipment – oil and gas exploration, production and processing equipment, chemical process facilities, pulp and paper.

There are a handful of parametric models (for instance, Weibull, Extreme Value, Lognormal, Gamma, Birnbaum-Saunders) that have successfully served as population models for failure times arising from a wide range of products and failure mechanisms. Sometimes there are probabilistic arguments based on the physics of the failure mode that tend to justify the choice of model. Other times the model is used solely because of its empirical success in fitting actual failure data. A large gap still needs to be bridged between the fatigue experimentalists and researchers who use probabilistic methods to study the fatigue crack growth problems.

REFERENCES

Barlow, R. E., Hunter, L. C., & Proschan, F. (1963). Optimum checking procedures. *Journal of the Society for Industrial and Applied Mathematics*, *11*(4), 1078–1095. doi:10.1137/0111080

Bogdanoff, J. L., & Kozin, F. (1985). *Probabilistic Models of Cumulative Damage*. New York: Wiley.

Ghonem, H., & Dore, S. (1987). Experimental study of the constant probability crack growth curves under constant amplitude loading. *Engineering Fracture Mechanics*, *27*(1), 1–25. doi:10.1016/0013-7944(87)90002-6

Itagaki, H., Ishizuka, T., & Huang, P. Y. (1993). Experimental estimation of the probability distribution of fatigue crack growth lives. *Probabilistic Engineering Mechanics*, *8*(1), 25–34. doi:10.1016/0266-8920(93)90027-S

Iyyer, N., Sarkar, S., Merrill, R., & Phan, N. (2007). Aircraft life management using crack initiation and crack growth models - P-3C Aircraft experience. *International Journal of Fatigue*, *29*(9-11), 1584–1607. doi:10.1016/j.ijfatigue.2007.03.017

Kapur, K. C., & Lamberson, L. R. (1977). *Reliability in Engineering Design*. New York: Wiley.

Lin, Y. K., & Yang, J. N. (1985). On statistical moments of fatigue crack propagation. *Engineering Fracture Mechanics*, *18*(2), 243–256. doi:10.1016/0013-7944(83)90136-4

Luss, H., & Kander, Z. (1974). Inspection policies when duration of checkings is non-negligible. *Operational Research Quarterly*, *25*(2), 299–309. doi:10.2307/3008371

Military Specification. (1974). *Airplane Damage Tolerance Requirements. MIL-A-83444*. USAF.

Munford, A. G. (1981). Comparison among certain inspection policies. *Management Science*, *27*(3), 260–267. doi:10.1287/mnsc.27.3.260

Munford, A. G., & Shahani, A. K. (1972). A nearly optimal inspection policy. *Operational Research Quarterly*, *23*(3), 373–379. doi:10.2307/3007891

Munford, A. G., & Shahani, A. K. (1973). An inspection policy for the Weibull case. *Operational Research Quarterly*, *24*(3), 453–458. doi:10.2307/3008129

Nechval, K. N., Nechval, N. A., Berzins, G., Purgailis, M., Rozevskis, U., & Strelchonok, V. F. (2009). Optimal adaptive inspection planning process in service of fatigued aircraft structures. In K. Al-Begain, D. Fiems, & G. Horvath (Eds.), Analytical and Stochastic Modeling Techniques and Applications, (LNCS) (vol. 5513, pp. 354–369). Berlin: Springer-Verlag doi:10.1007/978-3-642-02205-0_25

Nechval, N. A., Berzins, G., Purgailis, M., & Nechval, K. N. (2008a). Improved estimation of state of stochastic systems via invariant embedding technique. *WSEAS Transactions on Mathematics*, *7*, 141–159.

Nechval, N. A., Nechval, K. N., Berzins, G., Purgailis, M., & Rozevskis, U. (2008). Stochastic fatigue models for efficient planning inspections in service of aircraft structures. In K. Al-Begain, A. Heindl, & M. Telek (Eds.), Analytical and Stochastic Modeling Techniques and Applications (LNCS) (vol. 5055, pp. 114-127). Berlin: Springer-Verlag. doi:10.1007/978-3-540-68982-9_9

Nechval, N. A., Nechval, K. N., & Purgailis, M. (2011). Inspection policies in service of fatigued aircraft structures. In S.I. Ao & L. Gelman (Eds.), Electrical Engineering and Applied Computing, (LNEE) (vol. 90, pp. 459-472). Berlin: Springer Science+Business Media B.V. doi:10.1007/978-94-007-1192-1_37

Nechval, N. A., Nechval, K. N., & Purgailis, M. (2011a). Prediction of future values of random quantities based on previously observed data. *Engineering Letters*, *9*, 346–359.

Nechval, N. A., Nechval, K. N., & Vasermanis, E. K. (2003). Statistical models for prediction of the fatigue crack growth in aircraft service. In A. Varvani-Farahani & C. A. Brebbia (Eds.), Fatigue Damage of Materials 2003 (pp. 435-445). Southampton, UK: WIT Press.

Nechval, N. A., Nechval, K. N., & Vasermanis, E. K. (2004). Estimation of warranty period for structural components of aircraft. *Aviation*, *8*, 3–9.

Paris, R., & Erdogan, F. (1963). A critical analysis of crack propagation laws. *Journal of Basic Engineering*, *85*(4), 528–534. doi:10.1115/1.3656900

Sengupta, B. (1977). Inspection procedures when failure symptoms are delayed. *Operational Research Quarterly*, *28*(3-part-ii), 768–776. doi:10.1287/opre.28.3.768

Straub, D., & Faber, M. H. (2005). Risk based inspection planning for structural systems. *Structural Safety*, *27*(4), 335–355. doi:10.1016/j.strusafe.2005.04.001

Tadikamalla, P. R. (1979). An inspection policy for the gamma failure distributions. *Operational Research Quarterly*, *30*(1), 77–78. doi:10.1057/jors.1979.9

Virkler, D. A., Hillberry, B. M., & Goel, P. K. (1979). The statistic nature of fatigue crack propagation. *ASME Journal of Engineering Materials and Technology*, *101*(2), 148–153. doi:10.1115/1.3443666

Yang, J.N., His, W.H., & Manning, S.D. (1985). *Stochastic crack propagation with applications to durability and damage tolerance analyses.* Technical Report, Flight Dynamics Laboratory, Wright-Patterson Air Force Base, AFWAL-TR-85-3062.

Yang, J. N., & Manning, S. D. (1990). Stochastic crack growth analysis methodologies for metallic structures. *Engineering Fracture Mechanics*, *37*(5), 1105–1124. doi:10.1016/0013-7944(90)90032-C

Chapter 17
Stochastic Approach to Uncertainty Control in Multiphysics Systems:
Modeling of Carbon Balance and Analysis of GHG Emissions Using Satellite Tools

Yuriy V. Kostyuchenko
Scientific Centre for Aerospace Research of the Earth, National Academy of Sciences of Ukraine, Ukraine

Igor Artemenko
Scientific Centre for Aerospace Research of the Earth, National Academy of Sciences of Ukraine, Ukraine

Dmytro Movchan
Scientific Centre for Aerospace Research of the Earth, National Academy of Sciences of Ukraine, Ukraine

Ivan Kopachevsky
Scientific Centre for Aerospace Research of the Earth, National Academy of Sciences of Ukraine, Ukraine

ABSTRACT

The problem of uncertainty analysis in multi-component systems is considered. As an example a problem of decision making under uncertainty in task of modeling of carbon balance and analysis of greenhouse gases (GHG) emissions using satellite tools was considered. Approaches to decision making under uncertainty are described: interval, fuzzy, and stochastic assessments. Different approaches and algorithms to calculate carbon and GHG emissions are described. For every algorithm (deterministic inventory, ecological modeling, and satellite control of emissions) errors and uncertainties are analyzed and estimated. Algorithms for uncertainty analysis are presented. Algorithm for analysis of components of uncertainty of vegetation productivity assessment using satellite data is proposed. Uncertainty component analysis allows understand important properties of the system studied and its feedback to anthropogenic load and climate impact. It was demonstrated that the comprehensive analysis of uncertainties not only reduces errors but also obtains new knowledge about the systems studied.

DOI: 10.4018/978-1-5225-1639-2.ch017

INTRODUCTION

Decision Making Under Uncertainty: Problem Description and Task Definition

The problem of uncertainty analysis in multy-component systems and multy-physics processes is very complicated and urgent task, which cannot be reduced to measurements accuracy control. Management of anthropogenic, engineering and natural systems is aimed to secured and sustainable development, and so closely connected with multi-scale decision making.

In many problems of control and monitoring, engineering, communication, management and other practices, we should select a systems design, parameters of a process, so make a decision. This decision must be optimal, the best for the users, as well as industrial and public. In traditional operations research, we assume that we know the objective function $f(x)$ whose values describe what is best for the users. Problems of ecology, pollutions and greenhouse gases (GHG) emissions are most complicated for decision makers because it connected with industry and energy, public health, ecological safety and environmental services, climate and food production. So the instruments for decision support, including uncertainty analysis, should be comprehensive and enough complicated relating to complexity of the system studied.

For example, for a GHG analysis, this function $f(x)$ may represent the carbon stock rate (or calculated emissions) resulting from change ecosystems productivity represented by parameters x.

In such situations, the problem is to find parameters x that optimizes (e.g., maximizes) the given function $f(x)$ on the given range X. In work (Nguyen & Kreinovich, 2006) a good overview of existing engineering approaches and algorithm to decision making under uncertainty is presented.

Optimization of well-defined functions is what started calculus in the first place: once we know the objective function $f(x)$, we can use diferentiation to find its maximum, e.g., as the point x at which the derivative of f with respect to x is equal to 0. Sometimes, this equation $f0(x) = 0$ can be solved directly; sometimes, it is difficult to solve it directly, so we use gradient-based (i.e., derivatives-based) techniques to either solve this equation or to optimize the original objective function $f(x)$.

In many decision-making problems, instead of finitely many situations v, we have a continuum of possible situations. We may have one or more continuous variables that describe possible situations. In such problems, instead of finitely many probabilities $p(v)$, we have a probability distribution with a probability density $p(v)$, and instead of a sum, we represent the expected value as an integral $f(x) = \int_v p(v) \cdot f(x, v) dv$. In real situation, we often do not know the probabilities of different possible situations, or at least we only have partial and limited knowledge about these probabilities.

Decision making in environmental management and climate risk analysis requires carbon emission calculating under uncertainty. These uncertainties have different nature. Its connected with imperfection of emission account tools and methods, insufficient comprehensiveness of carbon cycle models, and incomplete account of variations of crop productivity. All these cases characterizing by different types of uncertainty and could be analyzed and reduced separately. But at the same time, the comprehensive analysis of uncertainties allow not only control and reduce errors, but also obtain new knowledge about the systems we study – multi-component ecosystem characterizing by multi-physics processes.

Decision Making Under Interval and Fuzzy Uncertainty

The simplest case is when we have tolerance-type (interval) uncertainty, when all we know is that the deviation of actual value x of the parameter from the nominal value x_0 cannot exceed the given tolerance Δ. In precise terms, this means that x belongs to the interval $[x_0-\Delta; x_0+\Delta]$, an we have no information about the probabilities within this interval.

In other situations, in addition to the interval that is guaranteed to contain x, experts can also provide us with narrower intervals that contain x with certain degree of confidence α. Such a nested family of intervals is also called a fuzzy set; because it turns out to be equivalent to a more traditional definition of fuzzy set (if a traditional fuzzy set is given, then different intervals from the nested family can be viewed as α-cuts corresponding to different levels of uncertainty α) (Moore & Lodwick, 2003).

The resulting interval and fuzzy computations techniques have been well developed, and they are still actively used in many application areas (Jaulin et al, 2001; Booker, Parkinson, & Ross, 2002).

When we dealing with interval uncertainty (in the ideal situation), when we know the exact values of $y(x)$ for every x, we can simply differentiate the corresponding dependence. In practice, however, the information on y comes from measurements, and measurements are never exact. For example, when we analyze a landscape, we measure the altitudes y_1; ..., y_n at different points $x_1 < ... < x_n$. Since the measurements are not exact, the measured values $\tilde{y}_i$ are, in general, slightly different from the the (unknown) actual altitudes y_i.

For measuring instruments we usually have an upper bound Δ_i on the measurement error $\Delta y_i \stackrel{def}{=} \tilde{y}_i - y_i$. This upper bound is usually provided by the manufacturer of this instrument: $\left|\Delta y_i\right| \leq \Delta_i$. Thus, after the measurement, the only information that we have about the actual (unknown) value y_i is that this value belongs to the interval $y_i = \left[\underline{y_i}, \overline{y_i}\right]$, where $\underline{y_i} \stackrel{def}{=} \tilde{y}_i - \Delta_i$ and $\overline{y_i} \stackrel{def}{=} \tilde{y}_i + \Delta_i$ (Jaulin et al, 2001).

Thus, the only information that we have about the actual dependence $y = f(x)$ of y on x is that the (unknown) function $f(x)$ belongs to the class

$$F \stackrel{def}{=} \{f(x) \mid f(x_i) \in y_i \forall i = 1, ..., n\}. \tag{1}$$

We also know that the (unknown) function $f(x)$ is smooth (differentiable), because otherwise, the notion of a slope does not make sense.

Since we only know the values of the unknown function $f(x)$ at finitely many points $x_1 < ... < x_n$, it is always possible that the derivative of the (unknown) function $f(x)$ attains the desired value s at some point between x_i and x_{i+1}. In other words, for every interval $[a; b]$ $(a < b)$, it is always possible to have a function f within the class F (defined by the equation (1)) for which $f_0(x) = s$ for some $s \in [a; b]$.

As we have mentioned, interval uncertainty is just the simplest possible case of non-probabilistic uncertainty. In many real life situations, instead of an interval, we have a fuzzy set - nested family of intervals corresponding to different levels of certainty α.

In this case, instead of an interval function, i.e., a finite sequence of pairs $\left\langle x_i, y_i \right\rangle (i = 1, 2, ..., n)$, where x_i is a real number and y_i is an interval, we have a fuzzy function - a finite sequence of pairs $\left\langle x_i, Y_i \right\rangle (i = 1, 2, ..., n)$, where Y_i is a fuzzy number.

Once we define the derivative of an interval-valued function, we can extend this definition to derivatives of fuzzy-valued function: for every degree of certainty α, we consider an interval function formed by the α-cuts of Y_i. We can then compute the derivative of this interval function.

So, for each level α, we have a derivative interval corresponding to this level α. The nested family of these derivative intervals forms a fuzzy number, which can be thus viewed as a derivative of the original fuzzy function.

In this context few important definitions and results could be formulated in framework of extending differentiation formalisms (Nguyen & Kreinovich, 2006).

1. By an interval function F, we mean a finite sequence of pairs $\langle x_i, y_i \rangle (i = 1, 2, ..., n)$, where for each i, x_i is a real number, y_i is a non-degenerate interval, and $x_1 < x_2 < ... < x_n$.
2. We say that a function $f : R \to R$ from reals to reals belongs to an interval function $F = \{\langle x_1, y_1 \rangle, ..., \langle x_n, y_n \rangle\}$ if *f(x)* is continuously differentiable and for every *i* from *1* to *n*, we have $f(x_i) \in y_i$.
3. Let F be an interval function, and let [a; b] be an interval. By a derivative $F'([a,b])$, we mean the intersection $F'([a,b]) \overset{def}{=} \bigcap_{f \in F} f'([a,b])$, where $f_0(x)$ denotes the derivative of a differentiable function *f(x)*, and $f'([a,b]) \overset{def}{=} \{f'(x) \mid x \in [a,b]\}$ is the range of the derivative $f_0(x)$ over the interval [a; b].
4. We say that a function *f(x)* is strongly increasing if $f'(x) > 0 \forall x$.
5. For every interval function *F*, the existence of a strongly increasing function $f \in F$ with $f'(x) > 0$ is equivalent to

$$\underline{y_i} < \overline{y_j} \forall i < j \tag{2}$$

6. According to the inequalities (2), all the differences $\overline{y_j} - \underline{y_i}, (i < j)$ are positive. Since all intervals are non-degenerate, the differences $\overline{y_i} - \underline{y_i}$ are also positive. Let us denote the smallest of these positive numbers by Δ. For every *i*, let us denote

$$y_i \overset{def}{=} \max(\underline{y}_1, ..., \underline{y}_i) + \frac{1}{2n} \cdot \Delta \tag{3}$$

7. We say that a function *f(x)* is strongly decreasing if $f'(x) < 0$ for all *x*.
8. For every interval function *F*, the existence of a strongly decreasing function $f \in F$ is equivalent to

$$\overline{y_i} < \underline{y_j} \forall i < j \tag{4}$$

9. For every interval function F and for every interval [a; b], the existence of a function $f \in F$ that is strongly increasing on the interval [a; b] is equivalent to

$$\underline{y_i} < \overline{y_j} \forall i < j \text{ for which } x_i, x_j \in [a,b] \tag{5}$$

10. For every interval function F and for every interval [a; b], $0 \in F'([a,b])$ if and only if neither conditions (5) not conditions (6) are satisfied.
11. Let $F = \{\langle x_1, y_1 \rangle, \ldots, \langle x_n, y_n \rangle\}$ be an interval function, and let v be a real number. Then, we define a new interval function $F - v \cdot x$ as follows: $F - v \cdot x = \{\langle x_1, y_1 - v \cdot x_1 \rangle, \ldots, \langle x_n, y_n - v \cdot x_n \rangle\}$, where, for an interval $y = [\underline{y}, \overline{y}]$ and for a real number c, the difference y-c is defined as $[\underline{y} - c, \overline{y} - c]$.
12. For every interval function F and for every interval [a; b], $v \in F'([a,b])$ if and only if $0 \in (F - v \cdot x)'([a,b])$.
13. For every interval function F and for every interval [a; b], let i_0 and j_0 be the first and the last index of the values x_i inside [a; b]. Then $F'([a,b]) = [\underline{F}_{i0,j0}, \overline{F}_{i0,j0}]$, where

$$\underline{F}_{i0jo} \stackrel{def}{=} \min_{i_0 \le i < j \le j_0} \overline{\Delta}_{ij}, \ \overline{F}_{iojo} \stackrel{def}{=} \max_{i_0 \le i < j \le j_0} \underline{\Delta}_{ij} \tag{6}$$

$$\underline{\Delta}_{ij} \stackrel{def}{=} \frac{\overline{y_i} - \underline{y_j}}{x_j - x_i}, \ \overline{\Delta}_{ij} \stackrel{def}{=} \frac{\overline{y_i} - \underline{y_j}}{x_j - x_i} \tag{7}$$

14. According to Kreinovich et al (2003) it possible to prove that with the propositions 1) – 13) there exists an algorithm that, given an interval function $F = \{\langle x_1, y_1 \rangle, \ldots, \langle x_n, y_n \rangle\}$, computes all possible values of the derivative $F'([a,b])$ in $O(n^2) = n \cdot (n+1) / 2$ computational steps.

In framework of task of uncertainty reducing for decision making optimization, task of analysis of observation and measurement data is the task of analysis distributions of x, and determination of Δ values.

Stochastic Model for Decision Making in GHG Control

Fundamentally different approach to decision making under uncertainty in field of environmental management was proposed by (Ermolieva et al, 2014), basing on robust cost-effective and environmentally safe carbon trading economic instruments for GHG emissions control.

This approach based on stochastic model of robust emission trading.

Proposed integrated robust multi-agent emission trading model seeks the goal that all the parties participating in emissions trading jointly achieve individual emission targets in a cost-efficient way under safety constraints by investing in emissions abatement, uncertainty reduction and by redistributing the emission permits through trading.

Emission trading affects and is being affected by emission uncertainties. In addition to asymmetric information, it implicitly contributes to reduction of other uncertainties for the two main reasons. First, trading is likely to lead towards more scrutinized emissions inventory compilation rules. Second, verifiability of trades requires that the reported emissions plus uncertainties are below the cap; therefore, trading creates incentives for parties to invest into uncertainty reduction prior to compliance. Proposed trading model exploits this phenomenon.

Different uncertainties differently affect emission trading causing market crashes and instabilities similar to financial markets. To limit the role of uncertainties, advocates of regulated trades argue in favor of uncertainty constraints distinguishing sources by their uncertainty levels (Godal et al., 2003). Market regulators may set restrictions on source category to be included in trading. Trading scheme may demand a Party to set source-specific targets depending on the level of uncertainty.

A model which includes uncertainty and risk-adjusted regulations into emission trading schemes is proposed following (Ermolieva et al, 2010). The model explores conditions of market's stability towards uncertainties by imposing appropriate safety constraints to control the level of admissible uncertainty which would guarantee cost efficiency of trades and safety levels of emission reduction targets.

These types of safety constraints are typical for pollution control, financial applications, stability regulations in the insurance industry and catastrophic risks management (Ermolieva & Ermoliev, 2005). In a sense, these constraints work as a probabilistic discounting mechanism which discounts the reported emissions to detectable levels overshooting uncertainty within a specified safety levels, i.e., portion of detectable emission changes.

Approach can be formally presented as follow (Ermolieva et al, 2014). Let denote the least costs $f_i(y_i)$ for party *i* to comply with imposed targets with given permits y_i and the target K_i. Formally, this function is defined by Equations 8 - 9. Denote the variability of reported emission x_i as random variable $\xi_i(x_i, \omega_i)$, where ω_i is a vector of all uncertainties (scenarios) affecting emissions of party *i*. A random variable $\xi_i(x_i, \omega_i)$ depends, in general, on reported emissions x_i. The uncertainty ξ_i can be reduced by investments in monitoring systems. For this purpose the variable u_i associated with monitoring and other technologies that may control the variability of emissions within the desirable safety level Q_i is introduced.

The individual optimization problem can be written as minimization of function $f_i(y_i)$ defined as the minimum of risk-adjusted expected emission reduction costs $c_i(x_i, \omega_i)$ and uncertainty reduction costs $d_i(x_i, \omega_i)$ for a given permit y_i to be defined through dynamic trading process:

$$f_i(y_i) = \min_{x_i u_i} E[c_i(x_i, \omega) + d_i(u_i, \omega)] \tag{8}$$

under quantile-based environmental safety constraints:

$$P[x_i + \xi(x_i, \omega_i) \leq K_i + y_i] \geq Q_i \tag{9}$$

for all parties *i*. Here Q_i denotes a safety level that ensures the probability that all potential emissions x_i and uncertainties $\xi_i(x_i, \omega_i)$ do not exceed emission target K_i adjusted by permits y_i. Safety level

Q_i is imposed by regulatory agency to ensure robust performance of the market. Uncertainties of cost functions c_i and d_i may be due to market performance, production shocks, and technological uncertainties, which are unknown in advance.

Safety constraints (9) are well known in financial applications as Value-at-Risk risk indicator. They are used for safety regulation of insurance companies, reliability of engineering structures, and catastrophic risk management (Ermolieva and Ermoliev, 2005).

Safety constraints (9) can be also written in the following form that in the case of analytically tractable distributions is reduced to deterministic nonlinear constraints (10). Let define quantile $z_i(x_i)$ as the minimal z such that

$$P[\xi_i(x_i, \omega_i) \leq z] \geq Q_i.$$

Then the following equivalent constraints can be substituted for the safety constraint (9):

$$x_i + u_i \leq K_i + y_i, u_i \leq z_i(x_i) \quad (10)$$

These equations show that reported emissions must undershoot targets K_i adjusted by uncertainties of emissions u_i and permits y_i. Equation 11 shows that safety constraints induce risk-related upper bounds $z_i(x_i)$ on uncertainty intervals dependent on reported emission level x_i. Therefore, it allows to introduce risk-based undershooting of emission targets defined by "critical" quantile $z_i(x_i)$. Functions $f_i(y_i)$ define costs of permits y_i for parties $i = 1,...,n$.

In (Ermolieva et al, 2014) solution of this task is proposed for different cases. For our study it is important that in this framework the task of uncertainty reducing for decision making optimization is the task of uncertainty u_i control in GHG emission x_i inventory. Further we demonstrate as it can be done using satellite tools.

GENERALIZED APPROACH TO ANTHROPOGENIC EMISSIONS ANALYSIS

Usually the task of emission analysis is reduced to task of assessment of quantity of carbon emissions (Intergovernmental Panel on Climate Change [IPCC], 1997).

However, the carbon emission coefficients are known with sufficient accuracy only for carbon dioxide emissions. In this case for emission inventory of other GHG such as CH_4, N_2O, NO_X, CO and non-methane hydrocarbons (NMVOC), which connected with corresponding fuel utilization, we use analysis and calculation in separate economical sectors with semi-empirical dependences and statistics on quantity of burned fuel and emission factors of these gases. Emissions of these gases determined by (IPCC, 1997):

$$E_x = mTK_x \quad (11)$$

where E_x – emission of x GHG, m – mass of utilized fuel; T – fuel calorific value, K_x – emission coefficient of the corresponding GHG.

The main types of fuel, which are sources of emissions of these gases are coal, natural gas, oil (gasoline, diesel) and biomass (wood, wood wastes, biofuel). Inventory for emissions can be carried out as well as by the separate types of fuel for all economy sectors, or by the separate sectors for all types of fuel.

Usually most attention is concentrated on the emissions of carbon dioxide and methane.

Mass of carbon in the utilized fuel m_C we could calculate using formula:

$$m_C = mTk_C \tag{12}$$

where m – mas of utilized fuel; T – calorific value of fuel, which is criteria of fuel thermal capacity; k_C – emission coefficient, which define actual content of carbon in utilized fuel. It should be noted, that during burning not all carbon is oxidized. Oxidization rate depends on equipment efficiency, so in mathematical models we should make correction to incomplete oxidization of carbon. For it we use coefficient of carbon oxidization fraction f_C – relation of quantity of burned carbon to total carbon content in the fuel:

$$E_{CO_2} = 44 m_C f_C / 12 \tag{13}$$

Also we should take into account that significant part of energy resources will not be utilized as fuel. Some part of it will be utilized for production of other industrial plastic, road bitumen etc. Carbon oxidation, and so, a direct emission in these processes is absent. This carbon calling accumulated and excluded from calculations of total emissions of carbon dioxide. IPCC methodology recommends make a correction to accumulated carbon for account of some types of fuel - m_{SC} (IPCC, 1997):

$$m_{SC} = f_{SC} m_C \tag{14}$$

where f_{SC} – is the fraction of accumulated carbon.

Therefore, a total emission of carbon dioxide reduced to accumulated carbon, we can calculate (according formulas (12) - (13)):

$$E_{CO_2} = 44\left(m_C - m_{SC}\right) f_C / 12 \tag{15}$$

This is the way to calculate anthropogenic emissions of carbon dioxide.

Emissions of CH_4, connected with relevant sources (such as coal mining) traditionally calculating as:

$$E_{CH_4} = 0,67 \cdot m k_{CH_4} \tag{16}$$

where m – mass of extracted product (for example, coal in tons), k_{CH_4} – emission coefficient, which defines volume of gas (m^3) emitted with extraction of one ton of product, and depends on production technology. To calculation mass of methane (Gg) from it volume usually using value of the methane density under the pressure of one atmosphere and with the temperature 20°C, which is equal 0,67•10^{-6} Gg/m^3.

Total anthropogenic emission in this case can be calculated using simple algorithm:

$$E_{tot} = \sum_{n=1}^{N} \sum_{m=1}^{M} k_{mn} \Delta x_{nm} \tag{17}$$

where M – sectors of economy, N – sites of territory, x – statistical data describing impacts and emissions.

Using algorithm (18) we can calculate GHG emissions from every fuels. Besides, data x (in particular, m as the reported mass of extracted fuel) in general case is the function of time, which describe technological and socio-economical development of regions. In this context algorithm (17) make a possible to forecast future emissions on the base of future development (see, for example, "emission scenarios" of IPCC).

NOTES ON UNCERTAINTY CONTROL FOR GENERALIZED APPROACH

Described approach has significant uncertainties, which depends on quality of used statistics on sectoral economy. These uncertainties are aleatoric and connected with inaccuracy of the data collected and with correctness of methodology of surveys and data processing.

In this case it is possible to apply a quite simple method to define errors and uncertainties of the relevant estimations. But we should also remember that distribution of x is not normal, so we should be careful with "average" assessments.

If we have two different (in ideal case - independent) assessment of emission $E^1_{tot,n}$ i $E^2_{tot,n}$ for certain region n from N and all existing sectors of economy m for studied types of GHG i. To estimate uncertainty of emissions assessments we can propose a form:

$$U_n = (| E^1_{tot,n} - E^2_{tot,n} |) / (E^1_{tot,n}) \tag{18}$$

And integrated uncertainty for a whole region N, and all types of GHG can be calculated as:

$$U_{tot} = \sqrt{\sum_n U_n^2} \tag{19}$$

This approach is quite primitive, and essentially limited. Particularly, it is essentially vulnerable toward quality of input data. Besides, using of methodologically disparate big sets of data may leads to high errors and reliability decreasing, especially on the long time periods. So strategic planning in field of ecological security, economic development, long-term emission reducing, etc., requires more sophisticated and complex algorithms to estimate uncertainty.

GHG AND CARBON CYCLE: KEY VARIABLES AND PARAMETERS

Correct analysis of aleatoric uncertainty will not be successful without analysis of epistemic uncertainty. Even estimation of measurement accuracy requires, in ideal case, understanding of what we are measuring. So we should recognize variables and parameters of processes, which determine transfer of carbon in Earth system.

So, key role in GHG emission uncertainty control may play ecological models, in particular, a correct analysis of carbon cycle. This issue, as well as some ecological models is presented in (White et al, 2011) These models are formalized reflection of our understanding of processes of accumulation and transformation of carbon compounds in ecosystems (Figure 1).

Model includes number of key elements, which are necessary to calculate correct output data for further analysis.

Mathematical model of carbon balance is usually integrated from regional or local components and links, which reflect mutual influence of neighboring sites. Integrated mathematical model of regional carbon cycle for region *r* may be presented as matrix equation:

$$\frac{d\mathbf{X}_r}{dt} = \mathbf{F}_r\left(\mathbf{X}_r, \mathbf{V}_r, y_r, t\right) \tag{20}$$

Figure 1. Integrated structural model of carbon cycle in ecosystems

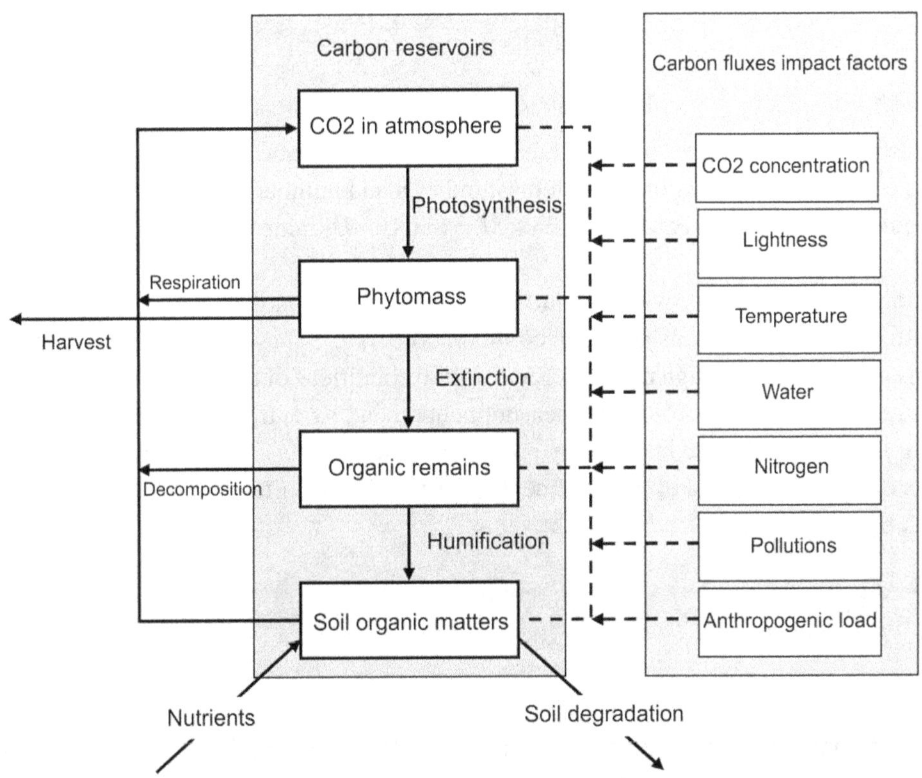

where $\mathbf{F}_r$ – column matrix of non-linear functions; $\mathbf{V}_r$ – column matrix of carbon fluxes between ecosystem components of region *r*; $\mathbf{X}_r = \left\| x_{ra}, \mathbf{X}_r^{cl}, \mathbf{X}_r^{gl}, \mathbf{X}_r^{bl}, \mathbf{X}_r^{df}, \mathbf{X}_r^{cf} \right\|^T$ – column matrix of unknown variables, and $\mathbf{X}_r^{cl} = \left\| \mathbf{X}_r^{clw}, \mathbf{X}_r^{cls}, \mathbf{X}_r^{clv}, \mathbf{X}_r^{clt}, \mathbf{X}_r^{clf} \right\|^T$; $\mathbf{X}_r^{gl} = \left\| X_{rp}^{glw}, X_{rl}^{glw}, X_{rs}^{glw} \right\|$; $\mathbf{X}_r^{bl} = \left\| X_{rp}^{glw}, X_{rl}^{glw}, X_{rs}^{glw} \right\|$; $\mathbf{X}_r^{df} = \left\| \mathbf{X}_r^{dfh}, \mathbf{X}_r^{dfs} \right\|$; $\mathbf{X}_r^{cf} = \left\| X_{rg}^{cf}, X_{rw}^{cf}, X_{rl}^{cf}, X_{rs}^{cf} \right\|$. Also $\mathbf{X}_r^{clw} = \left\| X_{rp}^{clw}, X_{rl}^{clw}, X_{rs}^{clw} \right\|$; $\mathbf{X}_r^{cls} = \left\| X_{rp}^{cls}, X_{rl}^{cls}, X_{rs}^{cls} \right\|$; $\mathbf{X}_r^{clv} = \left\| X_{rp}^{clv}, X_{rl}^{clv}, X_{rs}^{clv} \right\|$; $\mathbf{X}_r^{clt} = \left\| X_{rp}^{clt}, X_{rl}^{clt}, X_{rs}^{clt} \right\|$; $\mathbf{X}_r^{clf} = \left\| X_{rp}^{clf}, X_{rl}^{clf}, X_{rs}^{clf} \right\|$; $\mathbf{X}_r^{dfh} = \left\| X_{rg}^{dfh}, X_{rw}^{dfh}, X_{rl}^{dfh}, X_{rs}^{dfh} \right\|$; $\mathbf{X}_r^{dfs} = \left\| X_{rg}^{dfs}, X_{rw}^{dfs}, X_{rl}^{dfs}, X_{rs}^{dfs} \right\|$.

Variable y_r, included into this equation, corresponds to carbon fluxes in neighboring regions (with common borders). Value of this variable is proportional to the concentration (content) of carbon in atmosphere over these neighboring regions as:

$$y_r = \sum_{j=1, j \neq r}^{M} \theta_{rj} X_j^a \tag{21}$$

where θ_{rj} – proportional coefficients; X_j^a – carbon content in atmosphere over region *j*.

So, after substituting all expressions, the integrated model of carbon balance for all regions ($r = 1,2,\ldots,M$) may be written as:

$$\frac{d\mathbf{X}}{dt} = \mathbf{F}\left(\mathbf{X}, \mathbf{V}, t\right) \tag{22}$$

where $\mathbf{X} = \left\| \mathbf{X}_1, \mathbf{X}_2, \ldots, \mathbf{X}_M \right\|^T$ – column matrix of unknown parameters, $\mathbf{V} = \left\| \mathbf{V}_1, \mathbf{V}_2, \ldots, \mathbf{V}_M \right\|^T$ – column matrix of carbon fluxes. Number of equations and number of unknown variables in this model, in general case, is equal to the product of ecosystems number *n* and number of regions *M* (if all types of ecosystem are presented in every region): $n = 34 \text{ x } M = 850$ (for Ukraine quantity of ecosystems type is 34 and number of regions is 25).

Proportional coefficients θ_{rj}, which included into presented equations, determined by links of study region *r* with neighboring regions *j* through common borders L_{rj}. These links realized by wind transport of emissions and pollution though region borders. Initial conditions of transportation should be specified by analysis of long-term meteorological measurements: wind rose, frequency and speed of wind. Sum of vectors of regional wind rose in the study point will indicate as $\vec{w}$. This integrated vector is the function of time, depending of point of measurement. So, flux of transported carbon through border between regions will be calculated as:

$$y_{rj}(t) = X_j^a k_{j\theta} \int_0^{L_{rj}} w(l,t) \cos \psi(l,t) dl \tag{23}$$

where y_{rj} – external toward region *r* carbon flux from neighboring region *j*; $k_{j\theta}$ – conversion factor to calculate average concentration of carbon with height from integrated carbon content in atmosphere over region j; $\psi(l,t)$ – angle between the wind rose vector and perpendicular to the r region border. Form of equation for coefficients θ_{rj} is:

$$\theta_{rj} = k_{j\theta} \int_0^{L_{rj}} w(l,t) \cos \psi(l,t) dl \tag{24}$$

Integrated equation for atmospheric content of carbon may be presented as following:

$$\mathbf{Y} = \Theta \mathbf{X}^a \tag{25}$$

where $\mathbf{Y} = \left\| y_1, \ldots, y_M \right\|^T$ – column matrix of external carbon fluxes for all regions; $\mathbf{X}^a = \left\| X_1^a, \ldots, X_M^a \right\|^T$ – column matrix of carbon contents in atmosphere over all regions; $\Theta = \left\| \theta_{rj} \right\|$, $(r,j = 1,2,\ldots,M)$ – matrix of coefficients. Obviously, $\theta_{rj} \neq 0$ only if regions r and j are neighboring, also is obvious that є $\theta_{rj} = -\theta_{jr}$, and matrix Θ is skew-symmetric disperse matrix with zero diagonal. So for task solution we need only values of elements placed on one side of diagonal matrix Θ.

Initial Equation 21 describing carbon cycle, is the system with big dimension and big number of variables. From ecological viewpoint it is possible to divide studied natural system to sub-systems, in which separate components of studied processes has different velocity. In this case initial complicated system of equation may be presented as few more simple sub-systems, which describing separate processes (Bun & Dachuk, 1997):

$$\frac{d\mathbf{X}^a}{dt} = \mathbf{F}^a\left(\mathbf{X}, \mathbf{V}, \mathbf{Y}\right), \quad \frac{d\mathbf{X}^{cl}}{dt} = \mathbf{F}^{cl}\left(\mathbf{X}, \mathbf{V}, \mathbf{Y}\right),$$

$$\frac{d\mathbf{X}^{gl}}{dt} = \mathbf{F}^{gl}\left(\mathbf{X}, \mathbf{V}, \mathbf{Y}\right), \quad \frac{d\mathbf{X}^{bl}}{dt} = \mathbf{F}^{bl}\left(\mathbf{X}, \mathbf{V}, \mathbf{Y}\right), \tag{26}$$

$$\frac{d\mathbf{X}^{df}}{dt} = \mathbf{F}^{df}\left(\mathbf{X}, \mathbf{V}, \mathbf{Y}\right), \quad \frac{d\mathbf{X}^{cf}}{dt} = \mathbf{F}^{cf}\left(\mathbf{X}, \mathbf{V}, \mathbf{Y}\right),$$

where $\mathbf{F}$ with corresponding indexes is the column matrixes of non-linear functions of fluxes; $\mathbf{X}^a = \left\| X_1^a, \ldots, X_M^a \right\|^T$; $\mathbf{X}^{cl} = \left\| X_1^{cl}, \ldots, X_M^{cl} \right\|^T$. Therefore we analyze separately equations for carbon fluxes in atmosphere over all regions, separately for "arable land" ecosystems in all regions, and so on.

Solution of this task can be obtained as the set of discrete values $\tilde{X}_{rv}^a \approx X_r^a\left(t_v\right)$; $\tilde{\mathbf{X}}_{rv}^{cl} \approx \mathbf{X}_{rv}^{cl}\left(t_v\right)$ on the set of discrete values of time $t_v \in \left[t_0, t_N\right]$, where $v = 1,2,\ldots,N$ – number of integration step; t_0 i t_N – limits of simulation interval. Usually for modeling we use an algorithm based on calculation of carbon content in atmosphere over some regions. On the every step of modeling new values of unknown variables should be calculated, using algorithm for atmospheric carbon content calculation:

$$\tilde{X}^a_{rv+1} = X^a_{rv} + h_v \varphi^a_{rv}\left(\tilde{X}^a_{rv}\right),\ r = 1,2,\ldots,M\,, \tag{27}$$

where h_v – modeling step, φ^a_{rv} – non-linear functions defining by one-step method (Bun & Dachuk, 1997).

Taking into account low quality of data on atmospheric carbon content obtained from traditional measurements, it important to note, that using in (27) regional data from satellite on atmospheric concentration of GHG and on corresponding carbon fluxes could increase modeling reliability and simplify integrating procedure.

Other important notice concerns to necessity of correct problem-oriented land cover classification in study regions. Model shows (Figure 2) that land cover distribution is very important for correct estimation of carbon balance.

In this carbon balance model for different types of land covers used different variables: X^{cli}_p - carbon in phytomass in ecosystems "arable land" (index *i* indicates: *w* – winter crops, *s* – spring crops, *v* – vegetable, and *t* – technical crops), X^{gl}_p - carbon in phytomass in ecosystems "hayfields and pastures"; X^{cli}_l and X^{gl}_l – carbon in litter of these systems; X^{cli}_s та X^{gl}_s – carbon in soils of these systems; X_g – carbon in green phytomass (*cf* – conifers, *dfh* – hardwood and *dfs* – softwood), X_w – carbon in wood and root

Figure 2. Model of carbon fluxes in ecosystems

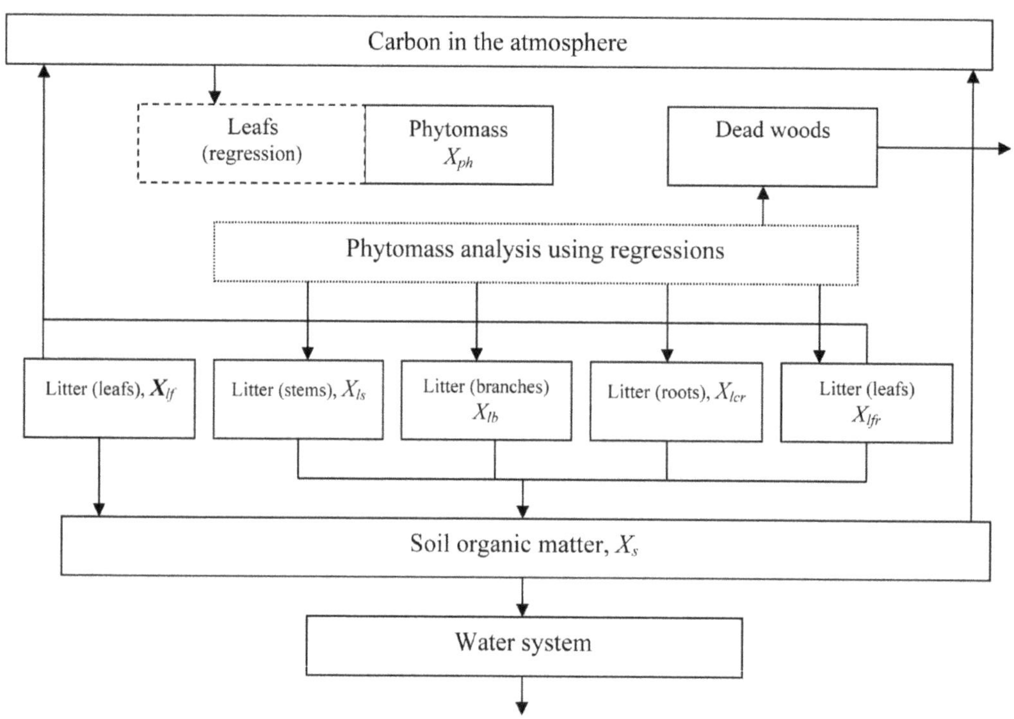

phytomass, X_s – carbon in forest soils. Let V_{ap} – carbon flux from atmosphere to sub-system "phytomass", which correspond to process of primary production, and V_{ag} – carbon flux from atmosphere to sub-system "green forest phytomass", which correspond to photosynthesis. It determinates as:

$$V_{ap} = \alpha_{ap} f_{ap}\left(C,T,P,t\right) \tag{28}$$

$$V_{ag} = \alpha_{ag} f_{ag}\left(C,T,P,t\right) \tag{29}$$

where α_{ap} and α_{ag} – empirical coefficients, determinate by calibration of model, f_{ap} and f_{ag} – non-linear functions describing dependencies between carbon flux and atmospheric carbon dioxide concentration *C*, temperature *T*, precipitation *P*, and depends on time *t*, (biophysical processes are mainly cyclical). Thus, implicitly introduced dependency of photosynthesis intensity from amount of green biomass.

This model shows that accuracy of classification of land covers and so reliability of technology of interpretation of satellite information aimed to classification is critical for task of analysis of regional carbon balance. Classification in this task means not only determination of vegetation types, but also analysis of photosynthesis, evapotranspiration, primary production, plant stress, and other parameters. Analysis of land cover distribution, vegetation condition control, and monitoring of atmospheric GHG concentrations using satellite data is the way to increase modeling reliability and correctness though utilization of more correct and comprehensive input data.

NOTES ON UNCERTAINTY ANALYSIS IN CARBON BALANCE MODELS

For error and uncertainty analysis in integrated models of carbon balance traditionally using data of statistically independent sources. For it may be used data of satellite observations.

In (Shvidenko et al, 2010) algorithm for determination of standard error m^2_y of carbon flux $Y=f(X_i)$, which defined as function of random distribution X_i with $i=1,2,\ldots k$ is proposed in form:

$$m_y^2 = \sum_{i=1}^{k}\left(\frac{dY}{dX_i}m_i\right)^2 + 2\sum_{i>j}\left(\frac{dY}{dX_i}\right)\left(\frac{dY}{dX_j}\right)r_{ij}m_{X_i}m_{X_j} \tag{30}$$

where r_{ij} – coefficients of correlation between distributions X_i and X_j.

Using this algorithm it is possible to estimate an error of carbon fluxes in all components of the system studied basing on all available data. Satellite observation data is also possible to integrate into this algorithm.

Relative integrated reliability of data can be estimated separately. It can be done through comparison of geospatial data of land cover distribution (obtained from satellite image classification), ecological modeling, geostatistics and inventory, as well as analysis of atmospheric concentration of GHG controlled from satellites. For this purpose could be used "agreement index of suitability" (Shvidenko et al, 2010) as the measure of corresponding of normalized distributions:

$$S_{nm} = \frac{1}{q}\sqrt{\sum_{j=1}^{q}(x_{nj}^{norm} - x_{mj}^{norm})^2}\ \text{, where } x_j^{norm} = \frac{x_j - x_{j\min}}{x_{j\max} - x_{j\min}} \tag{31}$$

where q – descriptive parameters; n – sites of studied area; m – sets of statistics (for example, departmental data); x – measured values of parameter j.

Mutual consistency of data distribution can be assessed using this method, and so we can more correctly combine disparate data, and minimize errors (aleatoric uncertainty).

Therefore, using of integrated models of carbon balance, especially regionalized with satellite data, allows calculate ecological and climate-related change more accurately. So modeling is suitable part of security management.

SATELLITE TOOL FOR GHG MONITORING

As an example of algorithm for atmosphere composition and conversion radiation parameters to separate gases concentration, we can use approach (Buchwitz et al, 2005; Schneising et al, 2009) developed for SCIAMACHY sensor. This algorithm (Weighting Function Modified Differential Optical Absorption Spectroscopy - WFM-DOAS)allows obtain concentrations of CO_2, CH_4 and O_2 using satellite observations data (Schneising et al, 2009).

This algorithm is based on least squares method for downscaling of selected vertical atmospheric profiles. Parameters are defining by direct observations of vertical characteristics of the atmosphere. Logarithm of linearized radiation transfer, which is defined by the atmosphere model and observation data, together with polynomial $P_i(\mathbf{a})$ is reducing to logarithm of relation of satellite measured radiation and solar radiation (i.e. to the normalized solar radiation I^{obs}). Integrated equation of WFM-DOAS algorithm can be presented as (Buchwitz et al, 2005):

$$\sum_{i=1}^{m}\left(\ln I_i^{\text{obs}} - \ln I_i^{\text{mod}}(\hat{\text{V}},\hat{\text{a}})\right)^2 \equiv \left\|\text{RES}\right\|^2 \to \min \tag{32}$$

In this equation the model of linear radiation transport is presented as:

$$\ln I_i^{\text{mod}}(\hat{\text{V}},\hat{\text{a}}) = \ln I_i^{\text{mod}}(\bar{\text{V}}) + \sum_{j=1}^{J}\left.\frac{\partial \ln I_i^{\text{mod}}}{\partial V_j}\right|_{\bar{V}_j} \times (\hat{V}_j - \bar{V}_j) + P_i(\hat{\text{a}}) \tag{33}$$

Index i refers a wavelength of image pixel i. components of vector V, denoted as V_j, determine vertical columns of all gases, which has absorption lines in selected spectral range. Factors of vertical distribution V_j and polynomial coefficients of vector **a** using as reduction parameters. Also the atmosphere temperature profile is used as parameter of smoothing.

Calculation parameters are determined as a minimum (on least squares method) between measured radiation ($\ln I^{obs}_i$) and modeled radiation ($\ln I^{mod}_i$), so by the length of vector **RES** (in particular, its components $\mathbf{RES}_i$ for every spectral measurement i).

For determination of content of carbon dioxide using average on observation depth dry air molar fraction of this gas, measured at same time with oxygen. So average on observation depth carbon dioxide concentration XCO_2 can be calculated as:

$$XCO_2 = \frac{CO_2^{col}}{\left(O_2^{col} / O_2^{mf}\right)} \tag{34}$$

where CO^{col}_2 is received absolute content of CO_2 (molecules on cm^2), O^{col}_2 the same for O_2, O^{mf}_2 – expected (averaged on observation depth) oxygen molar fraction reduced to dry air (0,2095).

This is the way to calculate tropospheric concentration of gas from radiation intensity in frame of its absorption band. Method in details is described in (Buchwitz et al, 2005).

The similar way we can use for determination of other gases concentration, for example methane. But concentration of carbon dioxide instead oxygen can be utilized in this algorithm (Schneising et al, 2009).

$$XCH_4 = \frac{CH_4^{col}}{\left(CO_2^{col} / CO_2^{mf}\right)} \tag{35}$$

where CH^{col}_4 is received absolute content of CH_4 (molecules on cm^2), CO^{col}_2 the same for CO_2, CO^{mf}_2 – expected (averaged on observation depth) carbon dioxide molar fraction reduced to dry air. For CO^{mf}_2 using constant 370 ppm.

On the Figure 3 result of application of described WFM-DOAS algorithm for processing of observations of AIRS (EOS satellite) and SCIAMACHY (ENVISAT satellite) and GOSAT (Ibuki satellite) sensors is presented.

SATELLITE TOOL FOR PLANT PRODUCTIVITY AND CARBON STOCK ASSESSMENT

Crop productivity control based on possibility of the photosynthesis measurements. Carbon is accumulating in terrestrial ecosystems by photosynthesis and emitting as CO_2 through autotrophic and heterotrophic respiration (Heimann & Reichstein, 2008). GPP (Gross Primary Production) is the total carbon assimilated by vegetation through photosynthesis.

Net Primary Production (NPP) is a residue of detent carbon after account of autotrophic respiration. Significant part of detent carbon can be accumulated in living crops and organic soils, and emission this carbon to atmosphere as CO_2 or methane can influencing to climate (Heimann & Reichstein, 2008).

Satellite instruments collect parameters distributions describing crop activity. In particular, the data from the Collection 5 of the MOD17A2 of MODIS satellite sensor can be used for vegetation analysis. The MOD17A2 is an 8-day summation of the GPP and photosynthetic production (PsnNet). Annual GPP and NPP for MOD17A determined as (Movchan & Kostyuchenko, 2015):

$$PsnNet = GPP - R_{ml} - R_{mr} \tag{36}$$

Figure 3. Temporal dynamics of carbon dioxide and methane atmospheric concentration over Ukraine 2003-2014
Source: Data Buchwitz et al, 2005 analyzed using algorithm (35)-(36), according to Groisman & Lyalko, 2012

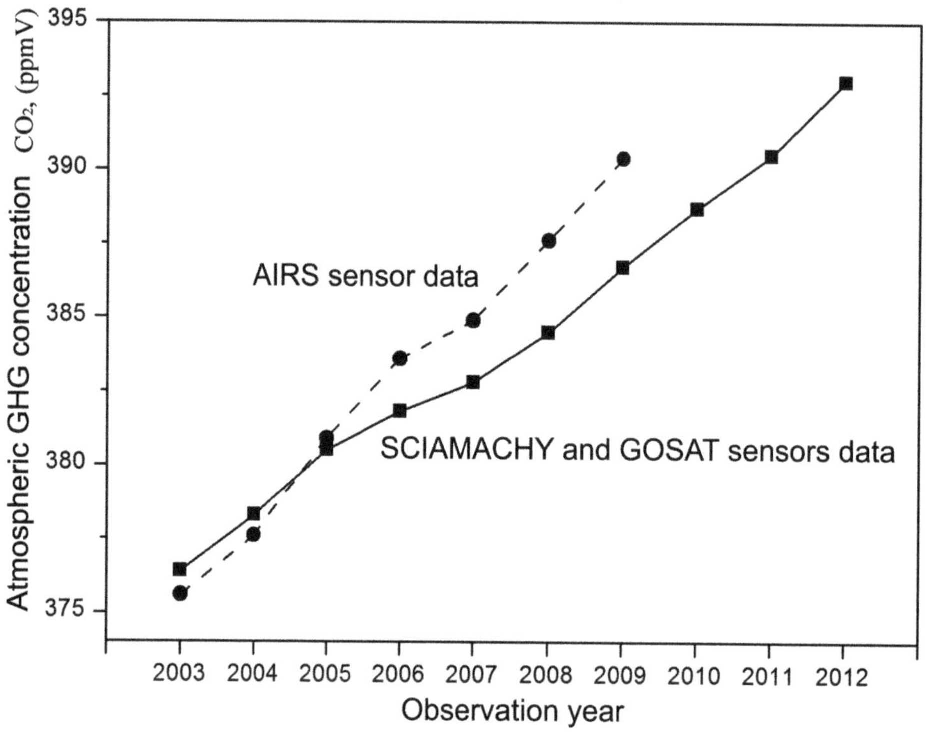

where R_{ml} and R_{mr} respiration of leafs and roots.

Annual *NPP* can be calculated as (Movchan & Kostyuchenko, 2015):

$$NPP = \sum_{i=1}^{365} PsnNet - (R_{mo} + R_g) = X_j \tag{37}$$

where R_{mo} respiration of all part of plant excluding roots and leafs, R_g – respiration of growth (Zhao et al, 2005), X_j – observed *NPP* in the year *j*.

Model MOD17 based on three sources of input data. For every pixel land cover data provided by MOD12Q1 product. Meteorological data provided by DAO (Data Assimilation Office) model. Data on spectral reflectance of surface FPAR i LAI provided from MOD15A2 product. Uncertainties of MOD12Q1, DAO, MOD15A2, as well as errors of algorithm (37)-(38) may reflect to correctness of output data MOD17.

Reliability of MOD12Q1 may be assessed in interval [0.7-0.8] for majority of land cover classes (Strahler et al, 2002; Movchan et al, 2014). Moreover, meteorological data from DAO is not direct measured data, but reanalysis data. This type of data has both systematic and random errors (Kostyuchenko, 2015). These errors seriously reflect on correctness of NPP calculation (Zhao, Running, & Nemani, 2006).

Algorithms MOD17, MOD15A2 are sensitive to accuracy of input meteorological data (temperature and precipitation), both absolute values and distribution. Regional monitoring and control of vegetation productivity using satellite tools are effective and useful, but require additional instruments to uncertainty control and errors reducing (Figure 4).

Data on vegetation productivity is necessary for models of carbon balance, and for for climate change control on regional scale (Movchan & Kostyuchenko, 2015). Many significant climate-related risks can be assessed using vegetation production data: from agricultural and food security to landscape fires. For example, fire risk dynamics may be assessed using forecasting of vegetation productivity (Janisch & Harmon, 2002).

In this context the task of uncertainty control of vegetation productivity estimation is a urgent problem for decision making in GHG emission control, environmental security, and climate adaptation planning.

UNCERTAINTY ASSESSMENT IN CROPS PRODUCTIVITY ANALYSIS

Annual sums of monthly productivity and annual productivity estimations, obtained from satellite observations using the same methodology, could be compared to estimate relative uncertainty of productivity assessments. Time-series of these assessments for various regions allow detect key trends of errors, analyze components of uncertainty, connected with variations of local climate, land-use parameters.

Figure 4. Regional dynamics of annual plant productivity (kgC/m²) and assessment of its error, measured by satellite sensor MODIS in Western part of Ukraine during 2000-2012

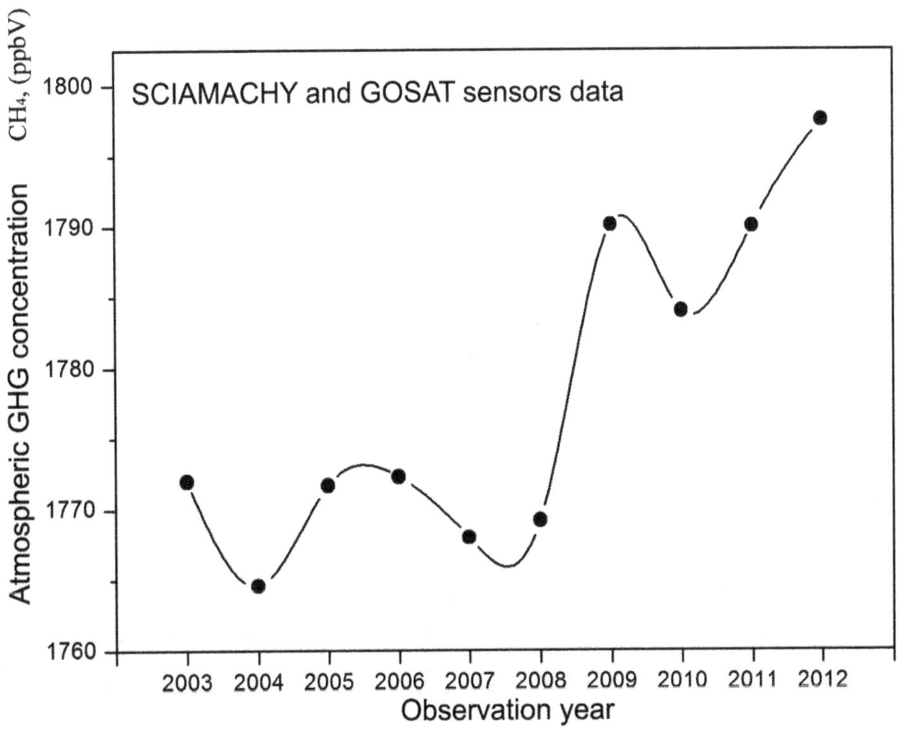

For separate region m in month i, during year j from interval k, an error of productivity estimation U_m can be determined as difference between observed annual productivity X_j and calculated monthly productivity x_{ij} according to:

$$U_m = \frac{\sum_j \frac{1}{k}((\sum_i x_{ij}) - X_j)}{< x_{ij} >} \tag{38}$$

In this case a total uncertainty of productivity assessment for all regions m of study area M may be assessed as:

$$U_{tot} = \sqrt{\sum_m U_m^2} \tag{39}$$

This algorithm has been utilized for analysis of productivity and calculation of uncertainty of vegetation of Western part of Ukraine during observation period 2000-2012. Average regional error of productivity estimation was assessed (Figure 4) according to (Movchan & Kostyuchenko, 2015).

Data shows that average annual error is between 15,7 and 18,2%. In separate regions it varied from 10,8% (regions with mainly homogeneous land-use structure, and low anthropogenic component), to 22,4% (landscapes and ecosystems with high productivity, high anthropogenic load, high heterogeneity of land covers).

Further consider an uncertainty as the aggregate of uncertainties, which connected with assessing the productivity. Thus, will assume that vectors of true state of systems $\boldsymbol{x}$ at the moment of observation k $(j, j \in \mathrm{k})$ is determined according (Aoki, 1967):

$$x_k = R_k x_{k-1} + A_k u_k + w_k , \tag{40}$$

where R_k – «system evolution matrix», e.g. formalized description of long-term managing impacts to productivity distributions (in our case – functional of climate processes) forming impacts to vector x_{k-1} at the moment k-1; A_k – «operating matrix», e.g. induced loads to system, influencing to productivity (such as land-use and land cover distribution), measured as impacts u_k to vector x; w_k – random process determining "noises" distribution, e.g. sensors, algorithms errors etc. In this case an uncertainty can be determined as:

$$\delta x = \delta \hat{x} \frac{\partial R}{\partial t} + \delta \hat{x} \partial A + \delta w , \tag{41}$$

where $\hat{x}$ - ensemble average values of measured parameter, t – time.

Formalized description of long-term managing impacts to productivity distributions will consider as functional:

$$R = f(s) \cdot F(T, W, R) + \Delta \tag{42}$$

where *f(s)* – function of density of vegetation (aggregate parameter of vegetation rarefaction, or degradation of crops related to bushiness or to the crown branching), *F(T, W, R)* – productivity functional, in general case depends of temperature, soil moisture and solar radiation (practically might be determined by remote observations), Δ – uncertainty coefficient (Kostyuchenko et al, 2013a).

If we assume the function of the density of vegetation the smooth homogeneous algebraic function, its behavior may be uniquely described using spectral vegetation indices. To describe operating impacts we can without generalization use a following form of equation (Kostyuchenko et al, 2013a):

$$\delta\hat{x}\frac{\partial R}{\partial t} \to \frac{\sum_i (x_i - \hat{x}_i)\sum_j (\delta T_j - \delta\hat{T}_j)}{\sqrt{(\sum_i x_i - \hat{x}_i)^2 (\sum_j \delta T_j - \delta\hat{T}_j)^2}} \frac{\partial(1 - \frac{\frac{1}{N}\sum_{n=1}^{N} T_n}{T^{\max}})(1 - \frac{1}{T^{\max} - \frac{1}{N}\sum_n T^{\max}})}{\partial t} \tag{43}$$

where T_n – air temperature during observation period; T^{max} – maximal air temperature during observation period; δT – «reduced temperature»,

$$\delta T = (1 - \frac{\frac{1}{N}\sum_{n=1}^{N} T_n}{T^{\max}})(1 - \frac{1}{T^{\max} - \frac{1}{N}\sum_{n=1}^{N} T^{\max}}).$$

Parameter «reduced temperature» is an optimal correlator of parameters of ecosystems vulnerability and risks, for example, productivity and climatic parameters. «Reduced temperature» was calculated basing on the analysis of long-term correlative relationships in multivariate parameters distributions using elliptic kopulas (Kostyuchenko et al, 2013b). Thus we approximate climatic impacts by changes of temperature. This is not absolutely correct, but measurements of temperature are most reliable measurement among all parameters.

Induced loads to ecosystem we will descried as impacts from land-use and land cover changes. Corresponding uncertainties connected with not correct classification and statistics (annual and seasonal), as well as significant variations of anthropogenic load to land on local scale.

This type of uncertainty (uncertainty connected with inaccuracy of classification of the observed system) we will through relative reliability of aggregated data. The «agreement index of suitability» (32) (Shvidenko, et al, 2010) may be used for it as:

$$\delta\hat{x}\partial A = \frac{\sum_i (x_i - \hat{x}_i)\sum_j (x_j^{norm} - \hat{x}_j^{norm})}{\sqrt{(\sum_i x_i - \hat{x}_i)^2 (\sum_j x_j^{norm} - \hat{x}_j^{norm})^2}} \frac{1}{n}\sqrt{\sum_{p,q} (\frac{x_p - x_p^{\min}}{x_p^{\max} - x_p^{\min}})(\frac{x_q - x_q^{\min}}{x_q^{\max} - x_q^{\min}})^2} \tag{44}$$

where q – study area sites; p – sets of statistical data (for example, field calibration measurements, statistics, cartography); x – measured value of investigated parameter; x^{norm} – «normalized agreement index of suitability», $x_{norm} = (\sum_{p,q} (\frac{x_p - x_p^{\min}}{x_p^{\max} - x_p^{\min}}) (\frac{x_q - x_q^{\min}}{x_q^{\max} - x_q^{\min}})^2)^{1/2}$. Using this approach we can estimate a mutual consistency of data distributions, most correctly combine disparate data, and so to minimize errors.

Distribution of uncertainties connected with methodical errors (aleatoric uncertainty) can be described as process of measurements of system state vectors, e.g. as variations of random process w_k determining set of sensor and processing errors. It can be assessing as:

$$\delta w = \sqrt{\sum_{i}^{n} (\Delta x_i \frac{\partial F}{\partial x})^2} + \nu' \tag{45}$$

where $F(x)$ – calculated value on the sequence of measured parameters x $(x_1, x_2, \ldots, x_n)$ e.g. $F=F(x_1, x_2, \ldots, x_n)$; ν' – unknown error, which we interpret as inherent uncertainty. Component $\frac{\partial F}{\partial x}$ and so $\sqrt{\sum_{i}^{n} (\Delta x_i \frac{\partial F}{\partial x})^2}$ can be calculated using assessments of accuracy of MOD17 FPAR i MOD15A2, according (Zhao et al, 2005; Wang et al, 2001). Using data and results (Wang et al, 2001; Morisette, Privette, & Justice, 2002), we can estimate this component in diapason [0,045; 0,065].

Therefore, integrated form of equation for assessment of components of uncertainty in task of vegetation productivity assessment could be presented as (Kostyuchenko et al, 2013a):

$$\frac{\sum_j \frac{1}{k}((\sum_i x_{ij}) - X_j)}{< x_{ij} >} = \frac{\sum_i (x_i - \hat{x}_i) \sum_j (\delta T_j - \delta \hat{T}_j)}{\sqrt{(\sum_i x_i - \hat{x}_i)^2 (\sum_j \delta T_j - \delta \hat{T}_j)^2}} \frac{\partial (1 - \frac{\frac{1}{N}\sum_{n=1}^{N} T_n}{T^{\max}})(1 - \frac{1}{T^{\max} - \frac{1}{N}\sum_{n=1}^{N} T^{\max}})}{\partial t} + \\ + \frac{\sum_i (x_i - \hat{x}_i) \sum_j (x_j^{norm} - \hat{x}_j^{norm})}{\sqrt{(\sum_i x_i - \hat{x}_i)^2 (\sum_j x_j^{norm} - \hat{x}_j^{norm})^2}} \frac{1}{n} \sqrt{\sum_{p,q} (\frac{x_p - x_p^{\min}}{x_p^{\max} - x_p^{\min}}) (\frac{x_q - x_q^{\min}}{x_q^{\max} - x_q^{\min}})^2} + \sqrt{\sum_{i}^{n} (\Delta x_i \frac{\partial F}{\partial x})^2} + \nu' \tag{46}$$

Separate components of uncertainty could be analyzed using this equation in task of assessment of vegetation productivity by satellite observation data.

It should be noted, that this equation may be reduced for calculation of carbon balance up to constant (Shvidenko et al, 2010), as an algorithm for determination of standard error m_y^2 of carbon flux $Y=f(X_i)$, determined as random distribution function X_i, $i=1,2,\ldots k$:

$$m_y^2 = \sum_{i=1}^{k} (\frac{dY}{dX_i} m_i)^2 + 2\sum_{i>j} (\frac{dY}{dX_i})(\frac{dY}{dX_j}) r_{ij} m_{X_i} m_{X_j} \quad (47)$$

where r_{ij} – coefficients of correlations between distributions X_i and X_j.

This algorithm allows estimate errors of carbon fluxes estimations in all components of studied system using all available data. Ii is important, that it is possible to incorporate into whole approach data of modeling, in-field ground measurements and satellite observations.

Calculated using this algorithm (47) component of uncertainty presented on the Figure 5.

As the calculations show, estimated uncertainty of assessment of vegetation productivity by satellite data is [0.138; 0.152]. Constant component (aleatoric uncertainty) of uncertainty is errors of sensor, processing of data (calculation of indexes, estimation of biophysical parameters from reflection, etc.) and monitoring methods (non-optimal time of observation, incorrect data integration, etc.). This uncertainty component is about [0.045; 0.065]. This constant component cannot be essentially reduced by data user.

Varied component, which correspond to regional variations of climatic parameters, lies in interval [0.02; 0.035]. This component may be reduced with account of correct regional climate trends.

Significant uncertainty component – [0.053; .074] is caused by incorrect classification of land cover and imperfect analysis of local land-use structure. This type of uncertainty (epistemic) is connected with lack of information about system studied, and requires more comprehensive models to reduce it. Besides ground calibration data and observation verification are required too.

Figure 5. Estimation of components of uncertainty of vegetation productivity detection using satellite data
Source: Kostyuchenko et al, 2013a

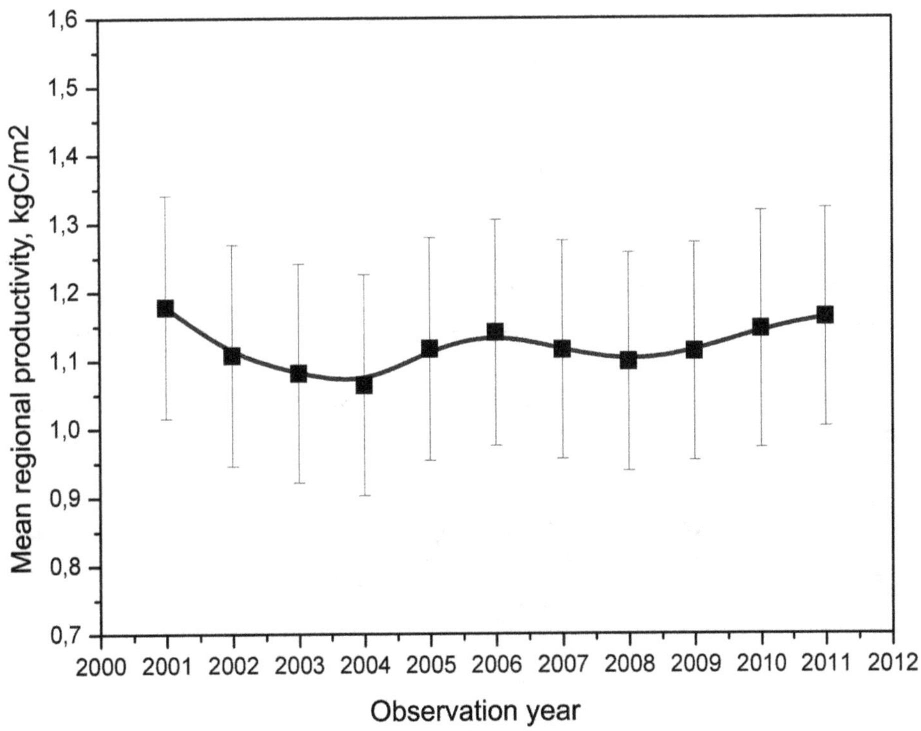

Besides, there is an uncertainty component about [0.01; 0.005] of unknown origin, which we identify as inherent uncertainty.

Therefore, it can be argued, that integration of Earth systems modeling, analysis of satellite data, and processing of ground measurements for calibration and verification allow to reduce 55-65% of error of vegetation productivity estimation.

EMISSION CALCULATION AND UNCERTAINTY ANALYSIS

Basing on measured atmospheric GHG concentrations it is possible to calculate corresponding GHG emissions.

The simplest way to calculate aggregate annual GHG emission by its concentrations, according to (Houghton, 2005, IPCC, 1997) is:

$$E_i = \sum_i \kappa_i \frac{(C_{current} - C_{previous})M_i}{C_{0i}} / t \tag{48}$$

here $C_{current}$ – current mean annual atmospheric concentration of gas i over certain observed territory; $C_{previous}$ – previous current mean annual atmospheric concentration of gas i over certain observed territory; C_{0i} - mean annual atmospheric concentration of gas i over same territory during pre-industrial era (according to geological data and historical records); M_i – historical mass of estimated GHG in the atmosphere (according to geological data and historical records) (Raynaud et al, 2000). For carbon dioxide and methane C_0 is 275 and 700 ppmv; and M_i - 220,89 x 10^{10} tons for CO_2 and 408,94 x 10^{10} tons for CH_4 (Petit et al, 1999) for whole Earth atmosphere. Coefficient κ_i determining with account of lifetime of GHG and its mass at the moment of estimation (for short-term forecasting it possible to assume κ_i = 1 for all GHG); t – averaging period (in years).

This algorithm allows to control national GHG emissions by independent tools (satellite observations). Analysis of identified differences (Figure 6) is a base for decision making if field of national emission management and environmental protection policy.

In particular, detected difference of assessments could indicate inadequateness the national inventory system during the observation period. This data also should focus attention to importance ecological processes. Input these processes into GHG balance could be 19-23% and it should adequately reflected in emission report data.

In the case of described distributions (44)-(46) we can present a modified with Equations 19 form of uncertainty analysis:

$$U_n = (| E^1_{tot,n} - E^2_{tot,n} |) / (\frac{\sigma_i \sum_{i,m} (\frac{| E^1_{tot,n} - E^2_{tot,n} |}{2} - \sum_i E^1_{tot,n})}{CoVar_m}) \tag{49}$$

Figure 6. Comparison of emission data from official inventory report (anthropogenic emissions), ecological models of carbon balance, and satellite observations for period 2003-2010
Source: Groisman & Lyalko, 2012

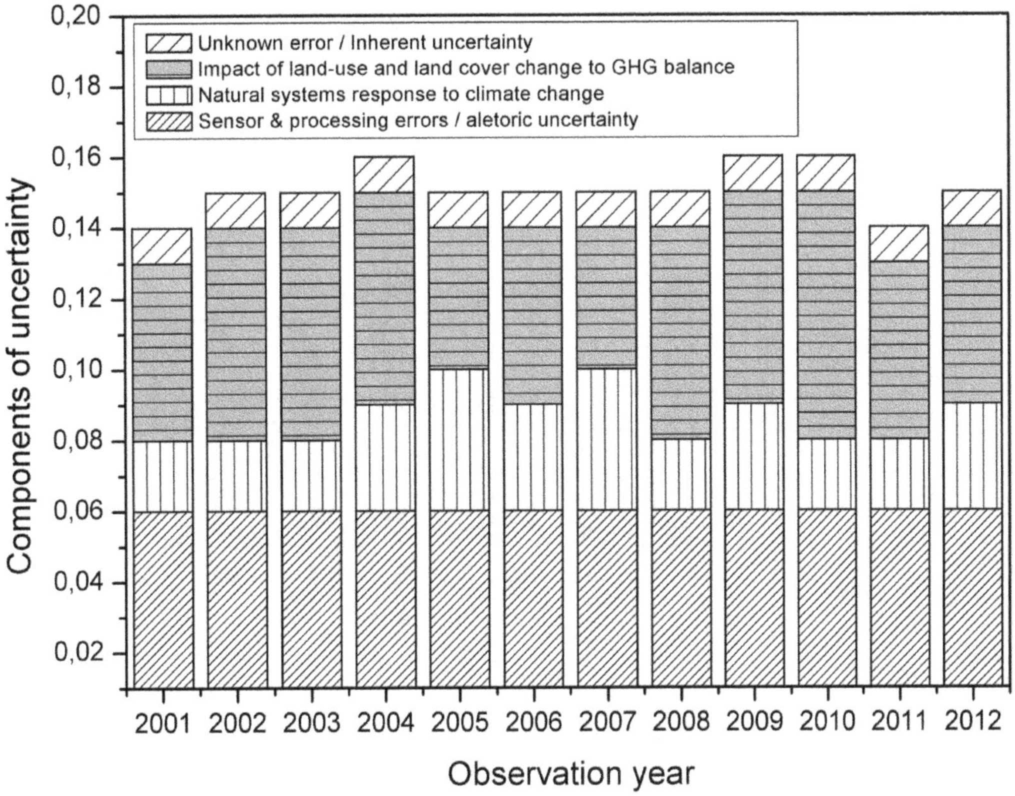

Application of (20) and (50) allow to estimate average change of data reliability (connected with reducing of relative uncertainty) depending on data source (offal statistics, ecological models, satellite observations) and with respect to each other (Figure 7).

Calculations show that assessments based on ecological models and observation data are more correct and accurate on all scales than statistics. It, at least, means that stochastic and fuzzy approaches to account both aleatoric and epistemic uncertainty in complex systems are more correct base for decision making, than interval aleatoric uncertainty control only based on deterministic approaches.

CONCLUSION

In this chapter the advantages of applied mathematics based approaches for ecology, environmental security, climate regional studies oriented to change adaptation, and for remote sensing technologies development is demonstrated.

Decision making usually focusing on optimization (and so risk minimization) through collected information completing and/or uncertainty reduction. In case of complex multi-physics systems and

Figure 7. Reliability of emission assessments from different data sources
Source: Groisman & Lyalko, 2012

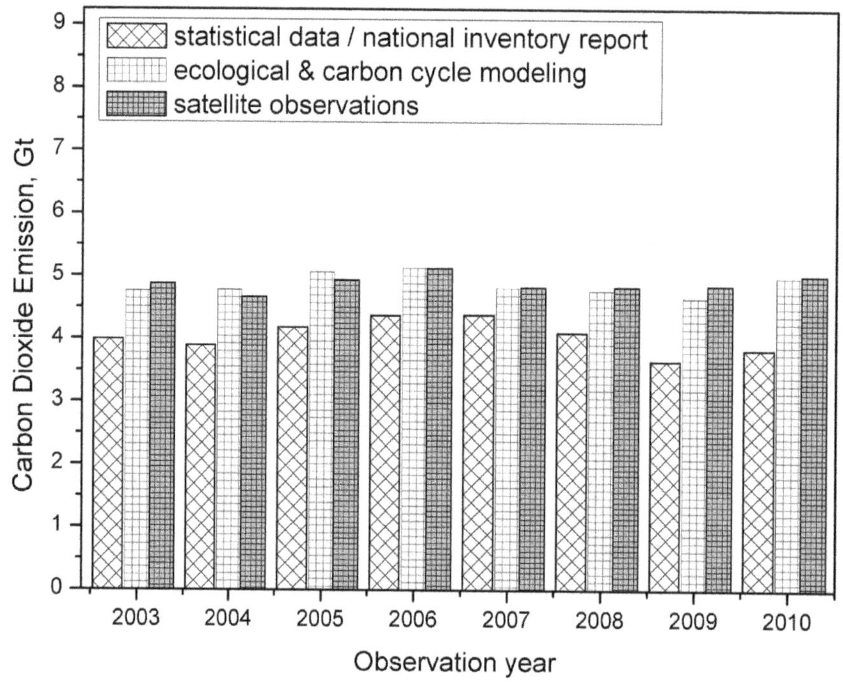

Figure 8. Ecological and carbon balance modeling

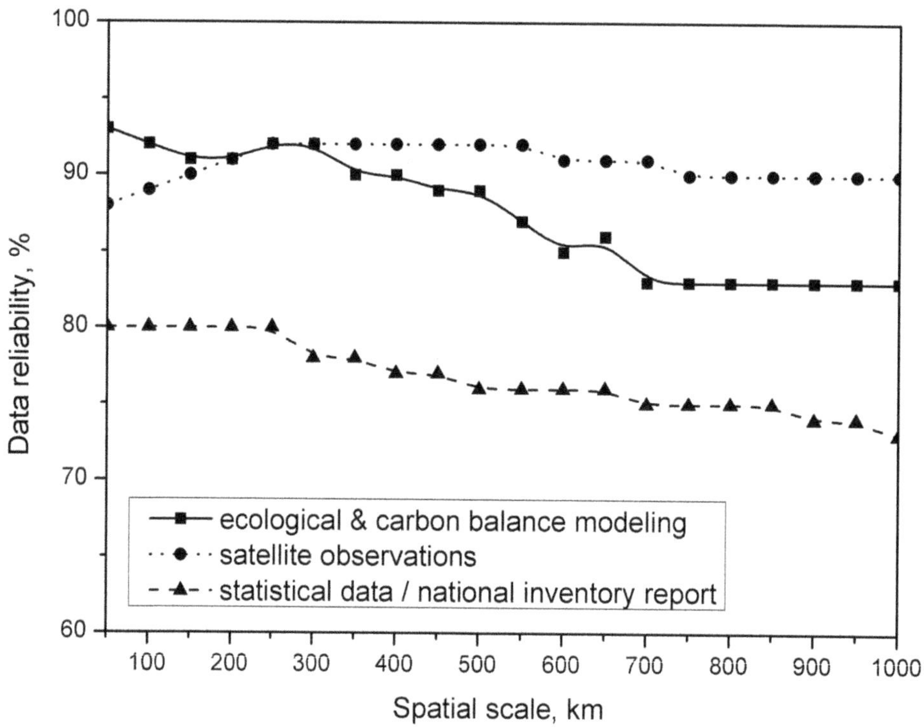

multivariate processes the analysis of uncertainty requires sophisticated tools. But at the same time it can be smart instrument for decision making and risk reduction.

The problem of uncertainty analysis in multy-component systems and multy-physics processes is very complicated and urgent task, which cannot be reduced to measurements accuracy control. On the base of applied mathematics approaches utilization we demonstrate that the comprehensive analysis of uncertainties allow not only control and reduce errors, but also obtain new knowledge about the systems we study.

In the framework of general task of decision making under uncertainties and risk the problem of uncertainty control in GHG emission inventory using satellite tools is described.

Deferent approaches to anthropogenic emissions inventory and analysis are described and discussed. Possible algorithms of uncertainty control for discussed approaches are proposed.

It is shown that a decisive role in uncertainty reducing in the tasks of GHG emission plays a correct analysis of carbon cycle, and so – mathematical models of ecosystems. For determination of key model variables, measured parameters and indicators for satellite observations the GHG and carbon cycle models have been analyzed. Uncertainties in carbon balance models are analyzed, and algorithms to reduce it are proposed.

Basing on the models described the satellite tool for GHG monitoring is defined and discussed. An algorithm for calculation of individual gas concentrations in atmosphere from radiation intensity within the separate absorption bands is analyzed.

As the important component of GHG monitoring system a satellite tool for plant productivity assessment is analyzed. Multicomponent stochastic model for uncertainty assessment in productivity analysis is proposed. As the separate, intercorrelated and multivariate components of uncertainty the instrumental errors, climate impact, and land-use change factors have been determined and analyzed. It was shown that quantitative assessments of these components in its temporal dynamics allow obtain new knowledge about investigated tools, climatic parameters and ecosystem we study.

Algorithm for emission calculation from atmospheric GHG concentration is proposed. Approach and algorithm to analyze and reduce an uncertainty in emission account are proposed and discussed.

Conclusions on the reliability and correctness of decisions based on the approach discussed are demonstrated. Calculations show that assessments based on ecological models and observation data are more correct and accurate on all scales than statistics. We can conclude than stochastic and fuzzy approaches to account both aleatoric and epistemic uncertainty in complex systems are more correct base for decision making, than interval aleatoric uncertainty control only based on deterministic approaches.

ACKNOWLEDGMENT

The authors are grateful to anonymous referees for constructive suggestions that resulted in important improvements to the chapter, to colleagues from the International Institute for Applied Systems Analysis (IIASA) as well as from the American Society for Civil Engineers (ASCE) and from American Meteorological Society (AMS) for their critical and constructive comments and suggestions, and for presented important data and models.

REFERENCES

Aoki, M. (1967). *Optimization of stochastic systems: topics in discrete-time systems*. New York, London: Academic Press.

Booker, J., Parkinson, J., & Ross, T. J. (Eds.). (2002). *Combined fuzzy logic and probability applications*. Philadelphia: SIAM.

Buchwitz, M., de Beek, R., Burrows, J. P., Bovensmann, H., Warneke, T., Notholt, J., & Schulz, A. et al. (2005). Atmospheric methane and carbon dioxide from SCIAMACHY satellite data: Initial comparison with chemistry and transport models. *Atmospheric Chemistry and Physics*, *5*(4), 941–962. doi:10.5194/acp-5-941-2005

Bun, R.A., Dachuk, V.S. (1997). Modeling of global and regional carbon cycles in biosphere. *Problems of Operating Research and Informatics*, *2,* 141-148.

Ermolieva, T., & Ermoliev, Y. (2005). Catastrophic risk management: Flood and seismic risks case studies. In S. W. Wallace & W. T. Ziemba (Eds.), *Applications of stochastic programming*. Philadelphia, PA: MPS-SIAM Series on Optimization. doi:10.1137/1.9780898718799.ch22

Ermolieva, T., Ermoliev, Y., Fischer, G., Jonas, M., & Makowski, M. (2010). Cost effective and environmentally safe emission trading under uncertainty. In K. Marti, Y. Ermoliev, & M. Makowski (Eds.), *Coping with uncertainty: Robust solutions* (pp. 79–99). Heidelberg, Germany: Springer-Verlag. doi:10.1007/978-3-642-03735-1_5

Ermolieva, T., Ermoliev, Y., Jonas, M., Obersteiner, M., Wagner, F., & Winiwarter, W. (2014). Uncertainty, cost-effectiveness and environmental safety of robust carbon trading: Integrated approach. *Climatic Change*, *124*(3), 633–646. doi:10.1007/s10584-013-0824-2

Godal, O., Ermoliev, Y., Klassen, G., & Obersteiner, M. (2003). Carbon trading with imperfectly observable emissions. *Environmental and Resource Economics*, *25*(2), 151–169. doi:10.1023/A:1023914324084

Groisman, P., & Lyalko, V. (Eds.). (2012). *Earth Systems Change over Eastern Europe*. Kiev: Akademperiodyka.

Heimann, M., & Reichstein, M. (2008). Terrestrial ecosystem carbon dynamics and climate feedbacks. *Nature*, *451*(7176), 289–292. doi:10.1038/nature06591 PMID:18202646

Houghton, J. (2005). Climate change and sustainable energy. Met Office.

IPCC. (1997). *Revised 1996 IPCC Guidelines for National Greenhouse Gas Inventories* (Vols. 1-3). Bracknell, UK: Intergovernmental Panel on Climate Change, Meteorological Office.

Janisch, J. E., & Harmon, M. E. (2002). Successional changes in live and dead wood carbon stores: Implications for net ecosystem productivity. *Tree Physiology*, *22*(2-3), 77–90. doi:10.1093/treephys/22.2-3.77 PMID:11830405

Jaulin, L., Kieffer, M., Didrit, O., & Walter, E. (2001). *Applied interval analysis: with examples in parameter and state estimation, robust control and robotics*. London: Springer. doi:10.1007/978-1-4471-0249-6

Kostyuchenko, Y. V., Bilous, Y., Kopachevsky, I., & Solovyov, D. (2013). Coherent risk measures assessment based on the coupled analysis of multivariate distributions of multisource observation data. In *Proceedings of 11-th International Probabilistic Workshop.*

Kostyuchenko, Yu. V. (2015). Geostatistics and remote sensing for extremes forecasting and disaster risk multiscale analysis. In S. Kadry & A. El Hami (Eds.), *Numerical Methods for Reliability and Safety Assessment: Multiscale and Multiphysics Systems* (pp. 404–423). Springer International Publishing. doi:10.1007/978-3-319-07167-1_16

Kostyuchenko, Y. V., Bilous, Y., Movchan, D., Lászlő, M., & Kopachevsky, I. (2013). Toward methodology of satellite observation utilization for agricultural production risk assessment. *IERI Procedia*, *5*, 21-27.

Kreinovich, V., Nguyen, H. T., Ferson, S., & Ginzburg, L. (2003). From computation with guaranteed intervals to computation with confidence intervals: a new application of fuzzy techniques. In *Proceedings of the 21st International Conference of the North American Fuzzy Information Processing Society NAFIPS'2002.*

Moore, R. E., & Lodwick, W. A. (2003). Interval analysis and fuzzy set theory. *Fuzzy Sets and Systems*, *135*(1), 5–9. doi:10.1016/S0165-0114(02)00246-4

Morisette, J. T., Privette, J. L., & Justice, C. O. (2002). A framework for the validation of MODIS land products. *Remote Sensing of Environment*, *83*(1-2), 77–96. doi:10.1016/S0034-4257(02)00088-3

Movchan, D., Kostyuchenko, Y., Marton, L., Frayer, O., & Kyryzyuk, S. (2014). Uncertainty Analysis in Crop Productivity and Remote Estimation for Agricultural Risk Assessment. In Vulnerability, Uncertainty, and Risk: Quantification, Mitigation, and Management (pp. 1008-1015). ASCE. doi:10.1061/9780784413609.102

Movchan, D., & Kostyuchenko, Yu. V. (2015). Regional Dynamics of Terrestrial Vegetation Productivity and Climate Feedbacks for Territory of Ukraine. *International Journal of Geographical Information Science*, *29*(8), 1490–1505. doi:10.1080/13658816.2015.1051985

Nguyen, H. T., & Kreinovich, V. (2006). Optimization and Decision Making under Interval and Fuzzy Uncertainty: Towards New Mathematical Foundations. *Studies in Fuzziness and Soft Computing*, *201*, 275–290. doi:10.1007/3-540-33517-X_10

Petit, J. R., Jouzel, J., Raynaud, D., Barkov, N. I., Barnola, J. M., Basile, I., & Stievenard, M. et al. (1999). Climate and atmospheric history of the past 420,000 years from the Vostok ice core, Antarctica. *Nature*, *399*(6735), 429–436. doi:10.1038/20859

Raynaud, D., Barnola, J. M., Chappellaz, J., Blunier, T., Indermuhle, A., & Stauffer, B. (2000). The ice record of greenhouse gases: A view in the context of future changes. *Quaternary Science Reviews*, *19*(1-5), 9–17. doi:10.1016/S0277-3791(99)00082-7

Schneising, O., Buchwitz, M., Burrows, J. P., Bovensmann, H., Bergamaschi, P., & Peters, W. (2009). Three years of greenhouse gas column-averaged dry air mole fractions retrieved from satellite – Part 2: Methane. *Atmospheric Chemistry and Physics*, *9*(2), 443–465. doi:10.5194/acp-9-443-2009

Shvidenko, A., Schepaschenko, D., McCallum, I., & Nilsson, S. (2010). Can the uncertainty of full carbon acconting of forest ecosystem be made acceptable to policymakers? *Climatic Change*, *103*(1-2), 137–157. doi:10.1007/s10584-010-9918-2

Strahler, A. H., Friedl, M., Zhang, X., Hodges, J., Cooper, C. S. A., & Baccini, A. (2002). The MODIS Land Cover and Land Cover Dynamics Products. Presentation at Remote Sensing of the Earth's Environment from TERRA, L'Aquila, Italy

Wang, Y., Tian, Y., Zhang, Y., El-Saleous, N., Knyazikhin, Y., Vermote, E., & Myneni, R. B. (2001). Investigation of product accuracy as a function of input and model uncertainties: Case study with SeaWiFS and MODIS LAI/FPAR Algorithm. *Remote Sensing of Environment*, *78*(3), 296–311. doi:10.1016/S0034-4257(01)00225-5

White, T., Jonas, M., Nahorski, Z., & Nilsson, S. (Eds.). (2011). *Greenhouse gas inventories. Dealing with uncertainties*. Heidelberg, Germany: Springer.

Zhao, M., Heinsch, F. A., Nemani, R. R., & Running, S. W. (2005). Improvements of the MODIS terrestrial gross and net primary production global data set. *Remote Sensing of Environment*, *95*(2), 164–176. doi:10.1016/j.rse.2004.12.011

Zhao, M., Running, S. W., & Nemani, R. R. (2006). Sensitivity of Moderate Resolution Imaging Spectroradiometer (MODIS) terrestrial primary production to the accuracy of meteorological reanalyses. *Journal of Geophysical Research*, *111*(G1). doi:10.1029/2004JG000004

Chapter 18
Journey from Order to Chaos and Returning

Anuraj Singh
Graphic Era University, India

ABSTRACT

In this chapter, different aspects have been described to investigate the ecological system with respect to several factors that may be responsible for emergence of complex behaviors. Some mathematical models incorporate time delay(s) such as delay due to maturation, gestation and other kinds of negative feedback. These nonlinear delay models have led to complexities in the system. The emphasis is to explore the complex dynamical behaviors including chaos in ecological models with respect to different control parameters. Motivation behind synchronization and stabilization of chaos in the system has also been emphasized. Hence, in this chapter attempts have been made to study order and chaos in variety of models applicable to multi-species ecological systems. Such a study is important since the ecological systems have all the necessary ingredients to be able to support chaos. The attempt is made to brief different mechanism that may bring order into chaotic systems or vice versa.

INTRODUCTION

Nonlinearity is all pervading in the ecological systems. In the recent times increasing attention has been focused on exploring real technological applications of nonlinear dynamics in ecology: emergence of chaos, controlling of chaos, synchronization of chaos to name but a few. Numerous mathematical ideas and techniques have been used to study nonlinear ecological systems and these, in turn have enriched the field of mathematics itself. Long term behaviors are investigated through stability of equilibrium states of underlying nonlinear dynamical models which consists of system of coupled differential or delay differential equations. The underlying system of delay differential equations are much more challenging to analyze compared to corresponding system of ODE's. The characteristic equation corresponding to the linearized system of delay differential equation is no longer a polynomial, rather, it is exponential polynomial or quasi-polynomial. Even for this simple linear equation, the stability of the trivial equilibrium requires an analysis of the roots of a transcendental equation. Analysis will be more involved in

DOI: 10.4018/978-1-5225-1639-2.ch018

case of two or more delays. Because of the complexity involved in obtaining the eigenvalues from the transcendental characteristic equation, the bifurcation problem will be further complicated due to presence of multiple delays. These nonlinear delay models have lead to complexities in the system (Gakkhar & Singh, 2010; Gakkhar, Singh, & Singh, 2012). The issues of existence, extinction and persistence in biological communities are of concern to ecologists. The ecological systems contain the ingredients for possible occurrence of chaos. The behavior of solution is found to be sensitive to initial conditions as well as the specific model parameter with respect to which chaos is observed.

As such the chaotic systems are not robust in contrast to ecological systems. Since then, many attempts have been made to study the complexity in ecological system. Owing to the fact that the ecosystems are subjected to variety of perturbation in the real world, it is indispensable to exert suitable control in order to maintain the balance in ecological models. Stabilization via feedback linearization is the fundamental idea of nonlinear control theory. As synchronization might play an important role in enhancing population persistence, the embarrassing situation caused by chaos can be changed and system can reach up to a dynamic balance. In ecological aspect, asynchrony allows for the global persistence of a population through rescue effects, even there are local extinctions. Indeed in certain circumstances, the highly synchronizing effects of chaos have been shown to enhance the global persistence of model populations even though large population swings often associated with chaotic dynamics increase the chance of local extinction. The mathematical techniques are vital for exploring rapid changes in controlling factors across threshold that cause abrupt changes between different states of an ecological system. These changes between different states are often accompanied by the appearance of behaviors such as limit cycles, chaos, extinction of species, or changes in top-down and bottom-up controls.

ORDER TO CHAOS

Over the last two decades there has been a race between theoretical and empirical population biologists to look for complex dynamics and chaos in real systems (Aziz-Alaoui, 2002; Gakkhar et al., 2012; Hassell, Lawton, & May, 1976; Hastings & Powell, 1991; Klebano & Hastings, 1994; Ruxton, 1999). It is well accepted that ecosystems possess all the necessary features such as nonlinearity, multidimensionality etc. to display chaotic behavior. Many theoretical studies have demonstrated that ecological chaos, if and when exists, can affect important ecosystem features such as predictability, species persistence and biodiversity (Allen, Schaffer, & Rosko, 1993; Aziz-Alaoui, 2002; Berryman & Millstein, 1989; Bhattacharya & Begum, 1996; Jorgensen & Bernardi, 1997; Jorgensen, 2002). Unpredictability is everywhere in ecological system. Although Hassell et al. (1976) found no evidence of chaotic behavior in any population, the notion that populations fluctuations in nature may be caused by deterministic chaos has persisted. It was discussed that the ecological systems have seeds of chaos (Hastings & Powell, 1991; Klebano & Hastings, 1994; Rai, 2004). Rai (2004) explained one of the reasons for non-occurrence of chaos lies in the organization of the ecological system. Chaos may be driven by human actions that increase growth rates or induce delays in the regulatory process. The chaotically fluctuating population is prone to extinction, with the consequence that group selection acts to eliminate species and chaos disappears.

ORDER/STABILITY

Let $X(t) \in R^n$ describes the state of a given system at a time t. The dynamics of the system is governed by a system of first order nonlinear differential equation:

$$\dot{\bar{X}} = F(X) \tag{1}$$

Let $F(X) = 0$, then X^* be an equilibrium point of the system. Consider X_1 be its neighboring point. The equilibrium point X^* is stable (Lyapunov stable) if for each $\varepsilon > 0$ there is a $\delta > 0$ such that

$$\left\| X(t) - X^* \right\| < \varepsilon, \text{whenever } \left\| X^* - X_1 \right\| < \delta$$

In other words, X^* is stable if solutions starting close to X^* at a given time remain close to X^* for all future times. More simply, nearby trajectory stay nearby. It is asymptotically stable if nearby solutions not only stay close, but also converge to X^* as *t* tends to infinity for all future time.

Mathematically, If X^* is stable and

$$\lim_{t \to \infty} X(t) = X^*$$

then the solution X^* is asymptotically stable.

Asymptotic Stability $\rightarrow$ Stability.

Stability of an equilibrium point is a local property.

An equilibrium X^* that is not stable is called unstable.

LOCAL STABILITY ANALYSIS

The equations governing interacting populations often taken the form

$$\frac{dX_i}{dt} = F_i(X_1, X_2, \ldots, X_n) \tag{2}$$

with $X_i(0) = X_{i0} \geq 0,\ 1 = 1, 2, \ldots, n$. It is assumed that F_i is such that solution of initial value problem is unique.

Let $X(t)$ be any other solution in the neighborhood of equilibrium point X^*

$$X = X^* + \xi \tag{3}$$

where $\xi = (\xi_1, \xi_2, ..., \xi_n)$ is a small perturbation from equilibrium point. Then the linear approximation for small perturbation can be written as

$$\frac{d\xi}{dt} = \left.\frac{\partial F}{\partial X}\right|_{X=X^*} \xi = A\xi. \tag{4}$$

The matrix $A = (a_{ij})$ is the Jacobian (variational) matrix at the equilibrium point X^*. Let $\xi(0)$ be the initial perturbation from the equilibrium point X^*, then the formal matrix solution to (4) can be given by

$$\xi(t) = e^{At}\xi(0) \tag{5}$$

The equilibrium point X^* of the system is stable when the perturbation $\xi(t)$ dies out as *t* tends to ∞. This is possible only if the real parts of the eigenvalues λ_i's of the matrix A namely, $\mathrm{Re}\{\lambda_i\}$, are less than zero for all *i*. If $\mathrm{Re}\{\lambda_i\} > 0$ for at least one value of *i* then $\|\xi(t)\|$ diverges as *t* increases and the equilibrium point is unstable. Since $\xi(t)$ is the solution of linearized system which is an approximate of nonlinear system, the stability is referred to local/linear stability only.

Therefore, the eigenvalues of the Jacobian matrix decide whether the equilibrium point is linearly stable or unstable. The characteristic equation for the variational matrix is obtained as

$$\det\left(A - \lambda I\right) = \lambda^n + a_1\lambda^{n-1} + ... + a_n = 0,\ a_0 \neq 0 \tag{6}$$

The coefficients $a_i; i = 1, 2, 3, ..., n$ of characteristic equation are all real. The system (2) is locally stable about the equilibrium point if all of the eigenvalues have negative real parts. On the other hand, the system is unstable if at least one of the eigenvalues has positive real part. In other words, all the eigenvalues of Jacobian matrix must lie in the left half of the complex plane. Accordingly, the necessary condition (not sufficient) for all eigenvalues to have negative real part is

$$\sum_{i=1}^{n} Trace(A) < 0.$$

In the special case of $Trace(A) = 0,$ either at least one eigenvalue must lie in the right half plane, or all eigenvalues must be purely imaginary (the pathological case of neutral stability). Another necessary, but not sufficient, condition is $\left(-1\right)^n \det | A |> 0.$

Routh-Hurwitz Criterion gives necessary and sufficient conditions to ensure that all eigenvalues lie in left half complex plane. These conditions, together with a_n, are

$$D_1 = a_1 > 0,\ D_2 = \begin{vmatrix} a_1 & a_3 \\ 1 & a_2 \end{vmatrix} > 0,\ D_3 = \begin{vmatrix} a_1 & a_3 & a_5 \\ 1 & a_2 & a_4 \\ 0 & a_1 & a_3 \end{vmatrix} > 0,$$

$$D_k = \begin{vmatrix} a_1 & a_3 & . & . & . & . \\ 1 & a_2 & a_4 & . & . & . \\ 0 & a_1 & a_3 & . & . & . \\ 0 & 1 & a_2 & . & . & . \\ . & . & . & . & . & . \\ 0 & 0 & . & . & . & a_k \end{vmatrix} > 0,\ k = 1, 2, ..., n$$

If the choice of parametric values is such that all the constraints are simultaneously satisfied, then the system will be locally asymptotically stable at X^*. On the other hand, violation of any one of these conditions implies instability of the equilibrium state.

GLOBAL STABILITY

An equilibrium is globally asymptotic stable if it is asymptotically stable and all initial values converge to the equilibrium (Alligood, Sauer, & Yorke, 1997). A system which is stable for all initial conditions in the state space is globally stable.

The global behavior of a nonlinear system can be different than its local behavior. In local stability, the system is stable at a particular point but in global stability, the system is stable in entire domain. A locally stable system may not be stable globally but globally stable system is always locally stable everywhere.

Asymptotic stability is a local property however, for linear systems this is the same as global asymptotic stability.

Persistence

From biological point of view, persistence of a system means the long term survival of all species of the system, irrespective of initial densities. Mathematically, persistence of a system means that strictly positive solutions do not have omega limit points on the boundary of the non-negative cone.

The system (2) is said to be persistent (Butler, Freedman, & Waltman, 1986; Freedman, 1980) if for every trajectory with positive initial conditions

$$\liminf_{t\to\infty} X_i(t) > 0, i = 1, 2, ..., n.$$

Further, the system (2) is said to be uniformly persistent (also known as permanence) if there exists a positive number $\delta > 0$ such that for every trajectory with positive initial conditions

$$\liminf_{t\to\infty} X_i(t) > \delta, i = 1, 2, ..., n.$$

A system is said to (uniformly) persist if each component (uniformly) persists.

PERIODIC SOLUTIONS

The solution $X(t)$ of autonomous system of differential equations (1) is called the periodic solution if there exists a constant T, such that $X(t+nT)=X(t)$) for all $t,\ n=1,2,3,\ldots$. The solution $X(t)$ is called T-periodic solution and T is called the period of the solution $X(t)$. Further, if no other periodic solution sufficiently close to it exists so that periodic orbit is an isolated one, then it is called a limit cycle. Every trajectory beginning sufficiently near a limit cycle approaches it either for $t\to\infty$ or $t\to-\infty$. If all nearby trajectories approach a limit cycle as $t\to\infty$, it is said to be stable. If they approach it as $t\to-\infty$, it is called unstable. A stable limit cycle is a period-T attractor or a periodic attractor.

BIFURCATION

Bifurcations are sudden qualitative changes in the nature of the motion that occur at critical control parameter values. In particular, fixed points can be created, destroyed or their stability can change with control parameter. As a parameter is varied, these qualitative changes in the behavior of the system can occur at certain critical value of the parameter. Such a behavior change is called bifurcation and the corresponding value of the parameter is called the bifurcation value.

Mathematically, dependence of the system (1) on the parameter is assumed as

$$\dot{X}=F(X,\mu) \tag{7}$$

Let X^* be a fixed point (equilibrium point) of the system. The eigenvalue $\lambda_i,\ i=1,2,\ldots,n$ of the Jacobian matrix about this fixed point gives the local dynamical behavior of the system. If the fixed point is stable, the eigenvalues $\lambda_i,\ i=1,2,\ldots,n$ must lie in the left- plane $\mathrm{Re}\,\lambda_i<0$. To destabilize the fixed point, one or more of the eigenvalues have to cross into the right half- plane as the parameter μ changes its value.

HOPF BIFURCATION

From (7), the critical value of the parameter for which the eigenvalues are purely imaginary is Hopf bifurcation point. The stability behavior of the system changes as the parameter crosses this critical value. It is characterized by a change of the real parts of the pair of complex conjugate eigenvalues associated with an equilibrium point from a negative to positive value while the imaginary part remains nonzero. The Hopf bifurcation can occur only in nonlinear systems with dimensions greater than one.

Supercritical Hopf bifurcation is a phenomenon in which the limit cycle is stable above the bifurcation point. In supercritical Hopf bifurcation, a small attracting limit cycle appears immediately when a stable node changes into an unstable node surrounded by a stable limit cycle.

Subcritical Hopf bifurcation is a Hopf bifurcation when bifurcated limit cycle will be unstable. Subcritical Hopf bifurcation is always much more dramatic, and potentially dangerous in engineering applications. After the bifurcation, the trajectories must jump to a distant attractor, which may be a fixed point, another limit cycle, infinity, or in three and higher dimensions-a chaotic attractor (Strogatz, 1994). The limit cycle is orbitally stable if a certain quantity called the first Lyapunov coefficient is negative, and the bifurcation is supercritical.

Otherwise it is unstable and the bifurcation is subcritical.

CHAOS

Chaos is the phenomenon of occurrence of bounded non-periodic evolution in completely deterministic nonlinear dynamical systems with high sensitive dependence on initial conditions.

- Non-periodic means that there are trajectories which do not settle down to fixed points, periodic orbits, or quasi periodic orbits as $t \to \infty$.
- Deterministic means their future dynamics are fully de ned by their initial conditions. That is, the system has no random or noisy inputs or parameters. The irregular behavior arises from the system's nonlinearity.
- Sensitive dependence on initial conditions means that nearby trajectories separate exponentially fast.

The necessary conditions for a system (1) to exhibit chaotic dynamics are:

- System must be non-linear,
- Order or dimension of the system must be 3.

The effect of nonlinearity often renders a periodic solution unstable for certain parametric choices. While these conditions do not guarantee chaos, they make existence possible.

DELAY MODELS

The principle of causality, i.e. future state of system is independent of the past and depends solely on the present state of the system is assumed in many applications. However, this assumption is merely an approximation to reality. Under this assumption, a dynamical system is represented by system of ODEs or PDEs. There are numerous studies involving mathematical models based on ordinary and partial differential equations in theoretical population dynamics. However, most natural and manmade processes in biology, medicine, chemistry, engineering or social sciences involve a time lag. In ecological problems delay occurs due to several reasons such as age structure of the population influencing the birth and death rates, maturation periods (thresholds), feeding times, reaction times, food storage and hunger coefficients in predator-prey interactions. Delay models are extensively used in epidemiology. Incubation period and latent period are the main sources of delay in epidemics. The consideration of delay in the model becomes indispensable when the order of time delay is of comparable magnitude with respect to the time span of the situation. By ignoring delay, one might miss the reality of the problem. The underlying system of delay differential equations are much more challenging to analyze compared to corresponding system of ODEs.

The delay models involving discrete delay(s) are represented by delay differential equation(s). A delay differential equation (DDE) is an equation where the evolution of the system at a certain time, t, depends on the state of the system at an earlier time, t. This is different from ordinary differential equations (ODEs) where the derivatives depend only on the current value of the independent variable. There are different kinds of delay differential equations, a general retarded delay differential equation (RDDE) or retarded functional differential equation (RFDE) is written as:

$$\dot{X} = F(X(t), X(t-\tau)) \tag{8}$$

To determine the rate of change of X at t_0, $X(t_0)$ and $X(t_0 - \tau)$ X must be known. Therefore, corresponding initial value problem involving a delay differential equation requires information not only at point t_0 but on the entire interval $[t_0 - \tau, t_0]$. Accordingly, the initial function $X(t) = X_0(t),\ t \in [t_0 - \tau, t_0]$ is associated with the delay differential equation (8). The function $X_0(t)$ describes the history.

CONTROL AND SYNCHRONIZATION OF CHAOS

One of the most basic and important problems in ecology concerns the survival of species. The notion that populations fluctuations in nature may be caused by deterministic chaos has persisted. The control and the management of population dynamics is one of the main objectives of mathematical modeling in ecology. With the fact of dynamic imbalance and extinction due to chaos, synchronization or control of chaos bring order into an otherwise chaotic system so as to maintain the ecological balance.

CONTROLLING OF CHAOS

It is not surprising that the man always seeks to introduce order in the laws that govern natural phenomenon. Controlling chaos is a focal issue in the nonlinear problems ranging from Physics and Chemistry to biology and economics. Chaos control concept has come to mean stabilization of unstable periodic orbits (including unstable equilibria) of a dynamical system (Chen & Dong, 1992). However, in a broader sense, controlling chaos can be understood as a process or mechanism which enhances existing chaos or creates chaos in a dynamical system when it is useful or beneficial, and suppresses it when it is harmful. The control of chaos involves eliminating and weakening chaos, it is usually a process of stabilizing unstable periodic orbits or reducing the leading positive Lyapunov exponent of the dynamics of the chaotic system. It can be achieved by small system perturbations.

Mathematically, let a n-dimensional nonlinear system be

$$\dot{X} = F(X, p, t), X \in R^n \tag{9}$$

p is a control parameter such that system is executing chaotic motion.

In existing control algorithms either the parameter p is suitably perturbed or an appropriate additional perturbation $C(X,t) = (c_1, c_{2,...,} c_n)^T$ is added to the R.H.S. Many of the existing chaos control methods can be broadly classified into feedback and non-feedback methods.

Feedback methods essentially make use of intrinsic properties of chaotic systems to stabilize orbits to equilibrium point. Non-feedback methods modify the underlying chaotic systems weakly so that stable periodic solutions appear. Feedback methods (Chen & Dong, 1992; Pyragas, 1992; Rajasekar & Lakshmanan, 1993) are primarily devised to control chaos by stabilizing a desired unstable periodic orbit to equilibrium points. The non-feedback methods (Ramirez, 1994) suppress chaotic orbits by converting the system dynamics to a periodic orbit.

The delayed feedback control (DFC) method is given by Pygras (1992) to control chaos in continuous dynamical systems. An advantage of this approach is that it does not require any other control force input, nor access to system parameters. Also, it may not need to exactly calculate the target trajectory. The DFC method allows one to stabilize unstable periodic orbits of a strange attractor over a large range of parameters. Another approach of stabilizing the unstable periodic orbits to the stable equilibrium point is feedback linearization.

SYNCHRONIZATION OF CHAOS

In general, synchronization is an adjustment of rhythms of oscillating objects due to their interaction.

Synchronization of chaos is the phenomenon that may occur when two or more chaotic oscillators are coupled or when a chaotic oscillator drives another chaotic oscillator. The possibility of driving two identical chaotic system to have same phase and amplitude through appropriate coupling is called synchronization of chaotic systems. Very strong coupling tends to make the states of both oscillators identical. It influences not only the mean frequencies but also the chaotic amplitudes. To synchronize a chaotic system, its all Lyapunov exponents should be made less than zero.

The variation of population abundances of interacting species is a well known ecological phenomenon; a good example is the hare-lynx cycle (Blasius, Huppert & Stone, 1999; Elton & Nicholson, 1942). A striking fact is that the abundances in different regions of Canada perfectly synchronize in phase, although the amplitudes are irregular and remain quite different. Blasius et al. (Blasius et al., 1999) assumed that the irregularity of the amplitudes is due to chaotic dynamics the predator-prey system, and the interaction between the populations in adjacent regions occurs because of the vial local migration of animals. They explained the phenomenon as synchronization in a lattice of coupled chaotic oscillators.

REFERENCES

Allen, J. C., Schaffer, W. M., & Rosko, D. (1993). Chaos reduces species extinction by amplifying local population noise. *Nature*, *364*(6434), 229–232. doi:10.1038/364229a0 PMID:8321317

Alligood, K. T., Sauer, T. D., & Yorke, J. A. (1997). *Chaos: An Introduction to Dynamical Systems. Newyork*. Springer. doi:10.1007/978-3-642-59281-2

Aziz-Alaoui, M. A. (2002). Study of a Leslie-Gower-type tritrophic population model. *Chaos, Solitons, and Fractals, 14*(8), 1275–1293. doi:10.1016/S0960-0779(02)00079-6

Berryman, A. A., & Millstein, J. A. (1989). Are ecological systems chaotic and if not, why not? *Trends in Ecology & Evolution, 4*(1), 26–28. doi:10.1016/0169-5347(89)90014-1 PMID:21227309

Bhattacharya, D. K., & Begum, S. (1996). Bionomic equilibrium of two-species system I. *Mathematical Biosciences, 135*(2), 111–127. doi:10.1016/0025-5564(95)00170-0 PMID:8768217

Blasius, B., Huppert, A., & Stone, L. (1999). Complex dynamics and phase synchronization in spatially extended ecological systems. *Nature, 399*(6734), 354–359. doi:10.1038/20676 PMID:10360572

Butler, G. J., Freedman, H. I., & Waltman, P. (1986). Uniformly persistent systems. *Proceedings of the American Mathematical Society, 96*(3), 425–430. doi:10.1090/S0002-9939-1986-0822433-4

Chen, G., & Dong, X. (1992). On feedback control of chaotic dynamic systems. *International Journal of Bifurcation and Chaos in Applied Sciences and Engineering, 2*(02), 407–411. doi:10.1142/S0218127492000392

Elton, C., & Nicholson, M. (1942). The ten-year cycle in numbers of the lynx in Canada. *Journal of Animal Ecology, 11*(2), 215–244. doi:10.2307/1358

Freedman, H. I. (1980). *Deterministic Mathematical Models in Population Biology*. New York: Marcel Dekker.

Gakkhar, S., & Singh, A. (2010). A delay model for viral infection in toxin producing phytoplankton and zooplankton system. *Communications in Nonlinear Science and Numerical Simulation, 15*(11), 3607–3620. doi:10.1016/j.cnsns.2010.01.010

Gakkhar, S. Singh, A., Singh, B.P. (2012). Effects of delay and seasonality on toxin producing phytoplankton-zooplankton system. *International Journal of Biomathematics, 5*(2).

Hassell, M. P., Lawton, J. H., & May, R. M. (1976). Patterns of dynamical behaviour in single-species populations. *Journal of Animal Ecology, 45*(2), 471–486. doi:10.2307/3886

Hastings, A., & Powell, T. (1991). Chaos in a three-species food chain. *Ecology, 72*(3), 896–903. doi:10.2307/1940591

Jorgensen, S. E. (2002). Ecosystem Management and Ecological Modeling. *Scientific World, 2*, 107–121. doi:10.1100/tsw.2002.80 PMID:12806044

Jorgensen, S. E., & Bernardi, R. D. (1997). The application of a model with dynamic structure to simulate the effect of mass fish mortality on zooplankton structure in Lago de Annone. *Hydrobiologia, 356*(1/3), 87–96. doi:10.1023/A:1003171012639

Klebano, A., & Hastings, A. (1994). Chaos in a three species food chain. *Journal of Mathematical Biology, 32*(5), 427–451. doi:10.1007/BF00160167

Pyragas, K. (1992). Control of chaos via extended delay feedback. *Physics Letters. [Part A]*, *170*, 421–428. doi:10.1016/0375-9601(92)90745-8

Pyragas, K. (1992). Continuous control of chaos by self-controlling feedback. *Physics Letters. [Part A]*, *170*(6), 421–428. doi:10.1016/0375-9601(92)90745-8

Rai, V. (2004). Chaos in natural populations: Edge or wedge? *Ecological Complexity*, *1*(2), 127–138. doi:10.1016/j.ecocom.2004.02.002

Rajasekar, S., & Lakshmanan, M. (1993). Algorithms for controlling chaotic motion: Application for the BVP oscillator. *Physica D. Nonlinear Phenomena*, *67*(1-3), 282–300. doi:10.1016/0167-2789(93)90211-I

Ramirez, J. A. (1994). Nonlinear feedback for controlling the Lorenz equation. *Physical Review E: Statistical Physics, Plasmas, Fluids, and Related Interdisciplinary Topics*, *50*, 2239–2342.

Ruxton, G. D. (1999). Chaos in a three-species food chain with a lower bound on the bottom population. *Ecology*, *77*(1), 317–319. doi:10.2307/2265680

Strogatz, S. H. (1994). *Nonlinear Dynamics and Chaos*. Perseus Books Publishing.

Chapter 19
Combustion Instability Analysis Using Wavelets in Conventional Diesel Engine

Rakesh Kumar Maurya
Indian Institute of Technology Ropar, India

Akhil Nekkanti
Indian Institute of Technology Ropar, India

ABSTRACT

Combustion instability affects the drivability, power, engine performance and efficiency. Elimination of combustion instabilities leads to increase in power for the same fuel economy. Effective controller can reduce combustion instabilities by understanding patterns in the cyclic variations. In this study, cyclic variations of combustion parameters were analyzed in a diesel engine using wavelet analysis. The experiments were conducted at varying loads and compression ratios at 1500 rpm. At steady state condition, cylinder pressure data of 2000 consecutive combustion cycles was measured. Continuous Wavelet Transform was used to capture non-stationary or transient features that might not have been detected using other transforms. Wavelet Power Spectrum (WPS) and Global Wavelet Spectrum were further used to determine the relationship between fluctuations in combustion parameters. Contour Plots were plotted based on WPS for visualizing the intensity and frequency of cyclic Variations. It was found that cyclic variability decreases with increase in engine load and compression ratio.

INTRODUCTION

The internal combustion (IC) engine is the key to the development and existence of modern society. Without the transportation by the millions of personalized vehicles on road and at sea, human society would not have reached the current living standards (in terms of physical facilities). Majority of combustion engines working in vehicle are reciprocating piston engines operating on combustion of fossil fuels. Mainly two types of reciprocating engines (compression ignition and spark ignition) are widely used by automotive industry. Due to economic growth and increase in energy requirement, the number

DOI: 10.4018/978-1-5225-1639-2.ch019

of engines (stationary and automotive) are increasing, which leads to higher fuel consumption and emissions. To reduce increasing emissions and meet the stringent emission requirement, efficient combustion control and after treatment technologies are required by industry. To design effective controller, nature and deterministic characteristics of combustion instabilities are needed. Main objective of this chapter is to understand the cycle to cycle variations in combustion using wavelets.

Combustion process in IC Engines are governed by various engine operating conditions and parameters such as engine load, speed, fuel-air properties, flow characteristics of the fuel air mixture in the cylinder, residual gas mixture, intake pressure and temperature etc. High cycle to cycle variations can be observed by even small variation in these factors depending on engine combustion mode. Cycle to cycle variations in combustion leads to undesirable affects on engine efficiency, power, and drivability and also affects the combustion noise (Heywood, 1988). Therefore it is important to characterize and control the cyclic variations in combustion process in IC engines. This chapter is mainly to understand the characteristics of cyclic variations in conventional compression ignition engine and accordingly develop the strategies to control cyclic to cycle variations.

Cycle to cycle variations is higher in spark ignition engine in comparison to compression ignition engines. The flame development and propagation have higher variations due to variations that are always present in the charge motion and flow which influences the flame growth and resulting in large cyclic variation in spark ignition engines (Heywood, 1988). Various researchers have analyzed the cycle to cycle variations in SI engines using statistical and chaotic methods (Ball, Raine, & Stone, 1998; Barton, Kenemuth, Lestz, & Meyer, 1970; Chen & Krieger, 1976; Ozdor, Dulger, & Sher,1994; Peters & Borman, 1970; Stone, Brown& Beckwith, 1996). However comparatively lowernumber of published papers are available on cyclic variations in diesel engine. Few researchers (Sagdeo, 1987; Strahle, 1977; Wolschendorf, Dürnholz, & Schmillen, 1991) have shown the relation between combustion randomness and the noise generated by these engines. Cycle to cycle variations in cylinder pressure of diesel engines has been reported in few early studies (Koizuml, Gyakushi, & Takamoto, 1977; Sczomak & Henein, 1979; Wing, 1975). Recently researchers have focused their attention to characterize the cycle to cycle variation in diesel engines due to stringent emission regulation and environmental concerns (Ajovalasit & Giacomin, 2003; Boguś & Merkisz, 2005; Macián, Luján, Guardiola, & Perles, 2006; Sen, Longwic, Litak, & Górski, 2008; Sen, Litak, Taccani, & Radu, 2008).

Wavelet transform provides good temporal as well as spectral resolution and therefore wavelets can be employed to estimate the magnitude as well as periodicities of cyclic variations in engines. Sen et al. (2008) demonstrated the use of continuous wavelet transform (CWT) to analyze the cycle to cycle variations at different engine speeds. They found that the cyclic variation decrease with increase in engine speed due to higher piston velocities which makes swirling more effective and enhance the combustion process. Few researchers have also used CWT to investigate the cycle to cycle variations in diesel engines (Ajovalasit et al., 2003; Boguś et al., 2005; Macián et al. 2006; Sen et al., 2008). Sen et al (2011) have applied CWT in SI-HCCI engines, which is considered alternative engine combustion mode. Recently Ali et al (2015) in their investigation of cyclic variability of alcohol addition to biodiesel have also made use of CWT. These studies have not explored the effect of compression ratio (CR)on cycle to cycle variations. In present study effect of compression ratio and engine load on cycle to cycle variation is investigated.

Diesel Engine Combustion

Diesel engine is typically a compression ignition type IC engine. Combustion in compression ignition engine is fundamentally different from spark ignition (SI) engines. Unlike SI engines, in the diesel engine pure air is drawn in to the combustion chamber during intake stroke of combustion cycle. The air is then compressed and towards the end of the compression stroke, shortly before TDC, the fuel is injected at high pressure into the hot air in the combustion chamber. The highly pressurized fuel is introduced into the combustion chamber via 5-8 fuel sprays, depending on the size of the cylinder. The injected fuel atomizes, evaporates and mixes with the hot compressed air and auto-ignites. In contrast of SI engines, time available for mixture formation is very short in diesel engines. Therefore a fast injection and the best possible atomization are required for intensive mixture formation. When fuel injection starts, some fuel atomizes, vaporizes and mixes partially with the air before auto-ignition occurs. Partially premixed mixture is then rapidly consumed when auto-ignition starts and the remainder combustion takes place under non premixed conditions. In the diesel engines, combustion process is very complex phenomenon and fuel injection, atomization, vaporization, mixing and combustion occurs at the same time in the combustion chamber. Mixing and vaporization are the slowest processes and therefore determine the combustion rate. Therefore fuel injection timing controls the start of combustion in diesel engine. The speed of diesel combustion is limited by the mixing between the injected fuel and the air in the combustion chamber, which also limits the maximum engine speed. Compression ignition engines have higher efficiency in comparison to SI engines due to a higher compression ratio.

Diesel engine combustion process consists of three different phases

1. The ignition delay period,
2. The premixed combustion phase, and
3. The mixing-controlled combustion phase(Lee, Jung, Kim, & van Nieuwstadt, 2013).

Duration between start of injection and start of combustion is known as ignition delay. The premixed combustion phase succeeds the ignition delay period and fuel air mixture formed in ignition delay period auto-ignites in short duration which leads to high heat release rates. The premixed phase combustion is followed by the mixing controlled combustion phase in which the combustion is controlled by the rate at which the mixture is available for burning after evaporation and mixing. Ignition delay period influences the rate of heat release and maximum in-cylinder pressure. As the Ignition delay increases the premixed combustion phase increases, which results into higher in-cylinder pressure and heat release rates. The fuel properties and spray characteristics affect the atomization of fuel, which in-turn affect the ignition delay. It was observed by Shoji(1997) that spray parameters such as spray cone angle, spray shape and spray direction are mainly responsible for cycle to cycle variations in diesels engine particularly at low speed and load conditions.

The cycle to cycle variations have been quantified by the using of statistical, chaotic and deterministic methods (Daw et al., 1996, 1998; Foakes & Pollard, 1993; DAILY, 1988). The statistical methods typically uses coefficient of variations (COV) to measure cyclic variations in combustion parameter. Indicated mean effective pressure (IMEP) and maximum cylinder pressure (P_{max}) are some of the commonly used combustion parameters in measuring cyclic variations. COV helps in characterizing the temporal variations present in the data series. The mathematical formula of estimating COV is as follows

$$COV_X = \frac{SD(X)}{\bar{X}} \text{x}100\%$$

where *SD* stands for standard deviation and *X* is the parameter for which variations are being measured (IMEP, THR and P_{max}). $\bar{X}$ denotes the mean value of the data series.

Statistical analysis only provides the temporal variations present in the data series, which is the main limitation of statistical methods. Statistical analysis is not able to account for the spectral characteristics of the time series (Ali et al, 2015). Hence wavelet transform is used for getting information regarding temporal as well as spectral characteristics of the time series. Though, short term Fourier transform (STFT) or windowed Fourier transform (WFT) could have also been used for this purpose but STFT and WFT exhibit poor temporal resolution. This motivated the researchers to use wavelet transform (Sen et al., 2008). Wavelets transform has been used for a wide range of applications such as for event related potentials in neurolinguistics (Samar, Swartz, & Raghuveer, 1995), two phase flow in pipe(Li, 2002), global climatic changes (Barbosa, Silva, & Fernandes, 2006; Oh, Ammann, Naveau, Nychka, & Otto-Bliesner, 2003), ultrasound testing of concrete structures(Hatanaka, Kawano, Ido, Hato, & Tagami, 2005)and vibrations of wind turbine(Basset, Rupp& Ting, 2010). Few researchers have successfully demonstrated the use of wavelet analysis for evaluating cyclic variations in IC engines (Sen et al., 2008; Sen et al., 2011, Sen, Litak, Finney, Daw, & Wagner, 2010; Sen, Zheng, & Huang, 2011; Sen, Wang, & Huang, 2011). The use of continuous wavelet transforms and power wavelet spectrum provides better view of periodicities and magnitude in cyclic variations. The basics of wavelets transform and its implementation is described in the following section.

WAVELET TRANSFORM

A wavelet is defined as quickly decaying small wave or an oscillation. Amplitude of wavelet is initially zero and increases with time and in small time decreases back to zero. A function can be characterized as a wavelet, when it has zero mean and finite energy. Admissibility condition of the wavelet is provided in Equation 1.

$$\int_{-\infty}^{\infty} \psi(t)\, dt = 0 \tag{1}$$

$$\int_{-\infty}^{\infty} \left|\psi(t)\right|^2 dt < \infty \tag{2}$$

Where $\psi(t)$ is any function. Wavelet is an exponential function, which tends to zero from both positive and negative directions of the time domain much faster than Fourier transform (sinusoidal function).

Wavelet transform is an integral transform that uses basis functions such as wavelet which are localized in both time and frequency domains. Wavelet transform decomposes a signal into frequency bands (or scales) with different resolutions by scaling the set of basis functions. The signal is projected on a set of basis functions known as mother wavelets (Barford, Fazzio, & Smith, 1992). The wavelets functions only differ with each other in terms of scale and translation. Morlet had initially considered the wavelets as a family of functions, which resulted from dilations and translations of a single function, called as mother wavelet (Sifuzzaman, Islam, & Ali, 2009). The equation (3) describes a daughter wavelet in terms of mother wavelet (Barford et al., 1992).

$$\psi_{a,b}(t) = \frac{1}{\sqrt{|a|}}\psi\left(\frac{t-b}{a}\right), \qquad a,b \in R, \quad a \neq 0 \tag{3}$$

Where a is the scaling factor and b is the translating factor, $\psi(t)$ is the mother wavelet with unit energy and $\psi_{a,b}(t)$ are known as daughter wavelets, which are obtained by scaling and translating of the mother wavelet. In case of $|a|$ is greater than 1 then the obtained wavelet is a compacted or compressed version of the mother wavelet and in case of $|a|$ is lesser than 1 then obtained wavelet is a widened or enlarged version of the mother wavelet. Parameter '*a*' is also known as the dilation parameter. Dilating is known as scaling of the function. The parameter '*b*' determines the time location of the wavelet and is the known as the translating parameter because it shifts the wavelet depending upon the sign of 'b'. Translating is basically shifting of the function on the real axis. The factor $\frac{1}{\sqrt{|a|}}$ is used, so that the function $\psi_{a,b}(t)$ has same energy for all scales '*a*'. Wavelets can change their time width by changing the scale 'a'. This indicates that the wavelets are adaptive of their frequencies. Scaling 'a' also helps the wavelets to narrow the width while focusing on higher frequencies and widen their width while focusing on smaller frequencies, similar to a zoom lens (Kaiser, 2010). Due to this property wavelets has a high frequency and time resolution. Wavelet transform can have combinations of time-frequency representations with different resolutions of the same signal, hence also termed as multi-resolution analysis. Due to these advantages, wavelet analysis is preferred over Fourier analysis.

Difference Between Wavelet Transform and Fourier Transforms

The Fourier transform decomposes a signal into its constituent frequencies. Inverse Fourier transform is used to convert the frequency domain into the time domain. Fourier transform breaks the waveform into sines and cosines functions, which means any signal can be represented as a combination of sinusoidal functions. Fourier transform can be applied to a wide range of applications such as Television signals, cell phone signals and sound waves. The Fourier transform of a function *f(t)* is given by Papoulis (2014) defined as

$$\hat{f}(\omega) = \int_{-\infty}^{\infty} f(t)e^{-i\omega t}dt \tag{4}$$

Where ω is angular frequency and the variable 't' represents time.

An inverse Fourier transform can be used to obtain the original function from the Fourier transform(Papoulis, 2014) and is represented by equation (5)

$$f\left(t\right) = \int_{-\infty}^{\infty} \hat{f}\left(\omega\right) e^{i\omega t} d\omega \tag{5}$$

The output function of Fourier transform is always complex unlike wavelet transform whose output function is both real as well as complex.

Fourier transform maps the time signal into frequency and phase, whereas wavelet transform maps the time series into time scale (frequency). Fourier transform provides information about the frequency of a signal. But if the frequency is changing with time then Fourier transform cannot provide the information on how frequency is changing (Harris, 1978). For example, if the Fourier transform is applied on a signal, which shows transient behaviour or is varying rapidly, then the Fourier transform provides information about the frequencies present in the signal but unable to provide any information regarding the time of occurrence of these frequencies and the duration for which these frequencies are present.

To overcome the drawbacks of Fourier transform, windowed Fourier transform (WFT) or short time Fourier transform (STFT) is used to extract the frequencies at different time locations of the signal. The procedure that WFT follows involves the divisions of the longer time signal into smaller segments of equal length, followed by the application of the Fourier transform on each of these smaller segments. A Fourier spectrum corresponding to each shorter segment is then obtained and plotted as a function of time (Kaiser, 2010). For localizing the time signal a windowed function $w(t)$ is defined. The windowed function is tapered at the ends for avoiding unnatural discontinuities. A continuous WFT is written in Equation 6 (Kaiser, 2010).

$$X(n,\omega) = \int_{-\infty}^{\infty} x\left(t\right) w\left(t-n\right) e^{-i\omega t} dt \tag{6}$$

Where $x\left(t\right)$ is the signal to be decomposed and $X(n,\omega)$ is the Windowed Fourier transform and $w(t)$ is the windowed function, ω is the angular frequency.

Similarly a discrete WFT (Kaiser, 2010)can be defined as Equation 7

$$X(m,\omega) = \sum_{n=-\infty}^{\infty} x\left[n\right] w\left[n-m\right] e^{-i\omega n} \tag{7}$$

Where $x\left[n\right]$ is the signal to be decomposed and $X(m,\omega)$ is the Windowed Fourier transform and $w\left[n\right]$ is the windowed function, ω is the angular frequency and is continuous whereas m is discrete.

A window function is a function that takes the value zero outside the chosen interval. If window function is constant inside the interval and zero outside then it is known as a rectangular window. The rectangular window exhibits good resolution for sinusoids of similar strength but shows poor resolution for the sinusoids having different strength. The sampling of signals may produce leakages, which are also known as aliases. Sampling may cause different signals to become indistinguishable and this effect is known as aliasing. If signal is a combination of two sinusoidal waves of different frequencies (one component stronger than the other) then the leakage from the stronger component can overshadow the weaker component. However when the windows have equal frequency, aliasing may make it difficult to resolve them even when the components are of similar strength. This is characterized as low dynamic range. Due to these difficulties, selection of a window function is critical decision. Generally moderate windows such as Hann and Hamming are selected as window functions. The mathematical formulation of Hann (Kumar & Foufoula-Georgiou, 1997) and Hamming windows (Kumar & Foufoula-Georgiou, 1997)is given by Equations 8 and 9 respectively.

$$w(n) = 0.5\left(1 - \cos\left(\frac{2\pi n}{N-1}\right)\right) \tag{8}$$

$$w(n) = 0.54 - 0.46\cos\left(\frac{2\pi n}{N-1}\right) \tag{9}$$

Where N is the total length of the sample.

The WFT uses a fixed window size resulting into fixed frequency as well as the time resolution, which may lead to poor spectral or temporal resolution. Aliasing of high and low frequency components that do not lie in the frequency range of window may result in inaccuracies (Torrence & Compo, 1998). If the signal has any features involving time intervals, which are shorter than the width of the window then these features are under localized in time, whereas if the features are longer than the width of the window then they are over localized. In both the cases of over-localization and under-localization, it is difficult to obtain the features of the signals and this makes WFT an inefficient tool (Kaiser, 2010). As wavelet transform uses a window size, which can be varied by scaling can solve both problems by capturing high and low frequencies. Hence wavelet transform is preferred over windowed Fourier transform.

Continuous Wavelet

The continuous wavelet transform (CWT) (Kaiser, 2010) with respect to a wavelet $\psi(t)$ is given by Equation 10.

$$CWT(a,b) = \frac{1}{\sqrt{|a|}}\int_{-\infty}^{\infty} x(t)\,\psi^*\left(\frac{t-b}{a}\right), \qquad a,b \in R, \quad a \neq 0 \tag{10}$$

Where $x(t)$ is continuous signal, $\psi(t)$ is a mother wavelet with unit energy, ψ^* indicates it's conjugate and a and b are scaling and translating parameter respectively. The Equation 10 can be re-written in terms of a daughter wavelet (refer to Equation 3) and shown in Equation 11

$$CWT\left(a,b\right)=\int_{-\infty}^{\infty}x\left(t\right)\psi_{a,b}^{*}\left(t\right) \tag{11}$$

Practical applications such as cyclic variations in engines involve a discrete time series signal. Continuous wavelet transform (CWT) (Kumar & Foufoula-Georgiou, 1997)on a discrete time series x_n as represented in Equation 12.

$$CWT_n\left(s\right)=\left(\frac{\delta t}{a}\right)^{\frac{1}{2}}\sum_{n'=0}^{N-1}x_{n'}\psi^{*}\left[\frac{\left(n'-n\right)\delta t}{a}\right] \tag{12}$$

Where $x_{n'}$ is the discrete time series, a is the scaling parameter, $\psi(t)$ is the wavelet function, which satisfies the Equations 1 and 2, ψ^* is its conjugate and n is the localized time index. From equation (12) N different signals are obtained, which need to combined for obtaining a single continuous wavelet transform (Torrence & Compo, 1998). Process of combining N different signals into one signal is known as convolution. N simultaneous convolutions can be carried out by the help of discrete Fourier transform in Fourier space (Torrence & Compo, 1998). The discrete Fourier transform is applied to time series $x_{n'}$ as in (Torrence & Compo, 1998)is shown in the Equation 13

$$\hat{x}_k=\frac{1}{N}\sum_{n=0}^{N-1}x_n e^{-2\pi ikn/N} \tag{13}$$

k ranges from 0,1,2,..., N -1
Combining the above Equation 13 with 12 we get

$$CWT_n\left(a\right)=\left(\frac{\delta t}{a}\right)^{\frac{1}{2}}\sum_{k=0}^{N-1}\hat{x}_k\hat{\psi}^{*}\left(s\omega_k\right)e^{i\omega_k n\delta t} \tag{14}$$

Where the angular frequency ω_k (Torrence & Compo, 1998) is given by Equation 15

$$\omega_k=\begin{cases}\dfrac{2\pi k}{N\delta t} k\leq\dfrac{N}{2}\\ -\dfrac{2\pi k}{N\delta t} k>\dfrac{N}{2}\end{cases} \tag{15}$$

Using the Equations 13, 14, and 15 a continuous wavelet transform at all the n time indices can be obtained.

Most commonly used continuous wavelets are Mexican hat wavelet and the Morlet wavelet. Morlet wavelet is a sinusoidal wave that is multiplied by a Gaussian function. The mathematical representation of morlet wavelet (Sen et al., 2008; Torrence & Compo, 1998) is represented in Equation 16.

$$\psi\left(\eta\right) = \pi^{-1/4} e^{i\omega_0\eta} e^{-\eta^2/2} \tag{16}$$

Where ω_0 is chosen as 6 to satisfy the admissibility condition, $\pi^{-1/4}$ is a normalizing factor. The chosen value of ω_0 helps in achieving good time and frequency localization and in this case Fourier period and the scale are equal. Similar value of ω_0 was also selected by Sen *et al.*(2008)for yielding good results.

Mexican Hat wavelet defined as the second derivative of Gaussian function (Daubechies, 1992) and is represented by the Equation 17

$$\psi\left(t\right) = \frac{2}{\sqrt{3\sigma}\pi^{1/4}}\left(1 - \frac{t^2}{\sigma^2}\right)e^{-t^2/2\sigma^2} \tag{17}$$

Where σ is the standard deviation. Mexican hat wavelet contains lesser oscillation than a morlet wavlet and is a real valued function, whereas Morlet wavelet is complex. Due to the real valued nature of mexican hat wavelet it captures both positive and negative oscillation of the signal as seperate peaks whereas morlet wavelet combines both the positive and negetive oscillations in a single broad peak (Torrence & Compo, 1998).

Discrete Wavelet

For a discrete time signal the resulting coefficients in the expansion series are called Discrete Wavelet Transform (DWT) of $s\left(n\right)$, where $s\left(n\right)$ is the discrete sequence that is defined for n =0,1,2...,M-1. The DWT is a collection of sub band signals. A discrete wavelet transform as described in (Barford et al., 1992; Harris,1978; Daubechies, 1992). is represented by Equation 18 and 19

$$W_\varphi\left(j_0,k\right) = \left(\frac{1}{M}\right)^{\frac{1}{2}} \sum_n s\left(n\right)\varphi_{j_0,k}\left(n\right) \tag{18}$$

$$W_\psi\left(j,k\right) = \left(\frac{1}{M}\right)^{\frac{1}{2}} \sum_n s\left(n\right)\psi_{j,k}\left(n\right) \tag{19}$$

Where $j \geq j_0$ and $s\left(n\right), \varphi_{j_0,k}\left(n\right), \psi_{j_0,k}\left(n\right)$ are functions of discrete variables and n =0,1,2...,M-1. The equation (18) computes the approximation coefficients whereas (19) computes for the detailed coefficients. M is generally chosen as a power of 2^j and $j_0 = 0$ .Haar wavelet is a basic type of DWT.

Haar wavelet is a square shaped function, which forms a wavelet family. It is the simplest possible wavelet. Main drawback of discrete wavelet is non differentiability due to discontinuity. The haar's wavelet function (Barford et al., 1992)is given by the Equations 20 and 21.

$$\psi\left(t\right) = \begin{cases} 1 & 0 \le t < \frac{1}{2} \\ -1 & \frac{1}{2} \le t < 1 \\ 0 & otherwise \end{cases} \tag{20}$$

The scaling function of haar wavelet is given as

$$\varphi\left(t\right) = \begin{cases} 1 & 0 \le t < 1 \\ 0 & otherwise \end{cases} \tag{21}$$

Small shifts in the input signal can lead to major variations in the distribution of energy of DWT, which is a disadvantage of DWT. It computes the coefficients of signals with respect to basis functions. However DWT is just able to provide an approximation of CWT using a discrete Fourier series. DWT is not able to capture non-stationary and transient features that would have been detected easily by CWT. Hence CWT is preferred in the analysis of transient cases. As the process taking place in IC engines is transient in nature, therefore CWT has been used in this study.

Wavelet Power Spectrum

The Wavelet Power Spectrum (WPS) gives information regarding the fluctuations of variances at different scales or frequencies. The magnitude of signal energy at a certain scale a and particular location n is obtained by the squared modulus of CWT. This is referred as wavelet power spectrum and is also known as scalogram. WPS is normalized by dividing with σ^2 so that power relative to white noise is obtained (Torrence & Compo, 1998)and is shown in Equation 22.

$$WPS = \left|CWT_n\left(a\right)\right|^2 \tag{22}$$

The normalized WPS (Torrence & Compo, 1998)is depicted in Equation 23

$$WPS_n = \frac{\left|CWT_n\left(a\right)\right|^2}{\sigma^2} \tag{23}$$

Where σ is the standard deviation. A CWT may generally be complex function (real and imaginary part), so a modulus in that case would actually mean the amplitude of CWT. WPS depends on time as well as scale (frequency) represented by a surface. The contours of the surface can be plotted on a plane to obtain a time scale representation of WPS. Through WPS important information can be obtained such as events with higher variances and their frequency of occurrences, and the duration of time for which they persist. This information can be used to modify and control the system.

The analyzed time series data is generally discrete and a finite length time series which would cause errors at the beginning and end of the WPS. The main reason for these errors is that Fourier transform assumes the data to be cyclic. To counter this problem the signal is padded with zeroes at the end of the time series and removed after the application of wavelet transform. To speed up the Fourier transform and to limit the edge effects the time series is padded with zeroes to make the total length of time series equal to next n^{th} power of 2 (Torrence & Compo, 1998). Padding of zeroes will introduce some discontinuities and will cause a decrease in amplitude at the edges. Cone of Influence (COI) is the region where edge affects become important and was defined by Torrence et al. (1998) as e-folding time for autocorrelation of WPS at each scale. The e-folding time is taken in such a way that at the edges where discontinuities exist, the magnitude of wavelet power reduces by a factor e^{-2}. Due of this factor the region inside the COI is considered and the region outside this is neglected (Torrence & Compo, 1998).

Global wavelet spectrum (GWS) is computed from WPS. It is the time average of the WPS and described by the following relation (Torrence & Compo, 1998).

$$GWS = W_s = \frac{1}{N}\sum_{n=1}^{N}\left|CWT_n\left(a\right)\right|^2 \tag{24}$$

The Global wavelet Spectrum is represented by W_s.The peak locations in global wavelet give an indication about the dominant periodicities.

EXPERIMENTAL METHODOLOGY

A four stroke, single cylinder conventional diesel engine coupled with eddy current dynamometer was used for present study. Experiments were conducted at constant engine speed of 1500 rpm at different engine operating loads and compression ratios. In-cylinder pressure signal was used for recording cycle to cycle variation in performance and combustion parameters. Piezoelectric pressure transducer was mounted in the cylinder head to measure the in-cylinder pressure. Crank angle encoder of resolution 1 CAD used for measurement of crank angle position. In-cylinder pressure of 2000 consecutive engine cycle was recorded by data acquisition system. Performance and combustion parameters were calculated on cycle to cycle basis form measured pressure data. Indicated mean effective pressure (IMEP), total heat release (THR), crank angle position at 50% heat release (CA_{50}) and the maximum in-cylinder pressure (P_{max}) are the calculated parameters used in the present study.

Maximum pressure (P_{max}) is calculated by estimating maximum pressure attained in each cycle form measured data. Indicated Mean effective Pressure (IMEP)is a performance parameter for comparing the design and output of different engines independent of the size of the engines. IMEP is calculated as the ratio of indicated work (W_{ind}) to displacement volume (V_d)(Heywood, 1988)and is represented in equation (25)

$$IMEP = \frac{W_{ind}}{V_d} \tag{25}$$

Total heat release (THR) is the total heat released during combustion. THR is calculated by integrating heat release rate from start of combustion to the end of combustion in the cylinder.

Detailed calculation of heat release rate is provided in reference (Heywood, 1988) and heat release rate equation used in this study is given by the equation (26)

$$\frac{\partial Q(\theta)}{\partial \theta} = \frac{\gamma}{\gamma - 1} P(\theta) \frac{\partial V(\theta)}{\partial \theta} + \frac{1}{\gamma - 1} V(\theta) \frac{\partial P(\theta)}{\partial \theta} \tag{26}$$

Where Q is heat release, γ is the ratio of specific heats, θ is the Crank Angle Degrees (CAD), P and V are pressure and volume respectively, which are a function of crank angle position. CA_{50} is the crank angle position at 50% of the total heat release rate and this gives an indication regarding the combustion phasing.

The effect of the cyclic variability on the aforementioned parameters can be quantified by analyzing the WPS of different parameters (IMEP, THR, CA_{50} and P_{max}).

RESULTS

The cycle to cycle variations affect the combustion characteristics of the engine and results into low performance, higher fuel consumption and drivability problems. Cyclic variations in four calculated parameters (THR, IMEP, CA_{50} and P_{max}) are analyzed using wavelet transform. Figure 1 shows the data series of 2000 consecutive engine cycles for IMEP and P_{max} at different engine loads for 17:1 compression ratio (CR). It can be noticed from figure 1 that mean values of IMEP and P_{max} increases with increase in engine load because higher amount of fuel is burnt in combustion chamber at higher engine loads. Figure 2 shows the data series of 2000 consecutive engine cycles for THR at different loads for CR= 17:1and at no load condition for three different compression ratio (CR= 16, 17 and 18). It is observed from Figure 2a that at constant compression ratio, total heat release increases with increase in engine load due to higher amount of fuel is burnt. It is observed that the variation in THR is higher at lowest load condition therefore at no load condition, effect of compression ratio is also analyzed and time series of THR at different compression ratio is shown in Figure 2b.

Presented time series of different combustion parameters are now analyzed using wavelets. Figure 3 shows wavelet power spectrum (WPS) and global wavelet spectrum (GWS) of THR at various engine loads for 17:1 compression ratio. The thick contour lines represent the 5% significance level which denotes the cone of influence (COI). It can be observed from Figure 3a that the strongest intensity of variance occurs in the periodic band 14-20 with the cycles ranging from 50-410, 570-880, 925-955, 1020-1070, 1180-1550 and 1655-1880. Other weaker bands are 128-157 band occurring in 630-770 cycles and 46-54 band spanning in the range of 145-160 cycles (Figure 3a). The presence of strong periodic band over a large number of cycles indicates a high cyclic variability in this case of no load (0%). Figure 3b depicts the WPS of THR at 25% load. It can be noticed from Figure 3b that a continuous

Figure 1. Data series of 2000 consecutive engine cycles for (a) IMEP and (b) P_{max} at different loads on CR= 17.

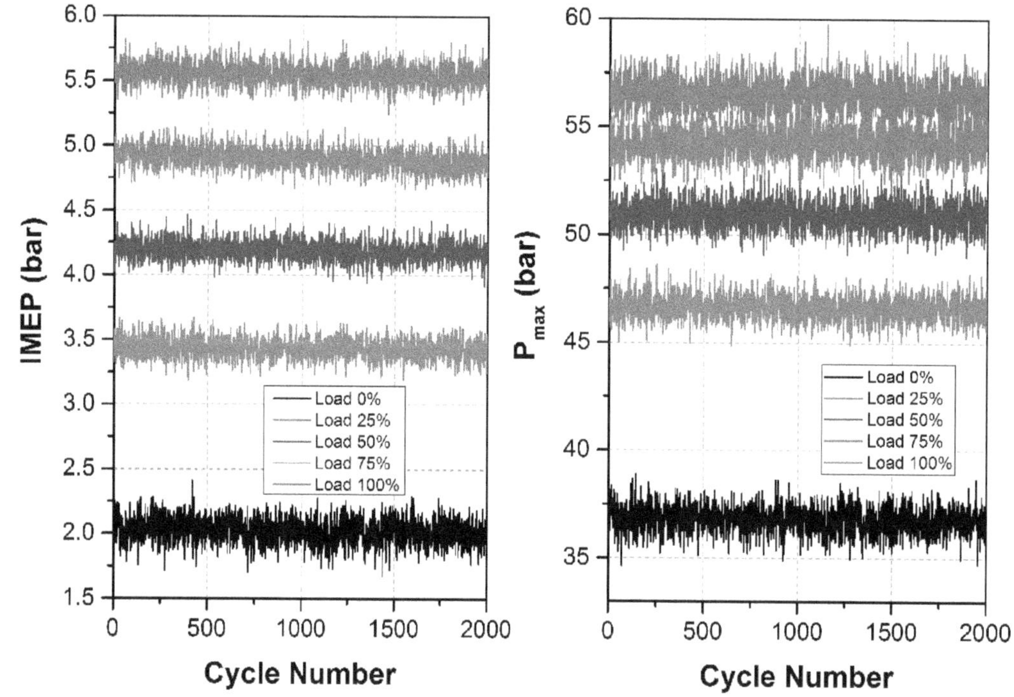

Figure 2. Data series of 2000 consecutive engine cycles for THR (a) at different loads for CR= 17 (b) at no load condition for CR= 16, 17 and 18.

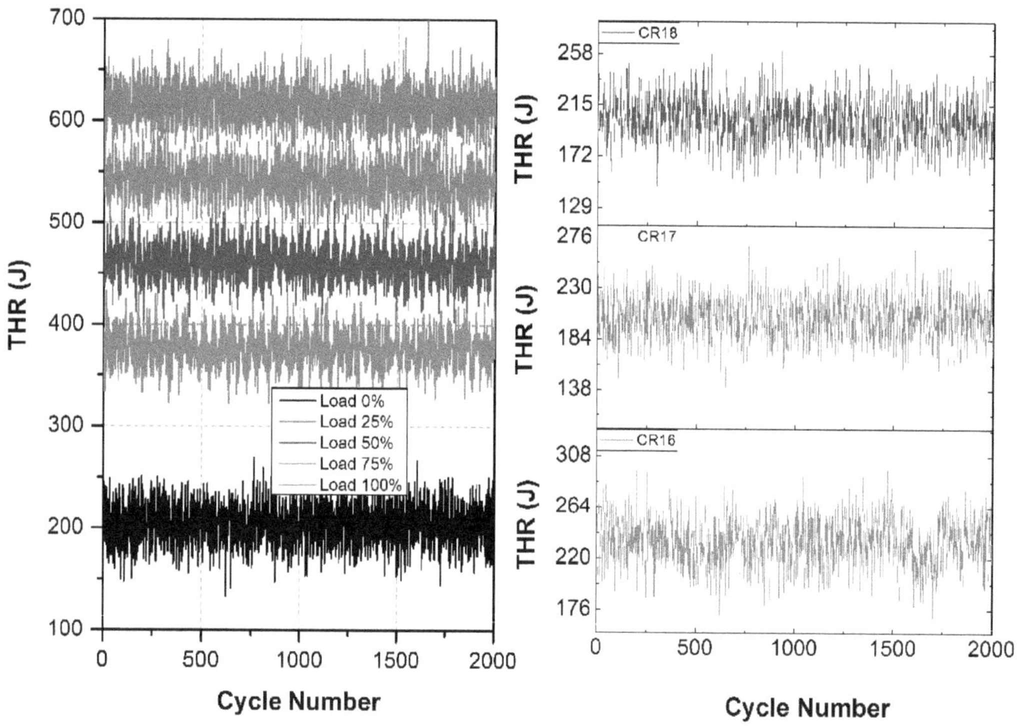

high intensity band within the period 32-64 in the cycles ranging from 130 to1920 cycles. This implies that cyclic variability exists in the complete range of cycles. On comparison of GWS of Figure 3b with 3a, it is inferred that the GWS increases in 3b indicating higher cyclic variations in the case of partial load than no load condition. A shift to a higher period indicates the cyclic variations are occurring with lower frequencies. On careful observation of GWS of Figure 3b with 3c shows that GWS decreases with increase in loads which means that the cycle to cycle variations decrease as the load increases. In this case a periodic band with strongest intensity was observed in the range of 32-64 cycles. This periodic band occurs almost continuously in cycles ranging from 70 to 1850 cycles but however this range consist of slightly lower intensity bands, which are indicated by the decrease in the intensity of red colour in the cycles 490-650 and 970-1100 cycles. The lower intensity bands result in a decrease of cycle to cycle variations. A decrease in cycle to cycle variations is observed because at high load combustion chamber temperature increases (higher fuel burnt) results into the increase in temperature of the residual gas and wall temperature, which will lead to better combustion leading to lower cyclic variations. The time scale representation of WPS for THR is shown in figure3d at 75% engine load conditions. A band with strong intensity was observed in the period of 16-28 for cycles 40-400, 500-770, 830-870, 960-1130 and 1220-1830. Some bands of weaker intensity were observed for the period 157-521 for cycles ranging in between 1160-1665, and in the period of 156-186 for cycles 310-525. The power in GWS is lower in comparison to that of 50% load, which indicates a decrease in the cyclic variations for increase in loads. The WPS of THR under full load (100%) conditions is shown in Figure 3e. The strongest intensity band is observed in the 10-16 periodic band intermittently in 141-253, 327-424,470-495, 550-582,640-704, 740-775, 1023-1194, 1266- 1494, 1510-1576 and 1891-1956 cycles. The GWS indicates lower power for this full load condition showing lower cycle to cycle variations due to the higher temperature achieved by the higher amount of fuel burnt.

Figure 4 represents the WPS and GWS for CA_{50} under different load conditions. CA_{50} under no load conditions are represented by WPS and GWS in Figure4a. The periodicity of strongest intensity is in the range of 6-10 ranging in the cycles 120-140,427-461,654-896, 1026-1661 and 1806-1921 intermittently. Other strong intensity bands are 64-110, 1036-180, 221-310 occurring in 1107-1216, 338-531 and 1110-1480 cycles respectively. GWS exhibits a power of 4.75 at a period of 10. Figure 4b represents the WPS for CA_{50} at 25% engineload. The colour bar shows the intensity of variance of the signals. The regions in WPS that are in white represent the value of intensity below the least specified intensity. The strongest intensity bands occur in 32-64 periodic band for cycles ranging in 55-512, 610- 1290,1300-1925 cycles. The GWS shows a highest power at a period of 47 and a value of 9.34 which is considerably higher than that of the previous case with no load condition. Hence the cyclic variations in CA_{50} increases from 0 to 25% load. At 50% engine load condition (Figure 4c) the GWS indicates a highest power of 7.85, which is smaller in comparison to the previous load. This indicates the decrease in cycle to cycle variations with increase in load from 25% to 50%. WPS in Figure 4c shows single strong intensity band in the period 28-64 cycles in the range of 46-302, 387-1070 and 1115-1940 cycles. Figure 4d shows that the GWS power has further decreased in comparison to the previous loads. The strong intensity bands are present in the period of 12-24 cycles and occur in 46-182, 285-1160 and 1295-1820 cycles. The strong periodic bands are intermittent in nature. Figure 4e depicts the WPS and GWS under full load conditions. There is only one strong intensity band which lies in the period of 10-16 cycles and ranges from 48-79, 207-253, 637-769, 837-1353 and 1702-1776 cycles intermittently. The power exhibited by GWS is 2.34 which further lower than 4.25 exhibited by 75% engine load. This observation shows that cyclic variations decrease on increasing load.

Figure 3. WPS and GWS of THR at various engine loads for 17:1 compression ratio.

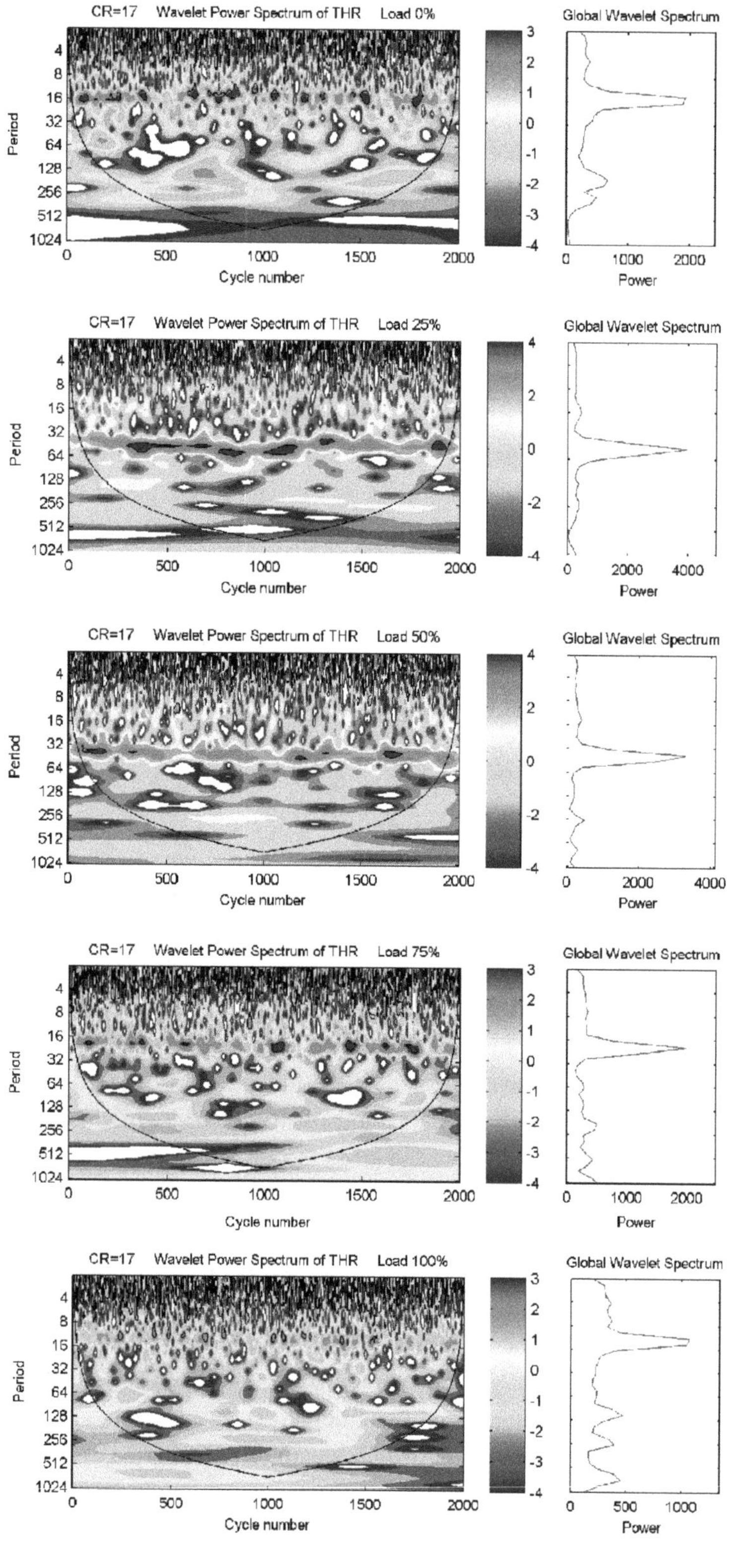

Figure 4. WPS and GWS of CA_{50} at various engine loads for 17:1 compression ratio.

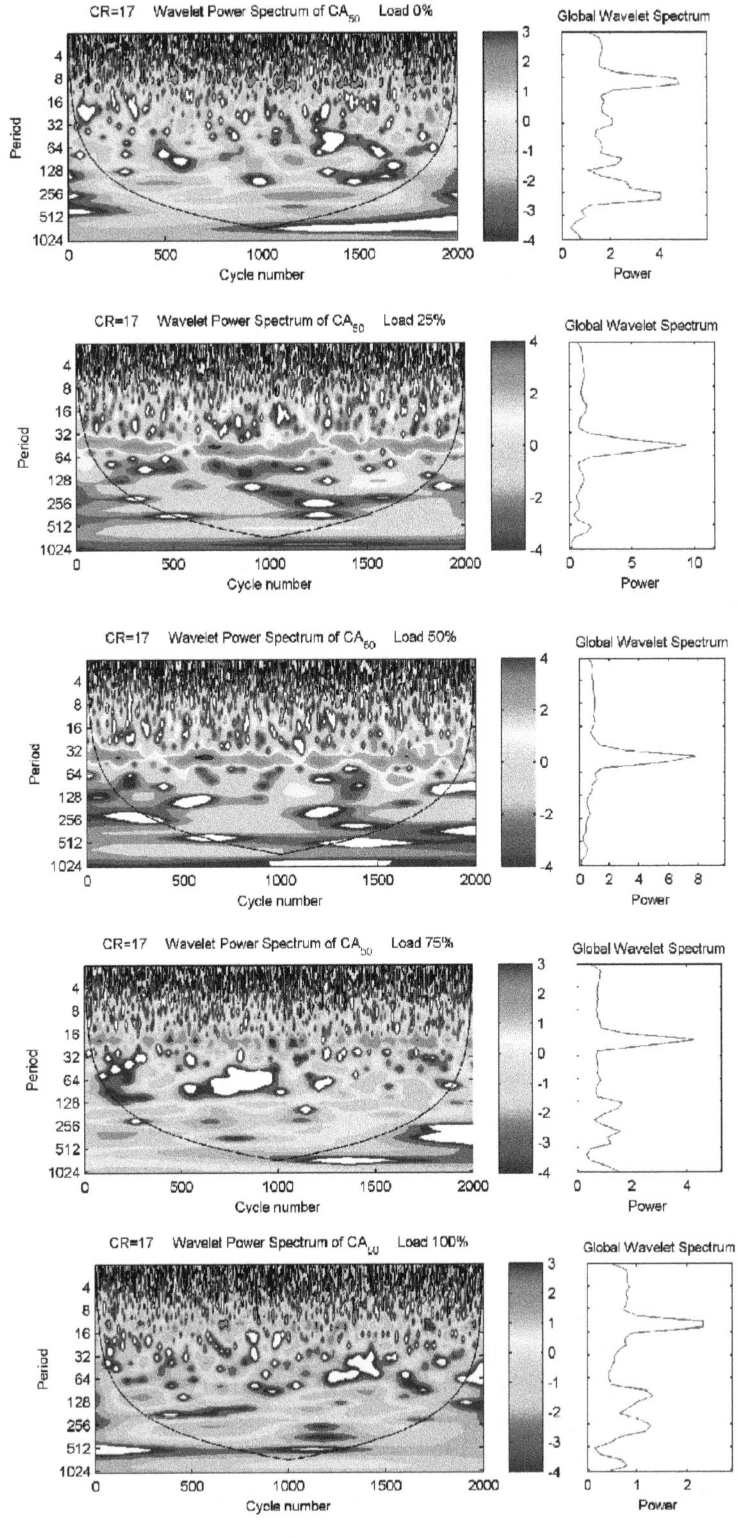

Figure 5 represents the WPS and GWS of IMEP at different loads. Figure 5a show that cycle bands of 11-16, 16-28, 55-93, 186-372 for cycles ranging between 597-637, 1130-1185, 1213-1400, 979-1642 cycles respectively have the highest variability. Some weaker bands of 32-64 period cycles ranging from 54-156 and 28-54 period cycles stretching from 227-336,406-996 cycles can also be observed. The presence of strong periodic bands suggests high cyclic variability in IMEP for no load condition. At 25% engine load (Figure 5b) one of strong periodic band present at 27-39 period occurring between 901-954 cycles. Other strong periodic band is 40-55 bands between 424-537 and 642-686 cycles. The power in GWS shows a decrease from 0.03 to 0.01 from no load to 25% load conditions, which indicates a decrease in cyclic variability. The GWS power further decreases for 50% load as shown in Figure 5c. WPS also shows presence of only a few strong intensity bands in the period of 20-32, 32-39 cycles ranging from 131-1368 and 1576-1691cycles respectively. It can be noticed from Figure 5d that GWS exhibits slight increase in power but the strongest intensity band is however outside the COI which indicates that it is not reliable. Other higher value of power in GWS is about 0.085, which is almost similar to that of the previous load condition. The WPS has randomly distributed strong intensity bands. Figure 5e represents the WPS and GWS of IMEP at full load (100%) conditions. The GWS power is slightly increased in this case of full load indicating higher variability. The strongest intensity bands in occur in period 93-128 and 47-108 with cycles spanning in the range 427-930 and 1466-1865 respectively. A comparatively weaker periodic band is noticed in 314-445 cycles and stretches in between 740-1562 cycles. However the GWS power is higher for low intensity period. A reason for this abnormality is the amplification of GWS power at higher periods (Wu & Liu, 2005).

Figure 6 shows WPS and GWS of P_{max}at various engine loads for 17:1 compression ratio. Figure6a shows strongest intensity periodic band 156-312, 39-55,55-93 and 55-78 occurring in the cycle range of 1143-1617, 224-380, 825-950 and 1307-1436 respectively. Maximum power of 1 at period 256 is shown by GWS in Figure6a. Figure 6b illustrates the decrease in GWS power showing decrease in cyclic variations with increase in engine load. Figure 6b shows only weak periodic bands such as 314-442 and 78-110 cycles occurring in range of 490-576 and 1192-1363 cycles respectively. For an increase in load to 50% shows GWS power to be 1 with corresponding strong intensity cycles outside the COI and hence the actual highest GWS power is 0.41. Very few strong intensity bands are observed in the periods 16-28, 20-28,30-38 cycles ranging from 820-840,1328-1363and 9417-1013 cycles respectively. With further increase in load the GWS power slightly increases indicating slightly higher cycle to cycle variations than the previous load condition (50% load). The WPS shows a single strong intensity band in the period 47-94 cycles spanning in the range of 1295-1493 cycles. Under full load conditions GWS exhibits a slightly higher power than the previous case. At 100% load (Figure 6e) only weak periodic bands are observed with 110-220 period band in cycles ranging in 644-1053 and 39-64 periodic band with cycles stretching from 965-1368.

It was found that lower load has higher cycle to cycle variations. Therefore effect of compression ratio (CR) on cycle to cycle variation is analyzed at low load condition. The WPS and GWS of THR data at CR =16, 17 and 18 are provided in Figure 7. Figure 7a (CR 16)shows that periodic cycle bands of 220-625 period have the highest cyclic variability in cycle ranging from 730-1680 along the COI and the strong intensity band are present both within as well as outside the COI. The maximum peak GWS power is considered 5250in this case. Some weaker bands are present in the period 128-313, 39-186 cycles which occur in the range from 519-968, 1425-1820 cycles respectively. As the CR increases from 16 to 17 the GWS power decreases considerably this indicates a decrease in cycle to cycle variations. The WPS of THR under no load conditions for a CR =17 is shown in figure7b. Figure 7b show the strongest

Figure 5. WPS and GWS of IMEP at various engine loads for 17:1 compression ratio.

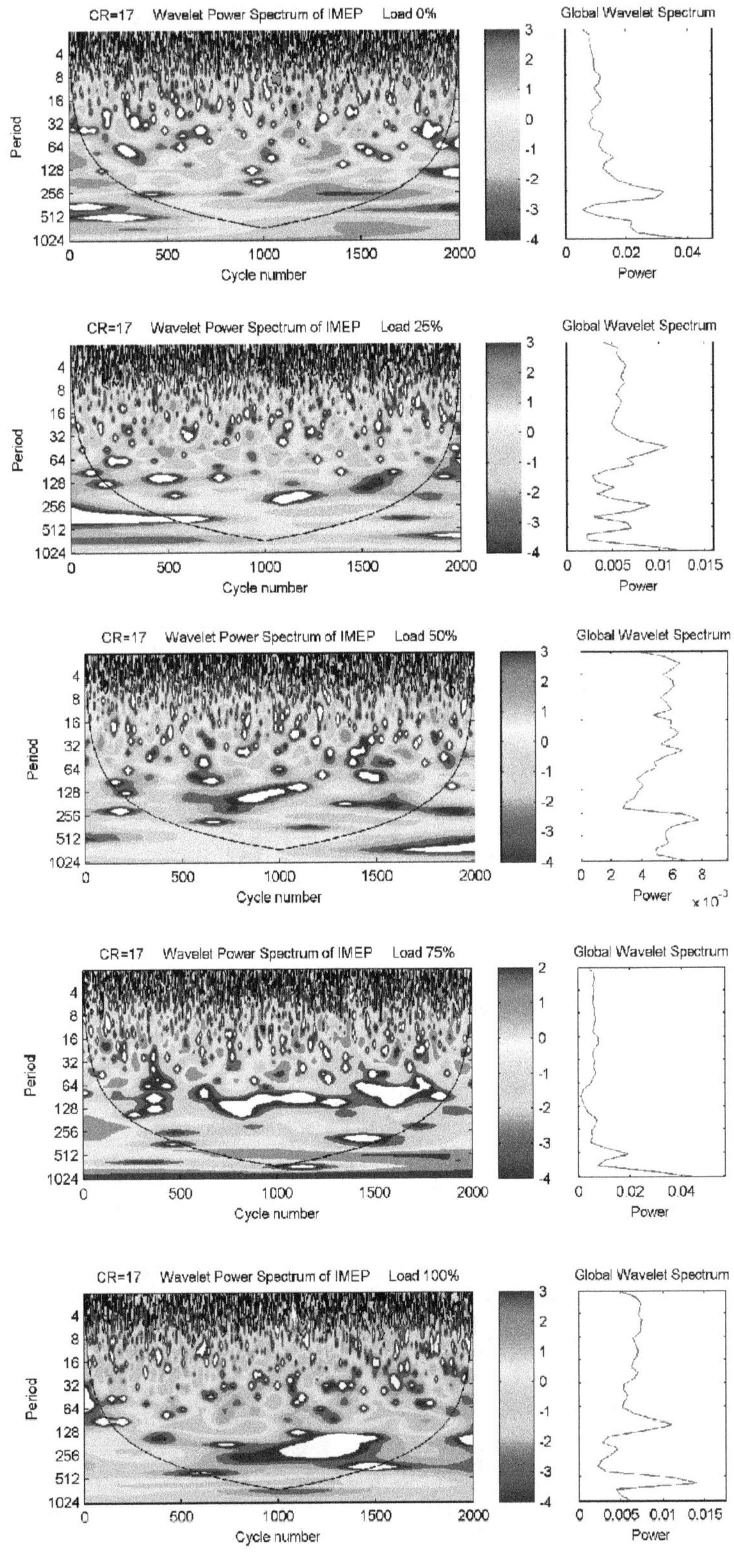

Figure 6. WPS and GWS of P_{max} at various engine loads for 17:1 compression ratio.

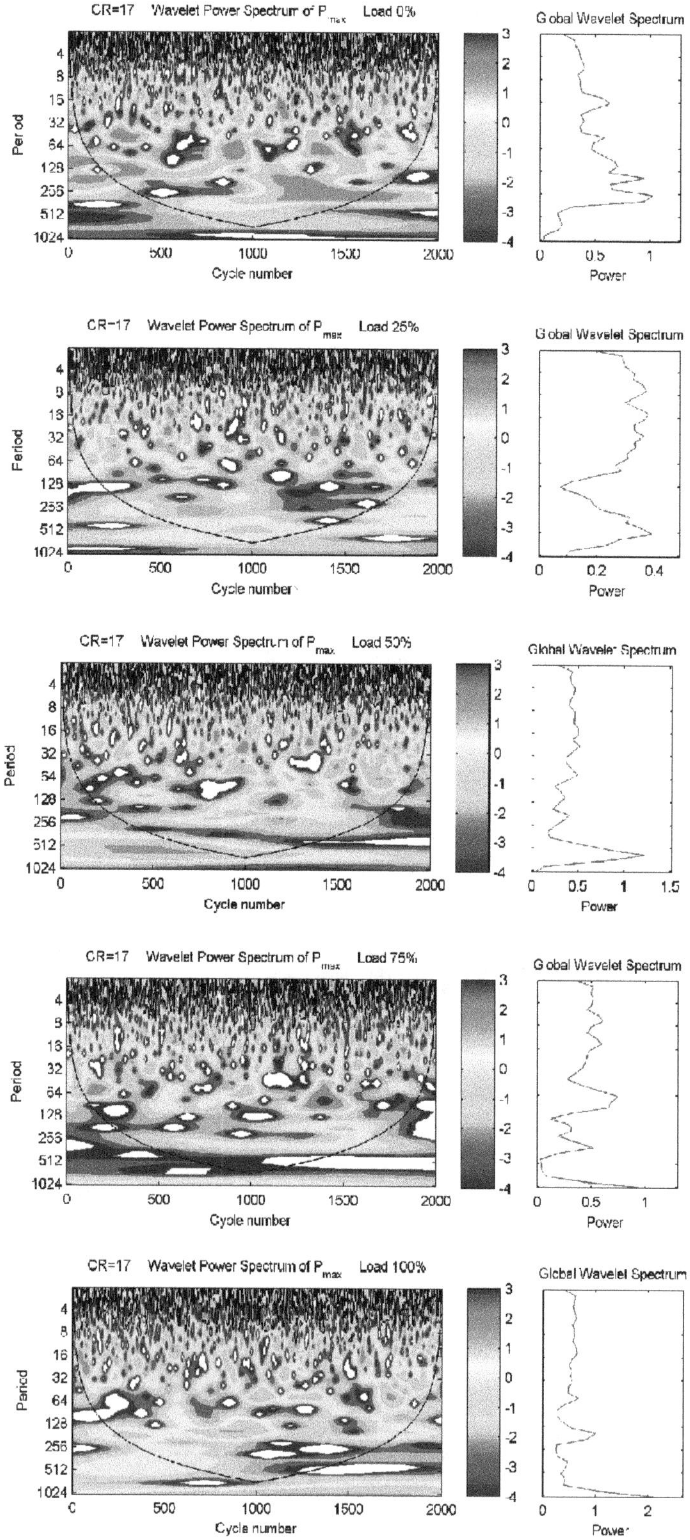

intensity of variance occurs in the periodic band 14-20 with the cycles ranging from 50-410, 570-880, 925-955, 1020-1070, 1180-1550 and 1655-1880. Other weaker bands are 128-157 band occurring in 630-770 and 46-54 band spanning in the range of 145-160 cycles (Figure 7b). Figure 7c shows the highest variance band is exhibited in the period 10-16 with the cycles ranging from 169-347, 466-714, 1052-1330, and 1400-1634. As CR increases to 18 (Figure 7c), the GWS power also decreases to 1031 from 1927 at CR17, which indicates a decrease in cyclic variability of the engine. It can be summarized that with increase in compression ratio cyclic variability decreases.

Figure 7. WPS and GWS of THR at different CR 16, 17, 18 for lowest load condition.

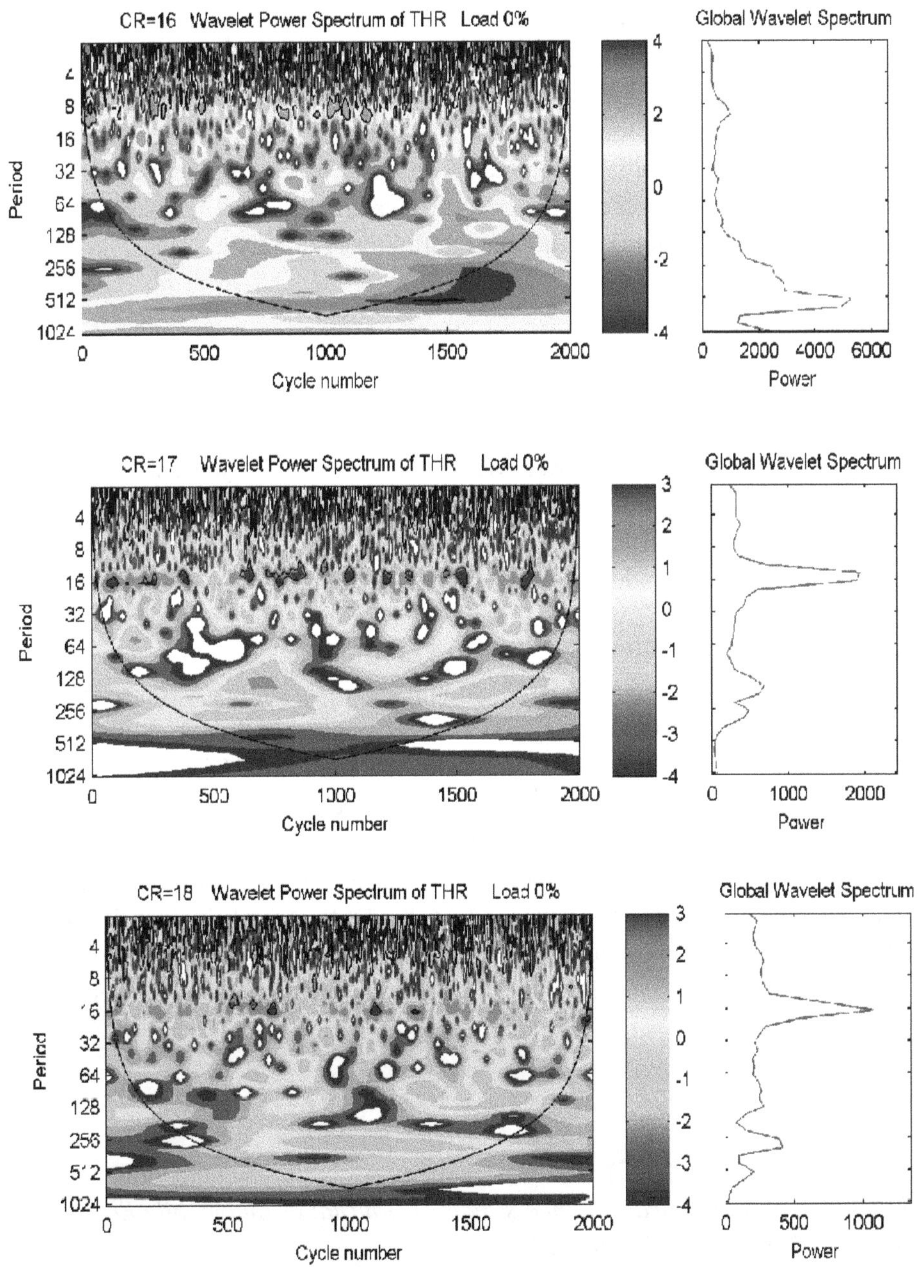

FUTURE RESEARCH DIRECTIONS

Cyclic variations in combustion affect engine performance and causes drivability problems. Wavelet transform is a good tool to analyze the characteristics of cyclic variations. Combustion instabilities using various renewable fuels in conventional as well as alternative combustion modes (such as HCCI, PPC and RCCI) can be analyzed using CWT. Wavelet analysis can also be used to understand the intermittent fluctuations in transition of different modes of engine combustion. Wavelet analysis can be used to validate the computational model and identify the dynamic features of computational models for wide range of applications. Wavelet transform and statistics require less computation time and efficiency. Therefore wavelets are utilized to develop effective combustion diagnostic feedback indicators for closed loop control of advanced engines. Wavelet transform can further be applied for estimating the engine knock intensity (Hou, Qiao, Wang, Liu, & Huang, 2010) for different engine combustion modes. The use of wavelets can be extended to get an estimation of relation between cycle to cycle variations of THR or IMEP with start of combustion or combustion duration. The cross wavelet spectrum would give the interrelation between the combustion parameters and the process or other parameters that affect it. Further Scale averaged wavelet power can also be used to examine the fluctuations in power over a range of scales and it is generally used for detecting the intermittency of a time series.

CONCLUSION

Wavelet analysis can be used for analysis of data with different timescales and transient nature of a time series. Wavelet transform has advantages over Fourier transform. In this study wavelet transform is used to study the combustion instability in conventional diesel engine. The cycle to cycle variations in different combustion parameters (THR, IMEP, P_{max} and CA_{50}) is analyzed using continuous wavelet transform. It was found that at lowest load conditions the GWS exhibited highest power signifying higher cycle to cycle variations in the parameters CA_{50}, IMEP, and P_{max}. GWS power decreased with increase in engine load implying a decrease in variability. Lower number of strong and moderate intensity periodic bands was found for higher engine load conditions. Maximum cyclic variability was distributed over a larger number of cycles for no load engine operating condition. Maximum cyclic variability occur in the cycles ranging from 50-410, 570-880, 925-955, 1020-1070, 1180-1550 and 1655-1880 in the periodic band of 14-20 cycles. At compression ratio 16 cycles with highest variability was found during 730-1680 cycles along the cone of influence. Whereas for a compression ratio 18 maximum cyclic variability was distributed over 169-347, 466-714, 1052-1330, and 1400-1634 cycles. It was found that decrease in CR from 18 to 16 drastically increased GWS power and resulting into severely high cyclic variability. Strong intensity THR variations are distributed over a large number of cycles whereas strong intensity variations of other parameters (IMEP, CA_{50}, P_{max}) are distributed over comparatively lower number of cycles.

REFERENCES

Ajovalasit, M., & Giacomin, J. (2003). Analysis of variations in diesel engine idle vibration. *Proceedings of the Institution of Mechanical Engineers. Part D, Journal of Automobile Engineering*, *217*(10), 921–933. doi:10.1243/095440703769683324

Ali, O. M., Abdullah, N. R., Mamat, R., & Abdullah, A. A. (2015). Comparison of the Effect of Different Alcohol Additives with Blended Fuel on Cyclic Variation in Diesel Engine. *Energy Procedia, 75*, 2357–2362. doi:10.1016/j.egypro.2015.07.490

Ball, J. K., Raine, R. R., & Stone, C. R. (1998). Combustion analysis and cycle-by-cycle variations in spark ignition engine combustion Part 1: An evaluation of combustion analysis routines by reference to model data. *Proceedings of the Institution of Mechanical Engineers. Part D, Journal of Automobile Engineering*, *212*(5), 381–399. doi:10.1243/0954407981526046

Barbosa, S., Silva, M. E., & Fernandes, M. J. (2006). Wavelet analysis of the Lisbon and Gibraltar North Atlantic Oscillation winter indices. *International Journal of Climatology*, *26*(5), 581–594. doi:10.1002/joc.1263

Barford, L. A., Fazzio, R. S., & Smith, D. R. (1992). *An introduction to wavelets. Hewlett-Packard Laboratories*, Technical Publications Department.

Barton, R. K., Kenemuth, D. K., Lestz, S. S., & Meyer, W. E. (1970). *Cycle-by-Cycle Variations of a Spark Ignition Engine-A Statistical Analysis* (No. 700488). SAE Technical Paper.

Basset, K., Rupp, C., & Ting, D. (2010). Vibration analysis of 2.3 MW wind turbine operation using the discrete wavelet transform. *Wind Engineering*, *34*(4), 375–388. doi:10.1260/0309-524X.34.4.375

Boguś, P., & Merkisz, J. (2005). Misfire detection of locomotive diesel engine by non-linear analysis. *Mechanical Systems and Signal Processing*, *19*(4), 881–899. doi:10.1016/j.ymssp.2004.06.004

Chen, K. K., & Krieger, R. B. (1976). A statistical analysis of the influence of cyclic variation on the formation of nitric oxide in spark ignition engines. *Combustion Science and Technology*, *12*(4-6), 125–134. doi:10.1080/00102207608946712

Daily, J. W. (1988). Cycle-to-cycle variations: A chaotic process? *Combustion Science and Technology*, *57*(4-6), 149–162. doi:10.1080/00102208808923950

Daubechies, I. (1992). Ten Lectures on Wavelets, 357 pp., Soc. for Ind. and Appl. Math., Univ. Sci. Cent., Philadelphia, PA.

Daw, C. S., Finney, C. E. A., Green, J. B., Kennel, M. B., Thomas, J. F., & Connolly, F. T. (1996). *A simple model for cyclic variations in a spark-ignition engine* (No. 962086). SAE Technical Paper.

Daw, C. S., Kennel, M. B., Finney, C. E. A., & Connolly, F. T. (1998). Observing and modeling nonlinear dynamics in an internal combustion engine. *Physical Review E: Statistical Physics, Plasmas, Fluids, and Related Interdisciplinary Topics*, *57*(3), 2811–2819. doi:10.1103/PhysRevE.57.2811

Foakes, A. P., & Pollard, D. G. (1993). Investigation of a chaotic mechanism for cycle-to-cycle variations. *Combustion Science and Technology*, *90*(1-4), 281–287. doi:10.1080/00102209308907615

Harris, F. J. (1978). On the use of windows for harmonic analysis with the discrete Fourier transform. *Proceedings of the IEEE*, *66*(1), 51–83. doi:10.1109/PROC.1978.10837

Hatanaka, H., Kawano, Y., Ido, N., Hato, M., & Tagami, M. (2005). Ultrasonic testing with advanced signal processing for concrete structures. *Nondestructive Testing and Evaluation*, *20*(2), 115–124. doi:10.1080/10589750500181084

Heywood, J. B. (1988). *Internal combustion engine fundamentals* (Vol. 930). New York: Mcgraw-hill.

Hou, J., Qiao, X., Wang, Z., Liu, W., & Huang, Z. (2010). Characterization of knocking combustion in HCCI DME engine using wavelet packet transform. *Applied Energy*, *87*(4), 1239–1246. doi:10.1016/j.apenergy.2009.06.014

Kaiser, G. (2010). *A friendly guide to wavelets*. Springer Science & Business Media.

Koizuml, I., Gyakushi, N., & Takamoto, Y. (1977). Study on the cycle-by-cycle variation in diesel engines. *Bulletin of the JSME*, *20*(145), 869–876. doi:10.1299/jsme1958.20.869

Kumar, P., & Foufoula-Georgiou, E. (1997). Wavelet analysis for geophysical applications. *Reviews of Geophysics*, *35*(4), 385–412. doi:10.1029/97RG00427

Lee, B., Jung, D., Kim, Y. W., & van Nieuwstadt, M. (2013). Thermodynamics-Based Mean Value Model for Diesel Combustion. *Journal of Engineering for Gas Turbines and Power*, *135*(9), 091504. doi:10.1115/1.4024757

Li, H. (2002). Application of wavelet multi-resolution analysis to pressure fluctuations of gas–solid two-phase flow in a horizontal pipe. *Powder Technology*, *125*(1), 61–73. doi:10.1016/S0032-5910(02)00012-8

Macián, V., Luján, J. M., Guardiola, C., & Perles, A. (2006). A comparison of different methods for fuel delivery unevenness detection in Diesel engines. *Mechanical Systems and Signal Processing*, *20*(8), 2219–2231. doi:10.1016/j.ymssp.2005.04.001

Oh, H. S., Ammann, C. M., Naveau, P., Nychka, D., & Otto-Bliesner, B. L. (2003). Multi-resolution time series analysis applied to solar irradiance and climate reconstructions. *Journal of Atmospheric and Solar-Terrestrial Physics*, *65*(2), 191–201. doi:10.1016/S1364-6826(02)00291-2

Ozdor, N., Dulger, M., & Sher, E. (1994). *Cyclic variability in spark ignition engines a literature survey* (No. 940987). SAE Technical Paper.

Papoulis, A. (2014). *The Fourier integral and its applications*. Academic Press.

Peters, B. D., & Borman, G. L. (1970). *Cyclic variations and average burning rates in a SI engine* (No. 700064). SAE Technical Paper.

Sagdeo, P. M. (1987). *Frequency spectra of diesel engine heat release rate, pressure development and engine noise: an experimental investigation of cause and effect relationship* (No. 871620). SAE Technical Paper.

Samar, V. J., Swartz, K. P., & Raghuveer, M. R. (1995). Multiresolution analysis of event-related potentials by wavelet decomposition. *Brain and Cognition*, *27*(3), 398–438. doi:10.1006/brcg.1995.1028 PMID:7626282

Sczomak, D. P., & Henein, N. A. (1979). *Cycle-to-cycle variation with low ignition quality fuels in a CFR diesel engine* (No. 790924). SAE Technical Paper.

Sen, A. K., Litak, G., Edwards, K. D., Finney, C. E., Daw, C. S., & Wagner, R. M. (2011). Characteristics of cyclic heat release variability in the transition from spark ignition to HCCI in a gasoline engine. *Applied Energy*, *88*(5), 1649–1655. doi:10.1016/j.apenergy.2010.11.040

Sen, A. K., Litak, G., Finney, C. E., Daw, C. S., & Wagner, R. M. (2010). Analysis of heat release dynamics in an internal combustion engine using multifractals and wavelets. *Applied Energy*, *87*(5), 1736–1743. doi:10.1016/j.apenergy.2009.11.009

Sen, A. K., Litak, G., Taccani, R., & Radu, R. (2008). Wavelet analysis of cycle-to-cycle pressure variations in an internal combustion engine. *Chaos, Solitons, and Fractals*, *38*(3), 886–893. doi:10.1016/j.chaos.2007.01.041

Sen, A. K., Longwic, R., Litak, G., & Górski, K. (2008). Analysis of cycle-to-cycle pressure oscillations in a diesel engine. *Mechanical Systems and Signal Processing*, *22*(2), 362–373. doi:10.1016/j.ymssp.2007.07.015

Sen, A. K., Wang, J., & Huang, Z. (2011). Investigating the effect of hydrogen addition on cyclic variability in a natural gas spark ignition engine: Wavelet multiresolution analysis. *Applied Energy*, *88*(12), 4860–4866. doi:10.1016/j.apenergy.2011.06.030

Sen, A. K., Zheng, J., & Huang, Z. (2011). Dynamics of cycle-to-cycle variations in a natural gas direct-injection spark-ignition engine. *Applied Energy*, *88*(7), 2324–2334. doi:10.1016/j.apenergy.2011.01.009

Shoji, T. (1997). Effect of Cycle-to-Cycle Variations in Spray Characteristics on Hydrocarbon Emission in DI Diesel Engines (Visualization of Sac Inner Flow, Needle Valve Motion and Cycle-to-Cycle Variations in Diesel Spray). *JSME International Journal Series B*, *40*(2), 312–319. doi:10.1299/jsmeb.40.312

Sifuzzaman, M., Islam, M. R., & Ali, M. Z. (2009). *Application of wavelet transform and its advantages compared to Fourier transform*. Academic Press.

Stone, C. R., Brown, A. G., & Beckwith, P. (1996). *Cycle-by-cycle variations in spark ignition engine combustion-part ii: Modelling of flame kernel displacements as a cause of cycle-by-cycle variations* (No. 960613). SAE Technical Paper.

Strahle, W. C. (1977). Combustion randomness and diesel engine noise: Theory and initial experiments. *Combustion and Flame*, *28*, 279–290. doi:10.1016/0010-2180(77)90033-5

Torrence, C., & Compo, G. P. (1998). A practical guide to wavelet analysis. *Bulletin of the American Meteorological Society*, *79*(1), 61–78. doi:10.1175/1520-0477(1998)079<0061:APGTWA>2.0.CO;2

Wing, R. D. (1975). The rotary fuel-injection pump as a source of cyclic variation in diesel engines, and its effect on nitric oxide emissions. *Proceedings - Institution of Mechanical Engineers*, *189*(1), 497–505. doi:10.1243/PIME_PROC_1975_189_061_02

Wolschendorf, J., Dürnholz, M., & Schmillen, K. (1991). *The IDI diesel engine and its combustion noise variations* (No. 910228). SAE Technical Paper.

Wu, S., & Liu, Q. (2005). Some problems on the global wavelet spectrum. *Journal of Ocean University of China*, *4*(4), 398–402. doi:10.1007/s11802-005-0062-y

Chapter 20
Numerical Simulation of High Temperature Deformation Behavior of Nickel-Based Superalloys Using Crystal Plasticity Models and Finite Element Method

M. K. Samal
Bhabha Atomic Research Centre, India & Homi Bhabha National Institute, India

ABSTRACT

Development of reliable computational models to predict the high temperature deformation behavior of nickel based super-alloys is in the forefront of materials research. These alloys find wide applications in manufacturing of turbine blades and discs of aircraft engines. The micro-structure of these alloys consists of the primary gamma-prime phase and the secondary and tertiary precipitates (of Ni_3Al type) are dispersed as gamma-prime phases in the gamma-matrix. It is computationally expensive to incorporate the explicit finite element model of the micro-structure in a crystal plasticity based constitutive framework to simulate the response of the polycrystalline micro-structure. Existing models in literature do not account for these underlying micro-structural features which are important for simulation of polycrystalline response. The aim of this chapter is to present a physically-motivated multi-scale approach for simulation of high temperature response of Nickel-based super-alloys. At the lower length scale, a dislocation density based crystal plasticity model is developed which simulates the response of various types of micro-structures. The micro-structures are designed with various shapes and volume fractions of gamma-prime precipitates. A new model for simulation of the mechanism of anti-phase boundary shearing of the gamma-prime precipitates, by the matrix dislocations, is presented in this chapter. The lower scale model is homogenized as a function of various micro-structural parameters and the homogenized

DOI: 10.4018/978-1-5225-1639-2.ch020

model is used in the next scale of multi-scale simulation. In addition, a new criterion for initiation of micro-twin and a constitutive model for twin strain accumulation are developed. This new micro-twin model along with the homogenized crystal plasticity model has been used to simulate the creep response of a single crystal nickel-based super-alloy and the results have been compared with those of experiment from literature. It was observed that the new model has been able to model the tension-compression asymmetry as observed in single crystal experiments.

INTRODUCTION

Nickel-based super-alloys are used for manufacturing of aircraft engine turbine blades, disk and other high temperature components such as combustion chambers, casings, liners, exhaust ducting and bearing housings etc (Furrer & Fecht, 1999). This is because of their superior mechanical properties such as excellent high temperature stability, oxidation resistance, creep resistance and mechanical strength. These properties are as a result of the two-phase structure (often called the $\gamma - \gamma'$ microstructure) (Sugui et al., 2011; Goff et al., 2004; Hopgood & Martin, 1986). The γ phase is mainly an alloy of Ni, Cr with the presence of small wt. % of other alloying elements. This phase is of face-centered cubic crystal structure and it remains stable up to its melting point. The γ' phase is a coherent ordered inter-metallic reinforcing phase of $L1_2$ crystal structure and of Ni_3A1 type. This phase often appears as a distribution of cuboidal precipitates embedded in a disordered Ni-Cr solid solution. However, the shape and size of these precipitates depends largely on the cooling rate (Babu et al., 2001; Epishin et al., 2001) as well as internal stress gradients (Ignat et al., 1993) encountered during casting and further heat treatment process of these alloys. Because of their shape and ordered structure, the γ' reinforcing precipitates act as effective obstacles to the motion of dislocations in the super-alloy.

When stresses are high, the dislocations will either bypass or shear the precipitates. The mechanism depends on the temperature and the stress level. The dependence of the super-alloy behavior on deformation conditions has been extensively analyzed in the last several years both for polycrystalline (Torster et al., 1997; Hong et al., 2009; Knobloch et al., 1997; Soula et al., 2009) and single crystal based super-alloys (Chatterjee et al., 2010; Cormier et al., 2007; Milhet et al., 2010). As is the case for other alloys reinforced with a dispersed hard second phase, the volume fraction of the γ' precipitates, their mean size and spacing have a major effect on the mechanical properties of these super-alloys. Another important micro-structural property controlling the behavior of super-alloys is related to high temperature changes in structure. When these super-alloys are deformed under creep conditions at temperatures above 850°C, their microstructure undergoes morphological changes. These changes have been described by models in literature (Manonukul et al., 2002; Ma et al., 2008) which take into account the intrinsic properties of a super-alloy (precipitate and matrix elastic moduli and volume fractions, misfit between the precipitate and matrix lattices, precipitate matrix interfacial energy), and also the experimental conditions (applied stress and temperature).

The creep deformation behavior of nickel-based single crystal super-alloys controls the service life of turbine blades used in modern aero-engines. Recently, there has been substantial effort to predict the life of the nickel based super-alloy components in literature (Brown et al., 1986; Koul & Castillo, 1991). Some of the modeling efforts for both single crystal and polycrystalline nickel based alloys are presented in literature (Coakley et al., 2011; Prasad et al., 2006; Kim et al., 2007; Zhan et al., 2008; Piard et al.,

2004). Though life prediction and failure analysis of all the high temperature components made of these super-alloys are very important, the failure of aircraft engine disc material, which is of polycrystalline micro-structure, is receiving the recent attention of researchers. The micromechanics of creep in these materials are very complex (Kovarik et al., 2009; Viswanathan et al., 2005). At lower temperature and high stresses, the creep mechanism is governed by different types of dislocation based shearing processes, whereas at higher temperatures, the creep deformation is controlled by Orowan looping and cross-slip mechanisms (Kovarik et al., 2009).

At intermediate temperatures and stresses, significant micro-twinning is observed in these alloys. At temperatures beyond 1000°C, the γ-γ' micro-structure structure forms the rafted morphology, which modifies the underlying dislocation based processes, leading to a strengthening effect. In the γ-γ' micro-structure, the γ' precipitates are sheared by dislocation ribbons which are dissociated into super-lattice partial dislocations separated by intrinsic and extrinsic stacking faults (Kovarik et al., 2009; Sarosi et al., 2007).

In order to simulate the deformation response of various types of alloys at the level of polycrystalline micro-structure, the crystal plasticity based finite element methods are most suitable. These models consider the effect of orientation and mis-orientation information of the grains in the simulation process which are obtained from the orientation image microscopy. It is hence possible to incorporate the various types of micro-structures which are obtained due to different heat treatment and mechanical working processes.

For simulation of response of polycrystalline microstructure, an experimentally validated crystal plasticity finite element model was developed in literature (Hasija et al., 2003; Deka et al., 2006; Venkataramani et al., 2006, 2007, 2008; Manchiraju et al., 2007). This model has been used for simulation of creep and fatigue responses of titanium based super-alloys. The effect of hold time (dwell) on fatigue response of Ti-6242 alloys has been studied in Venkataramani et al. (2006) using these crystal plasticity models and it was concluded that the dwell period has a significant deteriorating effect of the fatigue life of the components. Similar studies regarding the effect of hold time on low cycle fatigue behavior of HAYNES 230 nickel based super-alloy was also carried out in Lu et al. (2005).

The dislocation density based crystal plasticity models have received considerable attention in recent literature (Ma & Roters, 2004; Ma et al., 2006 a, b, 2008; Samal, 2015) due to their several advantages over the conventional crystal plasticity models. These dislocation density based models consider the evolution of statistically stored (SSD) and geometrically necessary dislocations (GND) in the matrix due to the plastic deformation. Various types of physically motivated constitutive models for generation and annihilation of SSDs can be easily incorporated into the model. Moreover, the plastic strain gradient at the geometrically incompatible locations such as matrix-precipitate interface and grain boundaries can lead to generation of GNDs (Ashby, 1970).

It is now well-known that the size effects observed in deformation response of various alloys are due to the presence of GNDs and hence, these must be considered explicitly (Ma & Roters, 2004; Ma et al., 2006 a, b, 2008; Samal, 2015) in the crystal plasticity constitutive framework. The geometry (shape and size) of the precipitates and their volume fraction play an important role in the polycrystalline response of the nickel based super-alloys as has been experimentally observed in Viswanathan et al. (2005). There is a lack of suitable models in research literature to explicitly consider the above micro-structural features and simultaneously perform a simulation for the polycrystalline microstructure. For this purpose, a multi-scale based approach has been developed in this work. The outline of the multi-scale framework is briefly described in the following paragraph.

In the lower scale, the geometrical features of the γ-γ' micro-structure are explicitly considered in the dislocation density based crystal plasticity model. The development of SSDs in the γ-channel due to various dislocation generation and annihilation mechanisms is considered. The evolution of GNDs at the locations with plastic in-homogeneity is considered through computation of the components of GNDs from the plastic strain gradient. The mechanism of anti-phase boundary (APB) shearing of the γ' precipitates by the matrix dislocations are also captured through a new model.

The lower scale model is homogenized in order to be used in the next level (i.e., crystal level) of simulation. There is no explicit consideration of the γ-γ' micro-structure at this level of simulation. At the crystal level of simulation, the homogenized crystal plasticity based model along with a new model for micro-twin initiation and propagation has been developed. The parameters of the homogenized model are calibrated through comparison of simulation of model response with single crystal experimental data. The parameters which are homogenized as a function of micro-structural features are:

1. Activation energy Q,
2. Passing stress τ_{pass} and
3. Activation volume V_{act}.

With the homogenized model and the calibrated micro-twin parameters, the creep response of a single crystal nickel based super-alloy has been simulated for loading in different directions such as [001] and [011] in both tension and compression. Significant tension-compression asymmetry has been observed in the single crystal experiments performed in Kakehi (1999) and Yamashita and Kakehi (2006). The new model has been used to simulate the creep response of this alloy used in Kakehi (1999). The creep experiment has been conducted at 700°C and 820 MPa of applied stress and the predicted responses compare very well with those of experiment in Kakehi (1999). For simulation of the response of the polycrystalline micro-structure, the actual micro-structure needs to be created from experimental data. A procedure to this effect has discussed in Ghosh et al. (2008) and Groeber et al. (2008 a, b). The model developed in this work can be being used for simulation of polycrystalline response of experimentally obtained micro-structures and this will be a scope of future research.

This chapter is divided into seven sections. The constitutive formulation for the dislocation density based crystal plasticity model is described in brief in Section-2. The criterion for onset of micro-twin due to the mechanism of dissociation of a dislocation into Shockley partials at the matrix-precipitate interface is discussed in Section-3 followed by the description of the new constitutive model for micro-twin strain accumulation in Section-4. The finite element implementation of the new model developed in this work is discussed in Section-5. The results of simulation of single crystal experiments have been discussed in Section-6. Initially, the parameters of the model have been calibrated by comparing the experimental response for a given micro-structure. The shape and volume fraction of the precipitates have been varied and the simulations have been conducted to predict the response of various types of micro-structures with the new model. The results of simulation have been homogenized for use in the next level of the multi-scale process. In addition, the homogenized model along with the micro-twin has been used to simulate the creep response of single crystal experiments and the results of simulation have been compared with those of experiment. The chapter ends with a summary in Section-7.

Constitutive Formulation for the Dislocation Density Based Crystal Plasticity Model

The constitutive formulation of the dislocation density based crystal plasticity model is based on the simulation of shear strain rate $\dot{\gamma}^{\alpha}$ on slip system α and it can be written as

$$\dot{\gamma}^{\alpha} = \rho_{\mathrm{m}}^{\alpha} b v^{\alpha} \tag{1}$$

where $\rho_{\mathrm{m}}^{\alpha}$ is the density of mobile dislocations on slip system α, b is the Burger's vector and v^{α} is the velocity of dislocations on slip system α. This velocity of dislocations is a function of applied shear stress τ^{α}, effective stress $\tau_{\mathrm{eff}}^{\alpha}$ in the slip system and other slip system resistances as discussed earlier and can be written mathematically as (Ma & Roters, 2004; Ma et al., 2006 a, b)

$$v^{\alpha} = \lambda v e^{\left(-Q_{\mathrm{act}}/k_B\theta\right)} \sinh\left(\frac{\tau_{\mathrm{eff}}^{\alpha} V_{\mathrm{act}}^{\alpha}}{k_B\theta}\right) \mathrm{sgn}\left(\tau^{\alpha}\right) \tag{2}$$

where λ is distance moved by the dislocation in each step (as it is a discontinuous process which depends upon probability of overcoming the barriers and presence of other mechanisms) and it is sometimes called dislocation jump, v is the frequency of oscillation of the dislocations, Q_{act} is the activation energy barrier to be overcome by the dislocations (which depends upon the mechanism of interaction of dislocations with matrix and other barriers, e.g., precipitates or other phases), k_B is Boltzmann's constant, θ is the applied absolute temperature in Kelvin, $V_{\mathrm{act}}^{\alpha}$ is the activation volume which again depends upon the mechanism of deformation being active at the applied stress and temperature and many other parameters. The slip system resistances are represented in terms of passing stress $\tau_{\mathrm{pass}}^{\alpha}$ and the activation volume $V_{\mathrm{act}}^{\alpha}$. The effective resolved shear stress $\tau_{\mathrm{eff}}^{\alpha}$ is the stress above the slip system threshold resistance (which is represented in terms of $\tau_{\mathrm{pass}}^{\alpha}$) and can be expressed as

$$\tau_{\mathrm{eff}}^{\alpha} = \begin{cases} \left|\tau^{\alpha}\right| - \tau_{\mathrm{pass}}^{\alpha} & \text{if } \left|\tau^{\alpha}\right| > \tau_{\mathrm{pass}}^{\alpha} \\ 0 & \text{otherwise} \end{cases} \tag{3}$$

The passing stress is due to interaction of mobile dislocations with other dislocations and dislocation networks in the same slip plane and can be expressed as

$$\tau_{\mathrm{pass}}^{\alpha} = c_1 G b \sqrt{\rho_P^{\alpha} + \rho_m^{\alpha}} \tag{4}$$

In the above Equation 4, c_1 is a material constant, G is the shear modulus, ρ_P^{α} is the density of dislocations which are immobilized and are in the plane of the slip system α and are also called as parallel dislocations. The expression for calculations of mobile and parallel dislocation densities will be discussed in the next sub-section. During the process of dislocation motion and their interaction with other dislocations and obstacles, the process of generation and annihilation of dislocations continues which gives rise to the net dislocation density. This is captured in the form of constitutive models as discussed below.

Constitutive Formulation for Hardening/Softening Due to SSDs and GNDs

The change in density of statistically stored dislocations can be thought of as net effect of some of the processes of dislocation multiplication and annihilation (Ma & Roters, 2004) and can be written as

$$\dot{\rho}^{\alpha}_{\mathrm{SSD}} = \dot{\rho}^{\alpha}_{\mathrm{SSD}_{\mathrm{lf}}} + \dot{\rho}^{\alpha}_{\mathrm{SSD}_{\mathrm{df}}} + \dot{\rho}^{\alpha}_{\mathrm{SSD}_{\mathrm{aa}}} + \dot{\rho}^{\alpha}_{\mathrm{SSD}_{\mathrm{ta}}} \tag{5}$$

where the subscripts stands for aa→lock formation, df→ dipole formation, aa→ athermal annihilation, ta→ thermal annihilation respectively. The increase in density of SSDs because of the mechanism of lock formation has been described in Ma & Roters (2004) as

$$\dot{\rho}^{\alpha}_{\mathrm{SSD}_{\mathrm{lf}}} = \frac{c_4}{b}\sqrt{\rho^{\alpha}_{F}}\,\dot{\gamma}^{\alpha} \tag{6}$$

where c_4 is a material constant. Similarly, the increase in SSDs due to the mechanism of dipole formation can be written as

$$\dot{\rho}^{\alpha}_{\mathrm{SSD}_{\mathrm{df}}} = \frac{c_5}{b}\frac{\sqrt{3}Gb}{16\pi\left(1-\nu\right)}\left(\left|\tau^{\alpha}\right| - \tau^{\alpha}_{\mathrm{pass}}\right)^{-1}\rho^{\alpha}_{m}\dot{\gamma}^{\alpha} \tag{7}$$

Two different mechanisms of dislocation annihilation have also been included in this model, i.e., thermal and athermal annihilation respectfully. The constitutive models for change in SSDs due to these mechanisms are described by the following equations.

The decrease in dislocation density due to the athermal process of dislocation annihilation is given as

$$\dot{\rho}^{\alpha}_{\mathrm{SSD}_{\mathrm{aa}}} = -c_6\rho^{\alpha}_{SSD}\dot{\gamma}^{\alpha} \tag{8}$$

Similarly, the decrease in dislocation density due to the thermal process of dislocation annihilation is given as

$$\dot{\rho}^{\alpha}_{\mathrm{SSD}_{\mathrm{ta}}} = -c_7\frac{D_0 b^3}{k_B\theta}\exp\left(\frac{-Q_{\mathrm{bulk}}}{k_B\theta}\right)\left(\rho^{\alpha}_{SSD}\right)^2\left|\tau^{\alpha}\right|\left(\frac{\dot{\gamma}^{\alpha}}{\dot{\gamma}_{\mathrm{ref}}}\right)^{c_8} \tag{9}$$

In Equation 8, c_6 is a material constant. In Equation 9, Q_{bulk} is the activation energy for the process of dislocation climb which is governs the rate of thermal annihilation of dislocations, $\dot{\gamma}_{\mathrm{ref}}$ is the reference shear rate, D_0 is the diffusion co-efficient, c_7 and c_8 are the material constants. In fact, there can be many other mechanisms being active in the matrix depending upon the condition of applied stress and temperature and these can be considered by formulation of suitable constitutive equations. Each term in the right hand side of Equation 5 can be constructed as a function of plastic shear strain increment $\dot{\gamma}^{\alpha}$, density of forest dislocations ρ_F (which depends upon density of SSDs and geometrically necessary dislocations GNDs), density of statistically stored dislocations (SSDs), applied shear stress τ^{α} and absolute temperature θ etc. as presented in Equations 6-9 respectively. Hence, these can be written in general as

$$\dot{\rho}^{\alpha}_{\mathrm{SSD}} = \dot{\rho}^{\alpha}_{\mathrm{SSD}}\left(\dot{\gamma}^{\alpha}, \rho^{\alpha}_{\mathrm{SSD}}, \rho_F, \tau^{\alpha}, \theta\right) \tag{10}$$

On the other hand, the increment of density of geometrically necessary dislocations depends upon the gradient of plastic strain and as it can be arbitrarily oriented with respect to the slip plane, the three components (screw, edge and normal) can be written as

$$\begin{aligned}
\dot{\rho}^{\alpha}_{\mathrm{GNDs}} &= \frac{1}{b}\left[\nabla_x \times \left(\dot{\gamma}^{\alpha} F_P^T n^{\alpha}\right)\right].d^{\alpha} \\
\dot{\rho}^{\alpha}_{\mathrm{GNDet}} &= \frac{1}{b}\left[\nabla_x \times \left(\dot{\gamma}^{\alpha} F_P^T n^{\alpha}\right)\right].t^{\alpha} \\
\dot{\rho}^{\alpha}_{\mathrm{GNDen}} &= \frac{1}{b}\left[\nabla_x \times \left(\dot{\gamma}^{\alpha} F_P^T n^{\alpha}\right)\right].n^{\alpha}
\end{aligned} \tag{11}$$

where the unit vectors d^{α}, t^{α} and n^{α} are in the directions of Burgers vector, tangent to edge dislocation direction and normal to the slip plane α respectively. The density of mobile dislocations, i.e., $\rho^{\alpha}_{\mathrm{m}}$ is computed as a function of forest (ρ^{α}_F) and parallel (ρ^{α}_P) dislocation density as

$$\rho^{\alpha}_{\mathrm{m}} = c\theta\sqrt{\rho^{\alpha}_F \rho^{\alpha}_P} \tag{12}$$

where c is a constant. This constant is often expressed in terms of shear modulus G, Boltzmann's constant k_B, Burger's vector b and other constant as

$$c = \frac{2k_B}{c_1 c_2 c_3 G b^3} \tag{13}$$

The density of forest and parallel dislocations are evaluated as functions of density of SSDs, GNDs and interaction coefficients $\chi^{\alpha\beta}$ among different slip system as

$$\begin{aligned}
\rho^{\alpha}_F &= \rho^{\alpha}_F\left(\rho^{\beta}_{\mathrm{SSD}}, \rho^{\beta}_{\mathrm{GND}}, \chi^{\alpha\beta}\right) \\
\rho^{\alpha}_P &= \rho^{\alpha}_P\left(\rho^{\beta}_{\mathrm{SSD}}, \rho^{\beta}_{\mathrm{GND}}, \chi^{\alpha\beta}\right)
\end{aligned} \tag{14}$$

The Equation 14 can be elaborated for calculation of forest and parallel dislocation densities as the following, i.e.,

$$\rho_F^\alpha = \sum_{\beta=1}^{N} \chi^{\alpha\beta} \begin{bmatrix} \rho_{SSD}^\beta \left|\cos\left(n^\alpha, t^\beta\right)\right| + \rho_{GNDs}^\beta \left|\cos\left(n^\alpha, d^\beta\right)\right| \\ +\rho_{GNDet}^\beta \left|\cos\left(n^\alpha, t^\beta\right)\right| + \rho_{GNDen}^\beta \left|\cos\left(n^\alpha, n^\beta\right)\right| \end{bmatrix} \tag{15}$$

$$\rho_P^\alpha = \sum_{\beta=1}^{N} \chi^{\alpha\beta} \begin{bmatrix} \rho_{SSD}^\beta \left|\sin\left(n^\alpha, t^\beta\right)\right| + \rho_{GNDs}^\beta \left|\sin\left(n^\alpha, d^\beta\right)\right| \\ +\rho_{GNDet}^\beta \left|\sin\left(n^\alpha, t^\beta\right)\right| + \rho_{GNDen}^\beta \left|\sin\left(n^\alpha, n^\beta\right)\right| \end{bmatrix} \tag{16}$$

The densities of forest and parallel dislocations evolve due to plastic deformation and the SSDs and GNDs also evolve due to plastic deformation and hardening mechanisms.

Criteria for Initiation of Micro-Twin

The criterion for initiation of micro-twin is based on the state of dissociation of the leading and trailing partials on a slip system α. Due to stress tensor σ_{ij} acting on a material point, the leading and trailing Shockley partials will experience stresses which can be denoted as τ_{lead} and τ_{trail} respectively. Based on the magnitudes of in-plane resolved shear stress and stress on leading and trailing Shockley partials, a criteria can be derived as to whether the partials will remain together or get dissociated at the interface of the γ and γ' phases of the super-alloy. The calculation procedure of τ_{lead} and τ_{trail} are described as follows. The traction vector on any slip plane α can be calculated as

$$T_i = \sigma_{ij}\eta_j \tag{17}$$

where σ_{ij} is the Cauchy stress tensor and η_j are the components of the unit vector η along the normal to the slip plane. The components of in-plane stresses can be calculated as

$$T_{Si} = T_i - \left(T_k\eta_k\right)\eta_i \tag{18}$$

The components of unit vector η_{Si} along the direction of resolved shear stress in the slip plane can be calculated from the vector T_{Si} as

$$\eta_{Si} = \frac{T_{Si}}{\left\|T_S\right\|} \tag{19}$$

Let η_{bi} be the components of the unit vector η_b along the slip direction. The angle θ between the resolved shear stress and slip direction is required for calculation of stresses on leading and trailing partials. This angle θ can be calculated as

$$\theta = \begin{cases} \cos^{-1}\left(\eta_S \cdot \eta_b\right) & \text{if } \left(\eta \times \eta_b\right) \cdot \eta_S \le 0 \\ -\cos^{-1}\left(\eta_S \cdot \eta_b\right) & \text{if } \left(\eta \times \eta_b\right) \cdot \eta_S > 0 \end{cases} \tag{20}$$

Stress on the Leading Partial

The stress on the leading partial is computed as

$$\tau_{\text{lead}} = \frac{1}{\left(\cos\theta_l + \cos\theta_t\right)}\left[\frac{Gbf_a}{\lambda} f_s + 2\tau_f\right] \tag{21}$$

where the angles θ_l and θ_t are written as $\pi/6 - \theta$ and $\pi/6 + \theta$ respectively where θ is obtained from Equation 20, G is the shear modulus, b is Burger's vector, f_s is the scaling factor (for dislocation line tension), λ is the width of γ channel, τ_f is the friction stress experienced by dislocations while moving through the matrix and f_a is a factor, the magnitude of which depends upon the angle θ as

$$f_a = \begin{cases} 1\Big/1 - \dfrac{\cos\theta_t}{\left(\cos\theta_l + \cos\theta_t\right)} & \text{if } \theta > 0 \\ 2 & \text{if } \theta \le 0 \end{cases} \tag{22}$$

Stress on the Trailing Partial

The stress on the trailing partial τ_{trail} is similarly computed as a function of applied stress and the intrinsic stacking fault energy Q_{ISF} as follows.

$$\tau_{\text{trail}} = \frac{-1}{2\lambda b\cos^2\theta_t}\left[\begin{array}{l} -\cos\theta_t \dfrac{Gb^2}{2\pi\left(1-\nu\right)} + \lambda Q_{ISF}\cos\theta_t - f_s Gb^2\cos\theta_t + \lambda b\cos\theta_t\left(-\tau_f + \dfrac{Q_{ISF}}{b}\right) \\ -\left(\begin{array}{l} \left[\dfrac{Gb^2}{2\pi\left(1-\nu\right)}\right]^2 \cos^2\theta_t - 2\lambda Q_{ISF}\dfrac{Gb^2}{2\pi\left(1-\nu\right)}\cos^2\theta_t \\ +2\cos\theta_t \dfrac{Gb^2}{2\pi\left(1-\nu\right)} f_s Gb^2\cos\theta_t + 2\cos^2\theta_t \dfrac{Gb^2}{2\pi\left(1-\nu\right)}\lambda b\left(-\tau_f + \dfrac{Q_{ISF}}{b}\right) \\ +\left(\lambda Q_{ISF}\cos\theta_t\right)^2 + 2\lambda f_s Q_{ISF} Gb^2\cos^2\theta_t - 2\lambda^2 b Q_{ISF}\cos^2\theta_t\left(-\tau_f + \dfrac{Q_{ISF}}{b}\right) \\ +\left(f_s Gb^2\cos\theta_t\right)^2 - 2 f_s Gb^3\lambda\cos^2\theta_t\left(-\tau_f + \dfrac{Q_{ISF}}{b}\right) + \left(\lambda b\cos\theta_t\right)^2\left(-\tau_f + \dfrac{Q_{ISF}}{b}\right)^2 \end{array}\right)^{0.5} \end{array}\right] \tag{23}$$

These Equations 22 and 23 are used for calculation of stresses on leading and trailing dislocation partials and the condition for dissociation of the partials at the precipitate-matrix interface depends upon the following criteria.

Criterion for Dissociation of Partials

Once the in-plane stress components are calculated according to Equation 18 and the stresses on leading and trailing partials are calculated. The criterion for dissociation of partials is laid down as follows.

1. If $\left\|T_S\right\| > \tau_{\text{lead}}$ and $\left\|T_S\right\| < \tau_{\text{trail}}$, then the leading partial passes through and hence the leading and trailing partials remain dissociated.
2. If $\left\|T_S\right\| > \tau_{\text{trail}}$ and $\left(\tau_{\text{trail}} - \tau_{\text{lead}}\right) > \tau_{\text{cutoff}}$, then both leading and trailing partials pass through and both remain dissociated, where τ_{cutoff} is the stress difference between the leading and trailing partials above which the two partials are treated to be remaining in the state of dissociation.
3. If $\left\|T_S\right\| > \tau_{trail}$ and $\left(\tau_{trail} - \tau_{lead}\right) \leq \tau_{cutoff}$, then both the partials stay together in un-dissociated state and both pass through the γ' precipitate.

The conditions 1 and 2 as discussed above can be treated as condition of initiation of twin at a material point.

Treatment of Plastic Strain Development in the Twinned Region

Once the condition of initiation of twin is reached following the conditions outlined in Section 3, the twinned region of the crystal is rotated by an angle with respect to the parent matrix. The rotation matrix is given as

$$R^{tw} = 2n^{\beta} \otimes n^{\beta} - 1 \tag{24}$$

where n^{α} is perpendicular to the plane which undergoes the twinning rotation. The elasticity tensor C of the twin region is calculated as

$$C^{tw}_{ijkl} = R^{tw}_{im} R^{tw}_{jn} R^{tw}_{ko} R^{tw}_{lp} C_{mnop} \tag{25}$$

The twin system vectors m^{β}_{tw} and n^{β}_{tw} are rotated as

$$m^{\beta}_{tw} = R^{tw} m^{\beta};\; n^{\beta}_{tw} = R^{tw} n^{\beta} \tag{26}$$

Constitutive Equations for Twin Deformation

The model is based the argument that the shearing of the γ' precipitates and subsequent re-ordering is the predecessor to the movement of partials causing the plastic straining. The thermal re-ordering rate by diffusion depends upon the activation energy barrier ΔG as

$$\mathrm{R}_{\mathrm{reorder}} \propto \exp\left(\frac{-\Delta G}{kT}\right) \tag{27}$$

where G is Gibb's free energy, k is Boltzmann's constant and T is absolute temperature. We can write ΔG as

$$\Delta G = \Delta E + p\Delta V - T\Delta S \tag{28}$$

where ΔE is the internal energy barrier, p is pressure, ΔV is the activation volume for pressure dependent diffusion, ΔS is the change in entropy of the system.

For solids, the pressure dependent activation volume is small and hence the corresponding term in Equation 28 can be ignored here. However, the change in entropy comes from irreversible processes of dissipation such as plastic work and damage. According to the principle of maximum plastic dissipation, the plastic strain tensor should be co-axial to the stress tensor and hence we can write plastic work as

$$W_p = \tau_{eff} A_p b_{tp} \tag{29}$$

where A_p is the area over which the shear force acts during the plastic deformation process. If other sources of dissipation are not accounted for, we can write

$$T\Delta S = W_p \tag{30}$$

In order to arrive at the strain rate equation, we state that the velocity of twin partials is written as

$$v_{tp} = P_{shear}\left(\tau_{eff}\right) f_{reorder} \lambda_{reorder} P_{reorder}\left(\Delta G\right) \tag{31}$$

where P_{shear} is the probability of the shearing process (by the partials) to take place which is a function of effective shear stress τ_{eff}, $f_{reorder}$ is the frequency of re-ordering, $\lambda_{reorder}$ is the reordering distance and $P_{reorder}$ is the probability of re-ordering against the energy barrier ΔG which is also a function of plastic dissipation and internal energy barrier. The internal energy barrier can be written as

$$\Delta E = \Gamma\left(t\right) A_{shear} \tag{32}$$

The effective shear stress considering the effect of tertiary γ' precipitates in the matrix may be written as

$$\tau_{eff} = \tau - \frac{f_3}{2}\frac{\Gamma_{pt}}{b_{tp}} \quad (33)$$

where f_3 is the area fraction of the tertiary γ' precipitates in the matrix. Hence the plastic shear strain rate due to micro-twinning can be written as

$$\dot{\gamma}\left(\tau_{eff}, T, t\right) = \rho_{tp}\left(\tau_{eff}, t\right) b_{tp} P_{shear}\left(\tau_{eff}\right) f_{reorder} \lambda_{reorder} \exp\left(\frac{-\Delta G}{kT}\right) \quad (34)$$

where ρ_{tp} is the density of partials which is a function of applied stress and time.

Finite Element Implementation of the Dislocation Density Based Crystal Plasticity Model

The above formulation has been implemented in an in-house finite element based code. The calculation of all the constitutive state variables are done at a material point except the geometrically necessary dislocation density $\rho^{\alpha}_{\text{GND}}$ which is a nonlocal parameter and depends upon the gradient of the product $\dot{\gamma}^{\alpha} F_P^T n^{\alpha}$, where F_P is the plastic part of the deformation gradient F. The deformation gradient is written as a product of elastic and plastic parts (i.e., F_e and F_P). The plastic velocity gradient L_p is calculated as summation of contribution of each slip system (the contributions being the product of shear rate $\dot{\gamma}^{\alpha}$ in each slip system and Schmidt's tensor $m^{\alpha} \otimes n^{\alpha}$), i.e.,

$$L_p = \sum_{\alpha=1}^{n} \dot{\gamma}^{\alpha} m^{\alpha} \otimes n^{\alpha} \quad (35)$$

and the plastic deformation gradient is updated based on the plastic shear accumulation in each slip system α as follows.

$$\dot{F}_p = L_P F_P \quad (36)$$

The parameter $\dot{\gamma}^{\alpha} F_P^T n^{\alpha}$ at each material point of integration of a finite element is extrapolated to the nodes of that element and in this way, the local variable are converted into a nodal variable so that the nonlocal parameter can be evaluated at the material integration point in the next iteration step.

For the nickel based super-alloys, the matrix γ and the secondary γ' phases are considered explicitly in the finite element model. The material properties including the elastic constants and the dislocation density coefficients are different for the matrix and secondary phases. Moreover, the secondary gamma-phase is an ordered phase and doesn't contain any dislocation initially. Due to the incompatibility, the dislocations pile up at the interface of the matrix and the gamma-phase. When a critical condition is reached, the dislocations cut through the precipitate and plastic deformation starts in the secondary gamma phase.

Criteria for APB Shearing in the γ' Phase

As the γ' phase is an ordered phase (with $L1_2$ microstructure), the matrix dislocations (in the disordered FCC phase) cannot enter the γ' phase. However, the matrix dislocations can form super-dislocations at the interface of matrix and precipitate and enter the ordered phase through the mechanism of anti-phase boundary (APB) shearing. APB shearing of γ' phase is assumed to take place when the resolved shear stress along the slip system α and the dislocation density at the interface exceed critical values, i.e.,

$$\tau^{\alpha} > \tau_{c} \text{ and } \rho_{\mathrm{m}}^{\alpha} > \rho_{c} \quad (37)$$

The flow rules for the γ' phase can be mathematically expressed as

$$\dot{\gamma}^{\alpha} = \mathbf{H}(\rho_{\mathrm{m}}^{\alpha} - \rho_{c})\rho_{\mathrm{m}}^{\alpha} b v^{\alpha} \quad (38)$$

with velocity of dislocations in the γ' phase is written as

$$v^{\alpha} = \lambda v e^{\left(-Q_{\mathrm{act}}/k_{B}\theta\right)} \sinh\left(\frac{\left\langle \tau^{\alpha} - \tau_{pass}^{\alpha} \right\rangle \mathbf{H}(\tau^{\alpha} - \tau_{c}) V_{\mathrm{act}}}{k_{B}\theta}\right) \mathrm{sgn}\left(\tau^{\alpha}\right) \quad (39)$$

where $\mathbf{H}(\ )$ is the Heaviside function, τ_c is the critical resolved shear stress and ρ_c is the critical density at the interface for onset of APB shearing in the γ' phase. The constitutive models for both the γ and γ' phases have been implemented in the FE code along with the criteria for onset of APB shearing in the γ' phase and has been used for simulation of constant strain rate and creep tests on single crystal experiments as discussed in the following section.

Results and Discussion

The dislocation density based crystal plasticity as model as described in Sections 2-4 and 5.1 has been used to simulate the deformation behavior of single crystal experiments of nickel based super-alloy CMSX-4. The experimental results have been taken from Fleury et al. (1996). This alloy contains around 70% of γ' precipitates and the precipitate are of cubical shape. The experiment has been conducted at 800 deg. C and the loading is along the [001] direction. The different slip systems in the FCC crystal of nickel-based super-alloy used in the simulation are presented in Table-1. The values of different parameters used in the dislocation density based crystal plasticity model are presented in Table-2.

In order to simulate the deformation behavior of single crystal experiments of nickel based super-alloy CMSX-4, a multi-scale approach has been adopted. The scheme of the multi-scale simulation procedure is presented in Figure 1. For simulation of response of polycrystalline experiments, the data

Table 1. Different slip systems in the FCC crystal of nickel-based super-alloy used in the simulation

Slip System No.	Slip Plane and Direction	Slip System No.	Slip Plane and Direction
1	(111) [-101]	7	(-1-11) [101]
2	(111) [01-1]	8	(-1-11) [0-1-1]
3	(111) [1-10]	9	(-1-11) [-110]
4	(-111) [01-1]	10	(1-11) [0-1-1]
5	(-111) [101]	11	(1-11) [-101]
6	(-111) [-1-10]	12	(1-11) [110]

Table 2. Values of different parameters used in the dislocation density based crystal plasticity model

Parameter	Value
Q_{act}	5.9 e-19 J
Q_{bulk}	4.8e-19 J
c_1	0.1
c_2	2.0
c_3	0.4
c_4	1.0 e-3
c_4'	2.5 e-3
c_5	1.0 e-4
c_6	10.0
c_7	10.0

regarding orientation of grains with respect to loading directions (from orientation imaging microscopy), distribution of orientation, relative orientation and misorientation among grains, grain size, structure within grains etc. are required. Modeling of all these details for simulation of deformation behavior of a polycrystalline material (with different mechanisms operating at different length scales) in a single scale will be computationally prohibitive. Hence, a multi-scale approach with homogenization schemes as presented in Figure 1 is necessary.

The term 'Homogenization' as used in this work refers to a single model for the grain without explicit consideration of matrix, precipitates and their interactions. The details of the mechanisms are

Figure 1. Schematic of the microstructure-sensitive multi-scale modeling approach for simulation of deformation behavior of nickel-based superalloys

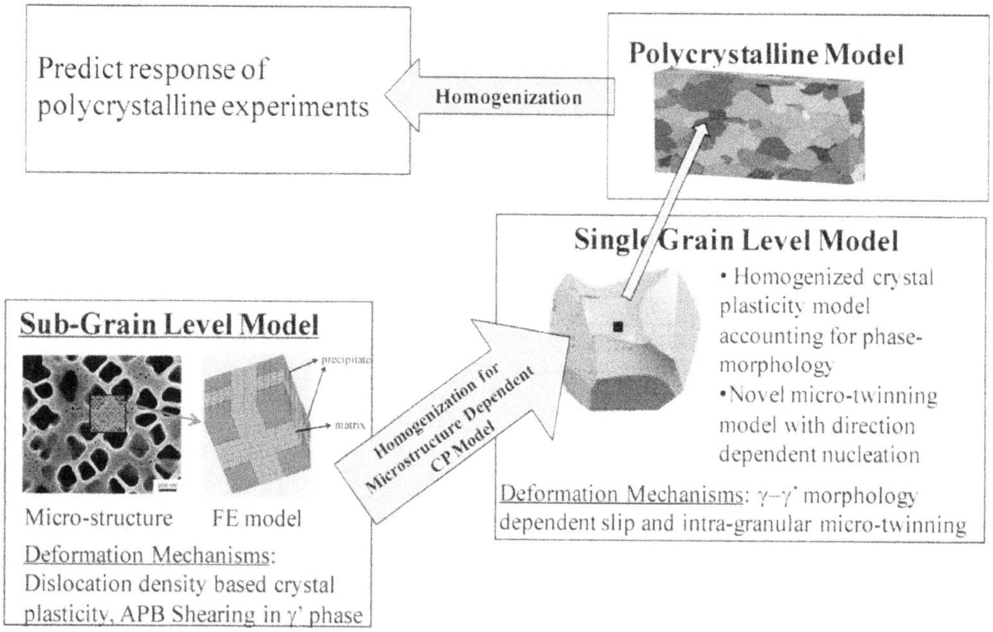

considered in the explicit lower length scale model. The response of different representative volume elements (in terms of overall shear rate in slip system α as a function of plastic work density) is derived from the lower length scale model as a function of different parameters like volume fraction and shape exponents of precipitates. These functional dependencies can be derived for parameters like homogenized activation energy, homogenized passing stress and homogenized activation volume so that these can be used in higher length scale model.

Initially, the mechanism of deformation in the single crystal level is modeled. It involves explicit consideration of matrix and precipitate geometry, their interaction as well as movement of dislocations in the slip planes of matrix and pile-up at the matrix-precipitate interface. The different mechanism of dislocation generation, annihilation and multiplication etc. are also modeled with dislocation density as state variables. Different types of dislocations, viz., mobile, immobile (forest and parallel), statistically stored as well as geometrically necessary dislocations, are considered in this dislocation density based grain scale crystal plasticity model. For shearing of precipitates at the matrix-precipitate interface, a new model for anti-phase boundary shearing has been developed and used in this scheme.

The response of the single crystal with various size, shape and distribution of precipitates have been homogenized in terms of global shear strain rate of the crystal as a function of plastic work density. This functional dependence of homogenized shear rate in a slip system has been derived as a function of shape exponent parameters and volume fraction of the precipitates so that the explicit modeling of matrix and precipitates can be avoided at the polycrystalline level simulation. In addition, plastic deformation by micro-twin as the grain level has been simulated with a new model which dictates initiation of micro-twin based on value of resolved shear stress in the twin system and stress acting on leading and trailing partials. Further evolution of twin strain is also modeled using a novel constitutive formulation.

From the knowledge about the microstructure of the alloy within a single grain, an explicit 3-D model of matrix-precipitate representative volume element is prepared as shown in Figure 2 for a typical precipitate shape embedded in the matrix. This model incorporates the precipitate shape, size, volume-fraction and the ligament length of the matrix between the precipitates into the finite element model. The responses of the lower length-scale model are homogenized into the grain-level model and the homogenized slip and micro-twin constitutive models are implemented at this length-scale. In this length-scale, the explicit sub-grain microstructure is not modeled and hence, it can expedite the computational simulation for polycrystalline experiments as schematically presented in Figure 1.

In this work, the sub-grain microstructure of CMSX-4 alloy has been explicitly modeled. This alloy contains around 70% of γ' precipitates and the precipitate are of primarily cubical shape. The tensile experiment has been conducted at 800 deg. C and the loading direction is along the [001] direction of the crystal. The experimental data have been taken from Fleury et al. (1996) for its comparison with the results of simulation.

The actual 3-D FE mesh of the cubical precipitate along with the matrix (Figure 3) as embedded in the representative volume element (RVE) is loaded in the vertical direction with symmetric boundary conditions in x, y and z-directions. The other faces use periodic boundary conditions except the loading face. The FE mesh consists of 8-noded 3D brick elements.

The results of the simulation (i.e., stress-strain response) with and without activation of APB shearing of the precipitates are shown in Figure 4. The experiment is carried out at a strain rate of 1e-4 per second. The experimental data is also plotted in Figure 4. As the precipitates have zero initial dislocation density and the volume fraction of the precipitates in the matrix is very high (i.e., of the order of 70%), the response is very stiff if the mechanism of APB shearing is not activated in the constitutive model.

The response of the γ phase only (i.e., without the precipitates) is also plotted in Figure 4. It can be noted that the yield stress as well the plastic hardening are very low for the RVE containing no γ' phases. This super-alloy gets its high yield strength and plastic hardening behavior from the presence of γ' phases in the matrix. However, with accumulation of dislocation density at the interfaces of matrix and the precipitates at high levels of applied stress, the critical stress and dislocation pile-up condition

Figure 2. FE model of a typical precipitate shape embedded in the matrix which is used in the lower length scale of the multi-scale approach

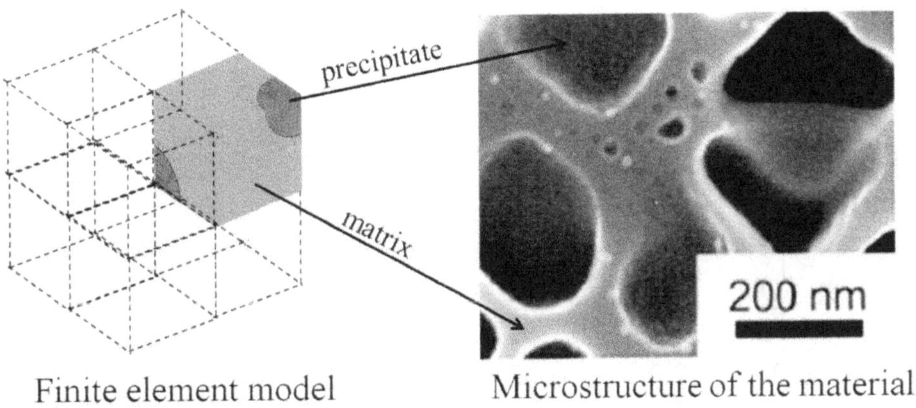

Figure 3. FE mesh of the matrix along with the cubic precipitates loaded along the vertical [001] direction

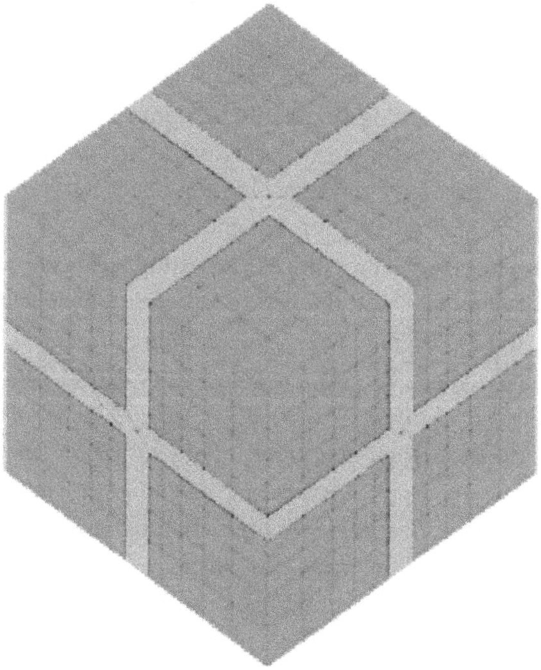

Figure 4. Stress-strain response of a single crystal nickel based super-alloy CMSX-4 for loading on [001] direction at 800 deg. C

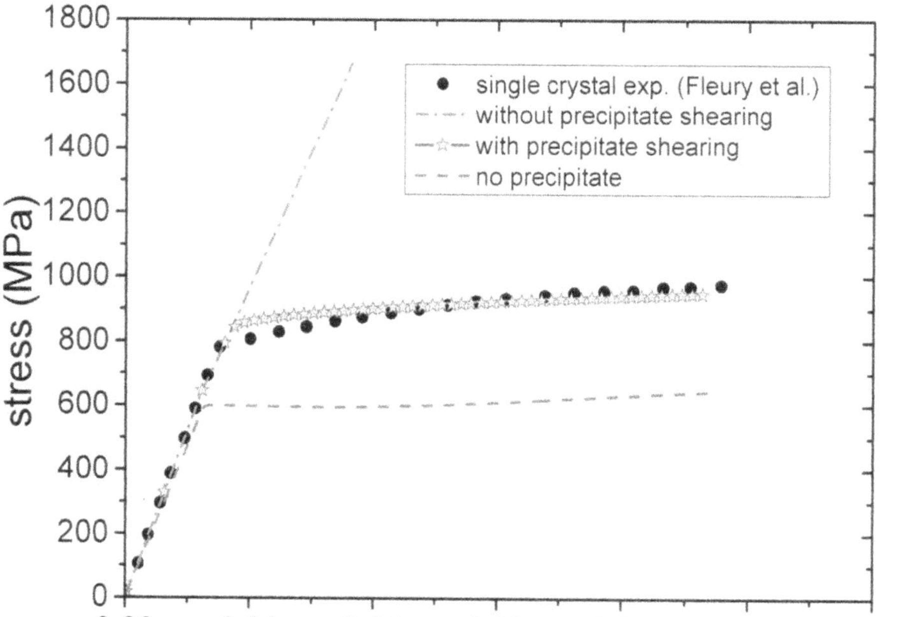

is satisfied so that the dislocations can form super-dislocations and enter the γ' phase. Hence, with incorporation of the criteria for APB shearing, the stress-strain response can be simulated satisfactorily as shown in Figure 4.

It can be observed from Figure 2 that the precipitates are modeled with symmetric boundary conditions and hence, these are located at the corners of the cube. Initially, there are no dislocations present in the precipitates and hence, plastic strain is zero before onset of APB shearing in precipitates by the matrix dislocations (Figure 5). The plastic strain distribution and its evolution with loading in the matrix as well as the precipitate can be very different for the two cases (i.e., case-(a): without, and case-(b): with APB shearing of γ' precipitates).

The plastic strain (for the [001] loading direction) distribution for the RVE is presented in Figure 5 for both the cases as mentioned above. As can be seen from the contour plots, maximum plastic strain is of the order of 4.5% at the end of numerical simulation and this strain is highly concentrated in the matrix in case-(a). The precipitates at the corners are not sheared in case-(a) and hence, there is no plastic strain. In case-(b), there is an evolution of plastic strain in both the precipitate and the matrix as can be seen from Figure 5. Hence, the maximum plastic deformation is less concentrated in the matrix in case-(b) as compared to case-(a).

The effect of loading direction on the stress-strain response of the crystal and the corresponding information regarding activation of different sets of slip systems is essential in order to build a constitutive model for homogenized plastic shear strain rate in the grain-level model. For this purpose, the RVE has been loaded in the [010] and [100] directions respectively. The slip rates as a function of plastic work

Figure 5. Contour of plastic strain in the loading direction in the matrix and precipitate of nickel-based super-alloy CMSX4 simulated with and without APB shearing model

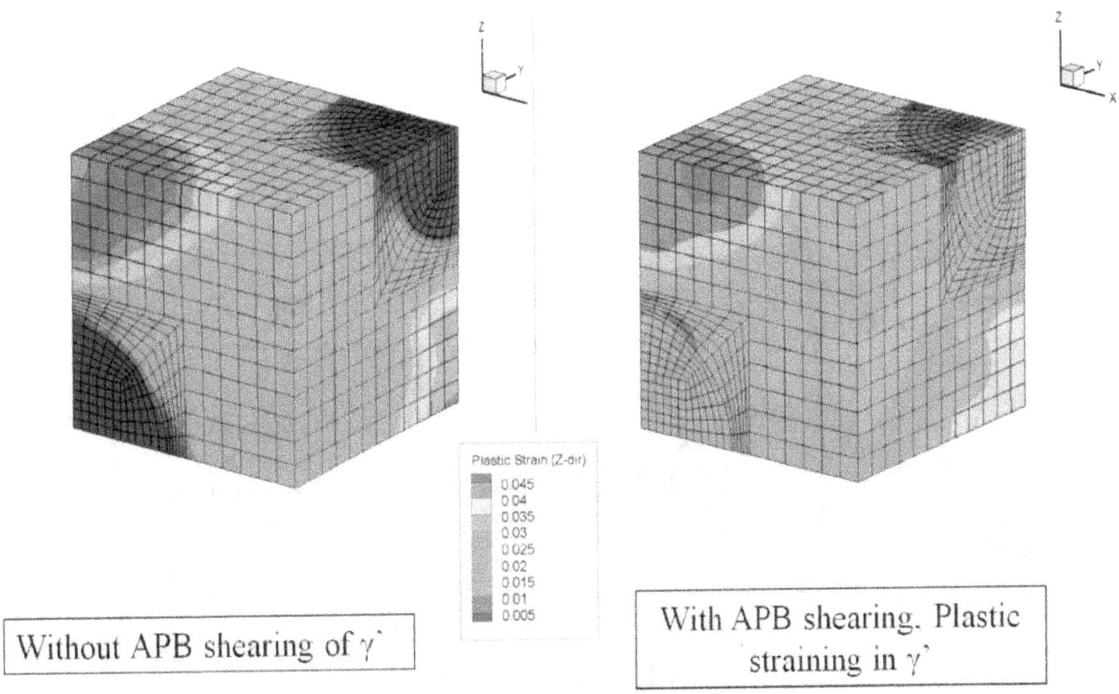

density has been presented in Figures 6 and 7 respectively for the different active slip systems when the RVE is loaded in two different orientations. The volume averaged plastic work density W_p can be evaluated from the FE results as follows.

$$W_p = \int_V \int_t \tau^\alpha \dot{\gamma}^\alpha dt dV \Big/ \int_V dV \tag{40}$$

where $\tau^\alpha, \dot{\gamma}^\alpha$ are the resolved shear stress and shear strain rate in slip system α, t is time and V is the volume of the RVE. It can be observed that 8 slip systems out of 12 become active in these two orientations. However, the slip rates are different for different active slip systems for these two orientations. It may be noted that after some amount of slip, one of the slip systems overtakes all other active slip systems and hence, slip rate increases monotonically for one or a pair of predominant slip system (3, 9 for loading in [010] direction and 6, 12 for loading [100] direction) and for others, it goes on reducing

When the slip rate activity of the dominant slip systems in different equivalent directions (i.e., [010] and [100]) are compared, it was observed that maximum homogenized slip rate variation with plastic work density almost remains same (Figure 8). This implies that [100], [010] and [001] directions are equivalent as FCC system is symmetric with respect to co-ordinate systems. However, for other loading directions, the slip rates will be different as the resolved shear stress on the slip systems will be different for a given value of applied stress.

Figure 6. Homogenized plastic slip rate as a function of plastic work density in different active slip systems for loading in [010] direction

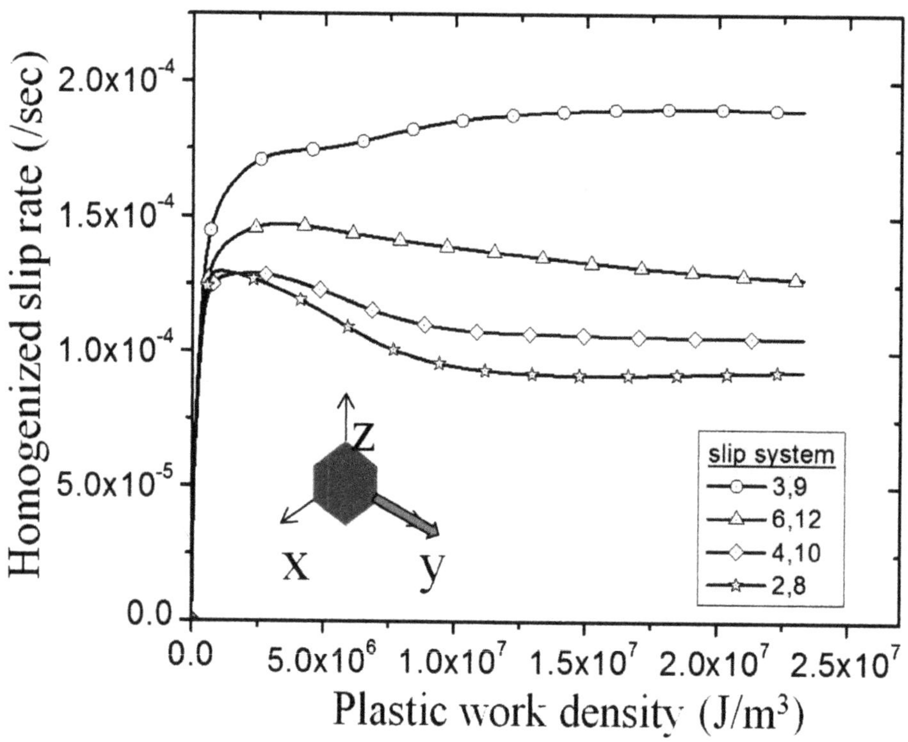

Figure 7. Homogenized plastic slip rate as a function of plastic work density in different active slip systems for loading in [100] direction

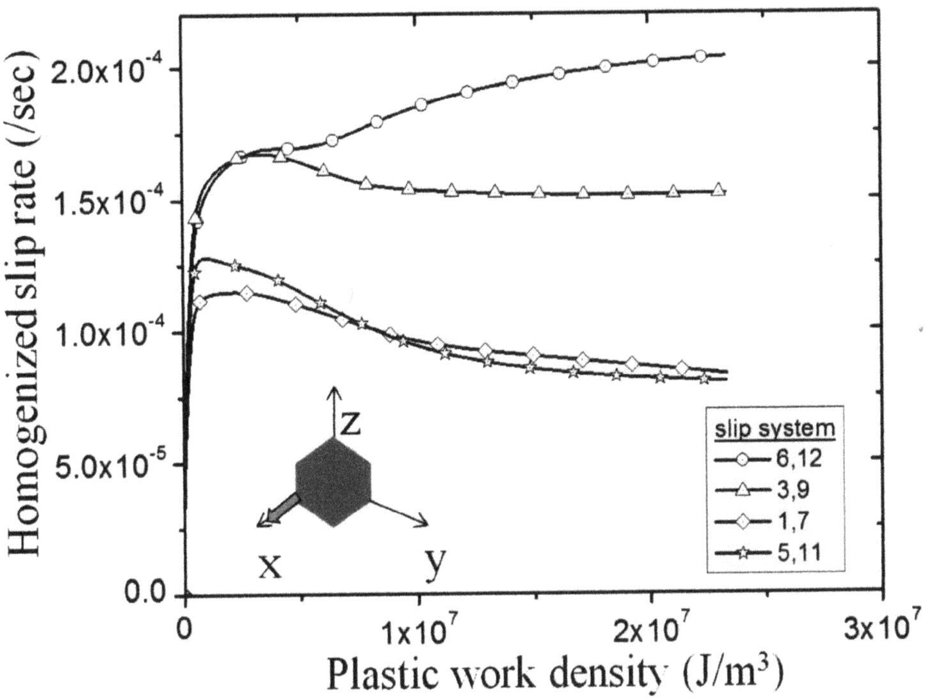

It can be noted from Figure 4 that the yield stress as well the hardening behavior of the matrix are very low. This super-alloy exhibits high yield stress and the hardening behavior and it is due to the presence of γ' phase in the matrix. However, the accumulation of dislocation density at the interfaces of matrix and precipitates at high stresses, the condition is satisfied so that dislocations can form super-dislocations and enter the γ' phase. Hence, with incorporation of the criteria for APB shearing and the corresponding constitutive relations for plastic flow in the γ' phase, the stress-strain response can be simulated satisfactorily as shown in Figure 4. The material constants can also calibrated from this single crystal experimental data.

Once the parameters are calibrated, these are used to simulate the other micro-structures with various shapes and volume fraction of the γ' precipitates. The precipitate shape is described by the parameter 'n' in the following equation for a generalized 3D ellipsoid.

$$\left(\frac{x}{a}\right)^n + \left(\frac{y}{b}\right)^n + \left(\frac{z}{c}\right)^n = 1 \tag{41}$$

where a, b, and c are the dimensions of the precipitate in three principal directions and 'n' is shape exponent. The shape exponent 'n' is equal to 2.0 for a sphere and tends to infinity for a cube. In order parameterize the shape exponent in the homogenized model, the shape exponent has been converted to the shape factor n_1 through

Figure 8. Comparison of maximum plastic slip rates as a function of plastic work density for loading in different directions

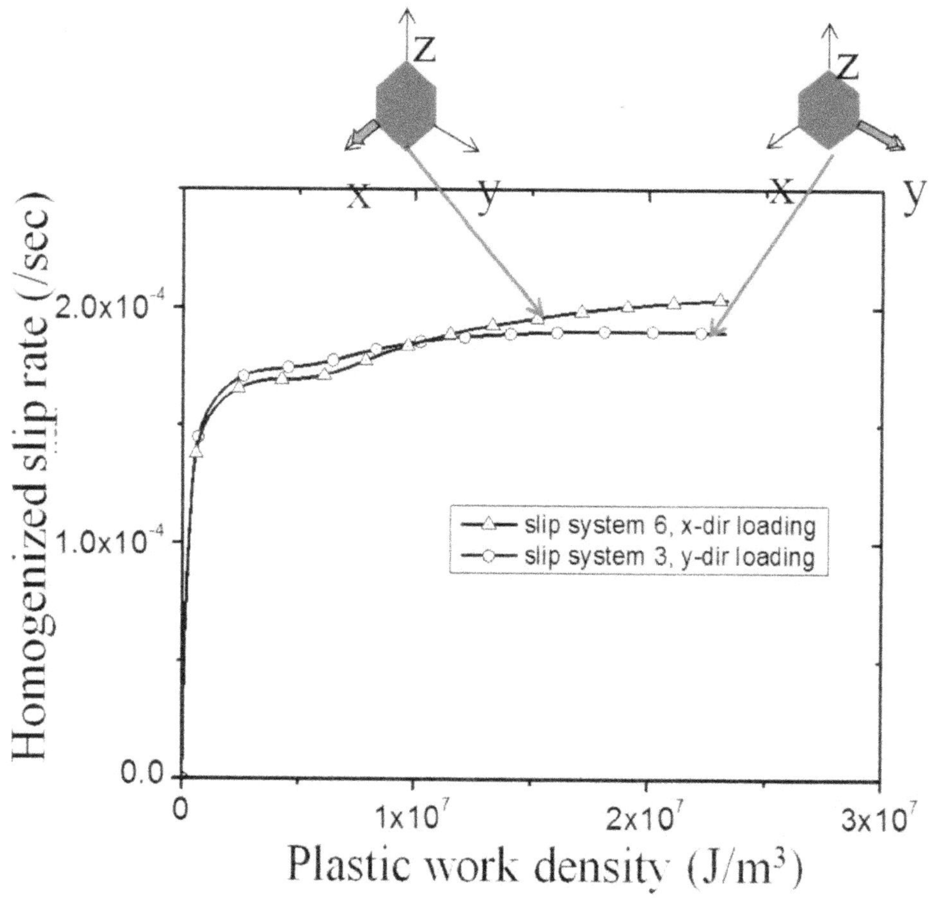

$$n_1 = \tan^{-1}(n) \tag{42}$$

The analysis has been carried out for various volume fractions of cubical precipitates ($n=\infty$) with volume fractions ranging from 12.5% to 75.1%. The stress-strain response of the micro-structure for loading in [001] direction at 800 deg. C is plotted in Figure 9 for these RVEs with various volume fractions of cubical precipitates.

It can be observed from the above figure that the yield stress increases with increasing volume fraction of precipitates. The increase in yield stress is initially small. However, it becomes significant at larger volume fractions of the γ' phase. The hardening response also changes with volume fraction of precipitates. Similarly, the results of analysis with shape exponent n equal to 1.5 are shown in Figure 10. It can be observed that hardening changes significantly with increasing volume fraction of precipitates. Hence, both the precipitate shape and the volume fractions affect the onset of yielding and plastic hardening response.

In order to validate the model, it has been used to predict the creep strain vs. time response of a single crystal nickel based super-alloy. The experimental data has been taken from Kakehi (1999). The

Figure 9. Variation of stress-strain response of the single crystal with various volume fractions of γ' precipitates with cubical shape ($n=\infty$)

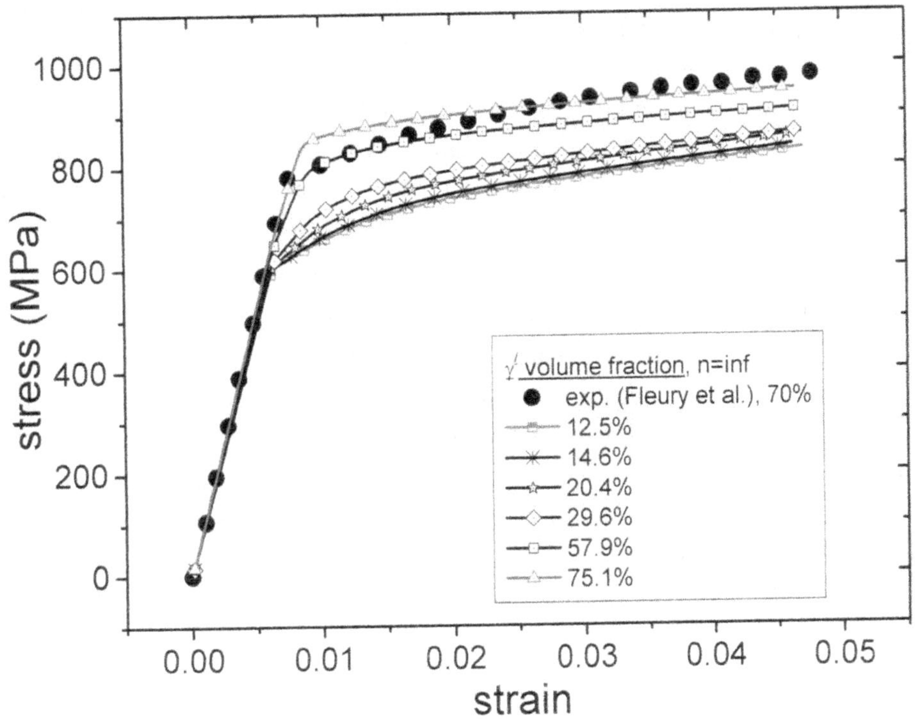

Figure 10. Variation of stress-strain response of the single crystal with various volume fractions of γ' precipitates with shape exponent n of 1.5

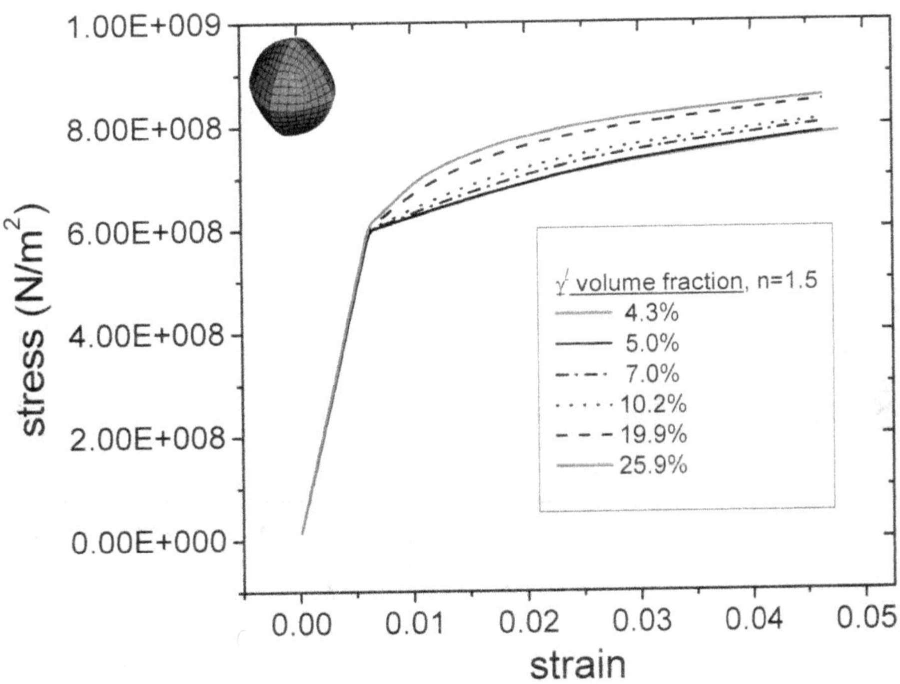

test has been conducted at 700 deg. C and loaded at 820 MPa of applied stress. The alloy contains 57.9% of γ' precipitates which are of cubical shape (n= ∞). The crystal is loaded in [001] direction. The results of simulation have been compared with that of experiment and the creep strain vs. time response is plotted in Figure 11. It can be observed that the dislocation density based crystal plasticity model has been able to predict the experimental creep response very satisfactorily.

In order to use this model in the higher polycrystalline length scale, the results of simulation with various micro-structural features have been homogenized. Various shapes of the precipitates have been modeled with the shape exponent 'n'. The shapes of the precipitates along with the values of parameter 'n' are shown in Figure 12. The parameter n_1 ($\tan^{-1}(n)$) along with the precipitates volume fraction (v_p) serve as the micro-structural parameters with which the homogenized variables will be expressed in suitable functional forms in the homogenized crystal plasticity model.

The volume averaged slip strain rate in each slip system has been evaluated and its evolution (maximum of all the slip systems) with plastic work density is presented in Figure 13 for various values of precipitate volume fraction (v_p) and shape factors (n_1). This data has been used to derive the following homogenized parameters, i.e., (a) homogenized activation energy, (b) homogenized passing stress and (c) homogenized activation volume. The variation of the difference of homogenized activation energy from the reference activation energy (for the γ phase alone) with shape factor (n_1) and volume fraction (v_p) is presented in Figure 14. The fitted surface is also shown in Figure 14 along with the 2D pro-

Figure 11. Simulation of creep response of CMSX-4 single crystal super-alloy CMSX-4 at 700 deg. C and 820 MPa of applied stress for tensile loading in [001] direction

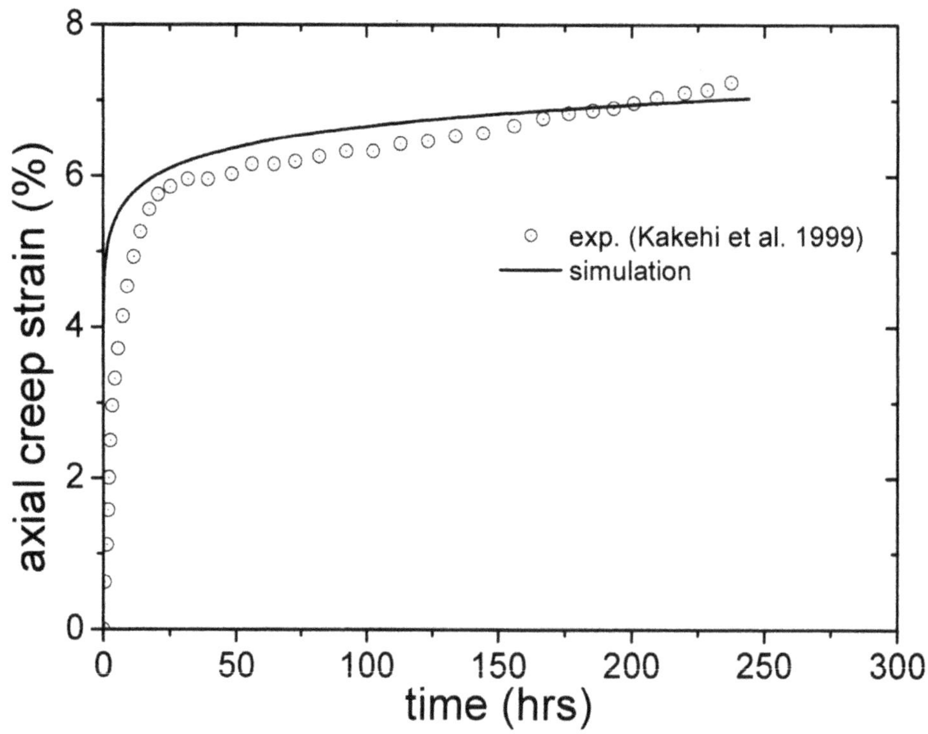

Figure 12. Variation of micro-structural parameter 'n' with shapes of precipitates

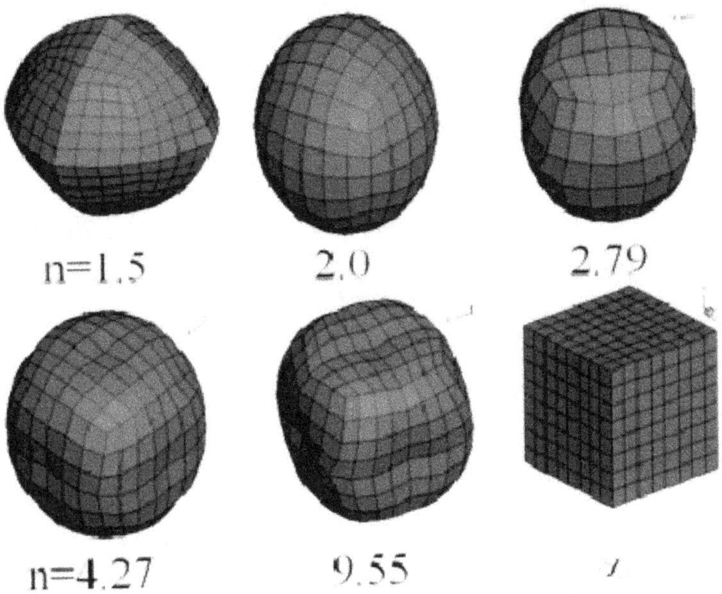

Figure 13. Variation of maximum homogenized slip rate with plastic work density as a function of volume fraction and shapes of precipitates

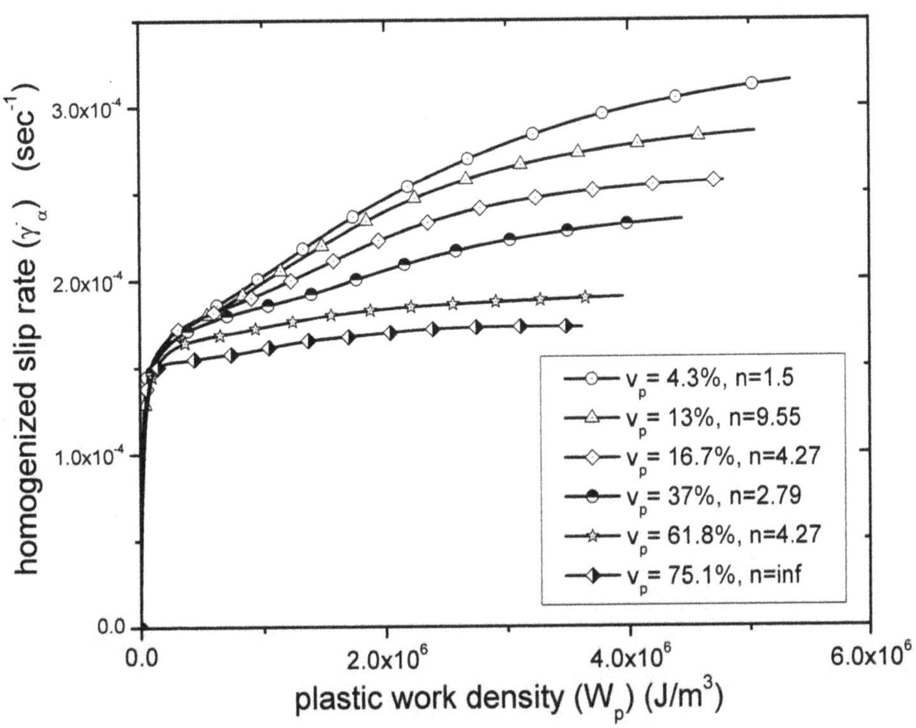

Figure 14. Variation of homogenized activation energy as a function of volume fraction and shape parameters of the precipitates

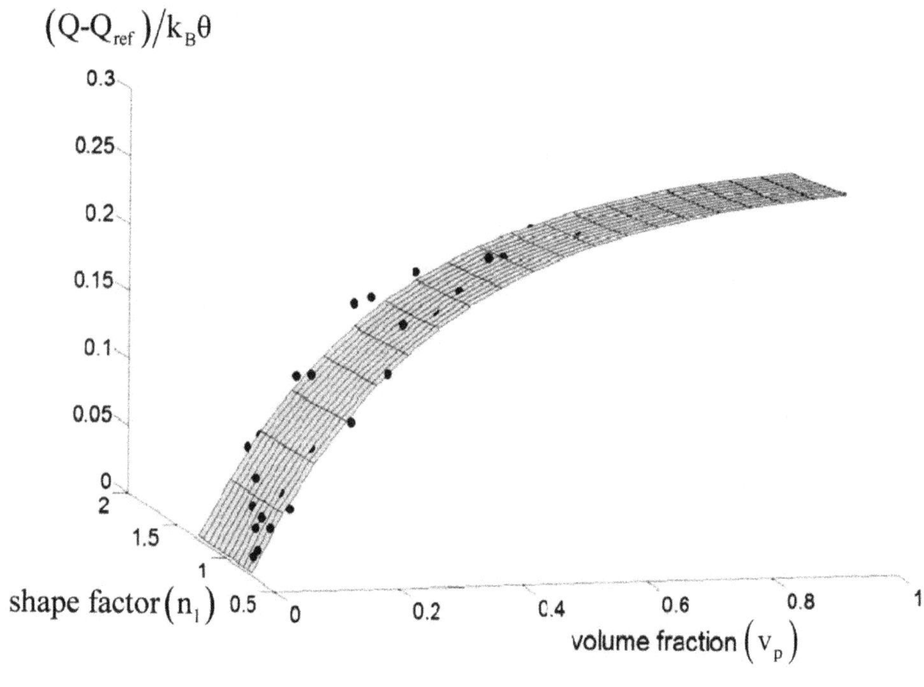

jected view in Figure 15. From this fitted surface, the parameters for the homogenized activation energy are derived. The homogenized activation energy can be expressed as

$$\frac{Q_{\text{hom}}}{k_B\theta} = \frac{Q_{\text{ref}}}{k_B\theta} - \ln\left(\frac{\dot{\gamma}_{\text{hom}}}{\dot{\gamma}_{\text{ref}}}\right) = \frac{Q_{\text{ref}}}{k_B\theta} + f\left(\mathrm{v_p},\mathrm{n_1}\right) \tag{43}$$

where the function $f\left(\mathrm{v_p},\mathrm{n_1}\right)$ is expressed as

$$f\left(\mathrm{v_p},\mathrm{n_1}\right) = \left[a_0 - a_1\exp(-a_2\mathrm{v_p})\right]\left(b_0 + b_1\mathrm{n_1} + b_2\mathrm{n_1^2}\right) \tag{44}$$

Similar exercise is carried out to evaluate the homogenized passing stress and the activation volume. The homogenized passing stress is expressed in terms of the function $\mathrm{f_a}\left(\mathrm{v_p},\mathrm{n_1}\right)$ as

$$\frac{\tau^{\alpha}_{\mathrm{p_{hom}}}}{\tau^{\alpha}_{\mathrm{p_{ref}}}} = 1 + \mathrm{f_a}\left(\mathrm{v_p},\mathrm{n_1}\right)\frac{\left(V^{\text{act}}_{\text{ref}}/k_B\theta\right)}{\tau^{\alpha}_{\mathrm{p_{ref}}}} \tag{45}$$

The homogenized activation volume is expressed in terms of the function $\mathrm{f_p}\left(\mathrm{v_p},\mathrm{n_1},\mathrm{W_p}\right)$ as

Figure 15. Projection of fitted functional form of homogenized activation energy as a function of volume fraction of precipitates

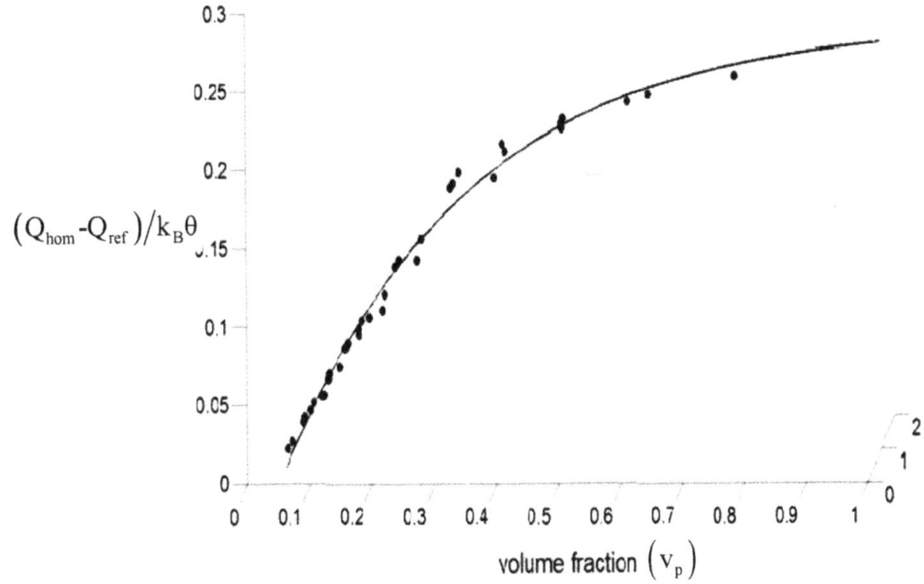

$$\frac{V_{ref}^{act}}{V_{hom}^{act}} = 1 + \left(\frac{V_{ref}^{act}}{k_B\theta \left(\tau_{ref}^{\alpha} - \tau_{p_{ref}}^{\alpha} \right)} \right) f_p \left(v_p, n_1, W_p \right) \tag{46}$$

where the volume averaged plastic work density can evaluated from the FE results as

$$W_p = \frac{\int_V \int_t \tau^{\alpha} \dot{\gamma}^{\alpha} dt dV}{\int_V dV} \tag{47}$$

It may be noted that the homogenized activation energy affects the onset of yielding of the homogenized medium. The passing stress affects the onset of yielding as well as subsequent plastic flow. However, it is not a function of the plastic work density W_p. However, the activation volume affects the plastic strain hardening and is a function of volume fraction, shape factor of precipitates as well as the plastic work density W_p. The variation of the functions $f_a(v_p, n_1)$ and $f_p(v_p, n_1, W_p)$ are plotted in Figs. 16 and 17 respectively. These have been evaluated from explicit simulation of different micro-structures and are expressed as functional forms of micro-structural parameters such as v_p and n_1 and plastic work density W_p. These homogenized parameters can be used in the next level of simulations where we don't need to consider the explicit FE discretization of the $\gamma - \gamma'$ phases.

Figure 16. Variation of functional form of homogenized passing stress as a function of volume fraction and shape factor of precipitates

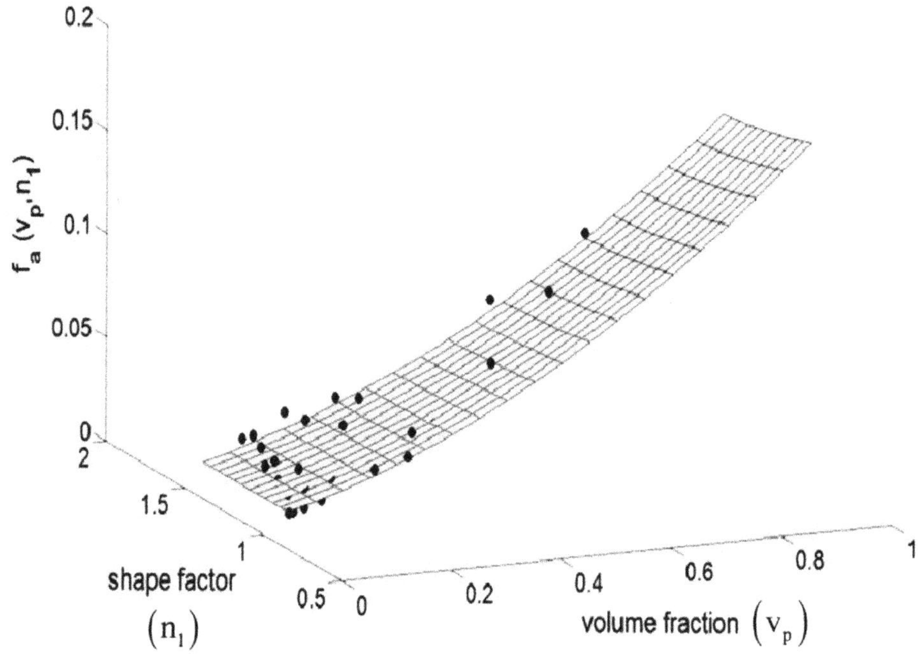

Figure 17. Variation of functional form of homogenized activation volume as a function of volume fraction and shape factor of precipitates

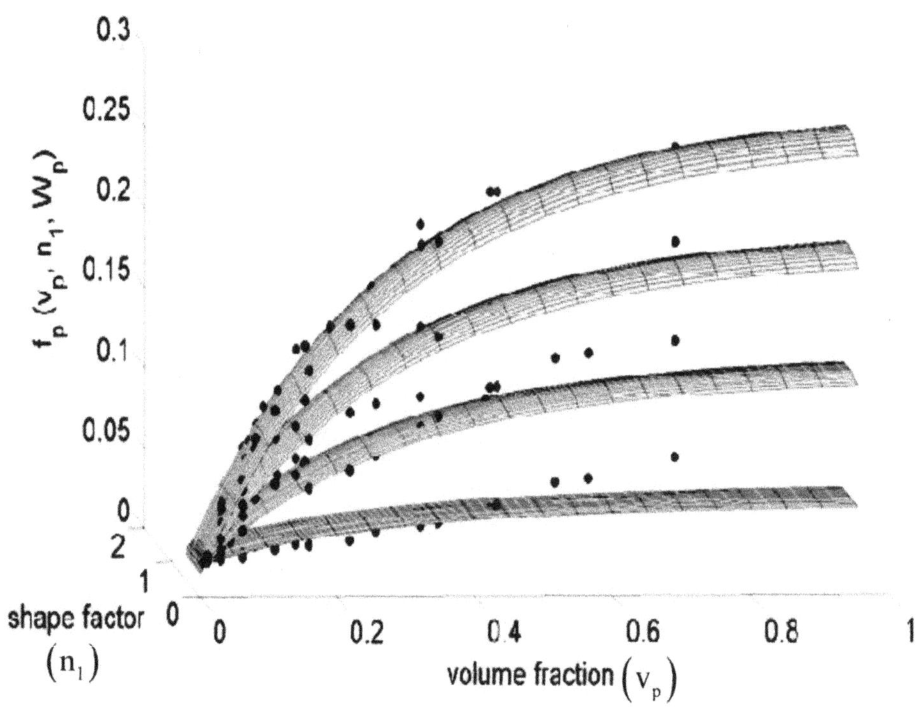

The homogenized dislocation density based crystal plasticity model has been used (along with the micro-twin nucleation and propagation model as discussed in Sections 3 and 4) to simulate the creep response of single crystal nickel based super-alloys with stress applied along different orientations of the crystal. The nickel based alloy is CMSX-4 and it contains 57.8% volume fraction of precipitates. The experimental data has been taken from Kakehi (1999). Significant tension-compression asymmetry is observed as can be seen from the experimental data presented in Figure 18 when the single crystal specimen is loaded along [001] as well as [011] direction. For loading in [001] tension, micro-twin formation is not observed as confirmed from transmission electron microscopy (TEM) observation in Kakehi (1999). However, when loaded in [001] compression, significant micro-twin development is observed in the alloy and the deformation is dominant by the strain accumulation due to micro-twin.

The experimental conditions have been simulated in the model and the predictions of the model are compared with those of experiment in Figure 18. It can be observed that the tension-compression asymmetry has been successfully modeled with the new formulation and the creep response with time as predicted by the model is very close to those of experiment both in [001] tension and [001] compression mode. Similar analysis has been carried out to simulate the single crystal experiment when loaded in [011] direction. Both tensile and compressive creep tests have been carried out. It can be observed the trend is just opposite to what is observed when the crystal is loaded in [001] direction. The presence of micro-twin in the deformed micro-structure has been observed through TEM observation for the [011] tension loading, whereas there was no micro-twin induced deformation for the [011] compression loading.

Figure 18. Prediction of tension-compression asymmetry in the creep response of the single crystal nickel based super-alloy CMSX-4due to loading in [001] direction in tension and compression.

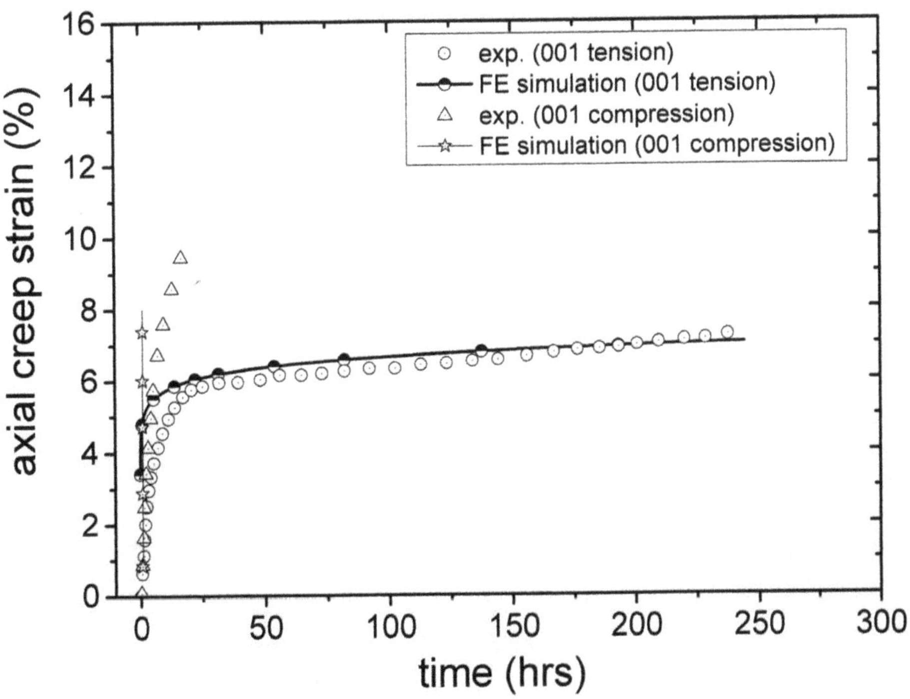

Because of presence of micro-twin, the creep strain accumulates faster for the [011] tension loading compared to [011] compression loading. The results of simulation have been compared with experimental data and are presented in Figure 19. It can be observed that the new model developed in this work has been very successful in satisfactorily predicting the single crystal experimental data. The micro-twin induced tension-compression asymmetry has also been satisfactorily predicted by the new model.

CONCLUSION

The aim of the project has been to develop a multi-scale approach for constitutive modeling of nickel based super-alloys which find applications in manufacturing of turbine blade and discs of aircraft engines. The γ' phase appears in both secondary and tertiary forms. For the numerical simulation, detailed information regarding the distribution of size, shape, spacing and volume fraction of these γ' phases of the material is required. These can be obtained from the analysis of the microstructure. In this work, a new model for APB shearing of the γ' precipitates has been developed. The dislocations density based constitutive relations for the γ phase, criteria for APB shearing of the γ' phase and flow rule for subsequent plastic deformation in the γ' phase (after APB shearing) are the essence of the new model. In order to avoid explicit discretization of all the phases in the $\gamma - \gamma'$ microstructure, a multi-scale model has been developed in this work. In the lower scale model, the explicit micro-structure has been modeled

Figure 19. Prediction of tension-compression asymmetry in the creep response of the single crystal nickel based super-alloy CMSX-4due to loading in [011] direction in tension and compression.

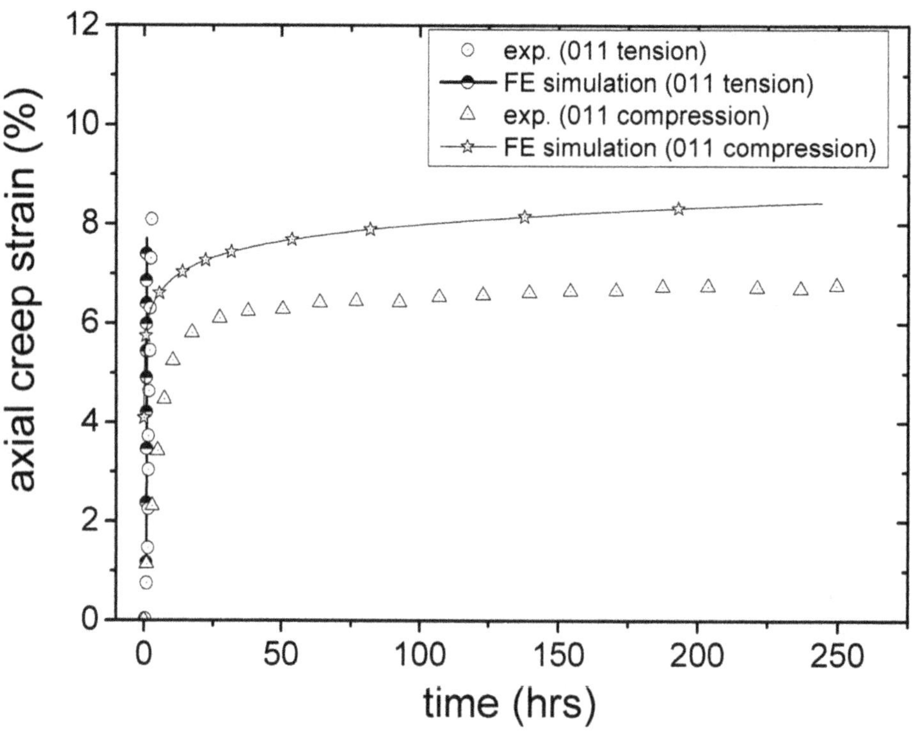

and the response has been homogenized so as to derive a homogenized slip based crystal plasticity model. Three parameters (viz. activation energy, passing stress and activation volume) has been expressed as functions of microstructural variables (i.e., v_p and n_1) and plastic work density W_p (representing hardening) so that the homogenized model can reproduce the response of the microstructure with different shapes and volume fractions of the γ' precipitates. The homogenized model as a function of micro-structural parameters have been used along with the model for micro-twin nucleation and propagation to simulate the observed tension-compression symmetry behavior in the creep responses of single crystal specimens. The results of the new model have been compared with those of experiment and it was observed that the simulated results compare well with those of experiment. The model also predicts tension-compression asymmetry (due to presence of micro-twin) very satisfactorily.

ACKNOWLEDGMENT

This work has been carried out under the XII-plan project of the Department of Atomic Energy, Government of India with Sub-project No. XII-N-R&D-45.08: Primary components integrity. The authors acknowledge the kind support of Mr. V. Bhasin, Associate Director, Nuclear Fuels Group, BARC for this work.

REFERENCES

Ashby, M. F. (1970). The deformation of plastically non-homogeneous materials. *Philosophical Magazine*, *21*(170), 399–424. doi:10.1080/14786437008238426

Babu, S. S., Miller, M. K., Vitek, J. M., & David, S. A. (2001). Characterization of microstructure evolution in a nickel base super-alloy during continuous cooling conditions. *Acta Materialia*, *49*(20), 4149–4160. doi:10.1016/S1359-6454(01)00314-7

Brown, S. G. R., Evans, R. W., & Wilshire, B. (1986). Creep strain and creep life prediction for the cast nickel-based super-alloy IN-100. *Materials Science and Engineering*, *84*, 147–156. doi:10.1016/0025-5416(86)90232-6

Chatterjee, D., Hazari, N., Das, N., & Mitra, R. (2010). Microstructure and creep behavior of DMS4-type nickel based super-alloy single crystals with orientations near <001> and <011>. *Materials Science and Engineering A*, *528*(2), 604–613. doi:10.1016/j.msea.2010.09.083

Coakley, J., Dye, D., & Basoalto, H. (2011). Creep and creep modelling of a multimodal nickel-base super-alloy. *Acta Materialia*, *59*(3), 854–863. doi:10.1016/j.actamat.2010.08.035

Cormier, J., Milhet, X., & Mendez, J. (2007). Non-isothermal creep at very high temperature of the nickel-based single crystal super-alloy MC2. *Acta Materialia*, *55*(18), 6250–6259. doi:10.1016/j.actamat.2007.07.048

Deka, D., Joseph, D., Ghosh, S., & Mills, M. J. (2006). Crystal plasticity modeling of deformation and creep in polycrystalline Ti-6242. *Metallurgical and Materials Transactions. A, Physical Metallurgy and Materials Science*, *37*(5), 1371–1388. doi:10.1007/s11661-006-0082-2

Epishin, A., Link, T., Brückner, U., & Portella, P. D. (2001). Kinetics of the topological inversion of the γ/γ-microstructure during creep of a nickel-based super-alloy. *Acta Materialia*, *49*(19), 4017–4023. doi:10.1016/S1359-6454(01)00290-7

Fleury, G., Schubert, F., & Nickel, H. (1996). Modelling of the thermo-mechanical behaviour of the single crystal super-alloy CMSX-4. *Computational Materials Science*, *7*(1-2), 187–193. doi:10.1016/S0927-0256(96)00079-1

Furrer, D., & Fecht, H. (1999). Ni-based super-alloys for turbine discs. *JOM Journal of the Minerals. Metals and Materials Society*, *51*(1), 14–17. doi:10.1007/s11837-999-0005-y

Ghosh, S., Bhandari, Y., & Groeber, M. (2008). CAD based reconstruction of three dimensional polycrystalline microstructures from FIB generated serial sections. *Journal of Computer Aided Design*, *40*(3), 293–310. doi:10.1016/j.cad.2007.11.003

Goff, S. D., Couturier, R., Guétaz, L., & Burlet, H. (2004). Effect of the microstructure on the creep behavior of PM Udimet 720 super-alloy—experiments and modeling. *Materials Science and Engineering A*, *387-389*, 599–603. doi:10.1016/j.msea.2004.01.094

Groeber, M., Ghosh, S., Uchic, M. D., & Dimiduk, D. M. (2008). A framework for automated analysis and simulation of 3D polycrystalline microstructures. Part 2: Synthetic structure generation. *Acta Materialia*, *56*(6), 1274–1287. doi:10.1016/j.actamat.2007.11.040

Groeber, M., Ghosh, S., Uchic, M. D., & Dimiduk, D. M. (2008). A framework for automated analysis and simulation of 3D polycrystalline microstructures. Part 1: Statistical characterization. *Acta Materialia*, *56*(6), 1257–1273. doi:10.1016/j.actamat.2007.11.041

Hasija, V., Ghosh, S., Mills, M. J., & Joseph, D. (2003). Modeling deformation and creep in Ti-6Al alloys with experimental validation. *Acta Materialia*, *51*, 4533–4545. doi:10.1016/S1359-6454(03)00289-1

Hong, H. U., Kim, I. S., Choi, B. G., Kim, M. Y., & Jo, C. Y. (2009). The effect of grain boundary serration on creep resistance in a wrought nickel-based super-alloy. *Materials Science and Engineering A*, *517*(1-2), 125–131. doi:10.1016/j.msea.2009.03.071

Hopgood, A. A., & Martin, J. W. (1986). The creep behaviour of a Nickel-based single-crystal superalloy. *Materials Science and Engineering*, *82*, 27–36. doi:10.1016/0025-5416(86)90092-3

Ignat, M., Buffiere, J.-Y., & Chaix, J. M. (1993). Microstructures induced by a stress gradient in a nickel-based super-alloy. *Acta Metallurgica et Materialia*, *41*(3), 855–862. doi:10.1016/0956-7151(93)90019-O

Kakehi, K. (1999). Tension/ Compression asymmetry in creep behavior of a Ni-based super-alloy. *Scripta Materialia*, *41*(5), 461–465. doi:10.1016/S1359-6462(99)00191-8

Kim, T.-W., Kang, D.-H., Yeom, J.-T., & Park, N.-K. (2007). Continuum damage mechanics-based creep–fatigue-interacted life prediction of nickel-based super-alloy at high temperature. *Scripta Materialia*, *57*(12), 1149–1152. doi:10.1016/j.scriptamat.2007.08.014

Knobloch, C., Saß, V., Siebörger, D., & Glatzel, U. (1997). Anisotropic creep behavior of a nickel-based super-alloy compared with single phase nickel solid solution and γ phase single crystals. *Materials Science and Engineering A, 234-236*, 237–241. doi:10.1016/S0921-5093(97)00142-1

Koul, A. K., & Castillo, R. (1991). A critical assessment of the θ projection concept for creep life prediction of nickel-based super-alloy components. *Materials Science and Engineering A, 138*(2), 213–219. doi:10.1016/0921-5093(91)90690-O

Kovarik, L., Unocic, R. R., Li, J., Sarosi, P., Shen, C., Wang, Y., & Mills, M. J. (2009). Microtwinning and other shearing mechanisms at intermediate temperatures in Ni-based super-alloys. *Progress in Materials Science, 54*(6), 839–873. doi:10.1016/j.pmatsci.2009.03.010

Lu, Y. L., Chen, L. J., Wang, G. Y., Benson, M. L., Liaw, P. L., Thompson, S. A., & Klarstrom, D. L. et al. (2005). Hold time effects on low cycle fatigue behavior of HAYNES 230 super-alloy at high temperatures. *Materials Science and Engineering A, 409*(1-2), 282–291. doi:10.1016/j.msea.2005.05.120

Ma, A., Dye, D., & Reed, R. C. (2008). A model for the creep deformation behaviour of single-crystal super-alloy CMSX-4. *Acta Materialia, 56*(8), 1657–1670. doi:10.1016/j.actamat.2007.11.031

Ma, A., & Roters, F. (2004). A constitutive model for fcc single crystals based on dislocation densities and its application to uniaxial compression of aluminium single crystals. *Acta Materialia, 52*(12), 3603–3612. doi:10.1016/j.actamat.2004.04.012

Ma, A., Roters, F., & Raabe, D. (2006a). A dislocation density based constitutive model for crystal plasticity FEM including geometrically necessary dislocations. *Acta Materialia, 54*(8), 2169–2179. doi:10.1016/j.actamat.2006.01.005

Ma, A., Roters, F., & Raabe, D. (2006b). On the consideration of interactions between dislocations and grain boundaries in crystal plasticity finite element modeling-theory, experiments and simulations. *Acta Materialia, 54*(8), 2181–2194. doi:10.1016/j.actamat.2006.01.004

Manchiraju, S., Asai, M., & Ghosh, S. (2007). A dual-time-scale finite element model for simulating cyclic deformation of polycrystalline alloys. *Journal of Strain Analysis in Engineering and Design, 42*(4), 183–200. doi:10.1243/03093247JSA233

Manonukul, A., Dunne, F. P. E., & Knowles, D. (2002). Physically-based model for creep in nickel-base super-alloy C263 both above and below the gamma solvus. *Acta Materialia, 50*(11), 2917–2931. doi:10.1016/S1359-6454(02)00119-2

Milhet, X., Cormier, J., & Organista, A. (2010). On the role of the internal stress during non-isothermal creep life of a first generation nickel based single crystal super-alloy. *Materials Science and Engineering A, 527*(9), 2280–2288. doi:10.1016/j.msea.2009.11.067

Piard, A., Gamby, D., Carbou, C., & Mendez, J. (2004). A numerical simulation of creep-fatigue crack growth in nickel-base super-alloys. *Engineering Fracture Mechanics, 71*(16-17), 2299–2317. doi:10.1016/j.engfracmech.2004.02.002

Prasad, S. C., Rajagopal, K. R., & Rao, I. J. (2006). A continuum model for the anisotropic creep of single crystal nickel-based super-alloys. *Acta Materialia, 54*(6), 1487–1500. doi:10.1016/j.actamat.2005.11.016

Samal, M. K. (2015). (in press). Simulation of material stress-strain curve and creep deformation response of nickel based superalloys using crystal plasticity based finite element models. *Transactions of the Indian Institute of Metals*. doi:10.1007/s12666-015-0591-z

Sarosi, P. M., Srinivasan, R., Eggeler, G. F., Nathal, M. V., & Mills, M. J. (2007). Observations of *a*<010> dislocations during the high-temperature creep of Ni-based super-alloy single crystals deformed along the [001] orientation. *Acta Materialia*, *55*(7), 2509–2518. doi:10.1016/j.actamat.2006.11.045

Soula, A., Renollet, Y., Boivin, D., Pouchou, J.-L., Locq, D., Caron, P., & Bréchet, Y. (2009). Analysis of high-temperature creep deformation in a polycrystalline nickel-base super-alloy. *Materials Science and Engineering A*, *510-511*, 301–306. doi:10.1016/j.msea.2008.04.122

Sugui, T., Jun, X., Xiaoming, Z., Benjiang, Q., Jianwei, L., Lili, Y., & Wuxiang, W. (2011). Microstructure and creep behavior of FGH95 nickel-base super-alloy. *Materials Science and Engineering A*, *528*(4-5), 2076–2084. doi:10.1016/j.msea.2010.11.038

Torster, F., Baumeister, G., Albrecht, J., Lütjering, G., Helm, D., & Daeubler, M. A. (1997). Influence of grain size and heat treatment on the microstructure and mechanical properties of the nickel-base super-alloy U 720 LI. *Materials Science and Engineering A*, *234-236*, 189–192. doi:10.1016/S0921-5093(97)00161-5

Venkataramani, G., Deka, D., & Ghosh, S. (2006). Crystal plasticity based FE model for understanding microstructural effects on creep and dwell fatigue in Ti-6242. ASME. *Journal of Engineering Materials and Technology*, *128*(3), 356–365. doi:10.1115/1.2204942

Venkataramani, G., Ghosh, S., & Mills, M. J. (2007). A size dependent crystal plasticity finite element model for creep and load-shedding. *Acta Materialia*, *55*(11), 3971–3986. doi:10.1016/j.actamat.2007.03.017

Venkataramani, G., Kirane, K., & Ghosh, S. (2008). Microstructural parameters affecting creep induced load shedding in Ti-6242 by a size dependent crystal plasticity FE model. *International Journal of Plasticity*, *24*(3), 428–454. doi:10.1016/j.ijplas.2007.05.001

Viswanathan, G. B., Sarosi, P. M., Whitis, D. H., & Mills, M. J. (2005). Deformation mechanisms at intermediate creep temperatures in the Ni-base super-alloy Rene 88 DT. *Materials Science and Engineering A*, *400*, 489–495. doi:10.1016/j.msea.2005.02.068

Yamashita, M., & Kakehi, K. (2006). Tension/compression asymmetry in yield and creep strengths of Ni-based super-alloy with a high amount of tantalum. *Scripta Materialia*, *55*(2), 139–142. doi:10.1016/j.scriptamat.2006.03.048

Zhan, Z., Fernando, U. S., & Tong, J. (2008). Constitutive modelling of viscoplasticity in a nickel-based super-alloy at high temperature. *International Journal of Fatigue*, *30*(7), 1314–1323. doi:10.1016/j.ijfatigue.2007.06.010

Compilation of References

Abdolshah, S., Abdolshah, M., & Tang, S. H. (2014). Trajectory Planning and Walking Pattern Generation of Humanoid Robot Motion. *IAES International Journal of Robotics and Automation, 4*(2).

Abu Farsakh, M. Y., & Titi, H. H. (2004). Assessment of direct cone penetration test methods for predicting the ultimate capacity of friction driven piles. *Journal of Geotechnical and Geoenvironmental Engineering, 130*(9), 935–944. doi:10.1061/(ASCE)1090-0241(2004)130:9(935)

Abu-Srhan, A., & Daoud, E. A. (2013). A Hybrid Algorithm Using a Genetic Algorithm and Cuckoo Search Algorithm to Solve the Traveling Salesman Problem and its Application to Multiple Sequence Alignment. *International Journal of Advanced Science and Technology, 61*, 29–38. doi:10.14257/ijast.2013.61.04

Adil, B. L., Gullu, H., Canakc, H., & Ozbakır, L. (2008). Prediction of compressive and tensile strength of limestone via genetic programming. *Expert Systems with Applications, 35*(1-2), 111–123. doi:10.1016/j.eswa.2007.06.006

Adil, B., Güllüb, H., Çanakçıb, H., & Özbakır, L. (2008). Prediction of compressive and tensile strength of limestone via genetic programming. *Expert Systems with Applications, 35*(1–2), 111–123.

Adjallah, K. H., Crespo Marquez, A., & Iung, B. (2007). A structured approach for the assessment of system availability and reliability using Monte Carlo simulation. *Journal of Quality in Maintenance Engineering, 13*(2), 125–136. doi:10.1108/13552510710753032

Agarwal, S. C., Sahani, M., & Bansal, S. (2010). Reliability Characteristic of cold-standby redundant system. *International Journal of Research and Reviews in Applied Sciences, 3*(2), 193–199.

Aggarwal, A., Singh, H., Kumar, P., & Singh, M. (2008). Optimizing power consumption for CNC turned parts using response surface methodology, Taguchis technique- a comparative analysis. *Journal of Materials Processing Technology, 200*(1-3), 373–384. doi:10.1016/j.jmatprotec.2007.09.041

Ahonen, H., de Souza Júnior, P. A., & Garg, V. K. (1997). A genetic algorithm for fitting Lorentzian line shapes in Mössbauer spectra. *Nuclear Instruments & Methods in Physics Research. Section B, Beam Interactions with Materials and Atoms, 124*(4), 633–638. doi:10.1016/S0168-583X(97)00107-9

Ajovalasit, M., & Giacomin, J. (2003). Analysis of variations in diesel engine idle vibration. *Proceedings of the Institution of Mechanical Engineers. Part D, Journal of Automobile Engineering, 217*(10), 921–933. doi:10.1243/095440703769683324

Ali, O. M., Abdullah, N. R., Mamat, R., & Abdullah, A. A. (2015). Comparison of the Effect of Different Alcohol Additives with Blended Fuel on Cyclic Variation in Diesel Engine. *Energy Procedia, 75*, 2357–2362. doi:10.1016/j.egypro.2015.07.490

Aljanaideh, O., Habineza, D., Rakotondrabe, M., & Janaideh, M. A. (2016). Experimental comparison of rate-dependent hysteresis models in characterizing and compensating hysteresis of piezoelectric tube actuators. *Physica B, Condensed Matter*, *486*, 64–68. doi:10.1016/j.physb.2015.10.021

Al-Kheer, A. A., El-Hami, A., Kharmanda, M. G., & Mouazen, A. M. (2011). Reliability-based design for soil tillage machines. *Journal of Terramechanics*, *48*(1), 57–64. doi:10.1016/j.jterra.2010.06.001

Allella, F., Chiodo, E., & Lauria, D. (2005). Optimal reliability allocation under uncertain conditions, with application to hybrid electric vehicle design. *International Journal of Quality & Reliability Management*, *22*(6), 626–641. doi:10.1108/02656710510604926

Allen, J. C., Schaffer, W. M., & Rosko, D. (1993). Chaos reduces species extinction by amplifying local population noise. *Nature*, *364*(6434), 229–232. doi:10.1038/364229a0 PMID:8321317

Alligood, K. T., Sauer, T. D., & Yorke, J. A. (1997). *Chaos: An Introduction to Dynamical Systems. Newyork.* Springer. doi:10.1007/978-3-642-59281-2

Aloufi, M., and Kazmierski, T. J. (2011). A response surface modelling approach to performance optimisation of kinetic energy harvesters IJRRCS Simulation. *Benchmarking and Modeling of Systems and Communication Networks,* 1-8.

Altinel, I. K. (1992). The design of optimum component test plans in the demonstration of a series system reliability. *Computational Statistics & Data Analysis*, *14*(3), 281–292. doi:10.1016/0167-9473(92)90039-I

Altinel, I. K. (1994). The design of optimum component test plans in the demonstration of system reliability. *European Journal of Operational Research*, *78*(3), 318–333. doi:10.1016/0377-2217(94)90043-4

Altinel, İ. K., & Özekici, S. (1998). Optimum component test plans for systems with dependent components. *European Journal of Operational Research*, *111*(1), 175–186. doi:10.1016/S0377-2217(97)00355-X

Altinel, İ. K., Özekici, S., & Feyzioğlu, O. (2001). Dynamic component testing of a series system with redundant subsystems. *IIE Transactions*, *33*(12), 1093–1108. doi:10.1080/07408170108936899

Altinel, I. K., Özekici, S., & Feyzioğlu, O. (2002). Component testing of a series system in a random mission. *Reliability Engineering & System Safety*, *78*(1), 33–43. doi:10.1016/S0951-8320(02)00047-9

Ameri, T. P. M., & Yousefi, M. Y. (2000). *Analysis of Affective Factors on Crisis Situation in Boilers of Steam Plants.* 15th International Conference Power Engineering Conference.

Andrews, J. D., & Moss, T. R. (2002). *Reliability and risk assessment.* Wiley-Blackwell.

Aoki, M. (1967). *Optimization of stochastic systems: topics in discrete-time systems*. New York, London: Academic Press.

Apeland, S., & Scarf, P. A. (2003). A fully subjective approach to capital equipment replacement. *The Journal of the Operational Research Society*, *54*(4), 371–378. doi:10.1057/palgrave.jors.2601505

Arnold, T. F. (1973). The concept of coverage and its effect on the reliability model of a repairable system. *IEEE Transactions on Computers*, *100*(3), 251–254. doi:10.1109/T-C.1973.223703

Ascher, H. E. (1968). Evaluation of repairable system reliability using the bad-as-old concept. *IEEE Transactions on Reliability*, *17*(2), 103–110. doi:10.1109/TR.1968.5217523

Ascher, H., & Feingold, H. (1984). *Repairable Systems Reliability: Modeling, Inference, Misconceptions and their Causes*. New York: Marcel Dekker, Inc.

Ashby, M. F. (1970). The deformation of plastically non-homogeneous materials. *Philosophical Magazine*, *21*(170), 399–424. doi:10.1080/14786437008238426

Asilturk, I., Neseli, S., & Ince, M. A. (2016). Optimisation of parameters affecting surface roughness of Co28Cr6Mo medical material during CNC lathe machining by using the Taguchi and RSM methods. *Measurement*, *78*, 120–128. doi:10.1016/j.measurement.2015.09.052

Atashgah, A. B., & Seifi, A. (2007). Application of semi-definite programming to the design of multi-response experiments. *IIE Transactions*, *39*(7), 63–769. doi:10.1080/07408170701245353

Avižienis, A., Laprie, J. C., Randell, B., & Landwehr, C. (2004). Basic concepts and taxonomy of dependable and secure computing. *IEEE Transactions on Dependable and Secure Computing*, *1*(1), 11–33. doi:10.1109/TDSC.2004.2

Avontuur, G. C. (2000). *Reliability analysis in mechanical engineering design: TU Delft*. Delft University of Technology.

Avontuur, G. C., & van der Werff, K. (2001). An implementation of reliability analysis in the conceptual design phase of drive trains. *Reliability Engineering & System Safety*, *73*(2), 155–165. doi:10.1016/S0951-8320(01)00041-2

Azaron, A., Furmans, K., & Modarres, M. (2010). Nurse Scheduling by Fuzzy Goal Programming. In D. Jones, M. Tamiz, & J. Ries (Eds.), *New Developments in Multiple Objective and Goal Programming* (pp. 154–163). Heidelberg, Germany: Springer.

Aziz-Alaoui, M. A. (2002). Study of a Leslie-Gower-type tritrophic population model. *Chaos, Solitons, and Fractals*, *14*(8), 1275–1293. doi:10.1016/S0960-0779(02)00079-6

Babu, S. S., Miller, M. K., Vitek, J. M., & David, S. A. (2001). Characterization of microstructure evolution in a nickel base super-alloy during continuous cooling conditions. *Acta Materialia*, *49*(20), 4149–4160. doi:10.1016/S1359-6454(01)00314-7

Bai, D. S., & Chun, Y. R. (1991). Optimum Simple Step-stress Accelerated Life Tests with Competing Causes of Failure. *IEEE Transactions on Reliability*, *40*(5), 622–627. doi:10.1109/24.106787

Bai, D. S., & Kim, M. S. (1993). Optimum simple step-stress accelerated life tests for Weibull distribution & Type-I censoring. *Naval Res. Logist.*, *40*(2), 193–210. doi:10.1002/1520-6750(199303)40:2<193::AID-NAV3220400205>3.0.CO;2-J

Bajeel, P. N., & Kumar, M. (2015). A component reliability test plan for a parallel system with failure rate as the exponential function of covariates. *Journal MESA, 6*(2), 177-190.

Balakrishnan, K. (1996). *Exponential distribution: theory, methods and applications*. CRC Press.

Ball, J. K., Raine, R. R., & Stone, C. R. (1998). Combustion analysis and cycle-by-cycle variations in spark ignition engine combustion Part 1: An evaluation of combustion analysis routines by reference to model data. *Proceedings of the Institution of Mechanical Engineers. Part D, Journal of Automobile Engineering*, *212*(5), 381–399. doi:10.1243/0954407981526046

Bansal, A. K. (2007). *Bayesian parametric inference*. Narosa Publishing House.

Bao, Y., Wang, H., & Wang, B. (2014). Short-term wind power prediction using differential EMD and relevance vector machine. *Neural Computing & Applications*, *25*(2), 283–289. doi:10.1007/s00521-013-1482-z

Barbosa, S., Silva, M. E., & Fernandes, M. J. (2006). Wavelet analysis of the Lisbon and Gibraltar North Atlantic Oscillation winter indices. *International Journal of Climatology*, *26*(5), 581–594. doi:10.1002/joc.1263

Barford, L. A., Fazzio, R. S., & Smith, D. R. (1992). *An introduction to wavelets. Hewlett-Packard Laboratories*, Technical Publications Department.

Barlow, R. E., & Hunter, L. C. (1960). Optimal preventive maintenance policies. *Operations Research*, *8*(1), 90–100. doi:10.1287/opre.8.1.90

Barlow, R. E., Hunter, L. C., & Proschan, F. (1963). Optimum checking procedures. *Journal of the Society for Industrial and Applied Mathematics*, *11*(4), 1078–1095. doi:10.1137/0111080

Barlow, R. E., & Proschan, F. (1965). *Mathematical Theory of Reliability*. New York: Wiley.

Barlow, R. E., & Proschan, F. (1975). *Statistical Theory of Reliability and Life Testing: Probability Models*. New York: Holt, Rinehart, and Winston.

Barlow, R. E., & Proschan, F. (1996). *Mathematical theory of reliability* (Vol. 17). Siam. doi:10.1137/1.9781611971194

Barriga, G. D. C., Lee Ho, L., & Cancho, V. G. (2008). Planning accelerated life tests under Exponentiated-Weibull-Arrhenius model. *International Journal of Quality & Reliability Management*, *25*(6), 636–653. doi:10.1108/02656710810881926

Barton, R. K., Kenemuth, D. K., Lestz, S. S., & Meyer, W. E. (1970). *Cycle-by-Cycle Variations of a Spark Ignition Engine-A Statistical Analysis* (No. 700488). SAE Technical Paper.

Barton, R. R., & Meckesheimer, M. (2006). Metamodel- Based Simulation Optimization. In S.G. Henderson & B.L. Nelson (Eds.), Handbooks in Operations Research and Management Science (pp. 535– 574). Amsterdam: Elsevier.

Basak, I., & Ghosh, A. (1992). Mechanism of material removal in electrochemical discharge machining: A theoretic model and experimental verification. *Journal of Materials Processing Technology*, *71*(3), 350–359. doi:10.1016/S0924-0136(97)00097-6

Basak, I., & Ghosh, A. (1996). Mechanism of spark generation during electrochemical discharge machining: A theoretical model and experimental investigation. *Journal of Materials Processing Technology*, *62*(1-3), 46–53. doi:10.1016/0924-0136(95)02202-3

Bashiri, M., & Hejazi, T. (2009). An extension of multi-response optimization in MADM view. *Journal of Applied Sciences*, *9*(9), 1695–1702. doi:10.3923/jas.2009.1695.1702

Bashiri, M., & Hejazi, T. H. (2012). A mathematical model based on principal component analysis for optimization of correlated multiresponse surfaces. *Journal of Quality*, *19*(3), 223–239.

Basset, K., Rupp, C., & Ting, D. (2010). Vibration analysis of 2.3 MW wind turbine operation using the discrete wavelet transform. *Wind Engineering*, *34*(4), 375–388. doi:10.1260/0309-524X.34.4.375

Beige, H., & Schmidt, G. (1982). Electromechanical Resonances for investigating linear and nonlinear properties of Dielectrics. *Ferroelectrics*, *41*(1), 39–49. doi:10.1080/00150198208210608

Berg, M. B. (1980). A marginal cost analysis for preventive maintenance policies. *European Journal of Operational Research*, *4*(2), 136–142. doi:10.1016/0377-2217(80)90023-5

Berryman, A. A., & Millstein, J. A. (1989). Are ecological systems chaotic and if not, why not? *Trends in Ecology & Evolution*, *4*(1), 26–28. doi:10.1016/0169-5347(89)90014-1 PMID:21227309

Bertsche, B. (2008). *Reliability in automotive and mechanical engineering: determination of component and system reliability*. Springer Science & Business Media.

Bharathi, R. S., & Baskar, N. (2010). Optimization techniques for machining operations: A retrospective research based on various mathematical models. *International Journal of Advanced Manufacturing Technology*, *48*(9-12), 1075–1090. doi:10.1007/s00170-009-2351-x

Bhattacharya, D. K., & Begum, S. (1996). Bionomic equilibrium of two-species system I. *Mathematical Biosciences*, *135*(2), 111–127. doi:10.1016/0025-5564(95)00170-0 PMID:8768217

Bhote, K. R., & Bhote, A. K. (2004). *World class reliability: Using multiple environment overstress tests to make it happen*. AMACOM Div American Management Association.

Billinton, R., & Allan, R. N. (1983). *Reliability Evaluation of Engineering Systems*. New York: Springer. doi:10.1007/978-1-4615-7728-7

Bischoff, W. (1993). On D-optimal designs for linear models under correlated observations with an application to a linear model with multiple response. *Journal of Statistical Planning and Inference*, *37*(1), 69–80. doi:10.1016/0378-3758(93)90080-P

Blasius, B., Huppert, A., & Stone, L. (1999). Complex dynamics and phase synchronization in spatially extended ecological systems. *Nature*, *399*(6734), 354–359. doi:10.1038/20676 PMID:10360572

Blischke, W. R., & Murthy, D. P. (2011). *Reliability: modeling, prediction, and optimization* (Vol. 767). John Wiley & Sons.

Bogdanoff, J. L., & Kozin, F. (1985). *Probabilistic Models of Cumulative Damage*. New York: Wiley.

Boguś, P., & Merkisz, J. (2005). Misfire detection of locomotive diesel engine by non-linear analysis. *Mechanical Systems and Signal Processing*, *19*(4), 881–899. doi:10.1016/j.ymssp.2004.06.004

Booker, J., Parkinson, J., & Ross, T. J. (Eds.). (2002). *Combined fuzzy logic and probability applications*. Philadelphia: SIAM.

Bose, G. K., & Mahapatra, K. K. (2014). Parametric study of Die Sinking EDM process on AISI H13 tool steel using statistical techniques. *Journal of Advances in Production Engineering and Management*, *8*(4), 168–180. doi:10.14743/apem2014.4.185

Bose, G. K., & Mahapatra, K. K. (2015). Experimental analysis of Die Sinking EDM Process on AISI P20 tool steel using response surface methodology. *Manufacturing Engineering*, 163–188.

Bose, G. K., & Mahapatra, K. K. (2015). Multi-criteria decision making of machining parameters for Die Sinking EDM process. *International Journal of Industrial Engineering Computations*, *6*(2), 241–252. doi:10.5267/j.ijiec.2014.10.005

Boser, B. E., Isabelle, M. G., & Vapnik, V. M. (1992). A Training Algorithm for Optimal Margin Classifiers. In *Proceedings of the 5th Annual Workshop on Computational Learning Theory (COLT'92)* (pp. 144-152). Pittsburgh, PA: ACM Press. doi:10.1145/130385.130401

Bouricius, W. G., Carter, W. C., & Schneider, P. R. (1969, August). Reliability modeling techniques for self-repairing computer systems. In *Proceedings of the 1969 24th national conference* (pp. 295-309). ACM. doi:10.1145/800195.805940

Box, G. E. P., & Draper, N. R. (2007). Introduction to Response Surface Methodology. In S. Quigley & S. S. Filler (Eds.), *Response Surfaces, Mixtures and Ridge Analyses* (pp. 1–16). Wiley and Sons. doi:10.1002/9780470072769.ch1

Box, G. E. P., & Wilson, K. B. (1951). On the experimental attainment of optimum conditions. *Journal of the Royal Statistical Society. Series B. Methodological*, *13*(1), 1–38.

Box, M. J., & Draper, N. R. (1972). Estimation and design criteria for multiresponse non-linear models with non-homogeneous variance. *Applied Statistics*, *21*(1), 13–24. doi:10.2307/2346599

Brezocnik, M., Kovacic, M., & Ficko, M. (2004). Prediction of surface roughness with genetic programming. *Journal of Materials Processing Technology*, *157–158*, 28–36. doi:10.1016/j.jmatprotec.2004.09.004

Brown, R. P., Kockott, D., Trubiroha, P., Ketola, W., & Shorthouse, J. (1995). *A Review of Accelerated Durability Tests*. VAMAS Technical Report No 18.

Brown, C., Liebovitch, L. S., & Glendon, R. (2007). Levy flights in Dobe Ju/'hoansi foraging patterns. *Human Ecology*, *35*(1), 129–138. doi:10.1007/s10745-006-9083-4

Brown, S. G. R., Evans, R. W., & Wilshire, B. (1986). Creep strain and creep life prediction for the cast nickel-based super-alloy IN-100. *Materials Science and Engineering*, *84*, 147–156. doi:10.1016/0025-5416(86)90232-6

BSI Standards Publication. (2013). *Methods for product accelerated testing*. BS EN 62506.

Buchwitz, M., de Beek, R., Burrows, J. P., Bovensmann, H., Warneke, T., Notholt, J., & Schulz, A. et al. (2005). Atmospheric methane and carbon dioxide from SCIAMACHY satellite data: Initial comparison with chemistry and transport models. *Atmospheric Chemistry and Physics*, *5*(4), 941–962. doi:10.5194/acp-5-941-2005

Bun, R.A., Dachuk, V.S. (1997). Modeling of global and regional carbon cycles in biosphere. *Problems of Operating Research and Informatics*, *2*, 141-148.

Bush, T. S., Catlow, C. R. A., & Battle, P. D. (1995). Evolutionary programming techniques for predicting inorganic crystal structures. *Journal of Materials Chemistry*, *5*(8), 1269–1272. doi:10.1039/jm9950501269

Butler, G. J., Freedman, H. I., & Waltman, P. (1986). Uniformly persistent systems. *Proceedings of the American Mathematical Society*, *96*(3), 425–430. doi:10.1090/S0002-9939-1986-0822433-4

Cady, W. G. (1964). *Piezoelectricity*. Dover.

Calabria, R., Pulcini, G., & Rapone, M. (1995). Optimal reliability and maintainability allocation in manufacturing systems. *Quality and Reliability Engineering International*, *11*(3), 183–188. doi:10.1002/qre.4680110308

Caruso, H., & Dasgupta, H. (1998). A Fundamental Overview of Accelerated Testing Analytical Models. *Journal of the IEST*, *41*(1), 16–20. doi:10.17764/jiet.41.1.3626j43p305q1978

Castillo, E. (2009). A Bayesian reliability approach to multiple response optimization with seemingly unrelated regression models. *Quality Technology & Quantitative Management*, *6*(4), 353–369. doi:10.1080/16843703.2009.11673204

Cayda U. (2010). Machinability evaluation in hard turning of AISI 4340 steel with different cutting tools using statistical techniques. *Proc. Inst. Mech. Eng. B-J Eng. Ma., 224*(7), 1043-1055. DOI: 10.1243/09544054JEM1822

Cempel, C., Natke, H. G., & Yao, J. T. P. (2000). Symptom reliability and hazard for systems condition monitoring. *Mechanical Systems and Signal Processing*, *14*(3), 495–505. doi:10.1006/mssp.1999.1246

Ceryan, N., Okkan, U., Samui, P., & Ceryan, S. (2013). Modeling of tensile strength of rocks materials based on support vector machines approaches. *International Journal for Numerical and Analytical Methods in Geomechanics*, *37*(16), 2655–2670.

Chaloner, K., & Larntz, K. (1992). Bayesian design for accelerated life testing. *Journal of Statistical Planning and Inference*, *33*(2), 245–259. doi:10.1016/0378-3758(92)90071-Y

Chaloner, K., & Verdinelli, I. (1995). Bayesian experimental design: A review. *Statistical Science*, *10*(3), 273–304. doi:10.1214/ss/1177009939

Chanas, S., & Kuchta, D. (2002). Fuzzy goal programming- one notion, many meanings. *Control Cybern*, *4*(31), 871–890.

Chang, S. (1997). An algorithm to generate near D-optimal designs for multiple response surface models. *IIE Transactions*, *29*(12), 1073–1081. doi:10.1080/07408179708966426

Chang, S. I. (1994). Some properties of multiresponse D-optimal designs. *Journal of Mathematical Analysis and Applications*, *184*(2), 256–262. doi:10.1006/jmaa.1994.1198

Chatterjee, D., Hazari, N., Das, N., & Mitra, R. (2010). Microstructure and creep behavior of DMS4-type nickel based super-alloy single crystals with orientations near <001> and <011>. *Materials Science and Engineering A*, *528*(2), 604–613. doi:10.1016/j.msea.2010.09.083

Chen, G., & Dong, X. (1992). On feedback control of chaotic dynamic systems. *International Journal of Bifurcation and Chaos in Applied Sciences and Engineering*, *2*(02), 407–411. doi:10.1142/S0218127492000392

Cheng, C. T., Chen, H. C., Hu, Y. Y., & Wong, C. C. (2009). Fuzzy balancing control of a small-size humanoid robot based on accelerometer. *International Journal of Fuzzy Systems*, *11*(3), 146–153.

Chen, K. K., & Krieger, R. B. (1976). A statistical analysis of the influence of cyclic variation on the formation of nitric oxide in spark ignition engines. *Combustion Science and Technology*, *12*(4-6), 125–134. doi:10.1080/00102207608946712

Chen, M. C., & Su, C. T. (1998). Optimization of machining conditions for turning cylindrical stocks into continuous finished profiles. *International Journal of Production Research*, *36*(8), 2115–2130. doi:10.1080/002075498192805

Cherkassky, V., & Mulier, F. (1998). *Learning from data: Concepts, theory, and methods*. New York: Wiley.

Chern, M. S. (1992). On the computational complexity of reliability redundancy allocation in a series system. *Operations Research Letters*, *11*(5), 309–315. doi:10.1016/0167-6377(92)90008-Q

Chernoff, H. (1962). Optimal Accelerated Life Designs for Estimation. *Technometrics*, *4*(4), 381–408. doi:10.1080/00401706.1962.10490020

Chevallereau, C., & Khalil, W. (1988). A new method for the solution of the inverse kinematics of redundant robots. *Proceedings. IEEE International Conference on Robotics and Automation*. IEEE Computer Society Press. doi:10.1109/ROBOT.1988.12020

Chiao, C. H., & Hamada, H. (2001). Analyzing Experiments With Correlated Multiple Responses. *Journal of Quality Technology*, *33*(4), 451–465.

Cho, T. M., & Lee, B. C. (2011). Reliability-based design optimization using convex linearization and sequential optimization and reliability assessment method. *Structural Safety*, *33*(1), 42–50. doi:10.1016/j.strusafe.2010.05.003

Chung, S. W., & Bai, D. S. (1998). Optimal Designs of Simple Step-stress Accelerated Life Tests for Lognormal Lifetime Distributions. *International Journal of Reliability Quality and Safety Engineering*, *5*(04), 315–336. doi:10.1142/S0218539398000285

Civicioglu, P., & Besdok, E. (2013). A conceptual comparison of the Cuckoo-search, particle swarm optimization, differential evolution and artificial bee colony algorithms. *Artificial Intelligence Review*, *39*(4), 315–346. doi:10.1007/s10462-011-9276-0

Coakley, J., Dye, D., & Basoalto, H. (2011). Creep and creep modelling of a multimodal nickel-base super-alloy. *Acta Materialia*, *59*(3), 854–863. doi:10.1016/j.actamat.2010.08.035

Coasta, A., Crespo, A., Navarro, J., Lizcano, G., Madsen, H., & Feitosa, E. (2008). A review on the young history of the wind power short-term prediction. *Renewable & Sustainable Energy Reviews*, *12*(6), 1725–1744. doi:10.1016/j.rser.2007.01.015

Comsol. (n.d.). Batteries Fuel Cells. *Comsol*. Retrieved from: http://www.comsol.co.in/products/batteries-fuel-cells/

Condra, L. W. (1994). *Accelerated Testing for Product Reliability Assurance. Hanse Environmental Test Report HETR 94-001*. Hans Environmental. Inc. Hopkins.

Contienetal-The Future in Motion. (2013). *Worldwide Emission Standards and Related Regulations*. Author.

Coolen-Schrijner, P., & Coolen, F. P. A. (2004). Nonparametric predictive inference for age replacement with a renewal argument. *Quality and Reliability Engineering International*, *20*(3), 203–215. doi:10.1002/qre.561

Cormier, J., Milhet, X., & Mendez, J. (2007). Non-isothermal creep at very high temperature of the nickel-based single crystal super-alloy MC2. *Acta Materialia*, *55*(18), 6250–6259. doi:10.1016/j.actamat.2007.07.048

Correia, A. E., & Davim, J. P. (2011). Surface roughness measurement in turning carbon steel AISI 1045 using wiper inserts. *Measurement*, *44*(5), 1000–1005. doi:10.1016/j.measurement.2011.01.018

Cox, D. R. (1955, July). The analysis of non-Markovian stochastic processes by the inclusion of supplementary variables. *Mathematical Proceedings of the Cambridge Philosophical Society*, *51*(3), 433–441. doi:10.1017/S0305004100030437

Cox, D. R., & Snell, E. J. (1989). *Binary response variables. Analysis of Binary Data* (2nd ed.). London, UK: Chapman & Hall.

Craig John, J. (1986). Introduction to Robotics: Mechanics and Control (3rd ed.). Addision-Wesley.

Crawley, E. F. (1994). Intelligent structures for aerospace: A technology overview and assessment. *AIAA Journal*, *32*(8), 1689–1699. doi:10.2514/3.12161

Cross, L. E. (1993). *Ferroelectric Ceramics* (N. Setter & E. Colla, Eds.). Basel, Switzerland: Birkhaeuser.

Daily, J. W. (1988). Cycle-to-cycle variations: A chaotic process? *Combustion Science and Technology*, *57*(4-6), 149–162. doi:10.1080/00102208808923950

Damjanovic, D. (1997). Stress and frequency dependence of the direct piezoelectric effect in ferroelectric ceramics. *Journal of Applied Physics*, *82*(4), 1788–1797. doi:10.1063/1.365981

Damjanovic, D., & Demartin, M. J. (1996). Article. *Physica D. Nonlinear Phenomena*, *29*, 2057–2069.

Damjanovic, D., Demartin, M., Chu, F., & Setter, N. (1996). *Proceedings of the Tenth IEEE International Symposium on Applications of Ferroelectrics*. Rutgers University Press.

Dasgupta, A., & Pecht, M. (1995). Material Failure Mechanisms & Damage Models. *IEEE Transactions on Reliability, 40*(5), 531-536.

Das, K., Mandal, N. K., & Sinha, B. K. (2003). Optimal experimental designs for models with covariates. *Journal of Statistical Planning and Inference*, *115*(1), 273–285. doi:10.1016/S0378-3758(02)00151-9

Das, S. K. (2005). *Process heat transfer*. Narosa Publishing House.

Das, S. K., & Basudhar, P. K. (2006). Undrained lateral load capacity of piles in clay using artificial neural network. *Computers and Geotechnics*, *33*(8), 454–459. doi:10.1016/j.compgeo.2006.08.006

Datta, R., & Majumder, A. (2010). *Optimization of Turning Process parameters Using Multi-Objective Evolutionary Algorithm*. IEEE; doi:10.1109/CEC.2010.5586296

Daubechies, I. (1992). Ten Lectures on Wavelets, 357 pp., Soc. for Ind. and Appl. Math., Univ. Sci. Cent., Philadelphia, PA.

David, I., & Robles, G. (2012). *PID Control Dynamics of a Robotic Arm Manipulator with Two Degrees of Freedom.* Retrieved from http://www.slideshare.net/popochis/pid-control-dynamics-of-a-robotic-arm-manipulator-with-two-degrees-of-freedom

Daw, C. S., Finney, C. E. A., Green, J. B., Kennel, M. B., Thomas, J. F., & Connolly, F. T. (1996). *A simple model for cyclic variations in a spark-ignition engine* (No. 962086). SAE Technical Paper.

Daw, C. S., Kennel, M. B., Finney, C. E. A., & Connolly, F. T. (1998). Observing and modeling nonlinear dynamics in an internal combustion engine. *Physical Review E: Statistical Physics, Plasmas, Fluids, and Related Interdisciplinary Topics, 57*(3), 2811–2819. doi:10.1103/PhysRevE.57.2811

Deb, K. (2001). *Multi-objective optimization using evolutionary algorithms.* Hoboken, NJ: Wiley.

Deb, K. (2002). *Multi-Objective Optimization using Evolutionary Algorithms.* Singapore: John Wiley & Sons.

Deb, K., Pratap, A., Agarwal, S., & Meyarivan, T. A. (2002). Fast and elitist multi-objective genetic algorithm: NSGA-II. *IEEE Transactions on Evolutionary Computation, 6*(2), 182–197. doi:10.1109/4235.996017

Deepak, A. G., & Badler, N. I. (2000). Real-Time Inverse Kinematics Techniques for Anthropomorphic Limbs. *Graphical Models, 62*(5), 353–388. doi:10.1006/gmod.2000.0528 PMID:12143897

Deka, D., Joseph, D., Ghosh, S., & Mills, M. J. (2006). Crystal plasticity modeling of deformation and creep in polycrystalline Ti-6242. *Metallurgical and Materials Transactions. A, Physical Metallurgy and Materials Science, 37*(5), 1371–1388. doi:10.1007/s11661-006-0082-2

Del Castillo, E. (1996). Multi response process optimization via constrained confidence regions. *Journal of Quality Technology, 28*(1), 61–70.

Del Castillo, E., & Montgomery, D. C. (1993). A nonlinear programming solution to the dual response problem. *Journal of Quality Technology, 25*(3), 199–204.

Denson, W., Chandler, G., Crowell, W., Clark, A., & Jaworski, P. (1994). *Nonelectronic Parts Reliability Data 1995.* DTIC Document.

Derringer, G., & Suich, R. (1980). Simultaneous optimization of several response variables. *Journal of Quality Technology, 12*(4), 214–219.

Dey, T., Dey, S., & Kundu, D. (2016). On progressively type-II censored two-parameter rayleigh distribution. *Communications in Statistics. Simulation and Computation, 45*(2), 438–455. doi:10.1080/03610918.2013.856921

Dhillon, B. S., & Singh, C. (1981). *Engineering Reliability: New Techniques and Applications.* New York: Wiley.

Dhillon, B. S., & Yang, N. (1993). Availability of a man-machine system with critical and non-critical human error. *Microelectronics and Reliability, 33*(10), 1511–1521. doi:10.1016/0026-2714(93)90105-8

Dhillon, B. S., & Yang, N. (1996). Reliability analysis of a repairable robot system. *Journal of Quality in Maintenance Engineering, 2*(2), 30–37. doi:10.1108/13552519610120432

Dinesh, M., John, F., & Canny. (1992). Real Time Inverse Kinematics for General 6R Manipulators. In *Proceedings. IEEE International. Conference on Robotics and Automation* (pp. 383-389).

Dodson, B., & Nolan, D. (1999). *Reliability engineering handbook.* Marcel Dekker New York.

Dorigo, M. (1992). *Optimization, learning and natural algorithms. Ph.D. book.* Politecnico di Milano.

Dube, G. P., Kapuria, S., & Dumir, P. C. (1996). Exact piezothermoelastic solution of simply supported orthotropic flat panel in cylindrical bending. *International Journal of Mechanical Sciences*, *38*(11), 1161–1177. doi:10.1016/0020-7403(96)00020-3

Dung, Kang, & Ro. (2010). Robot Manipulator Modeling in Matlab Simmechanics with PD Control and Online Gravity Compensation. *Proceedings of the IEEE.*

Easterling, R. G., Mazumdar, M., Spencer, F. W., & Diegert, K. V. (1991). System-based component-test plans and operating characteristics: Binomial data. *Technometrics*, *33*(3), 287–298. doi:10.1080/00401706.1991.10484835

Eberhart, R., & Kennedy, J. (1995). A new optimizer using particle swarm theory. In *Proceeding of International Symposium on Micro Machine and Human Science*, (pp. 39-43).

El-Hofy, H. (2005). *Advance Machining Process*. New York: McGraw Hill.

Elsayed, E. A., & Zhang, H. (2004). Design of optimum simple step-stress accelerated life testing plans. In *Proceedings of the Second IIEC.*

Elton, C., & Nicholson, M. (1942). The ten-year cycle in numbers of the lynx in Canada. *Journal of Animal Ecology*, *11*(2), 215–244. doi:10.2307/1358

Engelbrecht, A. P. (2007). *Computational intelligence: An introduction*. John Wiley & Sons Ltd. doi:10.1002/9780470512517

Epishin, A., Link, T., Brückner, U., & Portella, P. D. (2001). Kinetics of the topological inversion of the γ/γ-microstructure during creep of a nickel-based super-alloy. *Acta Materialia*, *49*(19), 4017–4023. doi:10.1016/S1359-6454(01)00290-7

Ermolieva, T., & Ermoliev, Y. (2005). Catastrophic risk management: Flood and seismic risks case studies. In S. W. Wallace & W. T. Ziemba (Eds.), *Applications of stochastic programming*. Philadelphia, PA: MPS-SIAM Series on Optimization. doi:10.1137/1.9780898718799.ch22

Ermolieva, T., Ermoliev, Y., Fischer, G., Jonas, M., & Makowski, M. (2010). Cost effective and environmentally safe emission trading under uncertainty. In K. Marti, Y. Ermoliev, & M. Makowski (Eds.), *Coping with uncertainty: Robust solutions* (pp. 79–99). Heidelberg, Germany: Springer-Verlag. doi:10.1007/978-3-642-03735-1_5

Ermolieva, T., Ermoliev, Y., Jonas, M., Obersteiner, M., Wagner, F., & Winiwarter, W. (2014). Uncertainty, cost-effectiveness and environmental safety of robust carbon trading: Integrated approach. *Climatic Change*, *124*(3), 633–646. doi:10.1007/s10584-013-0824-2

Ernst, B., Oakleaf, B., Ahlstrom, M., Lange, M., Moehrlen, C., Lange, B., & Rohrig, K. et al. (2007). Predicting the wind. *Power and Energy Magazine*, *5*(6), 78–89. doi:10.1109/MPE.2007.906306

Ernst, B., Reyer, F., & Vanzetta, J. (2009). Integration of wide-scale renewable resources into the power delivery system. In *CIGRE/IEEE PES Joint Symposium* (pp.1–9).

EURO Parliament. (2004), *EURO Emission Regulations*. Author.

Ezgi Aktar, D., Anagun, A. S., & Koksal, G. (2009). Determination of optimal product styles by ordinal logistic regression versus conjoint analysis for kitchen faucets. *International Journal of Industrial Ergonomics*, *39*(5), 866–875. doi:10.1016/j.ergon.2009.06.007

Fard, N., & Li, C. (2009). Optimal simple step stress accelerated life test design for reliability prediction. *Journal of Statistical Planning and Inference*, *139*(5), 1799–1808. doi:10.1016/j.jspi.2008.05.046

Fascio, V., Wüthrich, R., & Bleuler, H. (2004). Spark assisted chemical engraving in the light of electrochemistry. *Electrochimica Acta*, *49*(22-23), 3997–4003. doi:10.1016/j.electacta.2003.12.062

Faverjon, B., & Tournassoud, P. (1987). A local based approach for path planning of manipulators with a high number of degrees of freedom. *Robotics and Automation. Proceedings.IEEE International Conference* (*Vol. 4*). doi:10.1109/ROBOT.1987.1087982

Feinstein, C. H., & Thomas, M. (2002). Logit, probit, and tobit models. In J. Mackenzie (Ed.), *Making History Count: A Primer in Quantitative Methods for Historians* (pp. 384–434). Cambridge, UK: Cambridge University Press. doi:10.1017/CBO9781139164832.014

Fister, I. J., Xin-She, Y., Fister, D., & Fister, I. (2014). Cuckoo Search: A Brief Literature Review. *Studies in Computational Intelligence*, *516*, 49–62.

Fleury, G., Schubert, F., & Nickel, H. (1996). Modelling of the thermo-mechanical behaviour of the single crystal superalloy CMSX-4. *Computational Materials Science*, *7*(1-2), 187–193. doi:10.1016/S0927-0256(96)00079-1

Foakes, A. P., & Pollard, D. G. (1993). Investigation of a chaotic mechanism for cycle-to-cycle variations. *Combustion Science and Technology*, *90*(1-4), 281–287. doi:10.1080/00102209308907615

Fraley, S., Oom, M., Terrien, B., & Date, J. Z. (2006). *Design of Experiments via Taguchi Methods: Orthogonal Arrays*. The Michigan Chemical Process Dynamic and Controls Open Text Book.

Freedman, H. I. (1980). *Deterministic Mathematical Models in Population Biology*. New York: Marcel Dekker.

Freund, R. J., & Wilson, W. J. (1998). *Regression Analysis*. San Diego, CA: Academic Press.

Furrer, D., & Fecht, H. (1999). Ni-based super-alloys for turbine discs. *JOM Journal of the Minerals. Metals and Materials Society*, *51*(1), 14–17. doi:10.1007/s11837-999-0005-y

Gakkhar, S. Singh, A., Singh, B.P. (2012). Effects of delay and seasonality on toxin producing phytoplankton-zooplankton system. *International Journal of Biomathematics, 5*(2).

Gakkhar, S., & Singh, A. (2010). A delay model for viral infection in toxin producing phytoplankton and zooplankton system. *Communications in Nonlinear Science and Numerical Simulation*, *15*(11), 3607–3620. doi:10.1016/j.cnsns.2010.01.010

Gal, S. (1974). Optimal test design for reliability demonstration. *Operations Research*, *22*(6), 1236–1242. doi:10.1287/opre.22.6.1236

Gandini, A. (1990). Importance and sensitivity analysis in assessing system reliability. *IEEE Transactions on Reliability*, *39*(1), 61–70. doi:10.1109/24.52613

Gandomi, M., Soltanpour, M., Zolfaghari, M. R., & Gandomi, A. H. (2016). Prediction of peak ground acceleration of Irans tectonic regions using a hybrid soft computing technique. *Geoscience Frontiers*, *7*(1), 75–82. doi:10.1016/j.gsf.2014.10.004

Ganesan, H., & Mohankumar, G. (2013). Optimization of Machining Techniques in CNC Turning Centre using Genetic Algorithm. *Arab. J. Sci. Engg.*, *38*(6), 1529–1538. doi:10.1007/s13369-013-0539-8

Geem, Z. W., Kim, J. H., & Loganathan, G. V. (2001). A new heuristic optimization algorithm: Harmony search. *Simulation*, *76*(2), 60–68. doi:10.1177/003754970107600201

Ghaffarpour, M., Aziz, A., & Hejazi, T. H. (2015). Optimization of friction stir welding parameters using multiple response surface methodology. *Journal of Materials Design and Applications*, *0*(0), 1–13.

Ghonem, H., & Dore, S. (1987). Experimental study of the constant probability crack growth curves under constant amplitude loading. *Engineering Fracture Mechanics*, *27*(1), 1–25. doi:10.1016/0013-7944(87)90002-6

Ghosh, S., Bhandari, Y., & Groeber, M. (2008). CAD based reconstruction of three dimensional polycrystalline microstructures from FIB generated serial sections. *Journal of Computer Aided Design*, *40*(3), 293–310. doi:10.1016/j.cad.2007.11.003

Glasser, G. J. (1967). The age replacement problem. *Technometrics*, *9*(1), 83–91. doi:10.1080/00401706.1967.10490443

Godal, O., Ermoliev, Y., Klassen, G., & Obersteiner, M. (2003). Carbon trading with imperfectly observable emissions. *Environmental and Resource Economics*, *25*(2), 151–169. doi:10.1023/A:1023914324084

Goff, S. D., Couturier, R., Guétaz, L., & Burlet, H. (2004). Effect of the microstructure on the creep behavior of PM Udimet 720 super-alloy—experiments and modeling. *Materials Science and Engineering A*, *387-389*, 599–603. doi:10.1016/j.msea.2004.01.094

Goldberg, D. E. (1989). *Genetic algorithms in search, optimization and machine learning*. Boston: Addison Wesley.

Goldberg, D. E. (1989). *Genetic Algorithms in Search*. Addison-Wesley Professional.

Gopalan, M. N., & Nagarwalia, H. E. (1985). Cost benefit analysis of a one server two-unit cold standby system with repair and preventive maintenance. *Microelectronics and Reliability*, *25*(2), 267–269. doi:10.1016/0026-2714(85)90011-3

Goswami, D. & Chakraborty, S. (2015). A Study on the Optimization Performance of Fireworks and Cuckoo Search Algorithms in Laser Machining Processes. *Journal of The Institution of Engineers*, *96*(3), 215-229.

Goyal, N., Kaushik, A., & Ram, M. (2016). Automotive water cooling system analysis subject to time dependence and failure issues. *International Journal of Manufacturing, Materials, and Mechanical Engineering*, *6*(2), 1–22. doi:10.4018/IJMMME.2016040101

Groeber, M., Ghosh, S., Uchic, M. D., & Dimiduk, D. M. (2008). A framework for automated analysis and simulation of 3D polycrystalline microstructures. Part 1: Statistical characterization. *Acta Materialia*, *56*(6), 1257–1273. doi:10.1016/j.actamat.2007.11.041

Groeber, M., Ghosh, S., Uchic, M. D., & Dimiduk, D. M. (2008). A framework for automated analysis and simulation of 3D polycrystalline microstructures. Part 2: Synthetic structure generation. *Acta Materialia*, *56*(6), 1274–1287. doi:10.1016/j.actamat.2007.11.040

Groisman, P., & Lyalko, V. (Eds.). (2012). *Earth Systems Change over Eastern Europe*. Kiev: Akademperiodyka.

Gupta, S. & Tewari, P. C. (2009). Simulation modeling and analysis of a complex system of a thermal power plant. *Journal of Industrial Engineering and Management, 2*(2), 387-406.

Gupta, P., Lal, A. K., Sharma, R. K., & Singh, J. (2005). Numerical analysis of reliability and availability of the serial processes in butter-oil processing plant. *International Journal of Quality & Reliability Management*, *22*(3), 303–316. doi:10.1108/02656710510582507

Haggag, M. Y. (2009). Cost Analysis of Two-Dissimilar-Unit Cold Standby System with three States and Preventive Maintenance using Linear First Order Differential Equations. *Journal of Mathematics and Statistics*, *5*(4), 395–400. doi:10.3844/jmssp.2009.395.400

Hall, S. (2012). *Rules of Thumb for Chemical Engineers. Elsevier Science.*

Handbook, M. (1987). Reliability Test Methods, Plans, and Environments for Engineering Development, Qualification, and Production. MIL-HDBK-781.

Harrington, E. C. (1965). The desirability function. Industrial Quality Control. *American Journal of Operations Research*, *21*(10), 494–498.

Harris, F. J. (1978). On the use of windows for harmonic analysis with the discrete Fourier transform. *Proceedings of the IEEE*, *66*(1), 51–83. doi:10.1109/PROC.1978.10837

Hasija, V., Ghosh, S., Mills, M. J., & Joseph, D. (2003). Modeling deformation and creep in Ti-6Al alloys with experimental validation. *Acta Materialia*, *51*, 4533–4545. doi:10.1016/S1359-6454(03)00289-1

Hassani, V., & Tjahjowidodo, T. (2013). Structural response investigation of a triangular-based piezoelectric drive mechanism to hysteresis effect of the piezoelectric actuator. *Mechanical Systems and Signal Processing*, *36*(1), 210–223. doi:10.1016/j.ymssp.2012.10.011

Hassell, M. P., Lawton, J. H., & May, R. M. (1976). Patterns of dynamical behaviour in single-species populations. *Journal of Animal Ecology*, *45*(2), 471–486. doi:10.2307/3886

Hastie, T., Tibshirani, R., & Friedman, J. (2001). *The elements of statistical learning: Data mining, inference and prediction*. Berlin: Springer. doi:10.1007/978-0-387-21606-5

Hastings, A., & Powell, T. (1991). Chaos in a three-species food chain. *Ecology*, *72*(3), 896–903. doi:10.2307/1940591

Hatanaka, H., Kawano, Y., Ido, N., Hato, M., & Tagami, M. (2005). Ultrasonic testing with advanced signal processing for concrete structures. *Nondestructive Testing and Evaluation*, *20*(2), 115–124. doi:10.1080/10589750500181084

Hatogama, Y. (1997). Reliability analysis of three state system. *IEEE Transactions on Reliability*, *28*, 386–393.

Heimann, M., & Reichstein, M. (2008). Terrestrial ecosystem carbon dynamics and climate feedbacks. *Nature*, *451*(7176), 289–292. doi:10.1038/nature06591 PMID:18202646

Heisler, H. (2002). *Advanced vehicle technology*. Elsevier.

Hejazi, A. M. (2013). *Reliability Based Design of an Automotive Dry Friction Clutch* (Master of science thesis). Sahand University of Technology, Tabriz, Iran.

Hejazi, T. H., Bashiri, M., Díaz-García, J. A., & Noghondarian, K. (2012). Optimization of probabilistic multiple response surfaces. *Applied Mathematical Modelling*, *36*(3), 1275–1285. doi:10.1016/j.apm.2011.07.067

Hejazi, T. H., Seyyed-Esfahani, M., & Mahootchi, M. (2014). Optimization of degree of conformance in multiresponse–multistage Systems with a Simulation-based meta heuristic. *Quality and Reliability Engineering International*, *31*(4), 645–658. doi:10.1002/qre.1622

Henley, E. J., & Kumamoto, H. (1992). *Probabilistic Risk Assessment*. IEEE Press.

Heyliger, P., & Brooks, S. (1996). Exact solutions for laminated piezoelectric plates in cylindrical bending.*Transactions of ASME Journal of Applied Mechanics, 63*(4), 903-910. doi:10.1115/1.2787245

Heywood, J. B. (1988). *Internal combustion engine fundamentals* (Vol. 930). New York: Mcgraw-hill.

Hillier, F. S., & Lieberman, G. J. (2001). Other Algorithms for Linear Programming. In E.M. Munson. In M. Lorkovic (Ed.), *Introduction to the operational research* (pp. 332–340). New York, NY: Mc Graw- Hill.

Holland, J. H. (1962). Outline for a logical theory of adaptive systems. *Journal of the Association for Computing Machinery*, *9*(3), 297–314. doi:10.1145/321127.321128

Hong, H. U., Kim, I. S., Choi, B. G., Kim, M. Y., & Jo, C. Y. (2009). The effect of grain boundary serration on creep resistance in a wrought nickel-based super-alloy. *Materials Science and Engineering A*, *517*(1-2), 125–131. doi:10.1016/j.msea.2009.03.071

Hopgood, A. A., & Martin, J. W. (1986). The creep behaviour of a Nickel-based single-crystal super-alloy. *Materials Science and Engineering*, *82*, 27–36. doi:10.1016/0025-5416(86)90092-3

Hosmer, D. W., & Lemeshow, S. (2000). *Applied Logistic Regression* (W. A. Shewhart & S. S. Wilks, Eds.). New York, NY: John Wiley & Sons. doi:10.1002/0471722146

Houghton, J. (2005). Climate change and sustainable energy. Met Office.

Hou, J., Qiao, X., Wang, Z., Liu, W., & Huang, Z. (2010). Characterization of knocking combustion in HCCI DME engine using wavelet packet transform. *Applied Energy*, *87*(4), 1239–1246. doi:10.1016/j.apenergy.2009.06.014

Hsieh, K. L., Tong, L. I., Chiu, H. P., & Yeh, H. Y. (2005). Optimization of a multi-response problem in Taguchis dynamic system. *Computers & Industrial Engineering*, *49*(4), 556–571. doi:10.1016/j.cie.2005.08.002

Huang, B., Biro, F. M., & Dorn, L. D. (2009). Determination of Relative Timing of Pubertal Maturation through Ordinal Logistic Modeling: Evaluation of Growth and Timing Parameters. *The Journal of Adolescent Health*, *45*(4), 383–388. doi:10.1016/j.jadohealth.2009.02.013 PMID:19766943

Hwang, S. C., & McMeeking, R. M. (1998). A finite element model of ferroelectric polycrystals. *Ferroelectrics*, *211*(1), 177–194. doi:10.1080/00150199808232342

IEEE standards on piezoelectricity. (1988), *ANSI/IEEE standard*. The Institute of Electrical and Electronic Engineers.

Ignat, M., Buffiere, J.-Y., & Chaix, J. M. (1993). Microstructures induced by a stress gradient in a nickel-based super-alloy. *Acta Metallurgica et Materialia*, *41*(3), 855–862. doi:10.1016/0956-7151(93)90019-O

Ikeda, T. (1990). *Fundamentals of piezoelectricity*. New York: Oxford University Press.

Imhof, L. (2000). Optimum designs for a multi response regression model. *Journal of Multivariate Analysis*, *72*(1), 120–131. doi:10.1006/jmva.1999.1841

IPCC. (1997). *Revised 1996 IPCC Guidelines for National Greenhouse Gas Inventories* (Vols. 1-3). Bracknell, UK: Intergovernmental Panel on Climate Change, Meteorological Office.

Itagaki, H., Ishizuka, T., & Huang, P. Y. (1993). Experimental estimation of the probability distribution of fatigue crack growth lives. *Probabilistic Engineering Mechanics*, *8*(1), 25–34. doi:10.1016/0266-8920(93)90027-S

Iyyer, N., Sarkar, S., Merrill, R., & Phan, N. (2007). Aircraft life management using crack initiation and crack growth models - P-3C Aircraft experience. *International Journal of Fatigue*, *29*(9-11), 1584–1607. doi:10.1016/j.ijfatigue.2007.03.017

Jain, M., & Singh, C. J. (2000). Reliability of repairable multi-component redundant system. *International Journal of Engineering*, *3*, 107–114.

Jain, V. K., Dixit, P. M., & Pandey, P. M. (1999). On the analysis of electro chemical spark machining process. *International Journal of Machine Tools & Manufacture*, *39*(1), 165–186. doi:10.1016/S0890-6955(98)00010-8

Janisch, J. E., & Harmon, M. E. (2002). Successional changes in live and dead wood carbon stores: Implications for net ecosystem productivity. *Tree Physiology*, *22*(2-3), 77–90. doi:10.1093/treephys/22.2-3.77 PMID:11830405

Japan Automobile Standards Internationalization Center. ECE No.15. (2012). Exhaust Emission Standard.

Jaulin, L., Kieffer, M., Didrit, O., & Walter, E. (2001). *Applied interval analysis: with examples in parameter and state estimation, robust control and robotics*. London: Springer. doi:10.1007/978-1-4471-0249-6

Jeng, S. L., Lu, J. C., & Wang, K. (2007). A review of reliability research on nanotechnology. *IEEE Transactions on Reliability*, *56*(3), 401–410.

Jennings, D. E. (1986). Judging inference adequacy in logistic regression. *Journal of the American Statistical Association, 81*(394), 471–476. doi:10.1080/01621459.1986.10478292

Jones, D., & Tamiz, M. (2010). Practical goal programming. In F. S. Hillier (Ed.), *International Series in Operations Research & Management Science* (pp. 1–170). New York, NY: Springer.

Jones, T. L. (2010). *Handbook of reliability prediction procedures for mechanical equipment. Naval Surface Warfare Center*. West Bethesda, MD: Carderock Division.

Jorgensen, S. E. (2002). Ecosystem Management and Ecological Modeling. *Scientific World, 2*, 107–121. doi:10.1100/tsw.2002.80 PMID:12806044

Jorgensen, S. E., & Bernardi, R. D. (1997). The application of a model with dynamic structure to simulate the effect of mass fish mortality on zooplankton structure in Lago de Annone. *Hydrobiologia, 356*(1/3), 87–96. doi:10.1023/A:1003171012639

Kaiser, G. (2010). *A friendly guide to wavelets*. Springer Science & Business Media.

Kakehi, K. (1999). Tension/ Compression asymmetry in creep behavior of a Ni-based super-alloy. *Scripta Materialia, 41*(5), 461–465. doi:10.1016/S1359-6462(99)00191-8

Kalpakjian, S. (2001). *Manufacturing engineering and technology*. Pearson Education India.

Kamaruddin, S., Khan, Z. A., & Foong, S. H. (2010). Application of Taguchi method in the optimization of injection moulding parameters for manufacturing products from plastic blend. *IACSIT International Journal of Engineering and Technology, 2*(6), 574–580. doi:10.7763/IJET.2010.V2.184

Kanagaraj, G., Ponnambalam, S. G., & Jawahar, N. (2013). A hybrid cuckoo search and genetic algorithm for reliability–redundancy allocation problems. *Computers & Industrial Engineering, 66*(4), 1115–1124. doi:10.1016/j.cie.2013.08.003

Kanagaraj, G., Ponnambalam, S. G., & Lim, W. C. E. (2014). Application of a hybridized cuckoo search-genetic algorithm to path optimization for PCB holes drilling process. *2014 IEEE International Conference on Automation Science and Engineering (CASE)*. doi:10.1109/CoASE.2014.6899353

Kapur, K. C., & Lamberson, L. R. (1977). *Reliability in Engineering Design*. New York: Wiley.

Karl, S. (1992). *Heat transfer in boiling and condensation*. Berlin: Springer Verlag.

Kathleen, M.C., Natalia, Y.K., & Jeff, R. (2004). *Response surface methodology*. Center for computational analysis of social and organizational systems (CASOS), Technical Report.

Katić, D., & Vukobratović, M. (2003). Survey of intelligent control techniques for humanoid robots. *Journal of Intelligent & Robotic Systems, 37*(2), 117–141. doi:10.1023/A:1024172417914

Kaymaz, I., & McMahon, C. A. (2005). A response surface method based on weighted regression for structural reliability analysis. *Probabilistic Engineering Mechanics, 20*(1), 11–17. doi:10.1016/j.probengmech.2004.05.005

Kazemzadeh, R. B., Bashiri, M., Atkinson, A. C., & Noorossana, R. (2008). A general framework for multiresponse optimization problems based on goal programming. *European Journal of Operational Research, 189*(2), 421–429. doi:10.1016/j.ejor.2007.05.030

Ke, J. B., Lee, W. C., & Wang, K. H. (2007). Reliability and sensitivity analysis of a system with multiple unreliable service stations and standby switching failures. *Physica A: Statistical Mechanics and its Applications, 380*, 455-469.

Kennedy, J., & Eberhart, R. (1995). Particle swarm optimization. In *Proceeding of IEEE International Conference on Neural Networks,* (vol. 4, pp. 1942-1948). IEEE.

Khamis, I. H. (1997). Optimum M-step step-stress design with k stress variables. *Communications in Statistics. Simulation and Computation*, *26*(4), 1301–1313. doi:10.1080/03610919708813441

Khamis, I. H., & Higgins, J. J. (1996). Optimum 3-step Step-stress Tests. *IEEE Transactions on Reliability*, *45*(2), 341–345. doi:10.1109/24.510823

Khamis, I. H., & Higgins, J. J. (1998). A new model for step-stress testing. *IEEE Transactions on Reliability*, *47*(2), 131–134. doi:10.1109/24.722275

Khanna, O. P. (1993). *A Text Book of Welding Technology for Engineering Students*. Dhanpat Rai and Sons.

Khatib, O. (1985). Real-time obstacle avoidance for manipulators and mobile robots. In *Proceedings. IEEE International. Conference on Robotics and Automation* (pp. 500-505). doi:10.1109/ROBOT.1985.1087247

Khuri, A. I. (2006). Response Surface Methods and Their Application in the Treatment of Cancer with Drug Combinations: Some Reflections. In A. I. Khuri (Ed.), *Response Surface Methodology and Related Topics* (pp. 159–173). Singapore: World Scientific Printers. doi:10.1142/5915

Khuri, A. I., & Conlon, M. (1981). Simultaneous optimization of multiple responses represented by polynomial regression functions. *Technometrics*, *23*(4), 363–375. doi:10.1080/00401706.1981.10487681

Khuri, A. I., & Mukhopadhyay, S. (2010). Response Surface Methodology. *Wiley Interdisciplinary Reviews: Computational Statistics*, *2*(2), 128–149. doi:10.1002/wics.73

Kim, J. Y., Park, I. W., & Oh, J. H. (2007). Walking control algorithm of biped humanoid robot on uneven and inclined floor. *Journal of Intelligent & Robotic Systems*, *48*(4), 457–484. doi:10.1007/s10846-006-9107-8

Kim, K. J., & Lin, D. K. J. (1998). Dual response surface optimization: A fuzzy modeling approach. *Journal of Quality Technology*, *30*(1), 1–10.

Kim, K., & Lin, D. (2006). Optimization of multiple responses considering both location and dispersion effects. *European Journal of Operational Research*, *169*(1), 133–145. doi:10.1016/j.ejor.2004.06.020

Kim, T.-W., Kang, D.-H., Yeom, J.-T., & Park, N.-K. (2007). Continuum damage mechanics-based creep–fatigue-interacted life prediction of nickel-based super-alloy at high temperature. *Scripta Materialia*, *57*(12), 1149–1152. doi:10.1016/j.scriptamat.2007.08.014

Klebano, A., & Hastings, A. (1994). Chaos in a three species food chain. *Journal of Mathematical Biology*, *32*(5), 427–451. doi:10.1007/BF00160167

Kleijnen, J. P. C. (2008). Simulation optimization. In F. S. Hillier (Ed.), *Design and Analysis of Simulation Experiments* (pp. 101–138). New York, NY: Springer. doi:10.1007/978-0-387-71813-2_4

Klein, C., & Huang, C. (1983). Review on Pseudo inverse Control for Use with kinematically re- dundant manipulators. *IEEE Transactions on Systems, Man, and Cybernetics*, *13*(3), 245–250. doi:10.1109/TSMC.1983.6313123

Knobloch, C., Saß, V., Siebörger, D., & Glatzel, U. (1997). Anisotropic creep behavior of a nickel-based super-alloy compared with single phase nickel solid solution and γ phase single crystals. *Materials Science and Engineering A*, *234-236*, 237–241. doi:10.1016/S0921-5093(97)00142-1

Koizuml, I., Gyakushi, N., & Takamoto, Y. (1977). Study on the cycle-by-cycle variation in diesel engines. *Bulletin of the JSME*, *20*(145), 869–876. doi:10.1299/jsme1958.20.869

Kolahan, F., & Abachizadeh, M. (2008). Optimizing turning parameters for cylindrical parts using simulated annealing method. *Proceedings of World Academy of Science, Engineering and Technology.*, *36*, 436–439.

Kollár, B. (2014). *Credit value at risk and options of its measurement.* Paper presented at International Conference on Economics and Social Science (ICESS), Information Engineering Research Institute, Advances in Education Research, Shenzhen, China.

Kollár, B., & Cisko, Š. (2014). *Credit risk quantification with the use of Credit Risk.* Paper presented at International Conference on Management, Education, Business, and Information Science, Durgapur, India.

Kong, J. S., Lee, E. H., Lee, B. H., & Kim, J. G. (2008). Study on the real-time walking control of a humanoid robot using fuzzy algorithm. *International Journal of Control, Automation, and Systems*, *6*(4), 551–558.

Kopachevsky, I., Kostyuchenko, Y. V., & Stoyka, O. (2016). Land Use Drivers of Population Dynamics in Tasks of Security Management and Risk Assessment. *International Journal of Mathematical. Engineering and Management Sciences*, *1*(1), 18–24.

Korkut, I., Kasap, M., Çiftci, I., & Seker, U. (2004). Determination of optimum cutting parameters during machining of AISI 304 austenitic stainless steel. *Materials & Design*, *25*(4), 303–305. doi:10.1016/j.matdes.2003.10.011

Kostyuchenko, Y. V., Bilous, Y., Kopachevsky, I., & Solovyov, D. (2013). Coherent risk measures assessment based on the coupled analysis of multivariate distributions of multisource observation data. In *Proceedings of 11-th International Probabilistic Workshop.*

Kostyuchenko, Y. V., Bilous, Y., Movchan, D., László, M., & Kopachevsky, I. (2013). Toward methodology of satellite observation utilization for agricultural production risk assessment. *IERI Procedia*, *5*, 21-27.

Kostyuchenko, Yu. V. (2015). Geostatistics and remote sensing for extremes forecasting and disaster risk multiscale analysis. In S. Kadry & A. El Hami (Eds.), *Numerical Methods for Reliability and Safety Assessment: Multiscale and Multiphysics Systems* (pp. 404–423). Springer International Publishing. doi:10.1007/978-3-319-07167-1_16

Kotu, V., & Deshpande, B. (2015). Regression methods. In S. Elliot (Ed.), *Predictive Analytics and Data Mining* (pp. 165–193). Waltham, MA: Elsevier. doi:10.1016/B978-0-12-801460-8.00005-7

Koul, A. K., & Castillo, R. (1991). A critical assessment of the θ projection concept for creep life prediction of nickel-based super-alloy components. *Materials Science and Engineering A*, *138*(2), 213–219. doi:10.1016/0921-5093(91)90690-O

Kovach, J., & Cho, B. R. (2009). A D-optimal design approach to constrained multiresponse robust design with prioritized mean and variance considerations. *Computers & Industrial Engineering*, *57*(1), 237–245. doi:10.1016/j.cie.2008.11.011

Kovarik, L., Unocic, R. R., Li, J., Sarosi, P., Shen, C., Wang, Y., & Mills, M. J. (2009). Microtwinning and other shearing mechanisms at intermediate temperatures in Ni-based super-alloys. *Progress in Materials Science*, *54*(6), 839–873. doi:10.1016/j.pmatsci.2009.03.010

Koza, J. R. (1992). *Genetic Programming: On the Programming of Computers by Means of Natural Selection.* Cambridge, MA: MIT press.

Kreinovich, V., Nguyen, H. T., Ferson, S., & Ginzburg, L. (2003). From computation with guaranteed intervals to computation with confidence intervals: a new application of fuzzy techniques. In *Proceedings of the 21st International Conference of the North American Fuzzy Information Processing Society NAFIPS'2002.*

Kronmueller, H. (1970). Article. *Zeitschrift fur Angewandte Physik*, *30*, 9–15.

Kulkarni A.V. & Jain, V.K. (2015). Design and development of an electrochemical spark micro manufacturing equipment. *International Journal of Mechanical Engineering and Robotics Research*, *4*(4), 368-372. DOI: 10.18178/ijmerr.4.4

Kulkarni Anjali, V. (2013). *Performance investigations into microfabrication using electrochemical spark* (Doctoral dissertation). GBTU, Lucknow, India.

Kulkarni, A. V. (2012). Electrochemical Spark Micromachining Process. In M. Kahrizi (Ed.), Micromachining techniques for fabrication of micro and nano structures, (pp. 235-252). Academic Press.

Kulkarni, A.V., Jain, V.K., & Misra, K.A. (2013). Study of microstructure and transient, instantaneous current in different excitations during silicon micromachining. *J. Inst. Eng. India Ser. B.*

Kulkarni, A. V. (2009). Systematic analysis of electrochemical discharge process. *Int. J. Machining and Machinability of Materials*, *6*(3/4), 194–211. doi:10.1504/IJMMM.2009.027324

Kulkarni, A. V., Jain, V. K., & Misra, K. A. (2011a). Electrochemical spark micromachining microchannels and microholes of metals and non-metals. *Int. Journal of Manufacturing Technology Management*, *22*(2), 107–123. doi:10.1504/IJMTM.2011.038713

Kulkarni, A. V., Jain, V. K., & Misra, K. A. (2011b). Electrochemical spark micromachining: Present scenario. *IJAT*, *5*(1), 52–59. doi:10.20965/ijat.2011.p0052

Kumar, A., & Singh, S. B. (2008). Reliability analysis of an n-unit parallel standby system under imperfect switching using copula. *Computer Modeling and New Technologies*, *12*(1), 47–55.

Kumar, A., Varshney, A., & Ram, M. (2015). Sensitivity analysis for casting process under stochastic modelling. *International Journal of Industrial Engineering Computations*, *6*(3), 419–432. doi:10.5267/j.ijiec.2015.2.001

Kumar, D., & Singh, S. B. (2016). Stochastic Analysis of Complex Repairable System with Deliberate Failure Emphasizing Reboot Delay. *Communications in Statistics. Simulation and Computation*, *45*, 1–20. doi:10.1080/00949655.2016.1232723

Kumar, P., & Foufoula-Georgiou, E. (1997). Wavelet analysis for geophysical applications. *Reviews of Geophysics*, *35*(4), 385–412. doi:10.1029/97RG00427

Kuo, W. (2001). *Optimal reliability design: Fundamentals and applications*. Cambridge University Press.

Kuo, W., & Prasad, V. R. (2000). An annotated overview of system-reliability optimization. *IEEE Transactions on Reliability*, *49*(2), 176–187. doi:10.1109/24.877336

Kuo, W., & Wan, R. (2007). Recent Advances in Optimal Reliability Allocation. *IEEE Transactions on Systems, Man, and Cybernetics. Part A, Systems and Humans*, *37*(2), 143–156. doi:10.1109/TSMCA.2006.889476

Kuriakose, S., & Shunmugam, M. S. (2005). Multi-objective optimization of wire EDM process by non-dominated sorting genetic algorithms. *Journal of Materials Processing Technology*, *170*, 133–141. doi:10.1016/j.jmatprotec.2005.04.105

Lan, T. S. (2010). Fuzzy Deduction Material Removal Rate Optimization for Computer Numerical Control Turning. *American Journal of Applied Sciences*, *7*(7), 1026–1031. doi:10.3844/ajassp.2010.1026.1031

Larrosa, J. A. E. (2008). *New heuristics for global optimization of complex bioprocesses.* Universidade de Vigo.

Lasanovsky. (1982). Availability of two-unit cold standby system with degraded state. *IEEE Transactions on Reliability*, *31*, 123.

Latha, B., & Senthilkumar, V. S. (2010). Modeling and analysis of surface roughness parameters in drilling GFRP composites using fuzzy logic. *Materials and Manufacturing Processes*, *25*(8), 817–827. doi:10.1080/10426910903447261

Latombe, J. C. (1991). *Robot Motion Planning*. Norwell, MA: Kluwer Academic Publishers. doi:10.1007/978-1-4615-4022-9

Lawless, J. F. (2011). *Statistical models and methods for lifetime data* (Vol. 362). John Wiley & Sons.

Lawsonl, C., & Montgomery, D. C. (2006). logistic regression analysis of customer satisfaction data. *Quality and Reliability Engineering International*, *22*(8), 971–984. doi:10.1002/qre.775

Lee, B., Jung, D., Kim, Y. W., & van Nieuwstadt, M. (2013). Thermodynamics-Based Mean Value Model for Diesel Combustion. *Journal of Engineering for Gas Turbines and Power*, *135*(9), 091504. doi:10.1115/1.4024757

Leemis, L. M. (1995). *Reliability: Probabilistic models and statistical methods*. Prentice-Hall, Inc.

Leung, S. C. H. (2007). A non-linear goal programming model and solution method for the multi-objective trip distribution problem in transportation engineering. *Optimization and Engineering*, *8*(3), 277–298. doi:10.1007/s11081-007-9019-x

Levitin, G., Xing, L., & Dai, Y. (2015). Optimal backup frequency in system with random repair time. *Reliability Engineering & System Safety*, *144*, 12–22. doi:10.1016/j.ress.2015.06.014

Li, C., & Fard, N. (2007). Optimum bivariate step-stress accelerated life test for censored data. *IEEE Transactions on Reliability*, *56*(1), 77–84. doi:10.1109/TR.2006.890897

Li, H. (2002). Application of wavelet multi-resolution analysis to pressure fluctuations of gas–solid two-phase flow in a horizontal pipe. *Powder Technology*, *125*(1), 61–73. doi:10.1016/S0032-5910(02)00012-8

Lin, D. K. J., & Tu, W. (1995). Dual response surface optimization. *Journal of Quality Technology*, *27*(1), 34–39.

Lin, Y. K., & Yang, J. N. (1985). On statistical moments of fatigue crack propagation. *Engineering Fracture Mechanics*, *18*(2), 243–256. doi:10.1016/0013-7944(83)90136-4

Liu, Y. T., Chang, W. C., & Yamagata, Y. (2010). A study on optimal compensation cutting for an aspheric surface using the Taguchi method. *Journal of Manufacturing Science and Technology.*, *3*(1), 40–48. doi:10.1016/j.cirpj.2010.03.001

Li, W., & Pham, H. (2005). Reliability modeling of multi-state degraded systems with multi-competing failures and random shocks. *IEEE Transactions on Reliability*, *54*(2), 297–303.

Li, Z., & Liang, X. (2007). Vibro-acoustic analysis and optimization of damping structure with response surface method. *Materials & Design*, *28*(7), 1999–2007. doi:10.1016/j.matdes.2006.07.006

Luss, H., & Kander, Z. (1974). Inspection policies when duration of checkings is non-negligible. *Operational Research Quarterly*, *25*(2), 299–309. doi:10.2307/3008371

Lu, Y. L., Chen, L. J., Wang, G. Y., Benson, M. L., Liaw, P. L., Thompson, S. A., & Klarstrom, D. L. et al. (2005). Hold time effects on low cycle fatigue behavior of HAYNES 230 super-alloy at high temperatures. *Materials Science and Engineering A*, *409*(1-2), 282–291. doi:10.1016/j.msea.2005.05.120

Ma, A., Dye, D., & Reed, R. C. (2008). A model for the creep deformation behaviour of single-crystal super-alloy CMSX-4. *Acta Materialia*, *56*(8), 1657–1670. doi:10.1016/j.actamat.2007.11.031

Ma, A., & Roters, F. (2004). A constitutive model for fcc single crystals based on dislocation densities and its application to uniaxial compression of aluminium single crystals. *Acta Materialia*, *52*(12), 3603–3612. doi:10.1016/j.actamat.2004.04.012

Ma, A., Roters, F., & Raabe, D. (2006a). A dislocation density based constitutive model for crystal plasticity FEM including geometrically necessary dislocations. *Acta Materialia*, *54*(8), 2169–2179. doi:10.1016/j.actamat.2006.01.005

Ma, A., Roters, F., & Raabe, D. (2006b). On the consideration of interactions between dislocations and grain boundaries in crystal plasticity finite element modeling-theory, experiments and simulations. *Acta Materialia*, *54*(8), 2181–2194. doi:10.1016/j.actamat.2006.01.004

Macián, V., Luján, J. M., Guardiola, C., & Perles, A. (2006). A comparison of different methods for fuel delivery unevenness detection in Diesel engines. *Mechanical Systems and Signal Processing*, *20*(8), 2219–2231. doi:10.1016/j.ymssp.2005.04.001

Madić, M., & Radovanović, M. (2013). Application of Cuckoo Search Algorithm for Surface Roughness Optimization in CO2 Laser Cutting, Annals of Faculty Engineering Hunedoara. *International Journal of Engineering*.

Mahapatra, S. S., Patnaik, A., & Patnaik, P. (2006). Parametric analysis and optimization of cutting parameters for turning operations based on Taguchi method. *Proceedings of the Int. Conference on Global Manufacturing and Innovation*.

Malaiya, Y. K., Li, M. N., Bieman, J. M., & Karcich, R. (2002). Software reliability growth with test coverage. *IEEE Transactions on Reliability*, *51*(4), 420–426. doi:10.1109/TR.2002.804489

Manchiraju, S., Asai, M., & Ghosh, S. (2007). A dual-time-scale finite element model for simulating cyclic deformation of polycrystalline alloys. *Journal of Strain Analysis in Engineering and Design*, *42*(4), 183–200. doi:10.1243/03093247JSA233

Mandava, R. K., & Vundavilli, P. R. (2015). Design of PID Controllers for 4-DOF Planar and Spatial Manipulators. *International Conference on Robotics, Automation, Control and Embedded Systems*. Hindustan University.

Manonukul, A., Dunne, F. P. E., & Knowles, D. (2002). Physically-based model for creep in nickel-base super-alloy C263 both above and below the gamma solvus. *Acta Materialia*, *50*(11), 2917–2931. doi:10.1016/S1359-6454(02)00119-2

Marcus, E., & Stern, H. (2003). *Blueprints for high availability*. John Wiley & Sons.

Maria, G. D. G., Stefano, C., Antonio, F., & Paolo, M. C. (2014). Comparison between wind power prediction models based on wavelet decomposition with Least-Squares Support Vector Machine (LS-SVM) and Artificial Neural Network (ANN). *Energies*, *7*(8), 5251–5272. doi:10.3390/en7085251

Maugin, G. A. (1985). Nonlinear electromechanical effects and applications. World Scientific Publishing Company.

Mazumdar, M. (1977). An optimum procedure for component testing in the demonstration of series system reliability. *Reliability. IEEE Transactions on*, *26*(5), 342–345.

Mazumdar, M. (1980). An optimum component testing procedure for a series system with redundant subsystems. *Technometrics*, *22*(1), 23–27. doi:10.2307/1268379

Mazzuchi, T. A., & Soyer, R. (1996). A Bayesian perspective on some replacement strategies. *Reliability Engineering & System Safety*, *51*(3), 295–303. doi:10.1016/0951-8320(95)00077-1

Meeker, W. Q. (1984). A Comparison of Accelerated Life Test Plans for Weibull & Lognormal Distributions & Type I Censoring. *Technometrics*, *26*(2), 157–171. doi:10.1080/00401706.1984.10487941

Meeker, W. Q., & Hahn, G. J. (1977). Asymptotically Optimum Over-stress tests to Estimate the Survival Probability at a Condition with a Low Expected Failure Probability. *Technometrics*, *19*(2), 381–399. doi:10.1080/00401706.1977.10489577

Meeter, C. A., & Meeker, W. Q. (1994). Optimum Accelerated Life Tests with a Nonconstant Sigma. *Technometrics*, *36*(27), 71–83.

Mehrjoo, S., & Bashiri, M. (2013). An application of principal component analysis and logistic regression to facilitate production scheduling decision support systems: An automative industry case. *Journal of Industrial Engineering International*, *9*(9), 1–12.

Meng, Q., Arsecularatne, J. A., & Mathew, P. (2000). Calculation of optimum cutting condition for turning operation using a machining theory. *International Journal of Machine Tools & Manufacture*, *40*(12), 1709–1733. doi:10.1016/S0890-6955(00)00026-2

Merkuryeva, G. (2005). Response surface-based simulation meta modeling methods with applications to optimization problems. In A. Dolgui, J. Soldek, & O. Zaikin (Eds.), *Supply chain optimization product / process design, facility location and flow control* (pp. 205–215). Boston, MA: Springer.

Michalewicz, Z., & Fogel, D. B. (2004). *How to solve it: Modern heuristics*. Springer-Verlag. doi:10.1007/978-3-662-07807-5

Milhet, X., Cormier, J., & Organista, A. (2010). On the role of the internal stress during non-isothermal creep life of a first generation nickel based single crystal super-alloy. *Materials Science and Engineering A*, *527*(9), 2280–2288. doi:10.1016/j.msea.2009.11.067

Military Specification. (1974). *Airplane Damage Tolerance Requirements. MIL-A-83444*. USAF.

Military, U. (1992). *Reliability prediction of electronic equipment.* MIL-HDBK-217F Notice 1.

Miller, R. G. Jr. (2011). *Survival analysis* (Vol. 66). John Wiley & Sons.

Miller, R., & Nelson, W. B. (1983). Optimum Simple Step-stress Plans for Accelerated Life Testing. *IEEE Transactions on Reliability*, *32*(1), 59–65. doi:10.1109/TR.1983.5221475

Milligan, M., Porter, K., DeMeo, E., Denholm, P., Holttinen, H., Kirby, B., & Soder, L. et al. (2009). *Wind power myths debunked.* IEEE Power and Energy Society.

Minitab. (2010). *Minitab user manual release 16.* MINITAB Inc.

Misra, K. B. (2008). *Handbook of performability engineering*. Springer Science & Business Media. doi:10.1007/978-1-84800-131-2

Modarres, M. (2006). *Risk analysis in engineering: Techniques, tools, and trends*. CRC Press.

Modarres, M., Kaminskiy, M. P., & Krivtsov, V. (2009). *Reliability engineering and risk analysis: a practical guide*. CRC Press.

Mohan, C., & Shanker, K. (1988). Reliability Optimization of Complex Systems using Random Search Technique. *Microelectronics and Reliability*, *28*(4), 513–518. doi:10.1016/0026-2714(88)90133-3

Mokaddis, G. S., & Tawfek, M. L. (1996). Some characteristics of a two-dissimilar-unit cold standby redundant system with three modes. *Microelectronics and Reliability*, *36*(4), 497–503. doi:10.1016/0026-2714(94)00128-6

Montgomery, D. C. (2001). *Design and analysis of experiments*. New York: John Wiley.

Montgomery, D. C. (2005). *Design and Analysis of Experiments*. New York, NY: John Wiley & Sons, Inc.

Montgomery, D. C., Peck, E. A., & Vining, G. G. (2003). *Introduction to linear regression analysis* (3rd ed.). New York: Wiley.

Moore, R. E., & Lodwick, W. A. (2003). Interval analysis and fuzzy set theory. *Fuzzy Sets and Systems*, *135*(1), 5–9. doi:10.1016/S0165-0114(02)00246-4

Morisette, J. T., Privette, J. L., & Justice, C. O. (2002). A framework for the validation of MODIS land products. *Remote Sensing of Environment*, *83*(1-2), 77–96. doi:10.1016/S0034-4257(02)00088-3

Movchan, D., Kostyuchenko, Y., Marton, L., Frayer, O., & Kyryzyuk, S. (2014). Uncertainty Analysis in Crop Productivity and Remote Estimation for Agricultural Risk Assessment. In Vulnerability, Uncertainty, and Risk: Quantification, Mitigation, and Management (pp. 1008-1015). ASCE. doi:10.1061/9780784413609.102

Movchan, D., & Kostyuchenko, Yu. V. (2015). Regional Dynamics of Terrestrial Vegetation Productivity and Climate Feedbacks for Territory of Ukraine. *International Journal of Geographical Information Science*, *29*(8), 1490–1505. doi:10.1080/13658816.2015.1051985

Muljadi, E., & Butterfield, C. P. (2001). Pitch-controlled variable-speed wind turbine generation. *IEEE Trans. Ind. Application*, *37*(1), 240–246. doi:10.1109/28.903156

Munford, A. G. (1981). Comparison among certain inspection policies. *Management Science*, *27*(3), 260–267. doi:10.1287/mnsc.27.3.260

Munford, A. G., & Shahani, A. K. (1972). A nearly optimal inspection policy. *Operational Research Quarterly*, *23*(3), 373–379. doi:10.2307/3007891

Munford, A. G., & Shahani, A. K. (1973). An inspection policy for the Weibull case. *Operational Research Quarterly*, *24*(3), 453–458. doi:10.2307/3008129

Myers, A. F. (2007). *k*-out-of-*n*: G system reliability with imperfect fault coverage. *IEEE Transactions on Reliability*, *56*(3), 464–473. doi:10.1109/TR.2007.903229

Myers, R. H. (1999). Response Surface Methodology: Current Status and Future Directions (with discussion). *Journal of Quality Technology*, *31*(1), 30–44.

Myers, R. H., Douglas, C. M., & Anderson, C. M. (2009). *Process and Product Optimization Using Designed Experiments*. John Wiley & Sons.

Myers, R. H., Montgomery, D. C., & Anderson Cook, C. M. (2009). *Response Surface Methodology: Process and Product Optimization Using Designed Experiments* (3rd ed.). John Wiley and Sons.

Myers, R. H., Montgomery, D. C., Vining, G. G., Borror, C. M., & Kowalski, S. M. (2004). Response surface methodology: A retrospective and literature survey. *Journal of Quality Technology*, *36*(1), 53–77.

Naikan, V. A. (2008). *Reliability Engineering and life testing*. PHI Learning Pvt. Ltd.

Nair, J. H., & Sabnis, S. V. (2002). A reliability test-plan for series systems with components having stochastic failure rates. *Reliability. IEEE Transactions on*, *51*(1), 17–22.

Nakagawa, T. (1981). Summary of periodic replacement with minimal repair at failure. *Journal of the Operations Research Society of Japan*, *24*, 213–228.

Ne'el, N. (1942). Article. *Cahiers Phys.*, *12*, 1–6.

Nechval, K. N., Nechval, N. A., Berzins, G., Purgailis, M., Rozevskis, U., & Strelchonok, V. F. (2009). Optimal adaptive inspection planning process in service of fatigued aircraft structures. In K. Al-Begain, D. Fiems, & G. Horvath (Eds.), Analytical and Stochastic Modeling Techniques and Applications, (LNCS) (vol. 5513, pp. 354–369). Berlin: Springer-Verlag doi:10.1007/978-3-642-02205-0_25

Nechval, N. A., & Vasermanis, E. K. (2004). *Improved Decisions in Statistics*. Riga: Izglitibas soli.

Nechval, N. A., Nechval, K. N., & Purgailis, M. (2011). Inspection policies in service of fatigued aircraft structures. In S.I. Ao & L. Gelman (Eds.), Electrical Engineering and Applied Computing, (LNEE) (vol. 90, pp. 459-472). Berlin: Springer Science+Business Media B.V. doi:10.1007/978-94-007-1192-1_37

Nechval, N. A., Nechval, K. N., & Vasermanis, E. K. (2003). Statistical models for prediction of the fatigue crack growth in aircraft service. In A. Varvani-Farahani & C. A. Brebbia (Eds.), Fatigue Damage of Materials 2003 (pp. 435-445). Southampton, UK: WIT Press.

Nechval, N. A., Nechval, K. N., Berzins, G., Purgailis, M., & Rozevskis, U. (2008). Stochastic fatigue models for efficient planning inspections in service of aircraft structures. In K. Al-Begain, A. Heindl, & M. Telek (Eds.), Analytical and Stochastic Modeling Techniques and Applications (LNCS) (vol. 5055, pp. 114-127). Berlin: Springer-Verlag. doi:10.1007/978-3-540-68982-9_9

Nechval, N. A., Berzins, G., Purgailis, M., & Nechval, K. N. (2008a). Improved estimation of state of stochastic systems via invariant embedding technique. *WSEAS Transactions on Mathematics*, *7*, 141–159.

Nechval, N. A., Nechval, K. N., & Purgailis, M. (2011a). Prediction of future values of random quantities based on previously observed data. *Engineering Letters*, *9*, 346–359.

Nechval, N. A., Nechval, K. N., & Vasermanis, E. K. (2004). Estimation of warranty period for structural components of aircraft. *Aviation*, *8*, 3–9.

Negnevitsky, M., Mandal, P., & Srivastava, A. (2009). Machine learning applications for load, price and wind power prediction in power systems. *Intelligent System Applications to Power Systems*, 1–6.

Negrete, C.C, Nájera, C.J. & Valenzuela, M.J.C. (2013). *Optimization of Cutting Parameters in Turning of AISI 1018 Steel With Constant Material Removal Rate Using Robust Design for Minimizing Cutting Power*. Paper No. IMECE2013-62520.

Nelson, R. B. (2006). An introduction to Copulas (2nd ed.). Springer.

Nelson, W.B. (2008). *Accelerated Testing: Statistical Models, Test Plans, & Data Analysis*. John Wiley & Sons, Inc.

Nelson, W. B. (2005). A Bibliography of Accelerated Test Plans. *IEEE Transactions on Reliability*, *54*(2), 194–197. doi:10.1109/TR.2005.847247

Nelson, W. B., & Kielpinski, T. J. (1976). Theory for Optimum Accelerated Life Tests for Normal & Lognormal Life Distributions. *Technometrics*, *18*(1), 105–114. doi:10.2307/1267923

Nelson, W. B., & Meeker, W. Q. (1978). Theory for Optimum Accelerated Censored Life Tests for Weibull & Extreme Value Distributions. *Technometrics*, *20*(2), 105–114. doi:10.1080/00401706.1978.10489643

Neumann, N. (2002). *Nichtlineare Effekte bei Längsschwingungen axial polarisierter piezokeramischer Staabe: Experimentelle Untersuchungen und Parameteridentifikation. Diplomarbeit, Institut für mechanik*. Technische Universität Darmstadt.

Nguyen, H. T., & Kreinovich, V. (2006). Optimization and Decision Making under Interval and Fuzzy Uncertainty: Towards New Mathematical Foundations. *Studies in Fuzziness and Soft Computing*, *201*, 275–290. doi:10.1007/3-540-33517-X_10

O'Halloran, B. M., Hoyle, C., Stone, R. B., & Tumer, I. Y. (2012). *The early design reliability prediction method*. Paper presented at the ASME 2012 International Mechanical Engineering Congress and Exposition. doi:10.1115/IMECE2012-89592

Ochieng, P. O. (2014). Mathematical analysis of tip speed ratio of a wind turbine and its effects on power coefficient. *International Journal of Mathematics and Soft Computing*, *4*, 61–66.

O'Connor, A. N. (2011). *Probability distributions used in reliability engineering*. RiAC.

Oh, H. S., Ammann, C. M., Naveau, P., Nychka, D., & Otto-Bliesner, B. L. (2003). Multi-resolution time series analysis applied to solar irradiance and climate reconstructions. *Journal of Atmospheric and Solar-Terrestrial Physics*, *65*(2), 191–201. doi:10.1016/S1364-6826(02)00291-2

Oliveira, E. A., Alvim, A. C. M., & Melo, P. F. (2005). Unavailability analysis of safety systems under aging by supplementary variables with imperfect repair. *Annals of Nuclear Energy*, *32*(2), 241–252. doi:10.1016/j.anucene.2004.06.003

Ozdor, N., Dulger, M., & Sher, E. (1994). *Cyclic variability in spark ignition engines a literature survey* (No. 940987). SAE Technical Paper.

Pagano, N. J. (1970). Exact solutions for rectangular bidirectional composites and sandwich plates. *Journal of Composite Materials*, *4*(1), 20–34.

Palanikumar, K. (2010). Modeling and analysis of delamination factor and surface roughness in drilling GFRP composites. *Materials and Manufacturing Processes*, *25*(10), 1059–1067. doi:10.1080/10426910903575830

Pant, S., & Singh, S. B. (2011). Particle Swarm Optimization to Reliability Optimization in Complex System. In *Proceeding of IEEE Int.Conf. on Quality and Reliability*. Doi:10.1109/ICQR.2011.6031711

Pant, S., Kumar, A., Kishor, A., Anand, D., & Singh, S. B. (2015). Application of a Multi-Objective Particle Swarm optimization Technique to Solve Reliability Optimization Problem. In *Proceeding of IEEE Int.Conf. on Next Generation Computing Technologies*, (pp. 1004-1007). Doi:10.1109/NGCT.2015.7375272

Pant, S., Anand, D., Kishor, A., & Singh, S. B. (2015). A Particle Swarm Algorithm for Optimization of Complex System Reliability. *International Journal of Performability Engineering*, *11*(1), 33–42.

Papanicolaou, A. (2009). *Taylor approximation and the delta method.* Retrieved May 3, 2015, from http://web.stanford.edu/class/cme308/OldWebsite/notes/TaylorAppDeltaMethod.pdf

Papoulis, A. (2014). *The Fourier integral and its applications*. Academic Press.

Paris, R., & Erdogan, F. (1963). A critical analysis of crack propagation laws. *Journal of Basic Engineering*, *85*(4), 528–534. doi:10.1115/1.3656900

Park, S. H. (1998). *Robust Design and Analysis for Quality Engineering*. London: Chapman & Hall.

Park, S., Han, Y., & Hahn, H. (2009). Fuzzy controller based biped robot balance control using 3D image. *International Journal of Fuzzy Systems*, *11*(3), 202–212.

Parmar, R. S. (2010). *Welding engineering and technology*. Khanna Publishers.

Participants, O. (2002). *Offshore reliability data handbook*. Trondheim, Norway: DNV.

Paszkowicz, W. (1996). Application of the Smooth Genetic Algorithm for Indexing Powder Patterns–Tests for the Orthorhombic System. *Materials Science Forum*, *228*, 19–24. doi:10.4028/www.scientific.net/MSF.228-231.19

Pavlyukevich, I. (2007). Lévy flights, non-local search and simulated annealing. *Journal of Computational Physics*, *226*(2), 1830–1844. doi:10.1016/j.jcp.2007.06.008

PAYACLUTCH Corp, . (2013). *QC Laboratory Database*. Rasht, Iran: Author.

Payne, R. B., Sorenson, M. D., & Klitz, K. (2005). *The Cuckoos*. Oxford University Press.

Peng, W. Y., & Liao, Y. S. (2004). Study of electrochemical discharge machining technology for slicing non-conductive brittle materials. *Journal of Materials Processing Technology*, *149*(1-3), 363–369. doi:10.1016/j.jmatprotec.2003.11.054

Percy, D. F. (2002). Bayesian enhanced strategic decision making for reliability. *European Journal of Engineering and System Safety*, *51*, 295–303.

Peters, B. D., & Borman, G. L. (1970). *Cyclic variations and average burning rates in a SI engine* (No. 700064). SAE Technical Paper.

Petit, J. R., Jouzel, J., Raynaud, D., Barkov, N. I., Barnola, J. M., Basile, I., & Stievenard, M. et al. (1999). Climate and atmospheric history of the past 420,000 years from the Vostok ice core, Antarctica. *Nature*, *399*(6735), 429–436. doi:10.1038/20859

Phelps, R. I. (1983). Optimal policy for minimal repair. *The Journal of the Operational Research Society*, *34*(5), 425–427. doi:10.1057/jors.1983.95

Piard, A., Gamby, D., Carbou, C., & Mendez, J. (2004). A numerical simulation of creep-fatigue crack growth in nickel-base super-alloys. *Engineering Fracture Mechanics*, *71*(16-17), 2299–2317. doi:10.1016/j.engfracmech.2004.02.002

Popovic, P., & Ivanovic, G. (2005). Monograph Reliability design of mechanical systems. Belgrade: 'VINCA' Institute of Nuclear Sciences.

Popovic, P., & Ivanovic, G. (2007). A methodology for the design of reliable vehicles in the concept stage. *Strojniski Vestnik*, *53*(3), 173–185.

Popovic, P., & Ivanovic, G. (2010). Methodology design for reliability power transmission systems of vehicles. *International Journal of Vehicle Design*, *54*(3), 262–283. doi:10.1504/IJVD.2010.036130

Popovic, P., Ivanovic, G., Mitrovic, R., & Subic, A. (2011). Design for reliability of a vehicle transmission system. *Proceedings of the Institution of Mechanical Engineers. Part D, Journal of Automobile Engineering*, *226*(2), 194–209. doi:10.1177/0954407011416175

Porter, A. (2004). *Accelerated Testing & Validation: Testing, Engineering, & Management Tools for Lean Development*. Elsevier.

Powell, D., Martins, E., Arlat, J., & Crouzet, Y. (1995). Estimators for fault tolerance coverage evaluation. *IEEE Transactions on Computers*, *44*(2), 261–274. doi:10.1109/12.364537

Pradyut, K. M., Manas, R. D., Samui, P., & Sarat, K. D. (2013). Uplift Capacity of Suction Caisson in Clay Using Artificial Intelligence Techniques. *Marine Georesources and Geotechnology*, *31*(4), 375–390. doi:10.1080/1064119X.2012.690827

Prasad, S. C., Rajagopal, K. R., & Rao, I. J. (2006). A continuum model for the anisotropic creep of single crystal nickel-based super-alloys. *Acta Materialia*, *54*(6), 1487–1500. doi:10.1016/j.actamat.2005.11.016

Pyragas, K. (1992). Control of chaos via extended delay feedback. *Physics Letters. [Part A]*, *170*, 421–428. doi:10.1016/0375-9601(92)90745-8

Qin, Z., Yu, F., Shi, Z., & Wang, Y. (2006). Adaptive inertia weight particle swarm optimization. In *Proceeding of International conference on Artificial Intelligence and Soft Computing,* (pp. 450-459).

Qu, Z. (1994). Global Stability of Trajectory Tracking of Robot under PD Control. *Dynamics and Control*, *4*(1), 59–71. doi:10.1007/BF02115739

Raghavan, A., Seshu, P., & Gandhi, P. S. (2003). Hysteresis modeling in piezoceramic actuator systems. *Proceedings of the Society for Photo-Instrumentation Engineers*, *5062*, 560–568. doi:10.1117/12.514647

Ragheb, M., & Adam, M. R. (2011). Wind turbines theory - The betz equation and optimal rotor tip speed ratio, fundamental and advanced topics in wind power. In C. Rupp (Ed.), *Fundamental and Advanced Topics in Wind Power*. doi:10.5772/21398

Rai, V. (2004). Chaos in natural populations: Edge or wedge? *Ecological Complexity*, *1*(2), 127–138. doi:10.1016/j.ecocom.2004.02.002

Rajasekar, S., & Lakshmanan, M. (1993). Algorithms for controlling chaotic motion: Application for the BVP oscillator. *Physica D. Nonlinear Phenomena*, *67*(1-3), 282–300. doi:10.1016/0167-2789(93)90211-I

Rajgopal, J., & Mazumdar, M. (1988). A type-II censored, log test time based, component-testing procedure for a parallel system. *Reliability. IEEE Transactions on*, *37*(4), 406–412.

Rajgopal, J., & Mazumdar, M. (1993). Design of system based component test plans.*Proceedings of 2nd Industrial Engineering Research Conference.*

Rajgopal, J., & Mazumdar, M. (1995). Designing component test plans for series system reliability via mathematical programming. *Technometrics*, *37*(2), 195–212. doi:10.1080/00401706.1995.10484304

Rajgopal, J., & Mazumdar, M. (1996). A system-based component test plan for a series system, with type-II censoring. *Reliability. IEEE Transactions on*, *45*(3), 375–378.

Rajgopal, J., & Mazumdar, M. (1997). Minimum cost component test plans for evaluating reliability of a highly reliable parallel system. *Naval Research Logistics*, *44*(5), 401–418. doi:10.1002/(SICI)1520-6750(199708)44:5<401::AID-NAV1>3.0.CO;2-8

Rajgopal, J., & Mazumdar, M. (1997). System-based component test plans for reliability inferences. *Frontiers in Reliability*, *1998*, 295–302.

Rajgopal, J., Mazumdar, M., & Majety, S. R. V. (1999). Optimum combined test plans for systems and components. *IIE Transactions*, *31*(6), 481–490. doi:10.1080/07408179908969851

Rajgopal, J., Mazumdar, M., & Savits, T. H. (1994). Some properties of the Poisson distribution with an application to reliability testing. *Probability in the Engineering and Informational Sciences*, *8*(03), 345–354. doi:10.1017/S0269964800003454

Ram, M., Singh, S. B., & Singh, V. V. (2013) Stochastic Analysis of a Standby System with Waiting Repair Strategy. *IEEE Transactions on Systems, Man, and Cybernetics, 43*(3), 698-707.

Ramachandran, K. M., & Tsokos, C. P. (2015). Design of experiment. In J. Cetel & A. Sambamoorthy (Eds.), *Mathematical Statistics with Applications in R* (2nd ed., pp. 459–494). London, UK: Elsevier. doi:10.1016/B978-0-12-417113-8.00009-6

Ramezani, M., Bashiri, M., & Atkinson, A. C. (2011). A goal programming-TOPSIS approach to multiple response optimization using the concepts of non-dominated solutions and prediction intervals. *Expert Systems with Applications*, *38*(8), 9557–9563. doi:10.1016/j.eswa.2011.01.139

Ramirez, J. A. (1994). Nonlinear feedback for controlling the Lorenz equation. *Physical Review E: Statistical Physics, Plasmas, Fluids, and Related Interdisciplinary Topics*, *50*, 2239–2342.

Ram, M. (2013). On system reliability approaches: A brief survey. *International Journal of System Assurance Engineering and Management*, *4*(2), 101–117. doi:10.1007/s13198-013-0165-6

Ram, M., & Manglik, M. (2014). Performance evaluation of a multi-state system covering imperfect fault coverage. *Communications in Statistics. Simulation and Computation*, *45*(9), 3259–3280. doi:10.1080/03610918.2014.932802

Ram, M., & Manglik, M. (2014). Stochastic behaviour analysis of a Markov model under multi-state failures. *International Journal of System Assurance Engineering and Management*, *5*(4), 686–699. doi:10.1007/s13198-014-0234-5

Ram, M., & Manglik, M. (2016). Stochastic Biometric System Modelling with Rework Strategy. *International Journal of Mathematical. Engineering and Management Sciences*, *1*(1), 1–17.

Ram, M., & Singh, S. B. (2010). Availability, M. T. T. F. and cost analysis of complex system under preemptive repeat repair discipline using Gumbel-Hougaard family of copula. *International Journal of Quality & Reliability Management*, *27*(3), 576–595. doi:10.1108/02656711011043544

Ram, M., Singh, S. B., & Singh, V. V. (2013). Stochastic analysis of a standby system with waiting repair strategy. *IEEE Transactions on Systems, Man, and Cybernetics*, *43*(3), 698–707.

Ram, M., Singh, S. B., & Singh, V. V. (2013). Stochastic analysis of a standby system with waiting repair strategy. *IEEE Transactions on Systems, Man, and Cybernetics: Systems*, *43*(3), 698–707.

Ram, M., Singh, S. B., & Varshney, R. G. (2013). Performance improvement of a parallel redundant system with coverage factor. *Journal of Engineering Science and Technology*, *8*(3), 344–350.

Rani, K. A. N., Malek, M. F. A. & Siew-Chin, N. (2012). Nature-inspired Cuckoo Search Algorithm for Side Lobe Suppression in a Symmetric Linear Antenna Array. *Radio Engineering*, *21*(3), 865 – 874.

Rao, R. V. (2011). *Advanced modeling and optimization of manufacturing processes international research and development*. London: Springer. doi:10.1007/978-0-85729-015-1

Rao, S. S., & Tjandra, M. (1994). Reliability-based design of automotive transmission systems. *Reliability Engineering & System Safety*, *46*(2), 159–169. doi:10.1016/0951-8320(94)90133-3

Ravindran, A. R. (2008). Multiple criteria decision making. In A. R. Ravindran (Ed.), *operation research methodologies* (pp. 128–135). Washington, DC: CRC press.

Ravi, V., Murty, B. S. N., & Reddy, J. (1997). Nonequilibrium Simulated-Annealing Algorithm Applied to Reliability Optimization of Complex Systems. *IEEE Transactions on Reliability*, *46*(2), 233–239. doi:10.1109/24.589951

Rayleigh, R. (1887). XXV. Notes on electricity and magnetism. —III. On the behaviour of iron and steel under the operation of feeble magnetic forces. *Philosophical Magazine*, *23*(142), 225–232. doi:10.1080/14786448708628000

Ray, M. C. (1998). Closed-form solution for optimal control of laminated plate. *Computers & Structures*, *69*(2), 283–290. doi:10.1016/S0045-7949(97)00093-X

Ray, M. C., Bhattacharya, R., & Samanta, B. (1993a). Exact solutions for static analysis of intelligent structures. *AIAA Journal*, *31*(9), 1684–1691. doi:10.2514/3.11831

Ray, M. C., Bhattacharya, R., & Samanta, B. (1998). Exact solutions for dynamic analysis of composite plates with distributed piezoelectric layers. *Computers & Structures*, *66*(6), 737–743. doi:10.1016/S0045-7949(97)00126-0

Ray, M. C., Rao, K. M., & Samanta, B. (1992). Exact analysis of coupled electro-elastic behaviour of a piezoelectric plate under cylindrical bending. *Computers & Structures*, *45*(4), 667–677. doi:10.1016/0045-7949(92)90485-I

Ray, M. C., Rao, K. M., & Samanta, B. (1993b). Exact solution for static analysis of an intelligent structure under cylindrical bending. *Computers & Structures*, *47*(6), 1031–1042. doi:10.1016/0045-7949(93)90307-Y

Raynaud, D., Barnola, J. M., Chappellaz, J., Blunier, T., Indermuhle, A., & Stauffer, B. (2000). The ice record of greenhouse gases: A view in the context of future changes. *Quaternary Science Reviews*, *19*(1-5), 9–17. doi:10.1016/S0277-3791(99)00082-7

ReliaSoft Corporation. (2012). *Blocksim 8 User's Guide*. Retrieved from: http://www.reliasoft.com

ReliaSoft Corporation. (2012). *Reliasoft ALTA 8 User's Guide*. Retrieved from: http://www.reliasoft.com

Reynolds, A. M., & Frye, M. A. (2007). Free-flight odor tracking in Drosophila is consistent with an optimal intermittent scale-free search. *PLoS ONE*, *2*(4), e354. doi:10.1371/journal.pone.0000354 PMID:17406678

Ribeiro, J. S., Teofilo, R. F., Augusto, F., & Ferreira, M. M. C. (2010). Simultaneous optimization of the micro extraction of coffee volatiles using response surface methodology and principal component analysis. *Chemometrics and Intelligent Laboratory Systems*, *102*(1), 45–52. doi:10.1016/j.chemolab.2010.03.005

Rohatgi, V. K., & Saleh, A. M. E. (2011). *An introduction to probability and statistics* (Vol. 910). John Wiley & Sons.

Rosen, S. C., Harmonosky, C. M., & Traband, M. T. (2007). A simulation optimization method that considers uncertainty and multiple performance measures. *European Journal of Operational Research*, *181*(1), 105–110. doi:10.1016/j.ejor.2006.05.040

Ross, S. M. (2014). *Introduction to probability and statistics for engineers and scientists*. Academic Press.

Routara, B. C., Mohanty, S. D., Datta, S., Bandyopadhyay, A., & Mahapatra, S. S. (2010). Combined quality loss (CQL) concept in WPCA-based Taguchi philosophy for optimization of multiple surface quality characteristics of UNS C34000 brass in cylindrical grinding. *International Journal of Advanced Manufacturing Technology*, *51*(1-4), 135–143. doi:10.1007/s00170-010-2599-1

Roy, S., & Chaudhuri, S. S. (2013). Cuckoo Search Algorithm using Lèvy Flight: A Review. *International Journal of Modern Education and Computer Science*, *12*(12), 10–15. doi:10.5815/ijmecs.2013.12.02

Ruxton, G. D. (1999). Chaos in a three-species food chain with a lower bound on the bottom population. *Ecology*, *77*(1), 317–319. doi:10.2307/2265680

Ryan, T. P. (2007). *Reliability Analysis & Life Testing, Modern Engineering Statistics*. John Wiley & Sons, Inc.

Sabnis, S. V., & Agnihothram, G. (2006). Application of arithmetic-geometric mean inequality for construction of reliability test plan for parallel systems in the presence of covariates. *Economic Quality Control*, *21*(2), 219–230. doi:10.1515/EQC.2006.219

Sabnis, S. V., & Agnihothram, G. (2007). Reliability test plans for series systems in the presence of covariates. *Economic Quality Control*, *22*(2), 197–209. doi:10.1515/EQC.2007.197

Sagdeo, P. M. (1987). *Frequency spectra of diesel engine heat release rate, pressure development and engine noise: an experimental investigation of cause and effect relationship* (No. 871620). SAE Technical Paper.

Sahu, J., Acharya, J., & Meikap, B. (2009). Response surface modeling and optimization of chromium (VI) removal from aqueous solution using Tamarind wood activated carbon in batch process. *Journal of Hazardous Materials*, *172*(2-3), 818–825. doi:10.1016/j.jhazmat.2009.07.075 PMID:19748729

Said, K. M., & EL-Sherbeny, M. S. (2005). Profit Analysis of a Two Unit Cold Standby System with Preventive Maintenance and Random Change in Units. *Journal of Mathematics and Statistics*, *5*(4), 71-77.

Salazar, D., Rocco, C. M., & Galvan, B. J. (2006). Optimization of constrained multiple-objective reliability problems using evolutionary algorithms. *Reliability Engineering & System Safety*, *91*(9), 1057–1070. doi:10.1016/j.ress.2005.11.040

Salmasnia, A., Baradaran Kazemzadeh, R., & Niaki, S. T. A. (2012). An approach to optimize correlated multiple responses using principal component analysis and desirability function. *International Journal of Advanced Manufacturing Technology*, *62*(5-8), 1–12. doi:10.1007/s00170-011-3824-2

Saltelli, A., Ratto, M., Andres, T., Campolongo, F., Cariboni, J., Gatelli, D., & Tarantola, S. (2008). *Global sensitivity analysis: the primer*. John Wiley and Sons.

Samal, M. K. (2003). *Modelling of nonlinear behaviour of piezoelectric materials under weak electric fields* (Masters Thesis). Department of Mechanical Engineering, Indian Institute of Technology, Bombay, India.

Samal, M. K. (2015). (in press). Simulation of material stress-strain curve and creep deformation response of nickel based superalloys using crystal plasticity based finite element models. *Transactions of the Indian Institute of Metals*. doi:10.1007/s12666-015-0591-z

Samal, M. K., Seshu, P., & Dutta, B. K. (2007). An analytical formulation in 3D domain for the nonlinear response of piezoelectric slabs under weak electric fields. *International Journal of Solids and Structures*, *44*(14-15), 4656–4672. doi:10.1016/j.ijsolstr.2006.11.045

Samal, M. K., Seshu, P., Parashar, S. K., von Wagner, U., Hagedorn, P., Dutta, B. K., & Kushwaha, H. S. (2006a). Nonlinear Behaviour of Piezoceramics under Weak Electric Fields, Part-I: 3-D Finite Element formulation. *International Journal of Solids and Structures*, *43*(6), 1422–1436. doi:10.1016/j.ijsolstr.2005.06.084

Samal, M. K., Seshu, P., Parashar, S. K., von Wagner, U., Hagedorn, P., Dutta, B. K., & Kushwaha, H. S. (2006b). Nonlinear Behaviour of Piezoceramics under Weak Electric Fields, Part-II: Numerical results and validation with experiments. *International Journal of Solids and Structures*, *43*(6), 1437–1458. doi:10.1016/j.ijsolstr.2005.06.067

Samar, V. J., Swartz, K. P., & Raghuveer, M. R. (1995). Multiresolution analysis of event-related potentials by wavelet decomposition. *Brain and Cognition*, *27*(3), 398–438. doi:10.1006/brcg.1995.1028 PMID:7626282

Sankaya, M., & Guilu, A. (2015). Multi-response optimization of minimum quantity lubrication parameters using Taguchi-based grey relational analysis in turning of difficult-to-cut alloy Haynes 25. *Journal of Cleaner Production*, *91*, 347–357. doi:10.1016/j.jclepro.2014.12.020

Santos, P., Luisa, F. V., Aníbal, R., Andres, B., & Maudes, J. (2015). An SVM-Based Solution for Fault Detection in Wind Turbines. *Sensors (Basel, Switzerland)*, *15*(3), 5627–5648. doi:10.3390/s150305627 PMID:25760051

Sarkhel, Banerjee, & Hui. (2013). *Analysis and Control of a Six Link Serial Manipulator with Flexible Joints*. 142-Inacomm.

Sarosi, P. M., Srinivasan, R., Eggeler, G. F., Nathal, M. V., & Mills, M. J. (2007). Observations of *a*<010> dislocations during the high-temperature creep of Ni-based super-alloy single crystals deformed along the [001] orientation. *Acta Materialia*, *55*(7), 2509–2518. doi:10.1016/j.actamat.2006.11.045

Scheaffer, R. L. (1971). Optimum age replacement policies with an increasing cost factor. *Technometrics*, *13*(1), 139–144. doi:10.2307/1267081

Schenkelberg, F. (2012). *Effective reliability program traits and management*. Paper presented at the Reliability and Maintainability Symposium.

Schneising, O., Buchwitz, M., Burrows, J. P., Bovensmann, H., Bergamaschi, P., & Peters, W. (2009). Three years of greenhouse gas column-averaged dry air mole fractions retrieved from satellite – Part 2: Methane. *Atmospheric Chemistry and Physics*, *9*(2), 443–465. doi:10.5194/acp-9-443-2009

Scribner. (n.d.). *Electrochemical Impedance Spectroscopy (EIS): A Powerful and Cost-Effective Tool for Fuel Cell Diagnostics*. Scribner. Retrieved from: http://www.scribner.com/files/tech-papers/Scribner%20Associates%20-%20Electrochemical%20Impedance%20Spectroscopy%20for%20Fuel%20Cell%20Research.pdf

Sczomak, D. P., & Henein, N. A. (1979). *Cycle-to-cycle variation with low ignition quality fuels in a CFR diesel engine* (No. 790924). SAE Technical Paper.

Sen, A. K., Litak, G., Edwards, K. D., Finney, C. E., Daw, C. S., & Wagner, R. M. (2011). Characteristics of cyclic heat release variability in the transition from spark ignition to HCCI in a gasoline engine. *Applied Energy*, *88*(5), 1649–1655. doi:10.1016/j.apenergy.2010.11.040

Sen, A. K., Litak, G., Finney, C. E., Daw, C. S., & Wagner, R. M. (2010). Analysis of heat release dynamics in an internal combustion engine using multifractals and wavelets. *Applied Energy*, *87*(5), 1736–1743. doi:10.1016/j.apenergy.2009.11.009

Sen, A. K., Litak, G., Taccani, R., & Radu, R. (2008). Wavelet analysis of cycle-to-cycle pressure variations in an internal combustion engine. *Chaos, Solitons, and Fractals*, *38*(3), 886–893. doi:10.1016/j.chaos.2007.01.041

Sen, A. K., Longwic, R., Litak, G., & Górski, K. (2008). Analysis of cycle-to-cycle pressure oscillations in a diesel engine. *Mechanical Systems and Signal Processing*, *22*(2), 362–373. doi:10.1016/j.ymssp.2007.07.015

Sen, A. K., Wang, J., & Huang, Z. (2011). Investigating the effect of hydrogen addition on cyclic variability in a natural gas spark ignition engine: Wavelet multiresolution analysis. *Applied Energy*, *88*(12), 4860–4866. doi:10.1016/j.apenergy.2011.06.030

Sen, A. K., Zheng, J., & Huang, Z. (2011). Dynamics of cycle-to-cycle variations in a natural gas direct-injection spark-ignition engine. *Applied Energy*, *88*(7), 2324–2334. doi:10.1016/j.apenergy.2011.01.009

Sengupta, B. (1977). Inspection procedures when failure symptoms are delayed. *Operational Research Quarterly*, *28*(3-part-ii), 768–776. doi:10.1287/opre.28.3.768

Shah, H. K., Montgomery, D. C., & Carlyle, W. M. (2004). Response Surface Modeling and Optimization in Multiresponse Experiments Using Seemingly Unrelated Regressions. *Quality Engineering*, *16*(3), 387–397. doi:10.1081/QEN-120027941

Sharirli, M. (1985). *Methodology for system analysis using fault trees, success trees, and importance evaluations*. College Park, MD: Maryland Univ.

Shoji, T. (1997). Effect of Cycle-to-Cycle Variations in Spray Characteristics on Hydrocarbon Emission in DI Diesel Engines (Visualization of Sac Inner Flow, Needle Valve Motion and Cycle-to-Cycle Variations in Diesel Spray). *JSME International Journal Series B*, *40*(2), 312–319. doi:10.1299/jsmeb.40.312

Shvidenko, A., Schepaschenko, D., McCallum, I., & Nilsson, S. (2010). Can the uncertainty of full carbon acconting of forest ecosystem be made acceptable to policymakers? *Climatic Change*, *103*(1-2), 137–157. doi:10.1007/s10584-010-9918-2

Sifuzzaman, M., Islam, M. R., & Ali, M. Z. (2009). *Application of wavelet transform and its advantages compared to Fourier transform*. Academic Press.

Simens, V. (2003). *Worldwide emission standards and related regulations*. Siemens VDO.

Singh, H., & Kumar, P. (2006). Optimizing multi-machining characteristics through Taguchis approach and utility concept. *Journal of Manufacturing Technology Management*, *17*(2), 255–274. doi:10.1108/17410380610642304

Singh, V. V., Ram, M., & Rawal, D. K. (2013). Cost analysis of an engineering system involving subsystems in series configuration. *IEEE Transactions on Automation Science and Engineering*, *10*(4), 1124–1130. doi:10.1109/TASE.2013.2266134

Sitter, R. R., & Forbes, B. (1997). Optimal two-stage designs for binary response experiments. *Statistica Sinica*, *7*, 941–956.

Sitter, R. R., & Wu, C. F. J. (1999). Two-stage design of quantal response studies. *Biometrics*, *55*(2), 396–402. doi:10.1111/j.0006-341X.1999.00396.x PMID:11318192

Small, R. F., & Wagner, C. (2003). Economic Analysis of Boiler Tube Failures in NSW. *OMMI*, *2*(2), 1–20.

Smith, S. A., & Oren, S. S. (1980). Reliability growth of repairable systems. *Naval Research Logistics Quarterly*, *27*(4), 539–547. doi:10.1002/nav.3800270403

Smola, A., & Scholkopf, B. (1998). *A tutorial on support vector regression*. NeuroCOLT Technical Report NC-TR-98-030, Royal Holloway College, University of London, UK.

Smola, A., & Scholkopf, B. (1998). On a kernel based method for pattern recognition, regression, approximation and operator inversion. *Algorithmica*, *22*(1-2), 211–231. doi:10.1007/PL00013831

Soleimani, M., & Pourgol-Mohammad, M. (2014). Design for reliability of complex system with limited failure data; case study of a horizontal drilling equipment.*Proceedings of the Probabilistic Safety Assessment and Management PSAM*. doi:10.1155/2014/524742

Somashekhar, K. P., Ramachandran, N., & Mathew, J. (2010). Optimization of Material Removal Rate in Micro-EDM Using Artificial Neural Network and Genetic Algorithms. *Materials and Manufacturing Processes*, *25*(6), 467–475. doi:10.1080/10426910903365760

Sorkhel, S. K., Bhattacharyya, B., Mitra, S., & Doloi, B. 1996. Development of electrochemical discharge machining technology for machining of advanced ceramics.*International Conference on Agile Manufacturing*, (pp. 98-103).

Soula, A., Renollet, Y., Boivin, D., Pouchou, J.-L., Locq, D., Caron, P., & Bréchet, Y. (2009). Analysis of high-temperature creep deformation in a polycrystalline nickel-base super-alloy. *Materials Science and Engineering A*, *510-511*, 301–306. doi:10.1016/j.msea.2008.04.122

Spuchľáková, E., & Cúg, J. (2014). Lost Given Default and the Credit risk. In *Proceedings of ICMEBISInternational Conference on Management, Education, Business, and Information Science.* EDU Gait Press.

Srinivas, N., & Deb, K. (1995). Multi-objective optimization using non dominated sorting in genetic algorithms. *Evolutionary Computation*, *3*, 221–248.

Stamatios, V. K. (1995). *Understanding Neural Networks and Fuzzy Logic: Basic Concepts and Applications*. IEEE Press.

Stone, C. R., Brown, A. G., & Beckwith, P. (1996). *Cycle-by-cycle variations in spark ignition engine combustion-part ii: Modelling of flame kernel displacements as a cause of cycle-by-cycle variations* (No. 960613). SAE Technical Paper.

Strahler, A. H., Friedl, M., Zhang, X., Hodges, J., Cooper, C. S. A., & Baccini, A. (2002). The MODIS Land Cover and Land Cover Dynamics Products. Presentation at Remote Sensing of the Earth's Environment from TERRA, L'Aquila, Italy

Strahle, W. C. (1977). Combustion randomness and diesel engine noise: Theory and initial experiments. *Combustion and Flame*, *28*, 279–290. doi:10.1016/0010-2180(77)90033-5

Straub, D., & Faber, M. H. (2005). Risk based inspection planning for structural systems. *Structural Safety*, *27*(4), 335–355. doi:10.1016/j.strusafe.2005.04.001

Strogatz, S. H. (1994). *Nonlinear Dynamics and Chaos*. Perseus Books Publishing.

Sugui, T., Jun, X., Xiaoming, Z., Benjiang, Q., Jianwei, L., Lili, Y., & Wuxiang, W. (2011). Microstructure and creep behavior of FGH95 nickel-base super-alloy. *Materials Science and Engineering A*, *528*(4-5), 2076–2084. doi:10.1016/j.msea.2010.11.038

Suhail, A. H., Ismail, N., Wong, S. V., & Abdul, N. A. (2010). Optimization of cutting parameters based on surface roughness and assistance of workpiece surface temperature in turning process. *American Journal of Engineering and Applied Sciences.*, *3*(1), 102–108. doi:10.3844/ajeassp.2010.102.108

Suik, H., & Pihu, T. (2009). *Warranty reliability of CFB boiler burning oil shale*. Academic Press.

Sun, H., & Lee, S. (2005). Response surface approach to aerodynamic optimization design of helicopter rotor blade. *International Journal for Numerical Methods in Engineering*, *64*(1), 125–142. doi:10.1002/nme.1391

Swinehart, K., Yasin, M., & Guimaraes, E. (1996). Applying an analytical approach to shop-floor scheduling: A case study. *International Journal of Materials and Product Technology*, *1*, 98–107.

Tadikamalla, P. R. (1979). An inspection policy for the gamma failure distributions. *Operational Research Quarterly*, *30*(1), 77–78. doi:10.1057/jors.1979.9

Taguchi, G. (1990). *Introduction to Quality Engineering*. Tokyo: Asian Productivity Organization.

Taguchi, T., Ida, K., & Gen, M. (1997). Method for Solving Nonlinear Goal Programming Tasks with Interval Coefficients Using GA. *Computers & Industrial Engineering*, *33*(4), 597–600. doi:10.1016/S0360-8352(97)00201-5

Tahara, A., & Nishida, T. (1975). Optimal replacement policy for minimal repair model. *Journal of the Operations Research Society of Japan*, *18*, 113–124.

Tang, H., & Li, Y. (2015). Feed forward nonlinear PID control of a novel micromanipulator using Preisach hysteresis compensator. *Robotics and Computer-integrated Manufacturing*, *34*, 124–132. doi:10.1016/j.rcim.2014.11.006

Tang, L. C., & Xu, K. (2002). A unified approach for dual response optimization. *Journal of Quality Technology*, *34*(4), 437–447.

Teixeira, C. A. R., & Cavalca, K. L. (2008). Reliability as an added-value factor in an automotive clutch system. *Quality and Reliability Engineering International*, *24*(2), 229–248. doi:10.1002/qre.889

Terninko, J. (1997). *Step-by-Step QFD: Customer-Driven Product Design* (2nd ed.). Taylor & Francis.

Thakkar, D., & Ganguli, R. (2004). Helicopter vibration reduction in forward flight with induced shear based piezoceramic actuation. *Smart Materials and Structures*, *13*(3), 599–608. doi:10.1088/0964-1726/13/3/019

Thakkar, D., & Ganguli, R. (2006). Use of single crystal and soft piezoceramics for alleviation of flow separation induced vibration in smart helicopter rotor. *Smart Materials and Structures*, *15*(2), 331–341. doi:10.1088/0964-1726/15/2/013

Thakkar, D., & Ganguli, R. (2007). Induced shear actuation of helicopter rotor blade for active twist control. *Thin-walled Structures*, *45*(1), 111–121. doi:10.1016/j.tws.2006.11.001

Tillman, F. A., Hwang, C. L., & Kuo, W. (1980). Optimization of systems reliability. New York: Marcel Dekker.

Tillman, F. A., Hwang, C. L., & Kuo, W. (1977). Optimization Techniques for System Reliability with Redundancy: A Review. *IEEE Transactions on Reliability*, *26*(3), 148–155.

Tillman, F. A., Hwang, C. L., & Kuo, W. (1980). *Optimization of Systems Reliability*. New York: Marcel Dekker Inc.

Tipping, M. E. (2000). *The relevance vector machine.* In S. A. Solla, T. K. Leen, & K. R. Müller (Eds.), Advances in Neural Information Processing Systems (Vol. 12, pp. 652–658). Cambridge, MA: MIT Press.

Tipping, M. E. (2001). Sparse bayesian learning and the relevance vector machine. *Journal of Machine Learning Research*, *1*, 211–244.

Tong, L. I., Wang, C. H., & Chen, H. C. (2005). Optimization of multiple responses using Principal component analysis and technique for order preference by Similarity to ideal solution. *International Journal of Advanced Manufacturing Technology*, *27*(3), 407–414. doi:10.1007/s00170-004-2157-9

Torrence, C., & Compo, G. P. (1998). A practical guide to wavelet analysis. *Bulletin of the American Meteorological Society*, *79*(1), 61–78. doi:10.1175/1520-0477(1998)079<0061:APGTWA>2.0.CO;2

Torster, F., Baumeister, G., Albrecht, J., Lütjering, G., Helm, D., & Daeubler, M. A. (1997). Influence of grain size and heat treatment on the microstructure and mechanical properties of the nickel-base super-alloy U 720 LI. *Materials Science and Engineering A*, *234-236*, 189–192. doi:10.1016/S0921-5093(97)00161-5

Tu, J. V. (1996). Advantages and disadvantages of using artificial neural networks versus logistic regression for predicting medical outcomes. *Journal of Clinical Epidemiology*, *49*(11), 1225–1231. doi:10.1016/S0895-4356(96)00002-9 PMID:8892489

Tzeng, C. J., Lin, Y. H., Yang, Y. K., & Jeng, M. C. (2009). Optimization of turning operations with multiple performance characteristics using the Taguchi method and Grey relational analysis. *Journal of Materials Processing Technology*, *209*(6), 2753–2759. doi:10.1016/j.jmatprotec.2008.06.046

Tzikas, D., Wei, L., Likas, A., Yang, Y., & Galatsanos, N. (2006). A tutorial on relevance vector machines for regression and classification with applications. *EURASIP Newsletter*, *17*(2), 1–25.

Usher, T. & Sim, A. (2005). Nonlinear dynamics of piezoelectric high displacement actuators in cantilever mode. *Journal of Applied Physics, 98*(6).

Valdez-Flores, C., & Feldman, R. M. (1989). A survey of preventive maintenance models for stochastically deteriorating single-unit systems. *Naval Research Logistics*, *36*(4), 419–446. doi:10.1002/1520-6750(198908)36:4<419::AID-NAV3220360407>3.0.CO;2-5

Valian, E., Mohanna, S., & Tavakoli, S. (2011). a). Improved Cuckoo Search Algorithm for Global Optimization. *International Journal of Communications and Information Technology*, *1*(1), 31–44.

Valian, E., Mohanna, S., & Tavakoli, S. (2011). b). Improved Cuckoo Search Algorithm For Feedforward Neural Network Training. *International Journal of Artificial Intelligence & Applications*, *2*(3), 36–43. doi:10.5121/ijaia.2011.2304

Vapnik, V. N. (1998). *Statistical learning theory*. New York: Wiley.

Vapnik, V. N. (1999). *The nature of statistical learning theory* (2nd ed.). Berlin: Springer.

Vellaisamy, P., & Kumar, M. (2008). Optimal component test plans for a parallel system based on type-II censoring. *Statistical Methodology*, *5*(5), 454–461. doi:10.1016/j.stamet.2007.11.001

Vellaisamy, P., & Kumar, M. (2010). Design of optimal reliability test plan for a series system based on mixed-censoring. *Journal of Applied Statistical Science*, *17*, 1–10.

Venkataramani, G., Deka, D., & Ghosh, S. (2006). Crystal plasticity based FE model for understanding microstructural effects on creep and dwell fatigue in Ti-6242. ASME. *Journal of Engineering Materials and Technology*, *128*(3), 356–365. doi:10.1115/1.2204942

Venkataramani, G., Ghosh, S., & Mills, M. J. (2007). A size dependent crystal plasticity finite element model for creep and load-shedding. *Acta Materialia*, *55*(11), 3971–3986. doi:10.1016/j.actamat.2007.03.017

Venkataramani, G., Kirane, K., & Ghosh, S. (2008). Microstructural parameters affecting creep induced load shedding in Ti-6242 by a size dependent crystal plasticity FE model. *International Journal of Plasticity*, *24*(3), 428–454. doi:10.1016/j.ijplas.2007.05.001

Vining, G. G. (1998). A compromise approach to multiresponse optimization. *Journal of Quality Technology*, *30*(4), 309–313.

Virkler, D. A., Hillberry, B. M., & Goel, P. K. (1979). The statistic nature of fatigue crack propagation. *ASME Journal of Engineering Materials and Technology*, *101*(2), 148–153. doi:10.1115/1.3443666

Viswamurthy, S. R., & Ganguli, R. (2007). Modeling and analysis of piezoceramic actuator hysteresis for helicopter vibration control. *Sensors and Actuators. A, Physical*, *35*(2), 801–810. doi:10.1016/j.sna.2006.09.020

Viswamurthy, S. R., Rao, A. K., & Ganguli, R. (2007). Dynamic hysteresis of piezoceramic stack actuators used in helicopter vibration control: Experiments and simulations. *Smart Materials and Structures*, *16*(4), 1109–1119. doi:10.1088/0964-1726/16/4/020

Viswanathan, G. B., Sarosi, P. M., Whitis, D. H., & Mills, M. J. (2005). Deformation mechanisms at intermediate creep temperatures in the Ni-base super-alloy Rene 88 DT. *Materials Science and Engineering A*, *400*, 489–495. doi:10.1016/j.msea.2005.02.068

Volpe, R. A., & Khosla, P. K. (1990). Manipulator control with super quadric artificial potential functions: Theory and experiments. *IEEE Transactions on Systems, Man, and Cybernetics*, *20*(6), 1423–1436. doi:10.1109/21.61211

von Wagner, U. (2003). Non-linear longitudinal vibrations of piezoceramics excited by weak electric fields. *International Journal of Non-linear Mechanics*, *38*(4), 565–574. doi:10.1016/S0020-7462(01)00113-5

von Wagner, U. (2004). Non-linear longitudinal vibrations of non-slender piezoceramic rods. *International Journal of Non-linear Mechanics*, *39*(4), 673–688. doi:10.1016/S0020-7462(03)00108-2

von Wagner, U., & Hagedorn, P. (2002). Piezo-beam-systems subjected to weak electric field: Experiment and modelling of nonlinearities. *Journal of Sound and Vibration*, *256*(5), 861–872. doi:10.1006/jsvi.2002.5024

Vukobratovic, M., & Juricic, D. (1969). Contribution to the synthesis of biped gait. *Biomedical Engineering, IEEE Transactions on*, (1), 1-6.

Vukobratovic, M., & Timcenko, O. (1995). Stability analysis of certain class of bipedal walking robots with hybridization of classical and fuzzy control. In *Proc. International Conference on Advanced Robotics and Intelligent Automation* (pp. 290-295).

Wang, H., & Pham, H. (2006). *Reliability and optimal maintenance*. Springer Science & Business Media.

Wang, K. H., Yen, T. C., & Jian, J. J. (2013). Reliability and sensitivity analysis of a repairable system with imperfect coverage under service pressure condition. *Journal of Manufacturing Systems*, *32*(2), 357–363. doi:10.1016/j.jmsy.2013.01.004

Wang, Q.-M., Zhang, Q., Xu, B., Liu, R., & Cross, L. E. (1999). Nonlinear piezoelectric behavior of ceramic bending mode under strong electric fields. *Journal of Applied Physics*, *86*(6), 3352–3360. doi:10.1063/1.371213

Wang, Y., Tian, Y., Zhang, Y., El-Saleous, N., Knyazikhin, Y., Vermote, E., & Myneni, R. B. (2001). Investigation of product accuracy as a function of input and model uncertainties: Case study with SeaWiFS and MODIS LAI/FPAR Algorithm. *Remote Sensing of Environment*, *78*(3), 296–311. doi:10.1016/S0034-4257(01)00225-5

Wang, Z. H., Meng, J., & Fu, J. (2010). Novel method for evaluating surface roughness by grey dynamic filtering. *Measurement.*, *43*(1), 78–82. doi:10.1016/j.measurement.2009.06.008

Wasserman, G. S. (2002). *Reliability Verification, Testing, & Analysis in Engineering Design*. Marcel Dekker, Inc. doi:10.1201/9780203910443

Weldon, R. (1969, September). Paper 3: Boiler Feed Pump Design for Maximum Availability. In *Proceedings of the Institution of Mechanical Engineers, Conference Proceedings* (*Vol. 184*, No. 14, pp. 18-35). SAGE Publications. doi:10.1243/PIME_CONF_1969_184_421_02

White, T., Jonas, M., Nahorski, Z., & Nilsson, S. (Eds.). (2011). *Greenhouse gas inventories. Dealing with uncertainties*. Heidelberg, Germany: Springer.

Whitney, D. (1969). Resolved motion rate control of manipulators and human prostheses. *IEEE Transactions on Man-Machine Systems, 10*(2), 47–53. doi:10.1109/TMMS.1969.299896

Wing, R. D. (1975). The rotary fuel-injection pump as a source of cyclic variation in diesel engines, and its effect on nitric oxide emissions. *Proceedings - Institution of Mechanical Engineers, 189*(1), 497–505. doi:10.1243/PIME_PROC_1975_189_061_02

Wolf, K., & Gottlieb, O. (2001). *Nonlinear dynamics of a cantilever beam actuated by piezoelectric layers in symmetric and asymmetric configuration, Technical Report, No. ETR-2001-02*. Faculty of Mechanical Engineering, Israel Institute of Technology.

Wolschendorf, J., Dürnholz, M., & Schmillen, K. (1991). *The IDI diesel engine and its combustion noise variations* (No. 910228). SAE Technical Paper.

Wong, C. C., Cheng, C. T., Huang, K. H., & Yang, Y. T. (2008). Fuzzy control of humanoid robot for obstacle avoidance. *International Journal of Fuzzy Systems, 10*(1), 1.

Wong, C. C., Hwang, C. L., Huang, K. H., Hu, Y. Y., & Cheng, C. T. (2011). Design and implementation of vision-based fuzzy obstacle avoidance method on humanoid robot. *International Journal of Fuzzy Systems, 13*(1), 45–54.

Wong, W. K. (1994). Comparing robust properties of A, D, E and G-optimal designs. *Computational Statistics & Data Analysis, 18*(4), 441–448. doi:10.1016/0167-9473(94)90161-9

Wu, C. F. J. (2005). Optimization of correlated multiple quality characteristics using desirability function. *Quality Engineering, 17*(1), 119–126. doi:10.1081/QEN-200028725

Wu, S., & Liu, Q. (2005). Some problems on the global wavelet spectrum. *Journal of Ocean University of China, 4*(4), 398–402. doi:10.1007/s11802-005-0062-y

Xiao, N., Huang, H.-Z., Li, Y., He, L., & Jin, T. (2011). Multiple failure modes analysis and weighted risk priority number evaluation in FMEA. *Engineering Failure Analysis, 18*(4), 1162–1170. doi:10.1016/j.engfailanal.2011.02.004

Xie, L., Zhou, J., Wang, Y., & Wang, X. (2005). Load-Strength Order Statistics Interference Models. *International Journal of Performability Engineering, 1*(1), 23.

Xing, L. (2007). Reliability evaluation of phased-mission systems with imperfect fault coverage and common-cause failures. *IEEE Transactions on Reliability, 56*(1), 58–68. doi:10.1109/TR.2006.890900

Xiong, C. (1998). Inference on a Simple Step-stress Model with Type-II Censored Exponential Data. *IEEE Transactions on Reliability, 47*(2), 142–146. doi:10.1109/24.722278

Xiong, C., & Ji, M. (2004). Analysis of Grouped & Censored Data from Step-stress Life Testing. *IEEE Transactions on Reliability, 53*(1), 22–28. doi:10.1109/TR.2004.824832

Xiong, C., & Milliken, G. A. (1999). Step-stress Life-testing with R&om Stress-change Times for Exponential Data. *IEEE Transactions on Reliability, 48*(2), 141–148. doi:10.1109/24.784272

Xu, K. M., Noor, A. K., & Tang, Y. (1995). Three-dimensional solutions for coupled thermoelectroelastic response of multilayered plates. *Computer Methods in Applied Mechanics and Engineering, 126*(3-4), 355–371. doi:10.1016/0045-7825(95)00825-L

Yadava, V., Jain, V. K., & Dixit, P. M. (2002). Temperature Distribution during Electro-Discharge Abrasive Grinding. *Machinin. Science and Technology, 6*(1), 97–127.

Yaghoobi, M. A., Jones, D. F., & Tamiz, M. (2008). Weighted additive models for solving fuzzy goal programming problems. *Asia-Pacific Journal of Operational Research*, *25*(5), 715–733. doi:10.1142/S0217595908001973

Yaghoobi, M. A., & Tamiz, M. (2007). A method for solving fuzzy goal programming problems based on MINMAX approach. *European Journal of Operational Research*, *177*(3), 1580–1590. doi:10.1016/j.ejor.2005.10.022

Yalaoui, A., Chu, C., & Châtelet, E. (2005). Reliability allocation problem in a series–parallel system. *Reliability Engineering & System Safety*, *90*(1), 55–61. doi:10.1016/j.ress.2004.10.007

Yamashita, M., & Kakehi, K. (2006). Tension/compression asymmetry in yield and creep strengths of Ni-based superalloy with a high amount of tantalum. *Scripta Materialia*, *55*(2), 139–142. doi:10.1016/j.scriptamat.2006.03.048

Yang, J.N., His, W.H., & Manning, S.D. (1985). *Stochastic crack propagation with applications to durability and damage tolerance analyses.* Technical Report, Flight Dynamics Laboratory, Wright-Patterson Air Force Base, AFWAL-TR-85-3062.

Yang, X. S., & Deb, S. (2009). Cuckoo search via Levy flights. In *Proceeding of World Congress on Nature & Biologically Inspired Computing (NaBIC),* (pp. 210-214).

Yang, X. S., & Deb, S. (2009). Cuckoo Search via Lévy Flights. *World Congress on Nature & Biologically Inspired Computing (NaBIC 2009),* (pp. 210-214). IEEE Publications. doi:10.1109/NABIC.2009.5393690

Yang, G. (2007). *Life Cycle Reliability Engineering*. Wiley; doi:10.1002/9780470117880

Yang, G. B. (1994). Optimum Constant-stress Accelerated Life Test Plans. *IEEE Transactions on Reliability*, *43*(4), 575–581. doi:10.1109/24.370223

Yang, J. N., & Manning, S. D. (1990). Stochastic crack growth analysis methodologies for metallic structures. *Engineering Fracture Mechanics*, *37*(5), 1105–1124. doi:10.1016/0013-7944(90)90032-C

Yang, X. S., & Deb, S. (2010). Engineering optimization by cuckoo search. *Int. J. Mathematical Modelling and Numerical Optimization*, *1*(4), 330–343. doi:10.1504/IJMMNO.2010.035430

Yan, J. H., & Mazumdar, M. (1986). A comparison of several component-testing plans for a series system. *Reliability. IEEE Transactions on*, *35*(4), 437–443.

Yan, J. H., & Mazumdar, M. (1987). A comparison of several component-testing plans for a parallel system. *Reliability. IEEE Transactions on*, *36*(4), 419–424.

Yan, J., Yongqian, L., Shuang, H., & Meng, Q. (2013). Wind power grouping forecasts and its uncertainty analysis using optimized relevance vector machine. *Renewable & Sustainable Energy Reviews*, *27*, 613–621. doi:10.1016/j.rser.2013.07.026

Yeo, K. P., & Tang, L. C. (1999). Planning step-stress life-test with a target acceleration-factor. *IEEE Transactions on Reliability*, *48*(1), 61–67. doi:10.1109/24.765928

Yurdusev, M. A., Ata, R., & Cetin, N. S. (2006). Assessment of optimum tip speed ratio in wind turbines using artificial neural networks. *Energy*, *31*(12), 2153–2161. doi:10.1016/j.energy.2005.09.007

Yusoff, N., Ramasamy, M., & Yusup, S. (2011). Taguchis parametric design approach for the selection of optimization variables in a refrigerated gas plant. *Chemical Engineering Research & Design*, *89*(6), 665–675. doi:10.1016/j.cherd.2010.09.021

Zhang, G., Zhang, Z., Ming, W., Guo, J., Huang, Y., & Shao, X. (2014). The multi-objective optimization of medium-speed WEDM process parameters for machining SKD11 steel by the hybrid method of RSM and NSGA-II. *International Journal of Advanced Manufacturing Technology*, *70*(9-12), 2097–2109. doi:10.1007/s00170-013-5427-6

Zhan, Z., Fernando, U. S., & Tong, J. (2008). Constitutive modelling of viscoplasticity in a nickel-based super-alloy at high temperature. *International Journal of Fatigue*, *30*(7), 1314–1323. doi:10.1016/j.ijfatigue.2007.06.010

Zhao, J., & Badler, N. (1994). Inverse kinematics positioning using nonlinear programming for highly articulated figures. *ACM Transactions on Computer Graphics*, *13*(4), 313–336. doi:10.1145/195826.195827

Zhao, M., Heinsch, F. A., Nemani, R. R., & Running, S. W. (2005). Improvements of the MODIS terrestrial gross and net primary production global data set. *Remote Sensing of Environment*, *95*(2), 164–176. doi:10.1016/j.rse.2004.12.011

Zhao, M., Running, S. W., & Nemani, R. R. (2006). Sensitivity of Moderate Resolution Imaging Spectroradiometer (MODIS) terrestrial primary production to the accuracy of meteorological reanalyses. *Journal of Geophysical Research*, *111*(G1). doi:10.1029/2004JG000004

Zhao, X., Al-Khalifa, K. N., & Nakagawa, T. (2015). Approximate methods for optimal replacement, maintenance, and inspection policies. *Reliability Engineering & System Safety*, *144*, 68–73. doi:10.1016/j.ress.2015.07.005

Zhou, C., & Meng, Q. (2000). Reinforcement learning with fuzzy evaluative feedback for a biped robot. In *Robotics and Automation, 2000. Proceedings. ICRA'00. IEEE International Conference on* (Vol. 4, pp. 3829-3834). IEEE.

Zhou, F., Wu, D., Yang, X., & Jiao, J. (2008). Ordinal logistic regression for affective product design, In *Proceedings of Conference Industrial Engineering and Engineering Management conference*. Singapore: IEEE. doi:10.1109/IEEM.2008.4738219

Zhou, Y., Ouyang, X., & Xie, J. (2014). *A discrete cuckoo search algorithm for travelling salesman problem*. Inderscience Enterprises Ltd.

Ziegler, J. G., & Nichols, N. B. (1942). Optimum Setting for Automatic Controllers. *Trans. ASME*, *64*, 759–768.

About the Contributors

Mangey Ram received the B.Sc. degree in science from Chaudhary Charan Singh University, Meerut, India, in 2000, the M.Sc. degree in mathematics from Hemwati Nandan Bahuguna Garhwal University, Srinagar (Garhwal), India, in 2004, and the Ph.D. degree major in mathematics and minor in computer science from G. B. Pant University of Agriculture and Technology, Pantnagar, India, in 2008. He has been a Faculty Member for around eight years and has taught several core courses in pure and applied mathematics at undergraduate, postgraduate, and doctorate levels. He is currently a Professor with the Department of Mathematics, Graphic Era University, Dehradun, India. Before joining the Graphic Era University, he was a Deputy Manager (Probationary Officer) with Syndicate Bank for a short period. He is Editor-in-Chief of *International Journal of Mathematical, Engineering and Management Sciences*; Executive Associate Editor of the *Journal of Reliability and Statistical Studies* and the Guest Editor & Member of the editorial board of many journals. He is a regular Reviewer for international journals, including IEEE, Elsevier, Springer, Emerald, John Wiley, Taylor & Francis and many other publishers. He has published 72 research papers in IEEE, Springer, Emerald, World Scientific and many other national and international journals of repute and also presented his works at national and international conferences. His fields of research are reliability theory and applied mathematics. Dr. Ram is a member of the IEEE, Operational Research Society of India, Society for Reliability Engineering, Quality and Operations Management in India, International Association of Engineers in Hong Kong, and Emerald Literati Network in the U.K. He has been a member of the organizing committee of a number of international and national conferences, seminars, and workshops. He has been conferred with "Young Scientist Award" by the Uttarakhand State Council for Science and Technology, Dehradun, in 2009. He has been awarded the "Best Faculty Award" in 2011 and recently Research Excellence Award in 2015 for his significant contribution in academics and research at Graphic Era University.

J. Paulo Davim received his Ph.D. in Mechanical Engineering in 1997, M.Sc. degree in Mechanical Engineering (materials and manufacturing processes) in 1991, Licentiate degree (5 years) in Mechanical Engineering in 1986, from the University of Porto (FEUP), the Aggregate title from the University of Coimbra in 2005 and a D.Sc. from London Metropolitan University in 2013. He is Eur Ing and Senior Chartered Engineer by the Portuguese Institution of Engineers with a MBA and specialist title in Engineering and Industrial Management. Currently, he is Professor at the Department of Mechanical Engineering of the University of Aveiro. He has more than 30 years of teaching and research experience in Manufacturing, Materials and Mechanical Engineering with special emphasis in Machining & Tribology. Recently, he has also interest in Management/Industrial Engineering and Higher Education for Sustainability/Engineering Education. He has received several scientific awards. He has worked as evaluator of projects for international research agencies as well as examiner of PhD thesis for many universities. He is the Editor in Chief of several international journals, Guest Editor of journals, books Editor, book Series Editor and Scientific Advisory for many international journals and conferences. Presently, he is an Editorial Board member of 30 international journals and acts as reviewer for more than 80 prestigious Web of Science journals. In addition, he has also published as editor (and co-editor) more than 80 books and as author (and co-author) more than 10 books, 60 book chapters and 350 articles in journals and conferences (more than 200 articles in journals indexed in Web of Science/h-index 35+ and SCOPUS/h-index 43+).

* * *

Majid Abdolshah received his B.Sc. degree in Computer Science from Amirkabir University of Technology, Tehran, Iran, in 2013, and his M.Sc. degree in Computer Science from the University of Tehran, Tehran, Iran in 2016. Currently he is a Ph.D. student at University of Kansas, KS, USA. His main research interests are computer vision and machine learning approaches for solving the image-based problems. He is also pursuing the technological applications of statistical learning solutions for finding the relations among extracted features from images.

Mohammad Abdolshah received Bachelor's and Master's degrees in Industrial Engineering in 1998 and 2002, respectively, both from Amirkabir University of Technology, Tehran, Iran. He obtained Ph.D. from University of Putra Malaysia, Malaysia in 2011 in Industrial Engineering. Mohammad Abdolshah's current research interests are Industrial Engineering, Quality Engineering and human resource management.

Saeed Abdolshah received his B.Sc. degree in Mechanical Engineering from Semnan University, Semnan, Iran, in 2007, and his M.Sc. degree in Mechatronics Engineering from Sharif University of Technology, Tehran, Iran in 2012. Currently he is a Ph.D. student at University of Padua, Padua, Italy. His main research interests are humanoid robots, cable-driven parallel robots, assistive devices and rehabilitation.

Leilanaz Akbari is a master student of Industrial Engineering at Ferdowsi University of Mashhad. She received her B.Sc degree in Industrial Engineering from Sadjad University of Technology. Her research interests are focused on quality and reliability engineering, Transportation, and applied Operations Research. Award – 2013 for his innovative research on infrastructure project. He was elected Fellow of International Congress of Disaster Management in 2010. He served as a guest in disaster advance journal. He also serves as an editorial board member in several International journals.

Igor Artemenko received an MS in Aerospace Engineering in 2006, and Ph.D. in Remote Sensing in 2012. His studies directed to application of satellite observations to GHG atmospheric concentration control and GHG inventories. Igor Artemenko is author and co-author of more than 20 scientific papers and book chapters in fields of remote sensing and Earth sciences (ecology and climate change).

P. N. Bajeel received his MSc. (Tech) in Mathematics and Scientific Computing from National Institute of Technology Calicut in 2009. He is currently doing Ph.D in the Department of mathematics, National Institute of Technology (NIT) Calicut. His research interests are Reliability analysis, Estimation and Design of optimal reliability test plans.

Pradeep Reddy Bonikila received his B. Tech. in Mechanical Engineering from IIT Bhubaneswar, Odisha - 751013 in 2016. His research area includes robotics.

Goutam Kumar Bose is currently working as Head of the Department and Professor in Mechanical Engineering Department, Haldia Institute of Technology, Haldia, India. He has obtained Ph.D in Production Engineering from the Jadavpur University, Kolkata, India. He has obtained a Masters in Engineering in Mechanical Engineering from the Bengal Engineering & Science University, Shibpur, India. He has worked as an Assistant Professor in the Department of Mechanical Engineering of the College of Engineering & Management, Kolaghat for ten years. He was an Engineer in R & D Centre of M/s Hindustan Motors Ltd. West Bengal, India. His active areas of interests are Metal Cutting, Non-conventional machining, Industrial and Production Management and optimization techniques. He has published research papers in the Journals of International repute. He has presented papers at several international conferences in India and abroad.

Nupur Goyal received the bachelor's degree (BSc) in computer science from the Kurukshetra University, Kurukshetra, Haryana, India in 2009, the Master's degree (MSc) in mathematics from the H.N.B. Garhwal University, Srinagar, Uttarakhand, India in 2011. Currently, she is a research scholar in the department of mathematics at the Graphic Era University, Dehradun, Uttarakhand, India. Her research interests are in the area of reliability theory.

Seyyed Vahid Hashemi received his B.Sc. degree in Electrical Engineering from Semnan University, Semnan, Iran, in 2007. He has participated in many international robotic competitions as the team leader and electronic hardware designer. He had the key role in several successful projects related to mechatronic and control, and holds three patents in Iran. Currently he is working on some robotic projects mostly related to assistive devices and rehabilitation.

Amir-Mohsen Hejazi is a mechanical engineer whose main field of research is reliability engineering. He received his M. Sc. in Automobile Engineering and His B. Sc. in Mechanical Engineering respectively from Sahand University of Technology and Ferdowsi University of Mashhad. He has published several papers in the field of reliability and life prediction of mechanical complex systems.

Taha-Hossein Hejazi is an Assistant Professor at Sadjad University of Technology. He received his PhD, M.Sc, and B.Sc degrees in Industrial Engineering from Amirkabir University of Technology, Shahed University of Tehran, and Sadjad University of Techology respectively. His primary research areas include quality and reliability engineering, simulation, and multiple criteria analysis. He has also written four books in industrial engineering topics such as discrete and continuous system simulation, quality and reliability engineering, and multiple criteria decision making.

Jagan Jayabalan pursued the M.S. (By Research) at VIT University, Vellore and received B.E. Civil Engineering at Adhiyamaan College of Engineering, Hosur. He also worked as the Junior Research Fellow (JRF) under the sponsored research project by BRNS in Centre for Disaster Mitigation and Management at VIT University, Vellore. He is currently working as the Assistant Professor in School of Civil Engineering at Galgotias University, Uttar Pradesh. His research interests includes the application of artifical intelligence techniques in civil engineering. He also touches the area like slope stability, site characterization, pile foundation, etc.

Dookie Kim received his Bachelor degree in Civil and Environmental Engineering from Korea University in 1993, and he earned his Master and Ph.D. degrees in Civil Engineering from Korea Advanced Institute of Science and Technology (KAIST) in 1995 and 1999 respectively. He is as a post-doctoral researcher and at Korea Atomic Energy Research Institute (KAERI), University of California, and Irvine (UCI) and as a senior researcher at UNISON Co. Ltd in September 1999 – February 2003. Since March 2003, Dr. Kim has been teaching, researching, and consulting at the Department of Civil Engineering, Kunsan National University (KNU), Jeonbuk, Korea.

Ivan Kopachevsky received an MSc in Aerospace Engineering and Geodesy in 2008. His studies directed to application of satellite observations to risk analysis, uncertainty control, socio-ecological security assessment. Ivan Kopachevsky is author and co-author of more than 20 scientific papers and book chapters in fields of remote sensing, Earth sciences (ecology and climate change), engineering and social sciences.

Yuriy Kostyuchenko received an MS in Astrophysics and Stellar Astronomy from Moscow State University in 1993. After that he joined National Academy of Sciences of Ukraine as a Scientist. He received his PhD in Geophysics and Remote Sensing in 1997 from the Scientific Centre for Aerospace Research of the Earth of National Academy of Sciences of Ukraine. In 2011-2014 he served on position of Assistant Professor of the Geomorphology and Soil Science Department of the Faculty of Geography, Kiev University. Since 2000 he serves as Executive secretary of Ukrainian National Member Organization of International Institute for Applied Systems Analysis (IIASA). His research interests include remote sensing application for environmental and socio-ecological security analysis, risk analysis using non-parametric approaches. Yuriy V. Kostyuchenko is a member of Risk and Vulnerability Committee of American Society of Civil Engineers (ASCE), member of American Meteorological Society (AMS), American Statistical Society (ASA), and International Association for Promotion Geoethics (IAPG). He was honored by special award of Space Research Council of National Academy of Sciences of Ukraine in 2003 for fundamental research in remote sensing, and "Makarov" Medal of National Space Agency of Ukraine in 2012 for development of advanced technologies of remote sensing. Yuriy V. Kostyuchenko is author and co-author of more than 70 scientific papers and book chapters in fields of remote sensing, Earth sciences (ecology, climate change), engineering and social sciences.

Anjali Vishwas Kulkarni is working as 'Principal Research Engineer' at the Centre for Mechatronics, IIT Kanpur. She holds Ph.D. degree in Electrical Engineering. Advanced machining processes and Mechatronics are her areas of research interest. She has guided three M.Tech students under joint supervision. She has written a book chapter on, "Electrochemical Spark Micromachining Process" and has published several papers in 'Advanced machining processes' and in 'Mechatronics' areas of research. She has filed a group patent on "Process for drilling contoured deep hole in super alloys using STED to enhance cooling in Turbine blades". She has presented papers in many international and national conferences. She has to her credit five sponsored projects of national interest in the capacity of investigator and co-investigator. She has worked as the Manager of the prestigious project on 'Jugnu Nano Satellite', a joint venture of IITK and ISRO, which is launched on October, 2011. She has chaired technical sessions in national conferences and delivered invited talks on 'Mechatronics' and 'Advanced Machining' nation-wide. She has reviewed several papers for various international journals. She holds membership of IETE, ISAMPE, and RSI. Recently she has been nominated as the executive council member of Bundelkhand University, Jhansi, UP, India.

Anuj Kumar is an Assistant Professor of Mathematics at University of Petroleum & Energy Studies, Dehradun, India. He has obtained his Master's & doctorate degree in Mathematics from G. B. Pant University of Agriculture & Technology, Pantnagar, India. His area of interest is Reliability Analysis and Optimization.

Kaushik Kumar, B.Tech (Mechanical Engineering, REC (Now NIT), Warangal), MBA (Marketing, IGNOU) and Ph.D (Engineering, Jadavpur University), is presently an Associate Professor in the Department of Mechanical Engineering, Birla Institute of Technology, Mesra, Ranchi, India. He has 14 years of Teaching & Research and over 11 years of industrial experience in a manufacturing unit of Global repute. His areas of teaching and research interest are Quality Management Systems, Optimization, Non-conventional machining, CAD / CAM, Rapid Prototyping and Composites. He has 9 Patents, 3 Book, 8 Book Chapters, 97 international Journal publications, 18 International and 8 National Conference publications to his credit. He is on the editorial board and review panel of 7 International and 1 National Journals of repute. He has been felicitated with many awards and honours.

Mahesh Kumar received his Ph.D in Applied Statistics from Indian Institute of Technology (IIT) Bombay in 2006. Currently, he is working as Assistant Professor in the Department of Mathematics, National Institute of Technology (NIT) Calicut. His research interests include Reliability theory, Bayesian reliability analysis, System reliability estimation and design of test plans, Acceptance sampling plans, and Statistical quality control.

Pranshu Kumar has done his Bachelor of Technology in Mechanical Engineering from Graphic Era University, Dehradun. His research area is engineering based reliability theory.

Ravi Kumar Mandave received his B. Tech. in Mechanical Engineering from Acharya Nagarjuna University, Guntur-522510, India in 2008. He received his M.E in CAD/CAM from Andhra University, Visakhapatnam-530003, India. Presently, he is pursuing his Ph.D in Mechanical Engineering from School of Mechanical Sciences, IIT Bhubaneswar, Odisha – 751013, India. His research areas include legged robotics and soft computing.

Rakesh Kumar Maurya is currently working as Assistant Professor of Mechanical Engineering at Indian Institute of Technology, Ropar. He previously worked as Senior Research Associate (Pool Scientist- CSIR) at Indian Institute of Technology Kanpur. Dr. Maurya earned his B.Tech, M.Tech and PhD degrees in Mechanical Engineering from Indian Institute of Technology Kanpur. Dr. Maurya has authored/co-authored over 30 journal and conference publications and 4 book chapters. He is a member of various national and international societies in Mechanical and Automotive Engineering. He has served as a journal referee and committee member/session co-chair on multiple occasions. His areas of interest are HCCI combustion and control, engine instrumentation, engine management, combustion diagnostics, combustion and emissions in IC engines, alternative fuels.

Dmytro Movchan received an MS in Ecology in 2006. In 2009 he successfully completed his Ph.D study, and received an Ph.D. in Remote Sensing from the Scientific Centre for Aerospace Research of the Earth of National Academy of Sciences of Ukraine. His studies directed to application of geoinformation technologies, spatio-temporal approaches, and full carbon account for improving accuracy of GHG inventories. Dmytro Movchan is author and co-author of more than 20 scientific papers and book chapters in fields of remote sensing, Earth sciences (ecology, climate change) and social sciences.

Sayak Mukhopadhyay completed his B.Tech in Mechanical Engineering from Haldia Institute of Technology, Haldia, India under West Bengal University of Technology, Kolkata, India. Currently, he is working with Infosys Limited. His areas of interest are optimization techniques and modern machining. Apart from that he also likes to dwell in computer programming.

Beena Nailwal has done Ph. D. from Department of Mathematics, Statistics and Computer Science, G. B. Pant University of Agriculture and Technology, Pantnagar (India) under the guidance of Dr. S. B. Singh. She has done her M.Sc. (Mathematics) in 2007 with first division from Kumaun University, Nainital (India). Her area of research is Reliability Theory.

Konstantin N. Nechval (Ph.D. in Mechanical Engineering) is an Associate Professor in the Department of Applied Mathematics at the Transport and Telecommunication Institute, Riga, Latvia. He has authored and coauthored more than 75 papers. His research interests include stochastic processes, pattern recognition, operations research, statistical decision theory, adaptive control, and mechanical engineering. Dr. K. Nechval was awarded the "CASYS'07 Best Paper Award" for his paper: "Dual Control of Education Process" presented at the Eight International Conference on Computing Anticipatory Systems (Liege, Belgium, August 6-11, 2007) and the "MM2009 Best Paper Award" for his paper: "Optimal Statistical Decisions in a New Product Lifetime Testing" presented at the Fourth International Conference on Maintenance and Facility Management (Rome, Italy, April 22-24, 2009).

Nicholas A. Nechval (Ph.D. in Automatic Control and Systems Engineering, and Dr.Habil.Sc. in Radio Engineering) is a Professor in the Department of Mathematics at the Baltic International Academy, Riga, Latvia. Before his current position, from 1993 to 1999, Dr. Nechval was a Professor of Mathematics and Computer Science and the Head of the Research Laboratory at the Riga Aviation University. From 1999 to 2014, Dr. Nechval was a Professor and Head (Statistics Department and EVF Research Institute) at the University of Latvia, Riga, Latvia. He was also a Principal Investigator in the Institute of Mathematics and Computer Science at the University of Latvia. Dr. Nechval, in 1991, was awarded a Silver Medal of the Exhibition Committee (Moscow, Russia) for his research on the problem of Prevention of Collisions between Aircraft and Birds. His research interests include mathematics, econometrics, stochastic processes, pattern recognition, digital radar signal processing, operations research, statistical decision theory, adaptive control, and mechanical engineering. Professor Nechval has authored and coauthored more than 450 papers and 9 books, including the book: *Aircraft Protection from Birds* (Moscow: Russian Academy of Science, 2007) coauthored with V. D. Illyichev (Academician of the Russian Academy of Science), and the book: *Improved Decisions in Statistics* (Riga: Izglitibas Soli, 2004) coauthored with Prof. E. K. Vasermanis (University of Latvia). This book was awarded the "2004 Best Publication Award" by the Baltic Operations Research Society. Dr. Nechval is also an Editorial Board Member of several international journals.

Akhil Nekkanti was an undergraduate student working with Dr. Maurya on his B.Tech Project in mechanical engineering from IIT Ropar. He completed his B.Tech Degree from IIT Ropar in Mechanical Engineering in year 2016. His areas of interest include Internal Combustion Engines, Engine Instrumentation, HCCI Combustion, and Combustion Instability.

Pritam Pain is currently working as Professor in Mechanical Engineering Department, JLD College of Engineering and Management. He has obtained his M.Tech in the Mechanical Engineering department at Haldia Institute of Technology, Haldia, India. He completed his B.Tech in Mechanical Engineering from West Bengal University of Technology (Dr. Sudhir Chandra Sur Degree Engineering College), Kolkata, India. His areas of interest are non-traditional and modern machining and optimization by employing nature inspired techniques.

Sangeeta Pant received her doctorate from G. B. Pant University of Agriculture and Technology, Pantnagar, India. Presently she is working with the department of Mathematics of University of Petroleum & Energy Studies, Dehradun, as an Assistant Professor. Her area of interest is Numerical Optimization, Evolutionary Algorithms & Nature Inspired Algorithms

Hossein Poursabbagh is a Ph.D. Candidate of biosystems engineering at Ferdowsi University of Mashhad. He received his M.Sc, and B.Sc degrees in Reliability and Mechanical Engineering from Amirkabir University of Technology and Imam Hossein University of Tehran respectively. His primary research areas include reliability engineering, condition monitoring, neural networks and maintenance. He has also written 11 papers in reliability, condition monitoring or tolerancing topics. also he has worked more than 10 years in fields of NDT, technical inspection, thermography, reliability, etc.

Supriyo Roy, M.Sc (Statistics), M. Tech (ORI&BM), Ph. D (Business Model Optimization) has been working as Associate Professor, Birla Institute of Technology - Mesra, Ranchi. A Post Doctoral Fellow from Airbus Group Endowed (Formerly EADS-SMI) Center for Sourcing and Management, IIM - Bangalore, Dr. Roy has a rich experience of more than 5 years in Industry and 15 years of Post Graduate Teaching. His teaching interest comprises interdisciplinary areas of Management Science, Production and Operations Management, Decision Science, Supply Chain Management, Soft Computing for Manufacturing, etc. He has to his credit published 1 Book, 10 International book chapters, 12 Conference papers and more than 45 research papers in leading International as well as National Journals. Dr. Supriyo is associated as impaneled reviewer of many International and National Journals of repute. He has been facilitated with award of 'Outstanding Contribution to Teaching' in Management Science.

Mahendra Kumar Samal is a Senior Scientist at Reactor Safety Division of Bhabha Atomic Research Centre (BARC) and an Associate Professor in the divison of engineering sciences at Homi Bhabha National Istitute (HBNI), Mumbai, India. He has more than 20 years of research and teaching experience. He received his Ph.D. from University of Stuttgart, Germany and has carried out post-doctoral research work at Ohio State University, USA. His teaching and research interests include finite element analysis, multi-scale material modeling, smart structures, creep, fatigue, modeling of nolinear behavior of piezoelectric material, fracture and damage mechanics. He is credited with the implementation of the online creep and fatigue damage monitoring system at various chemical process and thermal power plants of India which helps in structural health assessment of critical components. He has developed finite element based packages such as DEFINE for damage mechanics and CRYSP for crystal plasticity simulations which are widely used by his students and colleagues for carrying out advanced nonlinear finite element analysis. He has 220 publications including 65 papers in various international journals.

Pijush Samui is working as an associate professor in civil engineering department at NIT Patna, India. He graduated in 2000, with a B.Tech. in Civil Engineering from Indian Institute of Engineering Science and Technology, Shibpur, India. He received his M.Sc. in Geotechnical Earthquake Engineering from Indian Institute of Science, Bangalore, India (2004). He holds a Ph.D. in Geotechnical Earthquake Engineering (2008) from Indian Institute of Science, Bangalore, India. He was a postdoctoral fellow at University of Pittsburgh (USA)(2008-2009) and Tampere University of Technology(Finland)(2009-2010). At University of Pittsburgh, he worked on design of efficient tool for rock cutting and application of Support Vector Machine (SVM) in designing of geostructure. At Tampere University of Technology, he worked on design of railway embankment, slope reliability and site characterization. In 2010, Dr. Pijush joined in the Center for Disaster Mitigation and Management at VIT University as an Associate Professor. He was promoted to a Professor in 2012. Dr. Pijush's research focuses on the application of Artificial Intelligence for designing civil engineering structure, design of foundation, stability of railway embankment, reliability analysis, site characterization, and earthquake engineering. Dr. Pijush is the recipient of the prestigious CIMO fellowship (2009) from Finland, for his integrated research on the design of railway embankment. He was awarded Shamsher Prakash Research Award (2011) by IIT Roorkee for his innovative research on the application of Artificial Intelligence in designing civil engineering structure. He was selected as the recipient of IGS Sardar Resham Singh Memorial. Dr. Pijush is active in a variety of professional organizations including the Indian Geotechnical Society, Indian Science Congress, Institution of Engineers, World federation of Soft Computing, and Geotechnical Engineering for Disaster Mitigation and Rehabilitation. He has organized numerous workshops and conferences on the applications of artificial intelligence in civil engineering.

Anuraj Singh is an Assistant Professor in Graphic Era University, Dehradun, Uttrakhand, India. She completed her Ph.D. from Indian Institute of Technology Roorkee, Roorkee, Uttrakhand, India in Mathematics in 2012. She has been NBHM postdoctoral fellow for two years from 2012 to 2014. She had been a visiting fellow at International Centre for Theoretical Physics (ICTP), Trieste, Italy. She has more than fifteen publications in international journals and in conference proceedings. Her research interests are mathematical biology, dynamical systems, complex systems and differential equations. Dr. Singh has membership of Society for Mathematical Biology (SMB) and Indian Academy for Mathematical Modeling and Simulation (IAMMS).

S. B. Singh is a Professor in the Department of Mathematics, Statistics and Computer Science, G. B. Pant University of Agriculture and Technology, Pantnagar, India. He has around 22 years of teaching and research experience to Undergraduate and Post Graduate students at different Engineering Colleges and University. Prof. Singh is a member of Indian Mathematical Society, Operations Research Society of India, National Society for Prevention of Blindness in India and Indian Science Congress Association. He is a regular reviewer of many books and International/ National Journals. He has been a member of organizing committee of many international and national conferences and workshops. He is an Editor of the 'Journal of Reliability and Statistical Studies'. He has authored and coauthored eight books on different courses of Applied/ Engineering Mathematics. He has been conferred with four national awards and a Best Teacher Award. He has published his research works at national and international journals of repute. His area of research is Reliability Theory.

Pandu Ranga Vundavilli received his B.Tech. in Mechanical Engineering from Jawaharlal Nehru Technological University, Kakinada-533003, India and M.Tech. in Computer Integrated Manufacturing from National Institute of Technology, Warangal- 506004, India in 2000 and 2003, respectively. He received his Ph.D. in Mechanical Engineering from the Indian Institute of Technology, Kharagpur, India in 2009. He is working at present, as Assistant Professor in the School of Mechanical Sciences of IIT Bhubaneswar, Odisha-751013. His research interests include robotics, modeling of manufacturing systems, genetic algorithms, differential evolution, particle swarm optimization, neural networks, fuzzy logic and others. He has published about 75 technical papers in various National, International Conferences and Journals. He is an active reviewer of many International Journals.

Dalkilic Yildirim (PhD) is a civil engineer by profession and has been working as a research associate in the Civil Engineering Department of DEU Faculty of Engineering since 1998 and assistant professor in the Civil Engineering Department of Erzincan University since 2012. He received his B.Sc. in civil engineering in 1997 and his M.Sc in 2001; his Ph.D. in 2010 on risk assessment in water resources management. His basic fields of research are hydrologic modeling, optimization of water power schemes, stochastic time series analysis, and water resources economics, planning and management. He is involved in a number of research projects in the relevant areas, supported by DEU, TUBITAK, UNDP, and EU.

Index

H

I

L

M

N

P

R

S

T

W

www.ingramcontent.com/pod-product-compliance
Lightning Source LLC
LaVergne TN
LVHW060351300926
849162LV00006B/1243